Water and Water Pollution Handbook

VOLUME 3

(*in four volumes*)

Water and Water Pollution Handbook

VOLUME 3

(*in four volumes*)

Edited by LEONARD L. CIACCIO

GTE LABORATORIES INCORPORATED
BAYSIDE, NEW YORK

MARCEL DEKKER, INC., New York   1972

MARCEL DEKKER, INC.
95 Madison Avenue, New York, New York 10016

LIBRARY OF CONGRESS CATALOG CARD NUMBER 78-134780
ISBN 0-8247-1117-3

PRINTED IN THE UNITED STATES OF AMERICA

Contributors to Volume 3

H. E. Allen, *Department of Environmental and Industrial Health, School of Public Health, University of Michigan, Ann Arbor, Michigan*

P. L. Brezonik, *Department of Environmental Engineering, University of Florida, Gainesville, Florida*

S. D. Faust, *Department of Environmental Sciences, Rutgers–The State University, New Brunswick, New Jersey*

R. A. Horne,* *Arthur D. Little, Inc., Cambridge, Massachusetts*

J. V. Hunter, *Department of Environmental Sciences, Rutgers–The State University, New Brunswick, New Jersey*

J. L. Lambert, *Department of Chemistry, Kansas State University, Manhattan, Kansas*

K. H. Mancy, *Department of Environmental and Industrial Health, School of Public Health, University of Michigan, Ann Arbor, Michigan*

H. C. Marks, *Pennwalt Corporation, King of Prussia, Pennsylvania*

A. H. Molof, *School of Engineering and Science, New York University, Bronx, New York*

D. A. Rickert, *Department of Environmental Sciences, Rutgers–The State University, New Brunswick, New Jersey*

V. T. Stack, Jr., *Director of Research and Development, Roy F. Weston Environmental Scientists and Engineers, West Chester, Pennsylvania*

I. H. Suffet, *Environmental Engineering and Science Program, Drexel University, Philadelphia, Pennsylvania*

*Present address: JBF Scientific Corp., Burlington, Massachusetts.

Contents

15. Biochemical Oxygen Demand Measurement   801

Vernon T. Stack, Jr.

16. Chemical Kinetics and Dynamics in Natural Water Systems 831

Patrick Lee Brezonik

17. Effect of Structure and Physical Characteristics of Water on Water Chemistry   915

R. A. Horne

18. The Use of Standards in the Quality Control of Treated Effluents and Natural Waters   949

Alan H. Molof

19. Design of Measurement Systems for Water Analysis   971

Herbert E. Allen and K. H. Mancy

20. Organic Analytical Chemistry in Aqueous Systems   1021

Joseph V. Hunter and David A. Rickert

21. Inorganic Analytical Chemistry in Aqueous Systems   1165

Jack L. Lambert

22. Residual Chlorine Analysis in Water and Waste Water   1213

Henry C. Marks

23. Analysis for Pesticides and Herbicides in the Water Environment   1249

Samuel D. Faust and Irwin H. Suffet

Contents of Volume 1

*Present address: Environomics, Houston, Texas.

ix

Contents of Volume 2

Contents of Volume 4

Chapter **15** **Biochemical Oxygen Demand Measurement**

Vernon T. Stack, Jr.
DIRECTOR OF RESEARCH AND DEVELOPMENT
ROY F. WESTON, ENVIRONMENTAL
SCIENTISTS AND ENGINEERS
WEST CHESTER, PENNSYLVANIA

I. Biochemical Oxygen Demand

A. Interest in a Biological Oxidation Parameter

For approximately 80 years, biochemical oxygen demand (BOD) has been used as a measure of the presence in aqueous solution of organic materials which can support the growth of microorganisms. By simple definition, expanded in Sec. 1.D, the interest is in oxygen utilization by microorganisms. The term BOD refers primarily to the results of a "standard" laboratory procedure(1,2). The procedure has been developed on the premise that the biological reactions of interest can be "bottled" in the laboratory and that kinetics and extent of reaction can be observed empirically. The attitude of many observers toward the laboratory procedure for BOD has been one of disenchantment from its inception, because the "bottled" system, on the basis of comparative kinetics, is

100% accurate in representing itself but is relatively unreliable as a representation of the system from which the sample was taken.

The existing standard laboratory procedures for observation of BOD are not defensible either as analytical or as empirical bioassay procedures. Therefore, this chapter is devoted to a discussion of the true worth of the standard procedures and of the necessity for a more fundamental approach to measurements of biological reactions.

1. MEASUREMENT OF WATER QUALITY

From the first realization that a relationship existed between microorganisms, organic materials, and utilization of oxygen in the waters of streams and lakes, techniques for measurement of the reactants have been of interest. Measurement of dissolved oxygen concentrations have not been overly difficult, but measurement of the concentration of organic materials which are biologically stabilized has been another story. Unfortunately, the only approach which has been applied to any significant extent is the measurement of oxygen utilization over an extended period of time, with correlation of the result to materials which existed at the start of the time period. The technique provides information, but only as an historical array displaying past water quality. For historical record and postmortem examination, the information is useful, but as direct input to water quality control, it is useless.

There would appear to be two approaches to the development of BOD information free of the time delay disadvantage: (a) Development of a BOD technique which would provide information in a few minutes instead of 5 or more days, or (b) capability for quick identification and analysis of all the specific organic compounds present and calculation of biochemical oxidation results from available BOD-organic compound correlations.

The analytical effort involved in specific determinations of organic materials appears to be impractical now or in the future. Therefore, the development of techniques for rapid BOD measurements is a more probable approach.

2. BOD AS A POLLUTION PARAMETER

Measurements of quantities of materials discharged to surface water from a source are required for detection and control of pollutional discharges. The merit of the BOD parameter is that it is supposed to reflect the presence of materials which will be oxidized biologically in the receiving water. The accuracy of the reflection is debatable, and the major question is: "Does the BOD reflect occurrence in the receiving stream?"

The correct answer is "... sometimes or partly." This is developed more fully in Sec.I.B.

The BOD as measured by "standard methods"(*1*) is a meaningful measurement within itself, but its usefulness as a parameter has been impaired by misinterpretation, misuse, and misunderstanding. The misinterpretation and misunderstanding begin when the closed system in the BOD bottle is prepared. The assumption is made that the closed system is parametrically related to the system from which the sample was taken; this assumption is not 100% correct and may be highly incorrect if the system which was sampled is physically unlike that in the BOD bottle and has inputs not reflected in the BOD bottle.

A second assumption is that consumption of dissolved oxygen is an absolute and complete parameter of biological reactions occurring in the BOD bottle. This assumption is reasonably accurate if the overall picture of oxygen consumption is developed, but can be highly inaccurate when only the accumulated oxygen consumption for 5 days is observed. The accuracy may or may not be improved by daily observations of oxygen consumption for 5 or 7 days and projection of the observations to an ultimate value via the fit of a first-order reaction curve. The oxygen consumptions which we observe are meaningfully related to reactions in the BOD bottle through the incubation period and at the time when the incubation is terminated; but, as discussed, the results may not be readily correlatable with conditions in the actual biological system being investigated.

3. BASIS FOR DECISIONS AND ACTIONS

Definition of water quality and the implementation of corrective measures are related in one significant aspect to oxygen consumption through biological oxidation reactions. A form of BOD measurement must be utilized as the basis for decisions and action. However, it must be recognized that the "standard" BOD measurement is not a truly quantitative entity in terms of water quality because (a) the "analytical" technique as normally applied does not qualify as a quantitative measurement, and (b) translation of the "analytical" information to the receiving waters is not a valid interpretation.

B. Theory

1. APPLICABLE TO A BOD BOTTLE

The basic difference between the BOD bottle and a stream is that the BOD bottle is a static system which is controlled to provide aerobic conditions, while the stream is a dynamic system which may include anaerobic areas(*3–5*). The BOD bottle is a batch operation in which the following operations take place: (a) A portion of the organic (and inorganic) material in the system is oxidized in chemical reactions: organic carbon to carbon dioxide; organic hydrogen to water; sulfide and organic sulfur to sulfates; organic nitrogen and ammonia to nitrites or nitrates, referred to as nitrification; and other reactions. The purpose of the oxidation reactions is to provide energy, and the oxygen consumed is referred to as energy oxygen. Energy oxygen is defined as that amount of oxygen required in energy reactions which support the synthesis of organic materials into new biological cells or biologically stable organic materials. Energy oxygen will be a fraction of that required for complete biological stabilization. For example, the energy oxygen for sucrose is approximately 10% of the total requirement. (b) Most of the energy produced does not appear as heat, but approximately 60–70% is utilized in coupled reactions which synthesize the remaining organic material into new microorganisms or convert it into stable compounds(*6*). The synthesis reactions generally do not consume oxygen. Thus, at the point where the original organic material no longer exists, oxygen consumption corresponds to energy oxygen and is equivalent to 10–40% of the chemical oxygen demand (COD) of the original organic material(*6, 7*). Elapsed incubation time is in the order of 1–2 days. When related to organic materials, COD refers to the oxidation of carbon and hydrogen to carbon dioxide and water. Ammonia or organic nitrogen is not oxidized. (c) As the synthesized microorganisms age, lysis of the cell wall occurs, and the cell contents are released(*8*). The remains of the cell wall are relatively stable materials, such as polysaccharides, which are not readily susceptible to further stabilization by biological reactions. Among the contents of the cells are biodegradable materials which are utilized in energy, synthesis and system maintenance reactions. Therefore, a cyclic degradation of the biological mass progresses through lysis-synthesis-lysis (endogenous reactions). The organic materials in the BOD bottle are eventually altered to the point where the materials remaining are chiefly the stable end products of the cell wall. Oxygen consumption at this point is normally calculated to be the long-term or first-stage carbon-aceous BOD and is in the range of 60–80% of the COD of the original organic material because of the remaining stable materials from the cell wall(*9*). Elapsed time is generally 10–20 days. (d) During the period of cell stabilization, degradation of amino acids and other nitrogenous compounds will release ammonia(*10*). The ammonia may or may not be oxidized to nitrates during this period through the action of nitrifying bacteria (nitrification). (e) If activity in the BOD bottle is allowed to continue for an extended period of time (beyond 100 days), the ultimate BOD development will represent essentially total oxidation of all the organic carbon, nitrogen, and hydrogen originally introduced into the system(*11*).

Past BOD theory(*12, 13*) has been based principally on the assumption that BOD development in the BOD bottle is a first-order progression and that the kinetics can be represented by a first-order rate constant and an ultimate BOD value. This assumption, historically, has not been substantiated, and as a result the BOD determination has not factually filled the quantitative need. In addition, application of the technique has been oversimplified by utilization of a fixed incubation period.

In order to establish the use of a BOD procedure which has quantitative meaning, it is necessary for our theoretical considerations to get into step with the observed facts, which basically are: energy oxygen, synthesis, and endogenous reactions.

2. APPLICABLE TO A SURFACE STREAM

Since prime interest is related to biological oxidation and to the resultant consumption of oxygen in rivers and lakes, a look at occurrences in surface waters is in order. Development of the oxygen demand along a flowing stream below the point of an organic discharge depends to a great extent upon the physical configuration of the stream(*3–5, 14, 15*).

a. A Deep Stream or Lake

In a relatively deep stream, the initial biological activity may be homogeneously dispersed if the organic materials in the discharge are soluble or colloidal. As microorganisms are synthesized, flocculation to some degree may occur, and the flocculated microorganisms and other material may settle to form a bottom deposit where anaerobic stabilization will occur(*4, 5*). Since the situation in the stream is dynamic in that the organic discharge, aerobic synthesis, and sludge deposition are continuous, the deposition of sludge can become significant.

The rate at which biological synthesis occurs is a function of the food-to-organism ratio(*16, 17*). Therefore, the rate of stabilization of organic material discharged to the stream is very much related to up-stream activity. If an established population of organisms already exists in the stream, an accelerated rate of synthesis and oxygen consumption can be anticipated.

There is little reason to think that occurrences in the BOD bottle are conveniently related to occurrences in the stream. The BOD bottle does not involve anaerobic stabilization, it does not involve the same degrees of agitation, it does not involve the upstream input of organic materials and organisms, and it is a static rather than a dynamic system.

b. A Shallow Stream

The physical geometry of a shallow stream provides excellent food-to-organism potential. The shallow configuration provides a large bottom surface area/stream volume ratio, which is enhanced if the stream bottom is rocky. In addition, a shallow stream normally has a high velocity, which provides turbulence in the flowing water. These characteristics of the shallow stream provide a high rate of stream reaeration and a suitable environment for growth of attached organisms. Immediately below an organic discharge, the attached organisms can grow luxuriantly. In effect, the stream becomes a trickling filter, and the rate of BOD removal as related to time of stream flow is accelerated compared to a deep stream (5, 14, 15).

In the zone of the stream where the attached growths are functioning, oxygen demand may not correlate with BOD removal in the same manner as for a deep stream, but it may be more related to energy oxygen. The attached growths may grow to a size where they become detached and are carried downstream to be deposited in a quiescent zone. Anaerobic stabilization in the sludge deposit will reduce the quantity of material which will eventually be aerobically stabilized, and the net result is that oxygen required for stabilization of organic material can be less than the BOD of the initial quantity of organic material discharged to the stream (4, 5, 14).

The major conclusion is that the BOD results obtained in a BOD bottle have a limited quantitative correlation with the oxygen demands which appear in a shallow stream. The statement is particularly true at this point in time because a definitive and fundamental evaluation has not been made of biological kinetics in a shallow stream.

C. Selected Results of Measurements (Standard BOD Methods)

Standard methods for the BOD determinations are presented in Refs. (1) and (2). The methods are basically similar and are directed toward the following: (1) a standardized laboratory technique, covering dilution procedure, incubation condition, and calculation of results; (2) a standardized dilution water which does not contain toxic materials, is essentially free of oxidizable materials, and contains an adequate level of nutrients; and (3) a viable culture of seed organisms which are capable of stabilizing the organic materials involved. Acclimation of the biological culture may or may not be necessary.

Once the procedure conforms to these elements of standardization, there is still the question of how to interpret the observed oxygen consumptions.

The BOD's of five organic compounds are shown in Figs. 1–5. The data were obtained by the "standard method." Changes in the concentration of oxygen were followed with great care in each dilution using a dissolved oxygen probe. Dilutions were incubated in a normal manner. When a dissolved oxygen measurement was to be made, the BOD bottle was opened, the determination was made, and the BOD bottle was resealed and returned to incubation. If the dissolved oxygen concentration in the BOD bottle was near depletion, the contents of the bottle were reaerated before resealing. An "acclimated seed" of organisms grown on a feed of each organic material was used in the BOD determination.

1. CARBONACEOUS COMPOUNDS

The BOD curve of glucose as observed in three dilutions is shown in Fig. 1. Oxygen consumption progressed in approximately 8 days to a

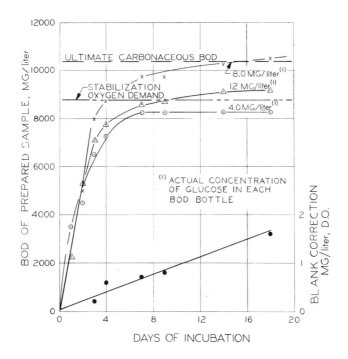

Fig. 1.   Biochemical oxygen demand, glucose, 10,000 mg/liter prepared sample.

point where synthesis and endogenous reactions had reduced biological cells and other materials in the system to relatively stable organic materials. Ammonia was not oxidized to nitrates.

For purposes of the following discussions, the *ultimate carbonaceous BOD* is defined as oxygen required for complete conversion of organic carbon and hydrogen to carbon dioxide and water. This is not the same ultimate BOD that is predicted by fitting a first-order curve through the BOD data points(*12, 13*). When successful, the ultimate value of the first-order projection is equivalent to the oxygen demand when relatively stable end products of cell walls of microorganisms remain.

The ultimate carbonaceous BOD for a pure compound can be arrived at in two ways. For pure compounds, such as glucose, the maximum oxygen demand when all organic carbon and hydrogen are converted to carbon dioxide and water corresponds to the chemical oxygen demand. A second approach, using glucose as an example, follows the biological reactions in the BOD bottle. Energy oxygen required in the synthesis of microorganisms from glucose is 0.34 g of oxygen per gram of glucose, as determined by the author from respiration measurements where endogenous respiration could be subtracted from the total to determine energy oxygen. Based on theoretical COD, the energy oxygen for glucose corresponds to oxidation of 32% of the material present and conversion of 68% into cell material or stable organic compounds.

The approximate molecular formula for cell material is $C_5H_7N_2O$, and the experimentally determined COD is approximately 1.42 g/per gram of cell material(*6*). On the basis of 1 g of glucose, the energy oxygen is 0.34 g which oxidizes 0.32 g of glucose. The remaining 0.68 g of glucose is synthesized into 0.50 g of cell material based on a carbon balance. Total oxidation of the cell material requires 0.72 g of oxygen, and total oxygen consumption is 1.06 g/per gram of glucose, which is essentially equivalent to the COD and the ultimate carbonaceous oxygen demand of glucose.

The approach to calculation of oxygen requirements can be adjusted to show oxygen requirements for any particular level of stabilization of cell material. For example, degradation of cell material to stable end products represents approximately 75% reduction in organic material (*6*). Thus, for glucose, the BOD in the bottle should approach energy oxygen of 0.34 plus 0.75×0.72 for oxidation of cell material, or 0.88 g of oxygen per gram of glucose. The corresponding oxygen consumption is shown in Fig. 1 as *stabilization oxygen demand*. The term stabilization is appropriate since the cell material has been reduced to relatively stable end products. Progression of oxygen demand from stabilization to ultimate is a slow process(*11*).

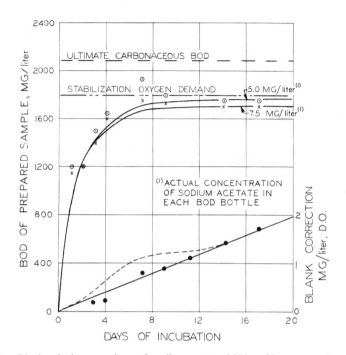

Fig. 2. Biochemical oxygen demand, sodium acetate, 2500 mg/liter prepared sample.

A BOD curve for sodium acetate is shown in Fig. 2. The curve development is quite similar to that observed for glucose and shows essentially the same relationship to ultimate carbonaceous BOD and the stabilization oxygen demand.

At the bottom of Figs. 1 and 2 the blank correction used during the calculation of BOD is shown. The reliability of this correction is open to question. Biological seed in a separate bottle does not necessarily undergo the same biological activity as does the seed in combination with organic material in a BOD determination. In the case of sodium acetate, the inadequacy of the blank correction is indicated in that the calculated BOD suggests values at about 8 days which are greater than BOD values at longer incubation periods. It appears that the blank provides too little correction for the earlier portion of the incubation period. The dashed adjustment of the blank correction may be a better representation of blank correction.

The glucose data in Fig. 1 suggests a second problem in blank correction in that the 4.0, 8.0 and 12 mg/liter concentrations appear to approach different values of stabilization oxygen demand. Great care was used in obtaining the data, and it appears that blanking of oxygen depletion by biological seed accounts for the differences in the calculated BOD curve for different dilutions. This conclusion is plausible because the presence of different concentrations of organic material in the BOD dilutions establishes different levels of biological activity in each bottle.

2. NITROGENOUS COMPOUNDS

BOD curves for acetonitrile (Fig. 3) show a rapid development of the stabilization oxygen demand. Lack of agreement between curves is attributed to blank variation. Nitrification has apparently started in the two

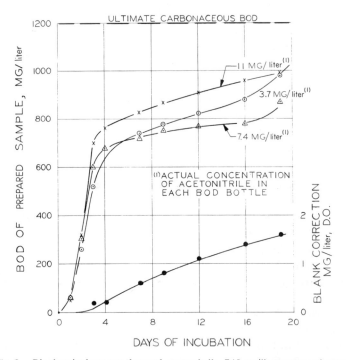

Fig. 3. Biochemical oxygen demand, acetonitrile, 740 mg/liter prepared sample.

lower dilutions on the 19th day, while it has not begun in the 11 mg/liter dilution. The biological seed was grown on a feed of acetonitrile, and thus, the existence of nitrifying organisms in the biological seed is probable.

BOD development for acrylonitrile is shown in Fig. 4. The biological seed was grown on a feed of acrylonitrile, but even with the acclimated culture, development of BOD lagged for 8 days and then progressed rapidly to the stabilization oxygen demand. From past experience, it

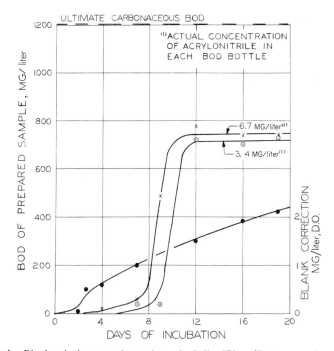

Fig. 4.   Biochemical oxygen demand, acrylonitrile, 670 mg/liter prepared sample.

is known that additional acclimation of the biological culture with an increase in the number of viable organisms will eliminate the lag in BOD development(*18*). Nitrification did not appear in the 20-day incubation period.

The BOD curves for monoethanolamine (Fig. 5) show nitrification in the lowest concentration of 3.3 mg/liter and a lesser degree of nitrification in the 5.0 mg/liter dilution. At the higher concentrations nitrification is not significant. The biological seed was developed on a feed of monoethanolamine, but there was a lag period of 3 days before BOD development started. These results demonstrate that nitrification is related to substrate concentration and is delayed at higher substrate concentrations.

3. IMPORTANT OBSERVATIONS AND CONCLUSIONS FROM THE BOD EXAMPLES

(a) The glucose and sodium acetate BOD values progressed along a curve which can be approximated roughly by a first-order equation. Oxygen consumption approached the estimated value representing a

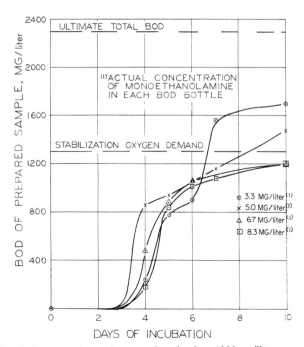

Fig. 5. Biochemical oxygen demand, monoethanolamine, 1000 mg/liter prepared sample.

condition where most of the biological cells had been reduced to stable
end products. This oxygen consumption value is referred to as the stabilization oxygen demand. (b) Two problems with blank correction were
indicated: (1) Oxygen demand in the blank may not provide an adequate
correction in the procedure, as in the case of sodium acetate, and (2)
the blank information may not apply equally to the different dilutions of
an organic material, as in the cases of glucose and acetonitrile. (c) The
BOD development of acrylonitrile, acetonitrile, and monoethanolamine
demonstrates the poor quantitative relationship of 5-day BOD values as
a meaningful representation of BOD development. (d) Nitrification was
encountered only when the initial carbon concentration in the BOD bottle
was low. In the cases of monoethanolamine and acetonitrile, dilutions
containing more than approximately 2.5 mg/liter of organic carbon did
not show nitrification during early stages of carbonaceous oxygen demand.

D. Definitions and Redefinitions

Standard Methods for Examination of Water and Wastewater(1)
defines the biochemical oxygen demand in the following manner:

The oxygen demand of sewage, sewage plant effluent, polluted water or industrial wastes is exerted by three classes of materials: (a) carbonaceous organic material usable as a source of food by aerobic organisms; (b) oxidizable nitrogen derived from nitrite, ammonia, and organic nitrogen compounds which serve as food for specific bacteria (e.g., Nitrosomonas and Nitrobacter); and (c) certain chemical reducing compounds (ferrous iron, sulfite, and sulfide) which will react with molecularly dissolved oxygen.

The *1967 Book of ASTM Standards, Part 23* (2) describes in Method D2329 the general classes of oxidizable material identically as quoted from *Standard Methods* (1). In addition, BOD is defined as "the quantity of oxygen required for the biological and chemical oxidation of waterborne substances under conditions of test."

The following definitions are given as an amplification of the quoted definitions:

1. *Biochemical oxygen demand* represents oxygen utilized in energy reactions which support the synthesis of organic material into new cell material. With time, the older cells die and organic materials associated with the cell are utilized in energy reactions and further synthesis. The cyclic lysis-synthesis-lysis reactions continue until the number of viable cells is greatly reduced and only relatively stable organic materials remain. The biochemical oxygen demand of a specific organic material is the summation of oxygen utilization in the energy reactions.

2. *Stabilization oxygen demand* is oxygen consumed in energy reactions through the point where death of microorganisms is essentially complete and only relatively stable organic materials (end products of energy reactions and of lysed cells) remain.

3. The relatively stable end products of lysed cells will continue to be degraded at a slow rate. Ultimately, almost complete oxidation to carbon dioxide, water, and nitrates will be achieved (11). This level of oxidation should be called *ultimate biochemical oxygen demand.*

The stated definitions are compatible with the intent of the definitions which have been used in the past, but in a practical sense the definitions differ. For example, the intent of the term "ultimate BOD" is oxygen consumption at the completion of biological oxidation reactions, but the analytical measurement has not met the intent. The approach of projecting a first-order representation of carbonaceous oxygen demand in some cases may lead approximately to an estimate of oxygen consumption corresponding to oxygen demand when the synthesized cells have been reduced to relatively stable end products. Therefore, the first-order estimation may be comparable to the stabilization oxygen demand rather than to the ultimate carbonaceous BOD. There is a definite advantage in the utilization of definitions which are related to practical occurrences which can be demonstrated.

E. Critique of Standardization Procedures

1. BACTERIOLOGICAL SEED

The objective of the BOD determination may be one of the following: An indication of oxidation and synthesis reactions accomplished by "natural" populations of organisms such as those obtained from surface streams or domestic sewage; or a detection of the presence of organic materials which can be utilized by an "acclimated" culture.

a. Natural Cultures

The important factor in the use of a "natural" culture is the assurance that an active population or organisms is present. *Standard Methods(1)* suggests the use of a glucose-glutamic acid check for BOD development by the seed in use. Since the check is a BOD determination, the results are after the fact for the particular culture. However, the glucose-glutamic procedures should be applied in order to establish confidence in the seed source. Many sources of domestic sewage contain industrial waste water and may be essentially sterile at particular times. If a domestic sewage treatment plant with secondary treatment is to be used as the seed source and the viability of organisms in the primary waste water is uncertain, the secondary effluent before chlorination is a better selection as a seed source.

The glucose–glutamic acid check procedure should be applied particularly to biological cultures maintained in the laboratory. Such cultures, if fed once a day, may be low in numbers of viable organisms and may require feeding at more frequent intervals or even semicontinuously to ensure acceptable viability.

b. Acclimated Cultures

The development of an acclimated culture of microorganisms capable of utilizing an organic material which is not normally utilized by "natural" cultures is basically a procedure of permitting nature to take its course (19). The absence of BOD development may be due to toxicity or to resistance to biological oxidation. The significance of toxicity should be determined prior to any attempt at acclimation of organisms. The toxicity test suggested in Appendix 5 of ASTM Method D2329-65T(2) can be used to determine acute toxicity.

The toxicity procedure is basically the BOD determination procedure extended over a broad range of dilutions. Interpretation of results is based on the assumption that the presence of acute toxicity will inhibit biological oxidation reactions, but it must be recognized that chemical reactions may utilize dissolved oxygen and mask biological reactions. Thus, determinations of immediate dissolved oxygen demands should be utilized in the interpretation of results.

If the toxicity test is applied where organic materials or waste waters do not show an oxygen demand, it may be modified by including a readily oxidized organic material as an internal control. The control should be introduced identically into each bottle in the dilution series and should produce an oxygen consumption of approximately 25–50% during the

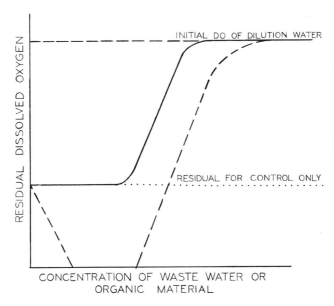

Fig. 6.   Idealized results from toxicity test.

incubation period(20). Idealized results are shown in Fig. 6 and may be interpreted as follows:

1. If oxygen depletions in all bottles are identical to that produced by the control, toxicity is absent and the organic material under examination was not biologically oxidized.

2. If the toxicity data follow the solid curve (which begins with an oxygen depletion equivalent to the control and shows decreased oxygen utilization at higher concentrations of the material in question), the waste water or material was acutely toxic to microorganisms and was not biologically oxidized.

3. If the results after incubation correspond to the dashed curve (which begins with an oxygen depletion equivalent to the control and shows

an increase in oxygen depletion at lower concentration of the material under test, but shows decreased oxygen depletion at higher concentration of the material under test), the material is oxidized biologically at lower concentrations and is toxic at higher concentration.

The initial appearance of acute toxicity can be termed a toxicity threshold to seed organisms. If acclimation is to be attempted, the concentration of the test material in the acclimation system should not exceed the toxicity threshold.

In Appendix 3 of ASTM Method D2329(2), a procedure is suggested for the development of acclimated microorganisms. The procedure as suggested generally will produce an acclimated culture in 2–3 weeks, but approximately 1 year was required in studies by the author in the acclimation of a culture for morpholine. Other nontoxic materials, such as 1,4-dioxane, appear to be entirely resistant to biological oxidation.

An alternative technique for the development of an acclimated culture in a potentially shorter period of time is to begin with an active biological culture and over a period of time progressively replace biodegradable organic material with the material under study(21). Proof of the development of an acclimated culture of microorganisms is the ability of the culture to function in the BOD procedure.

2. DILUTION WATER

Dilution water should be prepared by the addition of nutrients to highest quality water, which is free of both toxic and organic materials(1). The elimination of toxic inorganic cations is readily accomplished by distillation in suitable equipment or by demineralizing with suitable ion exchange resins. Removal of organic materials is imperative but not easily accomplished, particularly when volatile organics in approximately the same boiling range as water or below are involved.

Control of the problem of organic materials in dilution waters requires (a) a source of water for distillation or demineralization which does not contain organic materials, (b) a technique for removal of organic materials from the dilution water, or (c) the utilization of a blank correction for organic materials in the dilution water.

The quality of water to be used for BOD dilutions can be checked by incubation of seeded samples. The results, of course, are after the fact if the water is used immediately for BOD determinations. Alternatively, the total organic carbon analyzer, which will detect approximately 2 mg/liter of total carbon(22), may be used to determine the quality of water prior to use for dilution purposes. Whenever possible, the quality of dilution water should be determined before use.

The removal of low concentrations of organic materials from water in order to obtain suitable dilution water for BOD determinations is not easily accomplished. Adsorption procedures, such as activated carbon, are usually not effective. Corrective measures which may be applied include (a) seeding the dilution water and permitting an incubation period of 20 days or more before the water is used, or (b) for a quicker correction, boiling away a portion of the sample in an effort to strip off the organic materials.

Where the source of water to be distilled contains volatile organic materials, the distillation apparatus shown in Fig. 7 may be employed to

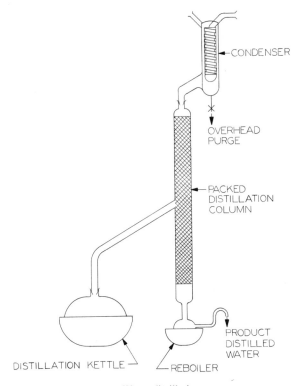

Fig. 7. Water distillation apparatus.

produce water low in organic materials. Water in a distillation kettle is vaporized into a packed column. Condensate from an overhead condenser is refluxed to the column, and condensate from the bottom of the column is reboiled to maintain rectification of organic material in the upper half of the column. Product-distilled water is removed from the reboiler, and approximately 10% of the water charged to the column is wasted in an overhead purge in order to remove organic materials from the system.

Neither *Standard Methods(1)* nor ASTM Method D2329(2) suggests that a blank correction can be made for organic materials in the dilution water, but there are occasions when such a correction must be made if BOD results are to be salvaged. The procedure for development of blank correction for the biological seed should consider the possibility that organic materials may be encountered in the dilution water. A suggested approach is to prepare a range of three or more dilutions of biological seed in dilution water such that the higher concentration of seed should produce at least a 50% depletion of dissolved oxygen.

Observed depletions of oxygen in the dilutions of biological seed require careful interpretation in order to derive suitable blanking values. Figure 8 illustrates characteristic depletions of oxygen by biological seed which have been observed by the author. The illustration presents some,

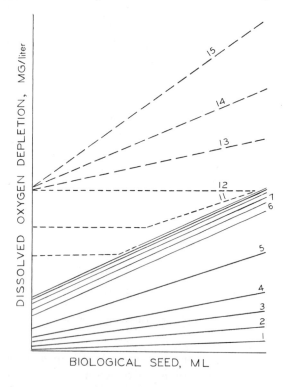

Fig. 8.   Oxygen depletion examples for the biological seed-dilution water system. Numbers refer to day of incubation. Oxygen depletion indicated by the curve, with (– – –) and without (——) concomitant nitrification.

but not all, potential relationships. Each line represents observed dissolved oxygen depletions for a specific incubation period. The numbers on the lines suggest days of incubation. The data points will usually fall on or near a straight line. Scatter of some points may occur, but the most reasonable location of the line can be determined by rationalization of data over several days of incubation.

Data such as those characterized in Fig. 8 can be interpreted to provide correction in BOD dilutions for (a) oxygen utilized because of organic material introduced with the biological seed and (b) oxygen utilized because of organic material introduced with the dilution water.

The slope of each line represents oxygen depletion per milliliter of seed and is, therefore, the proper correction to be made for organic material introduced with the biological seed.

If the dilution water used in the BOD procedure were free of organic material, there would be no depletion of oxygen at zero milliliters of biological seed; however, most distilled water contains organic material, and a positive intercept for dissolved oxygen depletion is a normal occurrence. The intercept at time zero for each line in Fig. 8 represents the depletion of oxygen due to organic material in the dilution water.

The blank correction for a BOD dilution is calculated as:

Blank correction, mg/liter = dissolved oxygen depletion in mg/liter/ml of seed × ml of biological seed + DO depletion by dilution water, mg/liter × fractional volume of dilution water

The corrections as discussed are applicable as long as nitrification is not involved.

In Fig. 8, the dashed lines illustrate conditions where nitrification occurs. Nitrification will occur earliest in the lower concentrations of biological seed. Oxygen depletion due to nitrification in lower concentrations of biological seed may be large enough to cause reversed slopes in the plot of data. As incubation time increases, nitrification will eventually occur at all concentrations of biological seed. When it is obvious that nitrification is essentially complete, the slope and intercept may be applied as a blank correction as previously suggested.

During the incubation period from beginning to completion of nitrification, the oxygen depletion observed for the biological seed cannot be interpreted as a blank correction. The problem involved is that nitrification in each BOD dilution is a function of biological seed and the quantity of organic material present. Therefore, interpretation of nitrification in a specific BOD dilution cannot be correlated to oxygen depletion in the blank data.

In summary, interpretation of oxygen depletion in dilutions of biological seed should determine (a) oxygen consumption related to the seed and associated organic materials, (b) oxygen consumption related to organic material in the dilution water, and (c) the occurrence of nitrification. Utilization of blank values obtained by the suggested procedure is an improvement in the determination of the "standard" biochemical oxygen demand.

Results may be observed where oxygen consumption is high in unseeded dilution water or where concentration of biological seed is small. Oxygen consumption under these conditions may not be limited to biological reactions (although organisms may be present) and may reflect autoxidation by free radical reactions(23). For example, the author has observed (unpublished data) that a dilute solution of 1,4-dioxane in sterile dilution water consumed oxygen. When the same dilution of dioxane was seeded with microorganisms, dissolved oxygen consumption other than the seed blank did not develop, and there was no evidence of oxidation of dioxane. The indicated results are attributed to free radical reactions (autoxidation) in the sterile solution; the introduction of biological seed and associated organic materials inhibited free radical reactions and dioxane was not utilized biologically. Similar free radical reactions have been observed by the author in relatively sterile samples of dilution water contaminated with organic materials, and it is not unusual to observe that unseeded samples of the water may have greater depletions of dissolved oxygen than the seeded sample. Therefore, if a correction is to be provided for organic materials present in the dilution water, it must be developed as an apparent correction observed in blanks containing a range of concentrations of biological seed where free radical reactions are inhibited.

3. INHIBITION OF NITRIFICATION

The oxidation of organic nitrogen or ammonia in the BOD determination has been a confusing factor(24). For example, in samples which have not been subjected to previous biological oxidation, carbonaceous BOD development can progress for possibly 20 days or longer without significant nitrification. In samples which have been partially stabilized biologically, nitrification may be significant from the start or before 5 days of incubation. Therefore, unadjusted values obtained by the BOD procedure and used to determine process efficiency of biological oxidation systems, or used to follow organic loading in a stream, are not truly comparable if nitrification is a variable. Determination of ultimate oxygen demands and the segregation of carbonaceous and nitrogenous demands become more important as emphasis on good water quality is increased.

Approaches to handling the problem of nitrification have utilized techniques for delay or inhibition of activity of organisms which oxidize ammonia, or have applied a correction for any nitrification which has occurred(24–32). Inhibition of nitrification has not been entirely successful because the agents or techniques utilized for specific control of nitrifying organisms have influenced the viability of other organisms.

Correction for nitrification through determination of nitrate formation during incubation is a logical solution(31), but application of the technique has been hampered by the fact that methods for nitrate determination have been time-consuming and have lacked the required accuracy. A further complication is that the nitrate determination must be performed on the same BOD bottle in which the oxygen consumption has been determined, because replicate bottles of the same sample may not develop the same degree of nitrification. The author incubated 20 replicates of a river water sample for 20 days and determined nitrification; the results (unpublished) ranged from very little to almost complete nitrification and showed essentially a normal statistical distribution. The technique involved an incubation volume of 500 ml. Three hundred ml were siphoned off for dissolved oxygen determination, and nitrates were determined on the remainder.

Corrections for nitrification through nitrate determinations are simplified by nondestructive determination of dissolved oxygen using a probe which leaves the original sample available for nitrate determination. Further simplification could be achieved if a suitably sensitive specific ion probe for nitrate were available.

Gaffney and Heukelekian(31) have suggested that nitrification can be delayed by providing nitrates in the dilution water as a source of nutrient, but the technique is limited to samples which require dilution and which do not contain high levels of ammonia nitrogen.

Sawyer and Bradney(25) suggest pasteurization, and Hurwitz et al. (26) suggest acidification, which essentially destroys the existing biological organisms, followed by reseeding with settled domestic sewage, which would contain few nitrifying organisms. The procedure would eliminate nitrification during the earlier days of incubation, but nitrification could be anticipated at longer incubation periods. In a situation where acclimation is important, the substitution of biological seed could produce erroneous results.

The more desirable objective is to inhibit the action of nitrifying bacteria without a detectable inhibition of other biological activity. A number of chemical agents have suggested: methylene blue(27), trichloromethyl pyridine (28), thiourea and allylthiourea (29), and ammonium ion $(30, 32)$.

The use of ammonium ion shows promise for convenient inhibition of nitrification without detrimental effects on other organisms in the biological culture (32).

The BOD curves in Figs. 1–5 demonstrate an alternative approach to the control of nitrification in the BOD bottle. In most cases, nitrification was basically absent through incubation periods of 20 days. The BOD development for monoethanolamine exhibited the most active nitrification, and the rate of nitrification was a function of the concentration of organic material in the bottle. At a concentration of 3.3 mg/liter of monoethanolamine, nitrification proceeded closely along with carbonaceous BOD; at a concentration of 5.0 mg/liter, nitrification was present but depressed; and at concentrations of 6.7 and 8.3 mg/liter, nitrification was not significant at any point during the 10-day incubation period.

These results indicate that the nitrification problem is inherent in the dilution procedure. The necessity to dilute the organic concentration so that demand for oxygen will not exceed that provided in the dilution water, results in concentrations of organic material which favor nitrification. This situation in some cases can be corrected by modifying the dilution procedure to include reaeration.

Certain dissolved oxygen probes can measure dissolved oxygen without displacement of the contents of the BOD bottle. Therefore, it is possible to monitor the dissolved oxygen concentration in a BOD bottle as frequently as desired. When the oxygen concentration approaches depletion, the BOD dilution can be reaerated. Reaeration can be accomplished readily and simply by utilizing a short piece of plastic pipe tapered on each end to fit into a BOD bottle. The piece of plastic pipe is inserted into the BOD bottle to be aerated, a clean BOD bottle is fitted to the other end of the pipe, and the entire assembly is tilted to permit a portion of the dilution to flow into the second BOD bottle. The entire system is shaken vigorously and set down in its original position to permit the BOD dilution to drain into the original bottle. The dissolved oxygen concentration established by reaeration is determined. The very small volume lost during the manipulation is replaced by dilution water, and the bottle is returned to the incubator.

The limited data in Figs. 1–5 suggest that utilization of a reaeration procedure which will permit higher concentrations of organic material in the BOD bottle can accomplish the development of carbonaceous BOD without significant interference from nitrification. However, the technique has no application to water samples which do not require dilution, such as well-oxidized effluents or surface water samples.

4. PRACTICALITY OF STANDARDIZATION IN THE STANDARD BIOCHEMICAL OXYGEN DEMAND

The attention to standardization in the BOD procedure as presented in Refs. (1) and (2) is meaningful as related to their defined terms and objectives. Most of the objectives are met, the only notable exception being the blank correction for biological seed. Endogenous respiration and oxidation of organic material introduced with the seed are complications which are variable and not readily corrected for in the blanking procedure.

Basically, the BOD method is based on the misconception that a first-order reaction is being observed and that standardization must establish conditions under which the reaction rate will be a constant(18,33–37). Within the misconception, a point in time such as the 5-day BOD value would have meaning. Actually, the 5-day BOD cannot stand alone and is valueless if it is not supported by other data such as COD; it has limited value under any circumstances.

Chemists or investigators who have had significant experience with the BOD procedure are aware of the mythical nature of the first-order reaction, and definitely the individuals who have devoted time to the writing of the standardized methods are aware of the myth, but it has been propagated because a more meaningful or more convenient representation of biological oxidation has not yet been developed.

5. REVISED APPROACH TO BIOCHEMICAL OXYGEN DEMAND

A continuing effort by environmental scientists and engineers to develop the theory, concept, and determination of biochemical oxidation is represented in the work of McCarty(6), Servizi and Bogan(7), Busch and Myrick(34–36), Bhatla and Gaudy(38), and others. In the light of better understanding, the concept of a first-order rate for development of biochemical oxygen demand essentially vanishes, and the obvious need for determinations of biochemical oxidations related to practical concepts is highlighted.

A convenient and more meaningful representation of the biochemical oxygen demand can be developed if the concept of the first-order reaction is discarded and the concept of actual occurrences in the BOD bottle is substituted. As discussed in Sec. I.B.2, oxygen consumption involves (a) oxygen utilized in energy reactions (energy oxygen) which oxidize part of the organic material to carbon dioxide, water, and other stable end products, and where the energy produced is used in the synthesis of the remaining utilizable organic materials into new biological cells; and (b) oxygen consumed in subsequent endogenous reactions which bring about a reduction of cell materials to relatively stable end products, such as polysaccharides. Therefore, the biochemical oxygen demand is the sum of energy oxygen and endogenous oxygen.

Progress is being made in techniques for the determination of energy oxygen, as discussed in Sec. II.B. Presumably, a method can be developed which will permit determination of energy oxygen in simple equipment within a few minutes (less than 1 hr). Biological cell production during the determination of energy oxygen can provide a basis for calculation of endogenous oxygen, and the sum of the two is the stabilization oxygen demand. Through this approach, the more meaningful stabilization oxygen demand could be determined in a relatively short time. The results would have a sound meaning directly related to occurrences in the system. The need is great for perfection of procedures to determine the stabilization oxygen demand as a replacement for the existing BOD procedures.

II. Measurements of Biochemical Oxidation

A. Respiration Measurements

Within the framework of the "standard" approach to BOD, the determination of oxygen consumption by almost any procedure can be considered. Two respiration measurement techniques which have been suggested and marketed are electrolytic regeneration of oxygen to replace that consumed (consumption detected by pressure change)(39), and manometric measurement of oxygen consumption from a closed gas space over the sample (40–44).

These techniques can function without dilution of the sample, and in a critical sense, the development of the BOD curve would not necessarily correlate with curves obtained by the "standard" dilution procedure. However, the values for stabilization oxygen demand should correspond for all procedures. Thus, if the concept of the 5-day BOD is discarded and the concept of the stabilization oxygen demand is utilized, essentially any approach to measurement of oxygen consumption is acceptable (see Chap. 27).

B. Measurements of Energy Oxygen

Measurements of energy oxygen have been used extensively for observation of biological activity(44, 45). A frequently applied technique utilizes the Warburg respirometer(44) to determine total oxygen consumption for biological seed and for the biological seed plus sample. Net oxygen consumption in the fed system, when endogenous conditions have been reached and a seed correction has been applied, corresponds to energy oxygen.

Vernimmen et al.(46) suggested a procedure for a short-term oxygen demand. In brief, the procedure utilizes a closed, stirred system containing a biological culture and fitted with a dissolved oxygen probe for measurement of dissolved oxygen concentrations. The rate of oxygen consumption for the biological culture under endogenous conditions is determined by plotting the decrease in dissolved oxygen concentration versus time. A sample of waste water is introduced and an increased rate of oxygen consumption results. The uptake rate in the system is observed until the rate returns to endogenous. Endogenous conditions can be reached within a few minutes if the feed of waste water has been suitably limited. The net consumption of oxygen from endogenous to endogenous is a measurement of energy oxygen.

In an article by Servizi and Bogan(7), energy oxygen determined by Warburg techniques was compared to energy oxygen values calculated from synthesis data (cell production determined by a calibrated turbidity measurement). Data obtained by the two techniques did not agree, and the difference was not resolved. It is probable that the energy-synthesis balance in a concentrated biological culture (Warburg system) is different from that in a dilute system functioning in the logarithmic growth phase. A potential complication is that energy required in system maintenance is a more pronounced factor in the concentrated culture and may be difficult to blank properly(6). Energy oxygen measurement is important to the definition of biological oxidation, and further work is required on procedures for determination of energy oxygen (see Chap. 27).

C. Measurements of Synthesis

A direct product of energy consumption is the synthesis of new cell material. Therefore, general knowledge of the magnitude of biological reactions requires that the quantity of synthesized cell material be determined. Potentially, biological growth can be determined by increase in weight or COD of suspended solids, as suggested by Busch and Myrick (34–36), but the determination can be complicated by the presence of suspended solids or precipitates or by low levels of synthesis in dilute solutions of organic material.

Measurements of deoxyribonucleic acid (DNA) are essentially directly related to the mass of biological cell material present(47). Since DNA is present in both dead and living cells, the measurement is total in nature. Synthesis can be represented by change in the quantity of DNA. The basic drawback is that procedures for the determination of DNA are tedious and time-consuming (48).

Measurements of dehydrogenase activity are a potential indirect measure of synthesis, in that dehydrogenase activity is directly related to the number of viable cells. Tetrazolium salts, especially 2,3,5-tri-phenyltetrazolium chloride (TTC), have been used to indicate cell viability (49). The reduced enzyme diphosphopyridine nucleotide (DPN) is oxidized via the reduction of TTC to formazan, which has a red color. When the reaction is carried out aerobically, there is a mole-for-mole relationship between DPN and formazan (50). Ford et al. (51) reported the use of TTC as a measure of activated sludge activity.

The TTC procedure is not complicated and holds promise as a method for measurement of biological activity and cell synthesis. Refinement of the procedure and increased sensitivity are desired so that observations of biological activity in dilute systems can be made.

Adenosine triphosphate (ATP) is the common intermediate in energy transfer reactions in bacterial growth. Production of ATP is indicated to be directly related to growth of bacteria (52,53). The suggested correlation is approximately 10.5 g of cells (dry weight) per mole of ATP. Therefore, determination of ATP production is also potentially applicable to calculations of synthesis (6) (see Chap. 27).

III. BOD and Related Parameters

It is improbable that a chemical or physical measurement of waste water content can in every case provide an understanding of biological stabilization reactions. Knowledge of biological activity can be determined only through measurement of parameters related to that function, such as oxygen consumption and cell production. There is a relationship between the organic materials and the biological activity which will utilize these materials, but the task of quantitative identification and determination of all organic materials in a surface water or complex waste water is prohibitive. Therefore, gross parameters such as chemical oxygen demand (COD) and total organic carbon (TOC) are utilized to establish the presence of organic material.

For any waste water stream, a statistical relation between the oxygen demand ratio (OD ratio) of BOD to COD or TOC of samples can be established. If only one organic material is present, the correlation will be good and will have standard deviations which reflect analytical error.

As the numbers, types, and occurrence of organic materials in the waste water become variable, a higher standard deviation in the BOD to COD or TOC correlation may appear because of the difference in OD ratio for different organic materials. For very complex waste waters which contain a broad spectrum of organic materials at all times, the OD ratio may show good precision and a low standard deviation because of the averaging effect.

In a biological oxidation process or in a surface stream below a point of pollution, the OD ratio will decrease as biological stabilization produces end products which are biologically stable organic materials, and which show COD or TOC but little or no BOD. The conclusion to be drawn is that BOD and COD or TOC data do not replace one another but are complementary in the determination of water quality (see Chaps. 19 and 27).

IV. Summary

There is a basic need for a biological oxidation parameter to be applied in the monitoring and control of water quality in waste water, surface streams, and lakes. The "standard" BOD procedure does not adequately satisfy this need because it has limited accuracy and problems in interpretation and requires long periods of time before results are obtained. Problems with the "standard" BOD determination are inherent in the inadequate kinetic theory from which it is derived. If the theory of first-order kinetics is disregarded and the BOD curve is carefully developed, useful information about carbonaceous oxygen demand may be obtained, but the technique requires 10–20 days and is complicated by the need for suitable seed organisms, the quality of dilution water, and the probability of nitrification.

A practical biological oxidation parameter is possible if observations are based on the fundamental reactions of energy oxygen, synthesis, and lysis-synthesis-lysis (endogenous respiration). Improved analytical procedures utilizing the dissolved oxygen probe make it possible to determine energy oxygen in less than 1 hour. Procedures for the measurement of synthesis through enzyme concentrations or activity may lead to convenient determinations of synthesis within a few minutes. Oxygen required during the stabilization of synthesized cell material can be calculated once synthesis is known. Therefore, it is possible in less than 1 hour to arrive at a fundamental measurement of biological oxidation by the summation of energy oxygen and endogenous requirements. The result has been termed *stabilization oxygen demand*.

REFERENCES

1. *Standard Methods for Examination of Water and Wastewater*, 12th ed., Am. Public Health Assoc., New York, 1965, p. 218.
2. *1967 Book of ASTM Standards, Part 23*, Am. Soc. Testing Materials, Philadelphia, 1967, p. 612.
3. H. W. Streeter, *Sewage Works J.*, **7**, 256 (1936).
4. C. C. Ruchhoft, *Sewage Works J.*, **13**, 542 (1941).
5. C. J. Velz and J. J. Gannon, *Intern. J. Air Water Pollution*, **7**, 587 (1963).
6. P. L. McCarty, *Proc. 2nd Intern. Water Pollution Res. Conf., Tokyo, 1964*, Pergamon, 1965, p. 169.
7. J. A. Servizi and R. A. Bogan, *J. Sanit. Eng. Div. Am. Soc. Civil Engrs.*, **89**, 17 (1963).
8. R. Y. Stanier, M. Doudoroff, and E. A. Adelberg, *The Microbial World*, 2nd ed. (H. Steinbach, ed.), Prentice-Hall, Englewood Cliffs, N.J., 1964, pp. 48, 333, 357–358.
9. W. W. Umbreit, *Modern Microbiology*, W. H. Freeman Co., San Francisco, 1962, p. 69.
10. Ref. (*12*), p. 537.
11. A. L. H. Gameson and A. B. Wheatland, *Inst. Sewage Purif. J. Proc.*, **5**, 106 (1958).
12. H. A. Thomas, Jr., *Sewage Works J.*, **9**, 425 (1937).
13. E. W. Moore, H. A. Thomas, Jr., and W. B. Snow, *Sewage Works J.*, **22**, 1343 (1950).
14. W. A. Cawley, *Sewage Ind. Wastes*, **30**, 1174 (1958).
15. O. Jaag and H. Ambuehl, *1st Intern. Conf. Water Pollution Res., London, 1962*.
16. C. C. Ruchhoft, J. F. Kachmar, and W. A. Moore, *Sewage Works J.*, **12**, 27 (1940).
17. R. F. Weston and W. W. Eckenfelder, *Sewage Ind. Wastes*, **27**, 802 (1955).
18. C. S. ReVelle, W. R. Lynn, and M. A. Rivera, *J. Water Pollution Control Federation*, **37**, 1679 (1965).
19. E. J. Mills, Jr., and V. T. Stack, Jr., *Sewage Ind. Wastes*, **27**, 1061 (1955).
20. V. T. Stack, Jr., *Ind. Eng. Chem.*, **49**, 913 (1957).
21. H. G. Schwartz, Jr., *J. Water Pollution Control Federation*, **39**, 1701 (1967).
22. C. E. Van Hall, J. Safranko, and V. A. Stenger, *Anal. Chem.*, **35**, 315 (1963).
23. E. S. Gould, *Mechanism and Structure in Organic Chemistry*, Holt, Rinehart and Winston, New York, 1964, p. 705.
24. C. N. Sawyer, *Chemistry for Sanitary Engineers*, (R. Eliassen, ed.), McGraw-Hill, New York, 1960, p. 274.
25. C. N. Sawyer and L. Bradney, *Sewage Works J.*, **18**, 1113 (1946).
26. E. Hurwitz, G. R. Barnett, R. E. Beaudoin, and H. P. Kramer, *Sewage Works J.*, **19**, 995 (1947).
27. W. E. Abbott, *Water Sewage Works*, **95**, 424 (1948).
28. G. Goring, *Soil Sci.*, **93**, 211 (1962).
29. H. A. Painter and K. Jones, *J. Appl. Bacteriol.*, **26**, 471 (1963).
30. J. H. Quastel and P. G. Scholefield, *Bacteriol. Rev.*, **15**, 1 (1951).
31. P. E. Gaffney and H. Heukelekian, *Sewage Ind. Wastes*, **30**, 503 (1958).
32. R. H. Siddiqi, R. E. Speece, R. S. Engelbrecht, and J. W. Schmidt, *J. Water Pollution Control Federation*, **39**, 579 (1967).
33. S. R. Hoover, L. Jasewicz, and N. Porges, *Sewage Ind. Wastes*, **25**, 1163 (1953).
34. A. W. Busch, *Sewage Ind. Wastes*, **30**, 1336 (1958).
35. N. Myrick and A. W. Busch, *J. Water Pollution Control Federation*, **32**, 741 (1960).
36. A. W. Busch and N. Myrick, *J. Water Pollution Control Federation*, **33**, 897 (1961).
37. A. F. Gaudy, M. N. Bhatla, R. H. Follett, and A. Niaaj, *J. Water Pollution Control Federation*, **37**, 444 (1965).
38. M. N. Bhatla and A. F. Gaudy, *J. Water Pollution Control Federation*, **38**, 1441 (1966).

39. J. W. Clark, *New Mexico State Univ. Eng. Exp. Station Bull. No. 11*, 1959.
40. L. L. Falk and W. Rudolfs, *Sewage Works J.*, **19**, 1000 (1947).
41. D. H. Caldwell and W. F. Langelier, *Sewage Works J.*, **20**, 202 (1948).
42. F. Sierp, *Ind. Eng. Chem.*, **20**, 247 (1928).
43. I. Gellman and H. Heukelekian, *Sewage Ind. Wastes*, **23**, 1267 (1951).
44. E. W. Lee and W. J. Oswald, *Sewage Ind. Wastes*, **26**, 1097 (1954).
45. M. N. Bhatla, V. T. Stack, Jr., and R. F. Weston, *J. Water Pollution Control Federation*, **38**, 601 (1966).
46. A. P. Vernimmen, E. R. Henken, and J. C. Lamb, *J. Water Pollution Control Federation*, **39**, 1006 (1967).
47. A. C. Giese, *Cell Physiology*, 2nd ed., W. B. Saunders Co., Philadelphia, 1962, p. 119.
48. F. J. Agardy and W. C. Shepherd, *J. Water Pollution Control Federation*, **37**, 1236 (1965).
49. E. Kun and L. G. Abood, *Science*, **109**, 144 (1949).
50. A. F. Brodie and J. S. Gotz, *Science*, **114**, 40 (1951).
51. D. L. Ford, J. T. Young, and W. W. Eckenfelder, *Proc. 21st Ind. Waste Conf. Purdue Univ. 1965*, p. 534.
52. T. Bauchop and S. R. Elsden, *J. Gen. Microbiol.*, **23**, 457 (1960).
53. T. Bauchop, *J. Gen. Microbiol.*, **18**, vii (1958).

Chapter 16 Chemical Kinetics and Dynamics in Natural Water Systems

Patrick Lee Brezonik
DEPARTMENT OF ENVIRONMENTAL ENGINEERING
UNIVERSITY OF FLORIDA
GAINESVILLE, FLORIDA

I. Introduction

Attempts to explain the composition of and reactions occurring in natural waters on a rational basis inevitably involve the two cornerstones of physical chemistry — thermodynamics and kinetics. The equilibrium situation, i.e., the condition toward which the system is tending, is described by the laws and equations of thermodynamics, while the rate at which the system is moving toward a given condition is described by the formulations of kinetics. In the absence of human effects, the composition of natural waters can often be described by equilibrium models which, for example, relate composition of the water to weathering reactions of rocks and soils in the watershed.

In that the solutes in natural waters are largely ionic, knowledge of ionic equilibria is particularly essential to an understanding of the reactions which control the composition of natural waters. Several types of ionic reactions (and equilibria) are important in this regard: acid-base, precipitation, complexation, and redox. Many equilibria are complex and involve several types of reactions occurring simultaneously. For example, carbonate equilibria involve acid-base reactions and possibly precipitation (calcium carbonate) and complexation (ion pair formation of bicarbonate and carbonate with cations). Manganese and especially iron equilibria involve all four types of reactions. Although the fundamental laws of equilibrium are common to all these reactions, each reaction type differs sufficiently from the rest to require separate and rather lengthy treatment. Natural waters also contain numerous non-ionic substances ranging from dissolved gases to complex organic substances. Clays and other suspended matter, while having net charges, are not ions in the usual sense of the word, and their equilibria are not described by the ionic equilibrium formulations used for simple cations and anions. Full understanding of the various processes affecting the equilibrium situation of water systems requires a broad knowledge of most classical areas of physical chemistry, ranging from ionic equilibria, phase equilibria, gas laws, surface and colloid behavior, to the laws of thermodynamics and thermochemistry. Clearly, an adequate treatment of all these topics would be far too lengthy for the present chapter.

The fundamentals of these topics are covered adequately in standard texts of physical chemistry (*1–4*). The subject of ionic equilibria has been covered in depth in several monographs (*5–9*). Furthermore, application of chemical equilibria to natural water descriptions has been the subject of several books and symposia (*10–13*) and the subject of many papers.

Weber and Stumm(*14*) have thoroughly discussed acid-base equilibria in natural waters; Garrels and Christ(*10*) similarly treat complexation, solubility, and redox equilibria. Models describing the composition of marine and fresh waters in terms of acid-base, solubility, complexation, and redox equilibria have been derived by Sillen(*15–17*). Stumm(*18*), Kramer(*19*), and Garrels and his co-workers(*20,21*). Thus, natural water equilibria have received extended treatments in readily available literature. On the other hand, there has been no comprehensive review of kinetic studies in water chemistry; it is hoped that the present chapter will remedy this situation.

While thermodynamic considerations provide the basis for calculations of the equilibrium composition of natural water systems, it is probably seldom that complete equilibrium is attained in natural water situations. Equilibrium calculations for some chemical species compare favorably with the actual concentrations in certain waters, but natural waters are open, dynamic systems in which true equilibrium is never reached for many species. In general, abiotic reactions in natural waters tend toward equilibrium; however, nonequilibrium conditions can be maintained because natural waters are not closed but can receive influxes of matter and energy. The adequacy of equilibrium calculations in describing the behavior of natural water solutes depends on the influx and reaction rates for the substance. If the influx is small and a substance reacts rapidly, equilibrium descriptions are probably adequate. The composition of most natural waters changes temporally, implying the inadequacy of equilibrium calculations. But in any case, the time-invariant state for open systems like natural waters is the steady state rather than the equilibrium state. Implied in this fact then is the importance of dynamic and kinetic considerations in descriptions of natural water systems. The biogenic elements (e.g., carbon, nitrogen, phosphorus, trace metals) are perhaps the most obvious examples of nonequilibrium distributions which require kinetic data for complete descriptions. However, kinetic considerations are essential to accurately describe the behavior of many inorganic compounds and elements in abiotic reactions.

The subject of kinetics can be conveniently divided into four main areas for the present discussion: (1) homogeneous chemical kinetics, i.e., abiotic reaction rates of substances dissolved in water, (2) heterogeneous reaction kinetics, e.g., reaction rates of dissolved substances with particulate matter, (3) biochemical kinetics, i.e., the dynamics of biogenic elements and compounds in reactions mediated by aquatic organisms, and (4) analytical implications of chemical kinetics.

II. Chemical Kinetics

A. General Considerations; Order and Molecularity

Rates of chemical reactions in general depend on the nature of the reactants, their concentrations, and on temperature. With the exception of certain (zero-order) reactions, increasing concentration of the reactant(s) results in higher reaction rates. In a closed system where no more reactants are added during the reaction, the concentrations of reactants do not remain constant but decrease as the reaction proceeds. Consequently, the reaction rate decreases with time and asymptotically approaches zero. Theoretically, the dependence of reaction rate on concentration of reactants is given by the law of mass action: The reaction rate at any time is proportional to the concentrations of the reactants, each of which is raised to a power corresponding to the number of molecules of that reactant participating in the reaction.

If the rate is symbolized by $-dC/dt$, where C is the concentration of reactant (which decreases with time), then according to this law for any reaction of the type

$$aA + bB + cC \cdots \longrightarrow \text{products} \tag{1}$$

the rate would in general be given by

$$-dC/dt = kC_A^a C_B^b C_C^c \cdots \tag{2}$$

where k is a constant of proportionality or the rate constant for the reaction. However, stoichiometry (and hence the law of mass action) does not agree with the experimentally derived rate formulations for many reactions. Kinetics is to a great extent an experimental science. The number of exponents appearing in the empirical rate equation determines the *order* of a reaction. For example, reactions whose rate data fit the equation:

$$-dC/dt = dP/dt = kC_A \tag{3}$$

where C and P represent reactant and product concentrations and C_A symbolizes any reacting species, are said to be first order, no matter how many reactants or products appear in the stoichiometric expression. Similarly, reactions fitting the equation

$$-dC/dt = kC_A C_B \tag{4}$$

are said to be second order overall, but first order with respect to A and B individually. Many reactions of the type $A + B \rightarrow P$, or $2A \rightarrow P$, may fit first-order expressions, and alternatively some reactions of the type $A \rightarrow P$ follow second-order kinetics. Furthermore, the order of a reaction need not be a whole number but may be zero or fractional. Reaction order is determined solely by the best fit of experimental data with a rate equation, and does not necessarily agree with the reaction stoichiometry or mechanism.

For a given reaction, the value of the rate constant depends on temperature and, if the reaction occurs in solution, on ionic strength. The units of the rate constant depend on the order of the reaction; in general, for a reaction of nth order, the dimensions of the constant are given in units of $(time)^{-1} (concentration)^{1-n}$.

An explanation for the frequent discrepancies between order and stoichiometry lies in the fact that most reactions are not simple one-step processes but involve several elementary reactions and unstable or metastable intermediates between reactants and products. While mechanism defines order, the converse is not true. Although our knowledge concerning reaction mechanisms is based largely on kinetic studies and determinations of reaction order, a given reaction order does not uniquely specify a certain mechanism. In multistep reactions one step is generally much slower than the rest and thus limits the overall rate of reaction. In many cases the stoichiometry of the rate-limiting step defines the overall reaction order. For example, if the rate-limiting step involves two molecules, it is likely that the reaction will follow second-order kinetics. A classic example given in physical chemistry texts is the decomposition of hydrogen iodide, $2 HI \rightarrow H_2 + I_2$, which results when two HI molecules collide with sufficient energy to rearrange the two H—I bonds to an H—H and an I—I bond. The elementary process involves two molecules and is called bimolecular. The experimentally determined rate formulation is $-d[HI]/dt = k[HI]^2$, which is a second-order equation. Although there is but one reacting species (HI) in this simple case, the reaction follows a second-order expression, and from this fact we infer that the mechanism of the elementary (rate-limiting) step is bimolecular. In reactions with complex mechanisms consisting of reversible and irreversible steps, product inhibitions, and alternate pathways, the order of the reaction may not be apparent from the mechanism, and conversely, the empirically determined order may provide few clues concerning the reaction mechanism.

As implied in the preceding paragraph, the molecularity of a reaction is defined as the number of molecules involved as reactants. For simple one-step reactions, molecularity and stoichiometry are the same. Molecularity should be applied only to simple one-step reactions, and for complex reactions we refer to the molecularity of each step but not of the overall process. To do this requires knowledge of the reaction mechanism.

Kinetic studies are important means of determining reaction mechanisms. Our knowledge about complex organic reaction mechanisms has been obtained largely through studies of reaction kinetics under varying conditions [see Gould(22), for example]. Treatment of mechanistic studies is largely beyond the scope of this chapter, and the interested reader should consult specialized texts on this subject(22–24). However, the following brief discussion is offered to indicate the limitations of interpreting mechanisms from empirical rate data. Assuming we know the mechanism for a reaction (which is true a priori only for theoretical reactions we write on paper), it is a simple matter to determine the overall rate of product formation or reactant loss. (For multi-reactant, multi-product reactions, the equation may depend on which reactant or product is considered).

Table 1 presents two possible mechanisms for the simple reaction $2A \rightarrow C$. Rate equations are developed for the reactant, intermediate, and product from the law of mass action Eq. (2); the rate of concentration change for a species is the sum of the rates by which the species is formed, minus the rates by which it reacts to form another species. In general

$$dA/dt = \sum \text{(rates of formation of A)} - \sum \text{(rates of loss of A)}$$

For short-lived intermediates we can assume that rate of formation becomes equal to rate of reaction after a short period of reaction time, and after this the intermediate concentration remains constant until the reactants are nearly exhausted. Thus intermediates reach a steady state (d[intermediate]$/dt = 0$) and the differential equation becomes an algebraic equation which can be solved for steady state concentration of intermediates in terms of reactant concentration and rate constants. The overall rate of product formation is proportional to the intermediate concentration which usually cannot be measured. However, the steady-state intermediate concentration may be expressed as a function of reactant concentrations and substituted into the equation for product formation. This yields a rate equation describing product formation as a function only of reactants (which can be measured) and rate constants.

Thus from a known mechanism we can derive a rate equation, and by comparing possible values for the rate constants, we can derive apparently simpler rate equations which still fit the mechanism. For the first mechanism in Table 1, three different "limiting" rate equations are possible depending on the relative sizes of the rate constants. The second mechanism produces three second-order rate equations which differ only in the magnitude of the rate constant. It is pertinent to note that both mechanisms can give the same second-order rate equation under appropriate

TABLE 1
Rate Equations for Reaction $2\,A \to C$, as Function of Mechanism and Magnitude of Rate Constants[a]

Mechanism	Rate equations	Steady state assumption	Overall rate of product formation	Limiting rate equation
(1) $A \underset{k_2}{\overset{k_1}{\rightleftharpoons}} B$ $B + A \xrightarrow{k_3} C$	$\dot{A} = k_2B - k_1A - k_3AB$ $\dot{B} = k_1A - k_2B - k_3AB$ $\dot{C} = k_3AB$	$\dot{B} = 0$ $B = \dfrac{k_1A}{k_2 + k_3A}$	$\dot{C} = \dfrac{k_1k_3A^2}{k_2 + k_3A}$	$k_2 \gg k_3,\ \dot{C} = k'A^2$ $k_3 \gg k_2,\ \dot{C} = k_1A$ $k_3 \approx k_2,\ \dot{C} = \dfrac{k_1A_2}{k'' + A}$
(2) $A + A \underset{k_2}{\overset{k_1}{\rightleftharpoons}} B$ $B \xrightarrow{k_3} C$	$\dot{A} = k_2B - k_1A^2$ $B = k_1A^2 - k_2B - k_3B$ $\dot{C} = k_3B$	$\dot{B} = 0$ $B = \dfrac{k_1A^2}{k_2 + k_3}$	$\dot{C} = \dfrac{k_1k_3A^2}{k_2 + k_3}$	$k_2 \gg k_3,\ \dot{C} = k'A^2$ $k_3 \gg k_2,\ \dot{C} = k_1A^2$ $k_3 \approx k_2,\ \dot{C} = k''A^2$

[a] The dot notation is used to express a differential with respect to time; thus $\dot{A} = dA/dt$.

circumstances. Thus, a given set of data which fit a second-order equation may indicate either mechanism, but from this information alone, we have no way of choosing. Further, Mechanism (1) yields a simple first-order equation when $k_3 \gg k_2$. This suggests a common trap which must be avoided in interpreting kinetic data: complex mechanisms may yield simple rate equations under appropriate conditions. The fact that measured rate data fit a simple equation is not sufficient evidence to conclude the reaction mechanism is also simple.

B. First-Order Reactions

A chemical reaction involving a single chemical species spontaneously reacting to form one or more products is unimolecular and should follow first-order kinetics. The rate of reaction at any time is proportional to the concentration of the substance. If the substance is symbolized by A, the initial concentration of A is a, and the amount of A reacted per unit volume at any time is x, then

$$-dC_A/dt = k_1 C_A = k_1(a-x) = dx/dt \qquad (5)$$

The integrated form of Eq. (5) is

$$\ln \frac{a}{a-x} = k_1 t \qquad (6)$$

For graphical solution, Eq. (6) is rearranged to the form

$$\log (a-x) = \left[\frac{-k_1}{2.303} \right] t + \log a \qquad (7)$$

A plot of $\log (a-x)$ versus t yields a straight line for experiments in which first-order kinetics are followed (Fig. 1). The y intercept is $\log a$, and k_1 is found from the slope of the line by:

$$\text{slope} = \frac{-k_1}{2.303}, \qquad \text{or } k_1 = -2.303\,(\text{slope}) \qquad (8)$$

The units of k_1 are reciprocal time; $\min^{-1}$ and hr^{-1} are commonly used, but even day^{-1} may be used for slow reactions in natural waters. Concentrations in solution are usually expressed in moles per liter; however, it is obvious that concentration units do not affect the value of k_1. Since reaction rates are temperature variant, the temperature at which the reaction is conducted should be specified.

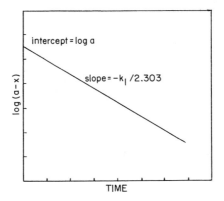

Fig. 1. Theoretical plot of log $(a-x)$ versus time for first-order reactions.

A useful term for first-order reactions is the half-life period. This is the time required for one-half of the reactant to decompose. From Eq. (6)

$$t_{1/2} = \frac{1}{k_1} \ln\left[\frac{a}{a/2}\right] = \frac{\ln 2}{k_1} = \frac{0.693}{k_1} \tag{9}$$

Equation (9) shows that the half-life of a first-order reaction is independent of the initial concentration. Thus, a first-order reaction takes as long to go halfway to completion when the initial concentration is high as when it is low. If $t_{1/2}$ is known or can be determined, it is a simple matter to compute k_1 from Eq. (9).

The decay of radioisotopes is a simple example of first-order reactions. Radioisotopes decompose spontaneously at a rate proportional to the amount of isotope left at any time. Thus, for the decomposition of carbon 14:

$$^{14}C \longrightarrow {}^{14}N + \beta^- \tag{10}$$

The decay rate is given by

$$-d(^{14}C)/dt = k(^{14}C) \tag{11}$$

Rates of radioactive decay cannot be modified by any chemical or physical means but are dependent only on the concentration of isotope. Radioisotope decomposition rates are usually expressed in terms of their half-life; for example, $t_{1/2}$ for ^{14}C is 5770 years(25). Many reactions of importance in natural waters are first order or follow first-order kinetics under certain circumstances. Examples are given in later paragraphs.

C. Second-Order Reactions

The rate expression for second-order kinetics depends on the concentration of two reactants. Bimolecular reactions of the general type

$$A + B \longrightarrow products \tag{12}$$

fit second-order reaction rate expressions. If a and b represent initial concentration of A and B, and x is the decrease in each at any time t, then the concentrations of A and B at any time will be $(a-x)$ and $(b-x)$, respectively. The rate equation for a second-order reaction then is

$$dx/dt = k_2(a-x)(b-x) \tag{13}$$

The integrated form of this equation, assuming $x = 0$ at $t = 0$, is

$$k_2 = \frac{2.303}{t(a-b)} \log \frac{b(a-x)}{a(b-x)} \tag{14}$$

Here k_2 has units of $t^{-1} C^{-1}$ and is equal to the reaction rate when A and B are unity. The magnitude of k_2 depends on the nature of the reaction, temperature, and the units in which t and C are expressed.

If A and B are initially present in equal concentrations or are the same compound, as in the reaction,

$$2A \longrightarrow products \tag{15}$$

Eq. (13) is simplified to

$$dx/dt = k_2(a-x)^2, \tag{16}$$

the integrated form of which is

$$\frac{1}{a-x} = k_2 t + 1/a \tag{17}$$

Graphical solutions for second-order reaction data are obtained by rearranging Eq. (14) in the form

$$\log \frac{(a-x)}{(b-x)} = \frac{k_2(a-b)}{2.303} t - \log b/a \tag{18}$$

Since a, b, and k_2 are constant for any given reaction, a plot of $\log (a-x)/(b-x)$ versus t gives a straight line for reactions following second-order kinetics, and k_2 is obtained from the slope of the line (Fig. 2):

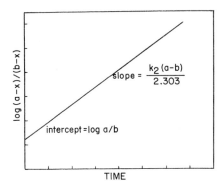

Fig. 2. Theoretical plot of log $(a-x)/(b-x)$ versus time for second-order reactions.

$$\text{slope} = \frac{k_2(a-b)}{2.303} \qquad (19)$$

For reactions of the type shown in Eq. (15), a plot of $1/(a-x)$ versus t [see Eq. (17)] should be linear with a slope equal to k_2.

The half-life period for second-order reactions is an ambiguous term if the concentrations of A and B are not equal, since they will then have different times for half-reaction. In reactions where the concentrations are the same or the two reacting species are identical, the half-life period (where $x = a/2$) is obtained from Eq. (17) as:

$$t_{1/2} = \frac{1}{k_2 a} \qquad (20)$$

Thus, for second-order reactions, the half-life period is inversely proportional to the first power of the initial concentration, and k_2 is easily obtained if $t_{1/2}$ and a are known.

A number of reactions important in aqueous environmental chemistry are second order. However, in most cases the reactions are complicated by consecutive or competing reactions or catalytic effects, and simple second-order formulations do not fully describe their kinetics. Several examples of second-order reactions (and their complicating factors) are described later in this discussion.

D. Third-Order Reactions

Reactions whose rates depend on the concentration of three reactants are less common in water chemistry. Termolecular reactions are of the general type

$$A + B + C \longrightarrow \text{products} \qquad (21)$$

If A, B, and C are all different molecules present at different initial concentrations a, b and c, respectively, the rate equation is

$$dx/dt = k_3 (a-x)(b-x)(c-x) \qquad (22)$$

The integrated form of Eq. (22) is rather complex; there are relatively few situations where these conditions prevail. More commonly, two molecules are identical, i.e.,

$$2A + B \longrightarrow \text{products} \qquad (23)$$

The concentrations at any time are $(a-2x)$ and $(b-x)$, and the rate equation is given by

$$dx/dt = k_3 (a-2x)^2 (b-x) \qquad (24)$$

the integrated form of which is

$$k_3 = \frac{1}{t(2b-a)^2} \left[\frac{2x(2b-a)}{a(a-2x)} + \ln \frac{b(a-2x)}{a(b-x)} \right] \qquad (25)$$

The simplest third-order reaction is of the type

$$3A \longrightarrow \text{products} \qquad (26)$$

or reaction (21) with $a = b = c$. The rate equation is

$$dx/dt = k_3 (a-x)^3 \qquad (27)$$

which is integrated to give

$$k_3 = \frac{1}{2ta^2} \left[\frac{x(2a-x)}{(a-x)^2} \right] \qquad (28)$$

E. Zero-Order Reactions

Reactions whose rates are independent of reactant concentration are said to be zero order. The rate equation for such reactions is given by

$$dx/dt = k \qquad (29)$$

which yields on integration

$$k = x/t \qquad (30)$$

The concentration independence for what appear to be zero-order reactions is often only apparent; many times the concentration at the reaction site is maintained at a constant level because of a substrate reservoir in another phase, thus yielding apparent zero-order kinetics. Examples include reactions of gases which adsorb strongly to solid surfaces and reactions in saturated solutions with a solid layer of the reactant at the bottom. These examples are heterogeneous (two-phase) reactions. However, many homogeneous reactions in natural waters can also be described by zero-order kinetics either because steady-state concentration is maintained by continuous influx of reactant or because the reaction is slow enough that reactant concentrations change only slightly during the period of observation. Rates of such slow reactions could be measured by sensitive tracer techniques.

F. Consecutive Reactions

Many reactions important in natural waters do not proceed in one step as the above kinetic descriptions imply, but rather pass through a series of intermediate steps before forming the final product. Such reactions as

$$A \xrightarrow{k_1} B \xrightarrow{k_2} C \tag{31}$$

are termed consecutive reactions. In any sequence of reactions of varying speed, the slowest step determines the rate of the overall reaction since each step depends on the previous reaction for its reactants. In a reaction sequence such as that in Eq. (31), the conversion of A to B will determine the rate of product (C) formation if $k_2 \gg k_1$. On the other hand, if $k_1 \gg k_2$, the conversion of B to C will control the rate of product formation. When k_1 and k_2 are comparable in magnitude, the overall reaction depends on both constants, resulting in complex mathematical descriptions. For example, the simple case in Eq. (31) results in the following rate expressions. If a is the initial amount of A, x the amount of A reacted (or B formed), and y the amount of C formed at any time t, then

$$dx/dt = k_1(a-x) \tag{32}$$

and

$$dy/dt = k_2(x-y) \tag{33}$$

The time-concentration course of A, B, and C is found by first integrating Eq. (32) and solving for x, and then substituting for x in Eq. (33) and integrating it. The following expressions are obtained for concentrations of A, B, and C (3):

$$[A] = (a - x) = a \exp(-k_1 t) \tag{34}$$

$$[B] = (x - y) = \frac{k_1 a}{k_2 - k_1}[\exp(-k_1 t) - \exp(-k_2 t)] \tag{35}$$

$$[C] = y = \frac{a}{k_2 - k_1}[k_2(1 - \exp(-k_1 t)) - k_1(1 - \exp(-k_2 t))] \tag{36}$$

G. Chain Reactions

A special type of consecutive reaction phenomenon, called a chain reaction, occurs when the reactants form intermediate(s) which react with more reactant to form product and more intermediate, for example

$$A_2 \rightleftharpoons 2A$$
$$A + B_2 \rightleftharpoons P + B$$
$$B + A_2 \rightleftharpoons P + A$$

overall
$$A_2 + B_2 \rightarrow 2P \tag{37}$$

A classic example of a chain reaction is the hydrogen–halogen reaction, $H_2 + X_2 \rightarrow 2HX$ which follows the mechanism described in Eq. (37) with $X_2 = Br_2$ or $Cl_2 = A_2$, $H_2 = B_2$ and $HX = P$. The chain is initiated by thermal dissociation or photolysis of X_2 to $2X$. Chain propagating steps form two molecules of HX and regenerate the halogen atom for another cycle. The reaction thus continues until the reactants are depleted or until the chain is broken by recombination of two X atoms to form X_2. The main reaction sequence consists of 4 steps:

Chain initiation (a) $X_2 \xrightarrow{k_1} 2X$

Chain propagation (b) $X + H_2 \xrightarrow{k_2} HX + H$ (38)

(c) $H + X_2 \xrightarrow{k_3} HX + X$

Chain termination (d) $2X \xrightarrow{k_4} X_2$

The overall reaction sequence is further complicated by side reactions when traces of moisture and oxygen are present and by the reversibility of the chain propagation reactions. The reaction rate equation can be derived from (38) using the steady-state treatment for the reactive atoms

$$d(X)/dt = 0 = k_1(X_2) + k_3(H)(X_2) - k_2(X)(H_2) - k_4(X)^2 \qquad (39)$$

$$d(H)/dt = 0 = k_2(X)(H_2) - k_3(H)(X_2)$$

Solving these equations for steady state concentrations of the atoms gives

$$(X) = [(k_1/k_4)(X_2)]^{1/2}$$

$$(H) = k_2(H_2)[(k_1/k_4)(X_2)]^{1/2}k_3^{-1}(X_2)^{-1} \qquad (40)$$

The rate of product (HX) formation is given by

$$d(HX)/dt = k_2(X)(H_2) + k_3(H)(X_2) \qquad (41)$$

Substituting for (X) and (H) and simplifying yields

$$d(HX)/dt = 2k_2K^{1/2}(H_2)(X_2)^{1/2} \qquad (42)$$

where $K = (k_1/k_4)$. The overall reaction rate equation for hydrogen–halogen reactions becomes further complicated when the reverse reactions for chain propagation steps are considered. However, the above treatment gives the general approach to deriving rate equations for chain reactions. It is interesting to note that Eq. (42) predicts the reaction to be first-order with respect to H_2 but only half-order with respect to X_2. Chain reactions characteristically (but not always) yield complex kinetic expressions, often with fractional orders.

All chain reactions have mechanisms involving the three stages of initiation, propagation and termination; the species or carriers that propagate the chain are atoms or free radicals — species with one (usually) or more unpaired electrons. Evidence that a particular reaction is a chain reaction can be obtained directly by detection of the electronic and other spectra of free radical intermediates and indirectly by several means. Chain termination occurs readily at the walls of the reaction vessel, or any solid surface, where radicals adsorb and recombine. Thus, if an increase in the ratio of surface area to volume of the vessel causes a decrease in reaction rate, a chain reaction with surface termination is probably taking place. A variety of substances also act as inhibitors of chain reactions, apparently by removing chain carrying radicals. In gas-phase reactions these inhibitors include nitric oxide and propylene; in liquid phase reactions a variety of Lewis bases — amines, alcohols, and other hydroxylated substances — act as effective chain breakers by forming less active radicals or nonradicals.

Chain reactions are common in organic chemistry and in many oxidation processes, e.g., of hydrocarbons, carbon monoxide, hydrogen sulfide and hydrogen. In atmospheric chemistry the oxidation of NO to NO_2 and formation of photochemical smog may involve free radical chain reactions(26), and in natural water chemistry, oxidation of H_2S and sulfite by molecular oxygen are thought to involve chain reactions.

H. Examples of Consecutive and Chain Reactions in Natural Waters

Consecutive reaction kinetics apply to the hydrolysis of some condensed phosphate compounds in natural waters. Tripolyphosphate hydrolyzes in a two step process with pyrophosphate as an intermediate:

$$M_5P_3O_{10} + H_2O \xrightarrow{k_1} M_3HP_2O_7 + M_2HPO_4 \qquad (43)$$

$$M_3HP_2O_7 + H_2O \xrightarrow{k_2} M_2HPO_4 + MH_2PO_4 \qquad (44)$$

where M symbolizes a monovalent metal ion, and k_1 and k_2 are first-order rate constants for hydrolysis of tripolyphosphate and pyrophosphate, respectively. Clesciri and Lee(27,28) determined hydrolysis rates for reactions (43) and (44) by measuring increases in orthophosphate with time. Reaction (44) was studied separately and k_2 was evaluated by standard first-order graphical plots. However, no simple method such as graphing the data can be used to evaluate k_1 since the only species that can be easily measured (orthophosphate) is also a product of the consecutive reaction of pyrophosphate hydrolysis. An equation to determine the rate constant (k_1) was derived, assuming reactions (43) and (44) are first order and independent(27):

$$C = -A_0 \left[\exp(-k_1 t) + \frac{2}{(k_2 - k_1)} (k_2 \exp(-k_1 t) + ck_1 \exp(-k_2 t)) \right]$$
$$+ A_0 \left[1 + \frac{2}{(k_2 - k_1)} (k_2 + ck_1) \right] + C_0$$

where

$$c = \frac{B_0(k_2 - k_1)}{A_0 k_1} - 1 \qquad (45)$$

Initial orthophosphate is given by C_0, and orthophosphate at time t is given by C; A_0 and B_0 are initial tripolyphosphate and pyrophosphate concentrations, respectively. The complexity of Eq. (45) implies the desirability of a computer solution for k_1. Clesciri and Lee[27] found k_1 in sterile lake water at 25°C to be 1.4×10^{-4} min^{-1} by this method; comparison of this value with results obtained in the presence of microorganisms under similar conditions[28] led these workers to conclude that tripolyphosphate hydrolysis in natural waters is controlled by enzymatic processes.

Another example of consecutive reaction kinetics in water chemistry is the chlorination of phenols and substituted phenols. The initial reaction of chlorine with phenol or substituted phenols conforms to a second-order rate expression. However, the chlorinated phenols can react to form more highly chlorinated products, resulting in a complex series of consecutive and competing reactions. Studies on this process have been extensively published elsewhere[29,30]. Morris[31] has also reviewed his and his co-workers' extensive kinetic studies on the chlorination of ammonia and amino compounds. These reactions are generally second order overall and first order in each reactant (chlorine and nitrogen compound).

The rate constant for formation of monochloramine (NH_2Cl) is about 10^4 times that for the formation of dichloramine ($NHCl_2$). The kinetics of each step can thus be studied separately, even though the overall sequence represents a consecutive reaction.

Oxidation of sulfide by dissolved oxygen is a common reaction in natural waters and generally occurs in transition zones between oxygenated and anoxic waters. Sulfide is produced by biological activity in anoxic waters (e.g., lake hypolimnia) and is oxidized via biological and chemical reactions when the anoxic water is mixed with oxygenated surface water. The mechanism of sulfide oxidation by dissolved oxygen is complex, involving consecutive and competing reactions, and a mixture of intermediates and products (elemental sulfur, sulfite, thiosulfate, and sulfate) are formed. Attempts to unravel the complicated kinetics have recently been reported. Cline and Richards[32] studied the kinetics of this reaction in seawater and found that the reaction fit a mixed second-order model (first-order in both total sulfide and dissolved oxygen) after the first few hours of reaction. During the initial period of reaction the reaction coefficient appeared to change continuously, thus obviating any simple kinetic interpretation. The reaction coefficient is given as the molar ratio (a/b) of hydrosulfide ion to oxygen consumed in the overall reaction: $a\mathrm{HS}^- + b\mathrm{O}_2 = $ products. The value of the reaction coefficient depends on the nature and relative proportions of the various products (mainly $S_2O_3^{2-}$, SO_3^{2-}, and SO_4^{2-}) being formed at any time. The authors[32] also reported a negative surface effect, i.e., reaction rate decreased as solid surface area increased. It was suggested that surfaces are active in degrading free radical intermediates.

Chen and Morris(33) further emphasized the complexity of aqueous sulfide oxidation by O_2. The initial process was found to exhibit considerable lag or autocatalytic character (see Sec. II.B.); further the stoichiometry was not constant with pH or time. The latter effect evidently results from changes in the amounts of the various partially oxidized sulfur species ($S_2O_3^{2-}$, etc.) formed as a function of time and pH. For example, increasing the pH of the reaction medium from 7.94 to 8.75 resulted in an increase in the fraction of total sulfur present as $S_2O_3^{2-}$ after 48 hr incubation from 0.25 to 0.34 and a decrease in the form present as SO_3^{2-} from 0.18 to 0.11. Elemental sulfur was not found to be a significant intermediate, except at higher sulfide concentration in solutions with pH less than 7. The order of the initial reaction with respect to sulfide was found to be fractional between first and second order within the pH range 6.90 to 10.30 [see Fig. 3(a)]. However, the dependency of reaction rate on O_2 concentration was found to be less than first order [Fig. 3(a)]. The authors fit the data for varied total sulfide and O_2 concentrations to an empirical rate equation which explained the observed rates at all pH values from 7 to 12.5 in spite of changes in stoichiometry and products

$$-d(\Sigma S^{2-})/dt = k_A(\Sigma S^{2-})(O_2)\{1 + k_B[(\Sigma S^{2-})/(O_2)]^{1/2}\} \qquad (46)$$

where (ΣS^{2-}) and (O_2) are molar concentrations of total sulfide and dissolved oxygen, respectively, and k_A and k_B are specific rate constants. While k_B was constant at 5.0 (dimensionless) for all pH values, k_A varied from 1.6 to 9.36 as a function of initial total sulfide and initial O_2 concentrations and pH. The effects of the first two parameters on k_A were relatively trivial compared to the effect of pH, where the observed specific rate increased with pH up to the range 8–8.5, decreased sharply near pH 9, then rose to a second maximum near pH 11 and finally declined under more alkaline conditions [Fig. 3(b)]. Over periods of several days the stoichiometry of the reaction indicated a net oxidation primarily to thiosulfate ($S_2O_3^{2-}$) since the reduction in oxygen per molecule of sulfide oxidized approached unity as the reaction

$$2\,H^+ + 2\,S^{2-} + 2\,O_2 \longrightarrow S_2O_3^{2-} + H_2O \qquad (47)$$

predicts.

Chen and Morris further found that the oxidation of sulfide by O_2 is catalyzed by a variety of metal ions, in order of activity $Ni^{2+} > Co^{2+} >$

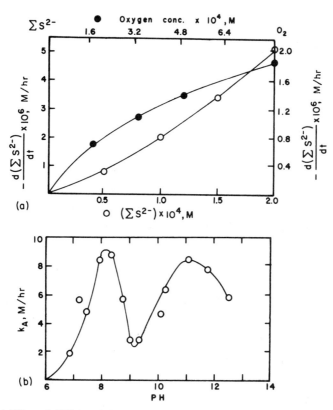

Fig. 3. (a) Effect of (ΣS^{2-}) and O_2 concentration on rate of reactant disappearance. Both curves were run at pH 8.34. The lack of linear response indicated fractional order for each reactant (less than first order for O_2, but greater than first order for (ΣS^{2-})). (b) Effect of pH on specific rate constant k_A for sulfide oxidation. [Both (a) and (b) are from Ref. *33* by courtesy of Pergamon Press.]

$Mn^{2+} > Cu^{2+} > Fe^{2+} > Ca^{2+}$ or Mg^{2+}. Nickel(II) ion is a particularly effective catalyst, reducing the time for oxidation of 0.01 M sulfide at pH 8.65 from several days to a few minutes. However, as Fig. 4(a) illustrates, nickel(II) ion also changes the apparent stoichiometry from 1.0 mole O_2 per g-atom S^{2-} toward 0.5 mole/g-atom and causes visible precipitation of elemental sulfur under pH conditions where this did not otherwise occur. The authors suggested that Ni(II) primarily catalyzes the first stages of oxidation to such a degree that later stages cannot keep pace. Chen and Morris also concluded that sulfide oxidation by O_2 is a chain reaction based on the fact that EDTA and a variety of other amine compounds inhibited the reaction rate [Fig. 4(b)]. This inhibition could

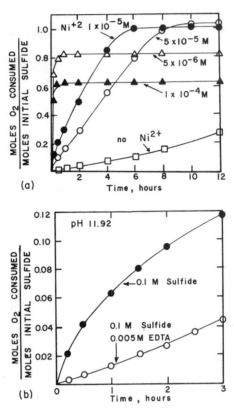

Fig. 4. (a) Effect of Ni(II) on stoichiometry and rate of sulfide oxidation by O_2. (b) Effect of EDTA on rate of sulfide oxidation by O_2. [Both (a) and (b) are reprinted from Ref. *33* by courtesy of Pergamon Press.]

not be explained solely by the possibility of the compounds complexing metal ion catalysts. EDTA and the other organic inhibitors were considered to act as chain breakers. A further complication in unraveling of the reaction mechanism is the authors' findings that glycerol, ethanol, and other hydroxylic materials, which are known chain breakers in sulfite oxidation by O_2, did not affect the rate of sulfide oxidation. Chen and Morris speculated that this indicates a different reaction chain in sulfite oxidation, but the reaction chains for both reactions have yet to be determined.

The kinetics of sulfite oxidation by dissolved oxygen was the subject of a separate study by Rand and Gale (*34*). These workers found that the rate data fit an equation involving fractional orders of sulfite and hydrogen ion ($dC/dt = 1.48 \times 10^{21} [SO_3^{2-}]^{2.33} [H^+]^{1.58}$) over the pH range 6.5–7.7. The oxidation rate was independent of dissolved oxygen at concentrations greater than 0.8 mg/liter. In the pH range 4.2–6.4 the rate was dependent on the first power of sulfite concentration and an undefined variable. The rate of sulfite oxidation is thus complex, and no single rate equation describes the kinetics over the whole range of variables. This indicates a complex mechanism which probably involves chain reactions or competing pathways.

I. Reversible Reactions

In all the above reactions, it was assumed that the processes proceed only in the direction indicated and that the products once formed have no tendency to react to form the original reactants. Such an assumption is valid only for reactions where the equilibrium lies far to the right (on the side of the products). If the equilibrium conditions for a reaction indicate appreciable concentrations of both products and reactants (that is, a reversible reaction), it is apparent that the reverse reaction will commence once the products begin to accumulate. Expressions describing the kinetics of such situations are thus complicated by introduction of a term to correct for the back reaction. For example, consider a simple first-order reaction, $A \rightarrow P$. If the reverse reaction is appreciable, the differential rate equation is given by

$$dx/dt = k_1(a-x) - k_{-1}(p+x) \tag{48}$$

where a and p are initial concentrations of reactant and product and k_{-1} is the first-order rate constant for the reverse reaction. The integrated form of this equation is

$$\ln \frac{[k_1 a - k_{-1} p]}{[k_1(a-x) - k_{-1}(p+x)]} = (k_1 + k_{-1})t \tag{49}$$

For higher order reversible reactions the rate expressions become considerably more complex, and the interested reader is referred to specialized texts (35–37) for further details. Frequently (but not always) problems of back reactions can be avoided in kinetic studies of reversible reactions by starting with an initial product concentration of zero and measuring the initial (forward) velocity.

J. Effect of Temperature on Reaction Rates; Theories of Reaction Rates

It is a well-known fact that chemical reactions occur more rapidly as temperature is increased. While the degree to which temperature affects reaction rates varies according to the nature of the reaction, it is in nearly all cases significant. The Arrhenius equation describes the temperature effect on reaction rates mathematically:

$$d \ln k / dT = E^{\neq} / RT^2 \tag{50}$$

In Eq. (50), k is the reaction rate constant, T the absolute temperature, R the gas constant in calories, and $E^{\neq}$ the activation energy for the reaction. Activation energy, a characteristic for each reaction, can be considered as the energy which reactants must absorb in order to become activated and react. Its significance will be further discussed presently. Assuming $E^{\neq}$ to be constant, Eq. (50) can be integrated:

$$\log k = \frac{-E^{\neq}}{2.303RT} + \text{constant}$$

or

$$k = (\text{constant}) \exp(-E^{\neq} / RT) \tag{51}$$

Between the limits of T_1 and T_2, Eq. (51) becomes

$$\log k_2 / k_1 = \frac{(T_2 - T_1) E^{\neq}}{2.303 \, RT_1 T_2} \tag{52}$$

From Eq. (51), if we plot $\log k$ vs. $1/T$, the result will be a straight line with a slope of $-E^{\neq}/2.303R$ if the Arrhenius equation is obeyed. From Eq. (52), we can determine $E^{\neq}$ from reaction rate constants at two different temperatures. Using Eq. (52) and the computed value of $E^{\neq}$ we can then estimate k values at other temperatures.

According to the concept of activation, reactants do not pass directly into products, but first go through an intermediate activated or transition state. The time-energy course for a reaction can be represented as shown in Fig. 5. The reactants (A) are at one energy state, while the products (B) are at a lower energy state (in an exergonic reaction). The energy difference between A and B is ΔH, the enthalpy or heat of reaction. In passing from A to B, the reactants pass through a transition state $A^{\neq}$ of higher energy, which requires them to absorb a certain amount of energy, $E^{\neq}$, the activation energy. Once the transition state is formed, it can proceed spontaneously to form B and release an amount of energy equal to $E^{\neq} + \Delta H$; thus, the net energy change for the reaction is ΔH.

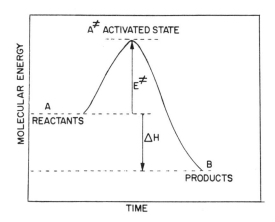

Fig. 5. Time-energy course of an exergonic reaction.

The concept of activation is valid for both theories of reaction mechanism, the collision theory of reaction velocity and the theory of absolute reaction rates. According to the collision theory the rate of reaction is determined by the number of collisions of reactants per unit time times the fraction of collisions whose reactants have sufficient energy for them to be activated. For gases, the number of collisions (Z) at a given temperature and pressure can be predicted from the kinetic theory of gases. The energy distribution among molecules is given by the Maxwell distribution law:

$$n^*/N = \exp(-E/RT) \tag{53}$$

where n^* is the number of molecules possessing an energy of E or greater at temperature T, and N is the total number of molecules. Obviously, for larger values of E, the fraction of molecules having (at least) that energy becomes rapidly smaller at constant T. If E is identified with $E^{\neq}$, Eq. (53) is seen to represent the second factor in the Arrhenius equation, Eq. (51). Then the constant in Eq. (51) is identified with Z, the frequency of molecular collisions at the stated temperature. Reaction rates at any given temperature are thus inversely dependent on activation energy. It is also apparent from Eqs. (51) and (53) that increasing temperature increases the fraction of molecules having an energy of $E^{\neq}$ or higher and therefore increases the rate of reaction.

Collision theory yields accurate predictions of reaction rates in only a few cases. Generally, predicted rates are larger than actual rates by a factor of $10-10^6$. Simple collision theory assumes that only activation energy and a rigid sphere collision frequency determine the rate of reaction. However, it is obvious that not all collisions between reactant molecules will lead to reaction; the molecules must have a proper spatial relationship during collision. In essence the atoms and bonds in the two molecules must be properly oriented so that energy and mass transfers can be effected. This steric factor and other complex factors such as entropy differences between reactants and transition states contribute to the general inadequacy of collision theory for predictive purposes. Collision theory of reaction rate is discussed in greater detail in physical chemistry texts(1–4).

The theory of absolute reaction rates is a more advanced approach based on statistical mechanics. In this case the transition state or activated complex is considered to exist in a certain equilibrium with the reactants. Based on this idea, Eyring(38) showed that the rate constant (k') for any reaction should be given by

$$k' = (kT/h)\, \exp\left(-\Delta G^{0\neq}/RT\right) \tag{54}$$

where k is Boltzmann's constant, h is Planck's constant, and $\Delta G^{0\neq}$ is the standard free energy of activation, or free energy difference between the activated complex and reactants. Derivation of Eq. (54) is beyond the scope of this discussion but can be found in several texts on physical chemistry(1–$4, 35, 37$). From the relationship between free energy, enthalpy, and entropy, i.e., $\Delta G^{0\neq} = \Delta H^{0\neq} - T\Delta S^{0\neq}$, Eq. (54) can be rewritten as

$$k' = (kT/h)\, \exp\left(\Delta S^{0\neq}/R\right) \exp\left(-\Delta H^{0\neq}/RT\right) \tag{55}$$

The enthalpy of activation ($\Delta H^{0\neq}$) and experimental activation energy ($E^{\neq}$) are nearly equivalent. Thus, the entropy of activation can be calculated from experimental k' and $E^{\neq}$. The entropy of activation is a measure of the freedom or disorder of the transition state compared to the reactants. The reactants are at least loosely bound in the transition state, with a consequent loss of freedom; thus, negative activation entropies are more common than positive values.

K. Kinetics of Reactions in Solution

Homogeneous gas phase kinetics is described rather well by the above laws; i.e., those derived from mass action concepts and the Arrhenius equation for temperature effects. The situation is rather more complicated for solution phase reactions, although the above laws are of course important in describing these reactions. Complete theoretical interpretation of reaction kinetics in solutions requires an understanding of such concepts as the Debye–Hückel theory for activity coefficients of ions and the theories of liquid dielectrics and polar molecules.

Salt effects are particularly important in dealing with kinetics of ionic reactions in water. The activity rate theory of Bronsted and Bjerrum was the first to present an adequate treatment of these reactions($1,36$). According to this theory, reactions between two ions are considered to proceed through an activated complex:

$$A^{Z_A} + B^{Z_B} \longrightarrow (A+B)^{Z(A+B)} \longrightarrow \text{products} \qquad (56)$$

The complex is considered to be in equilibrium with the individual ions, and the overall reaction rate is proportional to the concentration of complex. (This is in agreement with Eyring's theory of absolute reaction rates as discussed earlier.) The complex and free ions are related by an equilibrium expression, but since ions are involved, the constant must be expressed in activities rather than concentrations. Thus

$$K^{\neq} = \frac{a^{\neq}}{a_A a_B} = \frac{C^{\neq}}{C_A C_B} \frac{\gamma^{\neq}}{\gamma_A \gamma_B} \qquad (57)$$

where $K^{\neq}$ is the equilibrium constant for the formation of activated complex from free ions; $a^{\neq}$, a_A and a_B are the activities of the complex, A and B, respectively; C represents concentration; and γ the activity coefficient with proper subscripts for the species denoted. Activity coefficients can be expressed according to the Debye–Hückel theory as a function of solution ionic strength:

$$-\log \gamma_i = Z_i^2 A \sqrt{\mu} / [1 + \beta a_i \sqrt{\mu}] \qquad (58)$$

where A and β are constants characteristic of the solvent at a specified temperature and pressure [for water at 25°C and 1 atm pressure, A and β are 0.5085 and 0.3281×10^{-8}, respectively(10)], a_i is the effective diameter of the ion in solution, and μ is the ionic strength [$= \frac{1}{2}\Sigma m_i Z_i^2$].

For a given ionic strength (hence a given set of activity coefficients), the reaction rate for Eq. (56) is given by the bimolecular mass action equation:

$$-dC_A/dt = k_2 C_A C_B \qquad (59)$$

However, the reaction rate is also proportional to the concentration of activated complex, as suggested by Eq. (56), so that

$$-dC_A/dt = k_2^0 C^{\neq} \qquad (60)$$

which becomes

$$-dC_A/dt = k_2^0 K^{\neq} C_A C_B \gamma_A \gamma_B / \gamma^{\neq} \tag{61}$$

when Eq. (57) is used to replace $C^{\neq}$. Thus, k_2 in Eq. (59) is not a true constant but a product of the true rate constant k_2^0, the equilibrium constant for the activated complex, and the activity coefficients of the ions. Equating Eqs. (59) and (61), we find

$$k^2 = k_2^0 K^{\neq} \gamma_A \gamma_B / \gamma^{\neq}$$

or

$$\log k_2 = \log k_2^0 K^{\neq} + \log \gamma_A \gamma_B / \gamma^{\neq} \tag{62}$$

Using Eq. (58) to replace the activity coefficients, we obtain

$$\log k_2 = B - \frac{Z_A^2 A \sqrt{\mu}}{1 + \beta a_A \sqrt{\mu}} - \frac{Z_B^2 A \sqrt{\mu}}{1 + \beta a_B \sqrt{\mu}} + \frac{(Z_A + Z_B)^2 A \sqrt{\mu}}{1 + \beta a^{\neq} \sqrt{\mu}} \tag{63}$$

where $B = \log k_2^0 K^{\neq}$. From absolute reaction rate theory it can be shown that $B = \log (kT/h) K^{\neq}$, where k is Boltzmann's constant and h is Planck's constant. Thus, k_2^0 in Eqs. (60) to (62) represents the term kT/h, and $k_2 = (kT/h) K^{\neq}$.

Equation (63) can be considerably simplified at low ionic strengths. In this situation the limiting law of Debye and Hückel is applicable:

$$-\log \gamma_i = 0.51 Z_i^2 \sqrt{\mu} \tag{64}$$

Equation (64) is valid for aqueous systems at 25°C up to $\mu = 0.005$. Substituting this expression for γ_i into Eq. (62) results in the following expression after rearranging and simplifying:

$$\log k_2 = \log k_2^0 K^{\neq} + [-0.51 Z_A^2 - 0.51 Z_B^2 + 0.51 (Z_A + Z_B)^2] \sqrt{\mu}$$
$$= B + 1.02 Z_A Z_B \sqrt{\mu} \tag{65}$$

This last equation, a form of the Brønsted equation, predicts that a plot of $\log k_2$ versus $\sqrt{\mu}$ should be linear with a slope nearly equal to $Z_A Z_B$, the product of ionic charges. If the charges on both ions have the same sign, the product is positive and the rate constant increases with ionic strength. If the ion charges have different signs, the product is negative and the rate constant decreases with ionic strength. If one of the reactants is uncharged, the product is zero and ionic strength should not affect the rate constant. While the latter case is true for some ion-molecule reactions, it is not universally valid. For reactions involving a dipolar molecule and an ion or two dipolar molecules, there is an ionic strength effect(*36*). However, the effect is not as marked as in the case of two ions reacting. Equation (65) serves to illustrate one important point with regard to ionic reaction rate studies. In order to obtain reliable and reproducible values for reaction rate constants, it is necessary to work in media of constant or defined ionic strength. The problem is especially serious when pure water or very dilute natural waters are used for the reaction medium, since relative changes in ionic strength are great even for small (on an absolute basis) changes in ionic content. Changes in ionic strength as a reaction proceeds may thus lead to erratic rate constants.

L. Chemical Kinetics in Relation to Chemical Equilibrium

Kinetics is the study of change, especially the study of rates of change. From a philosophical viewpoint this definition is rather unsatisfactory, for how can we study or have knowledge about something that is in the process of changing into something else? According to Denbigh(*7*), many ancient Greek philosophers held that only changeless things could be made the subject of scientific study. The question of how a thing can cease to exist and become something else has long puzzled philosophers and natural scientists. Modern atomic theory provides at least a partial answer: According to this theory atoms themselves do not change during a chemical reaction, only their positions with respect to each other change. Current theories on the how and why of chemical change are still incomplete and unsatisfying; in general, equilibrium theory is more complete and advanced than kinetic theory. Kinetics has been and remains primarily an experimental science. This undoubtedly derives at least partly from the fact that chemical change is such an obvious occurrence. We can all see change (or at least the result of change), and in many cases it is a simple matter to measure the rate of change. The theoretical aspects of change remain somewhat more elusive.

In fact, there is no theory of reaction rates that stands alone; rather, all kinetic theories are founded on principles derived from equilibrium theories, i.e., theories concerning matter in an unchanging condition. Equilibrium predicts the probability and extent of a process occurring; kinetics indicates how fast the process will take place. The relationship of chemical equilibrium to chemical kinetics has been discussed by Denbigh(*7*). For example, the kinetic laws expressing reaction rates as functions of volume concentrations of reactants are merely extensions of the equilibrium law of mass action coupled with collision (kinetic) theory of molecules. The theory of absolute reaction rates is based on a transition state theory in which it is assumed that an activated complex or transition state exists in quasi-equilibrium with the reactants. Furthermore, activation energy, $\overrightarrow{E^{\neq}}$, is considered to be directly related to $\overrightarrow{\Delta U}$, the increase in internal energy in forming the complex from the reactants. The difference between $\overrightarrow{E^{\neq}}$ and $\overleftarrow{E^{\neq}}$ (the activation energies for the forward and reverse reactions) is equal to $n\Delta U$, where n is a small positive integer or its reciprocal and ΔU is the increase in internal energy from left to right of the reaction (7). Thus, the difference between two kinetic quantities is equal to a thermodynamic quantity. The role of equilibrium theories in kinetics is further seen in descriptions of reactions in solutions. Thermodynamic activities replace concentrations in the mass action laws; and activity coefficients derived from Debye–Hückel theory are used to relate measured concentrations to activities. With the philosophical state of affairs discussed at the beginning of this section, it is probable that kinetics will remain the stepchild of thermodynamics. Further advances in reaction rate theory very likely will find their basis in the laws of chemical equilibrium.

III. Heterogeneous Reaction Kinetics

Reactions in natural waters are not limited solely to interactions among dissolved substances but also include reactions between solutes and solid phases (sediments, suspended clays, detritus, organisms, etc.). These reactions are thus termed heterogeneous; their mechanisms and kinetics are generally more complex and less completely defined than those of homogeneous reactions. Heterogeneous reactions can be divided into at least five steps: diffusion of solutes to the solid surface, sorption of solute substrates onto the solid surface, reaction of the sorbed substrates, desorption of the products, and diffusion of products away from the surface (1). Diffusion in solution is usually rapid and not rate controlling, but intraparticle diffusion is sometimes rate limiting in reactions involving semiporous solids. In many cases no reaction occurs in the third step, and we are interested only in the adsorption and desorption of solutes on solid surfaces. This process is generally termed physical adsorption. Another possibility in the third step is chemical reaction between the sorbed substrate and sites on the solid surface. This phenomenon is called chemisorption. The nature of the solid surface changes irreversibly as chemisorption proceeds, and desorption is rather difficult and incomplete. Even in physical adsorption there is some interaction between the sorbate and solid surface, but only weak attractive forces like Van der Waal's and dipole-dipole forces are involved; chemisorption is characterized by strong interaction. However, a clear distinction between physical adsorption and chemisorption cannot always be made. A third possibility for the third step in sorption phenomena is the reaction between substrates on the solid surface to form products. Often, compounds do not measurably react with each other in homogeneous solutions, but they react rapidly if sorbed onto solid surfaces. This case is known as heterogeneous or contact catalysis.

A. Sorption Kinetics

Sorption from aqueous solution has received more attention in recent years because of the potential role of sediments and detritus in storing and transporting radionuclides and organic pollutants (e.g., pesticides) in rivers and streams. Most studies have been highly empirical and more concerned with equilibrium concentrations of sorbate than with kinetic descriptions of the process. However, a few recent kinetic studies can be mentioned.

The Langmuir adsorption model is commonly used in heterogeneous reaction kinetic formulations. This model was originally developed for sorption of gases onto solid catalysts and describes the fraction of surface covered by sorbate (gas) at equilibrium as a function of the gas pressure:

$$y = \frac{aP}{1 + bP} \qquad (66)$$

where y is the amount of gas sorbed per unit area or mass of the sorbent, P is gas pressure, and a and b are constants characteristic of the system under consideration. These constants, which are temperature dependent, are usually evaluated by graphical analysis of experimental data. If Eq. (66) is divided by P and the reciprocal is taken, the resulting equation is

$$P/y = (b/a)P + 1/a \qquad (67)$$

and a plot of P/y versus P should be linear with a slope of b/a and a y intercept of $1/a$ if the sorption follows the Langmuir isotherm. Equations (66) and (67) can be derived from the kinetics of condensation and evaporation (adsorption and desorption) of gas molecules on the solid surface (1–4). The rate of condensation (adsorption) according to the Langmuir model is given by

$$dA/dt = k_a P(1 - \theta) \qquad (68)$$

where A represents adsorption or condensation, k_a is the rate constant at a specified temperature, and θ is the fraction of total surface covered by adsorbed molecules. Thus, the rate of condensation is proportional to gas pressure and the fraction of uncovered surface. The rate of evaporation or desorption of gas from the surface is proportional to the fraction of surface covered:

$$dD/dt = k_d\theta \tag{69}$$

where D represents desorption or evaporation and k_d is the rate constant at a specified temperature. At equilibrium the rates of adsorption and desorption or condensation and evaporation are equal:

$$k_d\theta = k_aP(1-\theta) \tag{70}$$

Solving Eq. (70) for θ, the results are simplified in

$$\theta = bP/(1+bP) \tag{71}$$

where $b = k_a/k_d$. The Langmuir isotherm is often expressed in terms of y, the amount of gas (or sorbate) adsorbed per unit area or mass of adsorbent [Eq. (66)]. Since y is proportional to θ, the fraction of surface covered,

$$y = k\theta = kbP/(1+bP) \tag{72}$$

Equation (66) results from Eq. (72) when kb is replaced by the single constant a.

Sorption rates of gases are generally considered to be rapid, so that equilibrium is quickly obtained. The general applicability of Eq. (68) in describing the slower and more complex kinetics of sorption onto solids from aqueous solutions has not yet been verified. However, the Langmuir model does provide a foundation on which more sophisticated kinetic models can be derived to describe sorption from aqueous solution. Stöber(39) recently studied the dissolution of various silica particles (e.g., vitreous silica, quartz, cristo-balite) to form dissolved oligomeric silicic acid species. It was found that final dissolved silicic acid concentrations were influenced by adsorption of silicic acid onto surfaces of silica particles. A model was developed using the Langmuir adsorption isotherm to explain the kinetics of dissolution and the equilibrium concentrations found. Adsorbed silicic acid eventually stops further dissolution by forming an inhibitive adsorptive surface cover.

Sorption rates are temperature dependent but are generally less affected by temperature than homogeneous chemical reactions. This is not unexpected since physical adsorption involves only weak electronic interactions like hydrogen bonding and dipole–dipole and Van der Waals' forces. Thus, we would expect smaller activation energies (hence less temperature dependence) for these reactions than for homogeneous reactions involving strong electronic interactions. But this should not imply that temperature effects are insignificant in sorption. Carritt and Goodgal(40) found that an Arrhenius plot of rates of phosphate sorption onto estuarine sediments yielded activation energies of 6–8 kcal/mole. These values imply a 1.4- to 1.6-fold increase in rate per 10°C rise in temperature; homogeneous chemical reactions often show a 2- or 3-fold rate increase per 10°C rise. Similar low activation energies (3–5 kcal/mole) were recently found for sorption of 2,4-D (2,4-dichlorophenoxyacetic acid) onto some clays(41). Since the transition state in a sorption reaction would likely involve a lesser degree of freedom for the sorbate than its initial state, negative entropies of activation would be predicted. Haque et al.(41) found entropies of activation ranging from -7.0 to -7.6 e.u./mole at 25°C for sorption of 2,4-D onto three clays.

For sorbates exhibiting an acidic character (e.g., aquometal ions, organic acids), pH also affects the rate and extent of sorption. Morgan and Stumm(42) found that fractional pH changes significantly affected the rate of Mn(II) sorption onto freshly precipitated MnO_2. According to Weber(43), increasing hydrogen ion concentration increases both the rate and capacity for adsorption of alkylbenzene sulfonates onto activated carbon. The pH effect in the latter case results not only from changing the extent of ionization of the sorbate but also from neutralization of negative charges at the carbon surface with increasing hydrogen ion concentration. The effect of pH on rates of adsorption onto clays would also be considerable since negatively charged clays show a strong tendency to sorb hydronium ion(44).

Sorption from aqueous solution is frequently controlled by nonadsorptive processes such as diffusion within the interstices of the particle. For example, Carritt and Goodgal(40) found rapid initial sorption of phosphate onto estuarine sediments (50% was removed from solution in a few hours). However, equilibrium still was not reached after long (70 hr) contact times. These results suggested a mechanism of rapid phosphate adsorption on the sediment particle surface followed by slower intraparticle diffusion of phosphate. Gardner and Lee(45) found similar patterns for rates of oxygen uptake by lacustrine sediments. Intraparticle diffusion has been shown to control the rate of alkylbenzene sulfonate removal from aqueous solution by activated carbon(43,46,47). Rates of diffusion-controlled processes are described by Fick's first law

$$F = -D(\partial C/\partial x) \tag{73}$$

where F is the flux or mass transport through a unit cross section in unit time, C is mass concentration of the diffusing substance, x is the space coordinate in the direction of diffusion, and D is the coefficient of diffusion. Thus, in simple diffusion processes, the driving force is a concentration gradient $(\partial C/\partial x)$. However, as Weber and Rumer[46] point out, intra-particle transport through porous media (e.g., activated carbon) is likely to be affected by a number of molecular forces besides molecular diffusion. When the transport process is accompanied by surface adsorption, determination of characteristic diffusivities from experimental rate data requires that Fick's diffusion law be modified to include an adsorption term. The relationship between concentration of adsorbed solute and concentration of solute free to diffuse is often nonlinear; for example, the common Langmuir adsorption isotherm is nonlinear. In such cases the resulting modification of Fick's law is nonlinear, and numerical integration methods must be used to solve for the diffusion coefficient, D. The mathematics of diffusion-with-adsorption kinetics is beyond the scope of this discussion; the problem is discussed in detail by Weber and Rumer [46] and in other references[48–51].

Sorption kinetics may also be complicated by a dependence of activation energy on the amount of the substance sorbed[52]. Haque et al.[41] recently found that rates of 2,4-D sorption onto illite, kaolinite, and montmorillinite could be explained by a unimolecular equation modified to allow a variation in free energy of activation with the amount of pesticide sorbed:

$$-d(1-\phi)\, dt = k'(1-\phi) \tag{74}$$

where k' is the apparent first-order rate constant, and ϕ is the ratio of 2,4-D sorbed at any time to the 2,4-D sorbed at equilibrium. From absolute reaction rate theory, k' can be defined as

$$k' = \frac{kT}{h} \exp \frac{[-\Delta G_\alpha^{\neq} - \gamma(1-\phi)]}{RT} \tag{75}$$

where k is the Boltzmann constant, h is Planck's constant, $\Delta G_\alpha^{\neq}$ is the free energy of activation at equilibrium, and γ is a constant which specifies the magnitude of the change in free energy of activation as a function of ϕ. Rate constants for the three clays fell between 2 and 22×10^{-7} sec^{-1}, indicating long reaction times (several hundred hours to equilibrium). The low rate constants imply diffusion control since sorption on the clay surface itself is thought to be rapid.

An unusual approach to sorption onto clays was presented by Fruh[53] and Fruh and Lee[54]. They developed a model for sorption of cesium on stratified mica (vermiculite) based on steady-state enzyme kinetics (i.e., similar to the Michaelis–Menten equation presented in a later section). The model was not intended to imply that sorption occurs by a mechanism similar to enzyme catalysis, but the treatment is useful in describing the effects of various competing sorbate substances and inhibitors on the sorption kinetics of the ion (cesium) under study. Three basic patterns of cesium sorption were obtained when other solutes were present. These patterns were distinguishable by graphical analyses similar to the enzyme inhibition patterns discussed later in this chapter, and were described as complete general, partial general, and partial mutual displacement-type inhibition. In the general inhibition types, the presence of inhibitor prevented the maximum uninhibited sorption rate no matter how much sorbate was added. In complete general inhibition, a high inhibitor level caused a zero rate of sorption at low sorbate levels, while the sorption rate would not drop to zero under similar circumstances in partial general inhibition. In partial mutual displacement, the sorbate can replace the inhibitor at a sorption site and the maximum sorption rate can be restored at saturating concentrations of sorbate even though the inhibitor level may be high. This approach has not been applied to other sorption systems, but the general technique may be useful in describing the interactions of various solutes in sorption onto clays and other natural sestonic particles.

Reversible reaction kinetics have also been used to describe sorption from aqueous solutions. Cookson and North(55) found that the rate of *Escherichia coli* bacteriophage T_4 sorption onto activated carbon fit a reversible second-order equation, first-order with respect to virus and carbon concentration:

$$tk_1m = \ln\left[\frac{0.5(a_0+b_0+K^{-1}+m)-x}{0.5(a_0+b_0+K^{-1}-m)-x}\right]\left[\frac{a_0+b_0+K^{-1}-m}{a_0+b_0+K^{-1}+m}\right] \tag{76}$$

where a_0 and b_0 are initial virus and carbon site concentrations, respectively; k_1 and k_2 are forward and reverse rate constants, respectively; K is the equilibrium constant $(=k_1/k_2)$, t is time; x is concentration of adsorbed virus at any time t; and $m = [(a_0+b_0)^2+K^{-1}(2a_0+2b_0+K^{-1})]^{1/2}$. Kinetic data on phage sorption were linearized by plotting the right side of Eq. (76) as the ordinate versus t as the abscissa; the forward rate constant, k_1, was determined from the slope of the resulting line. Virus sorption at equilibrium followed the Langmuir isotherm and was temperature independent; K and b_0 were evaluated from Langmuir plots. Phage concentrations were determined by a plaque assay using *E. coli* cells. Sorption did not inactivate the virus. It was felt that the size of the

T_4 phage excluded it from pore areas and sorption occurred only on the outer surfaces of the particles. Cookson(56) further reported on the mechanism of T_4 phage sorption onto activated carbon and found that the forward rate constant for sorption varied markedly with pH (Fig. 6).

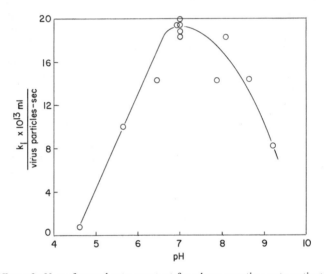

Fig. 6.   Effect of pH on forward rate constant for phage sorption onto activated carbon. [Reprinted from Ref. 56 by courtesy of the American Water Works Association.]

The maximum rate was found at pH 7; the rate dropped to zero below pH 4.5 and above pH 10. These and other results led Cookson to conclude that the mechanism for viral sorption onto activated carbon is provided by amino groups on the viral tail fibers which bind to carboxyl or lactone groups on the carbon by electrostatic attraction.

B. Catalysis and Autocatalysis

A catalyst is defined as a substance which changes the rate of a chemical reaction but which remains unchanged itself at the end. The reaction may be either accelerated or decelerated; the latter case is called negative catalysis and agents inducing such an effect are commonly known as inhibitors or catalyst poisons.

Catalysis can occur either homogeneously or heterogeneously. Homogeneous catalysis in liquid solutions is common and, according to Moore (1), is the rule rather than the exception. Catalysis by acids and bases is undoubtedly the most significant example of homogeneous catalysis in water chemistry. In homogeneous catalysis, the reaction rate depends on catalyst concentration. The full rate expression for a first-order reaction such as the acid-catalyzed inversion of sucrose then becomes:

$$-d(\text{sucrose})/dt = k(\text{sucrose})(\text{H}^+) \tag{77}$$

For any given experiment, (H^+) is constant and Eq. (77) degenerates to a simple first-order expression. The dependence of reaction rates on the concentration of acid or base catalysts can be easily determined by measuring the reaction rate at varying pH values. For example, Stumm and Lee[57] found that the oxygenation of ferrous iron at any given pH in solution can be described by

$$[-d\text{Fe(II)}/dt] = k[\text{Fe(II)}P_{O_2}] \tag{78}$$

where P_{O_2} is the partial pressure of oxygen. However, the reaction is highly pH dependent, as shown in Fig. 7(a). When the logarithm of the

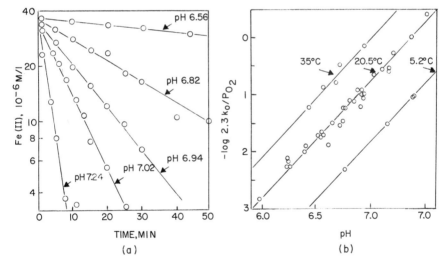

Fig. 7. (a) Oxygenation rates of ferrous iron are first-order with respect to Fe(II) and are strongly affected by pH. (b) Increase in pH by one unit increases oxygenation rate 100-fold. [Both (a) and (b) reprinted from Ref. 57 by courtesy of the American Chemical Society.]

rate constants (computed from Fig. 7(a)) divided by P_{O_2} is plotted versus pH as in Fig. 7(b), the resulting line has a slope of 2, indicating a 100-fold increase in reaction rate for an increase of one pH unit (10-fold increase in hydroxide concentration). The reaction rate thus has a second-order dependence on hydroxide and can be expressed as (57):

$$-dFe(II)/dt = kFe(II)P_{O_2}(OH^-)^2 \qquad (79)$$

Heterogeneous catalysis has been studied and described primarily for reactions of gases sorbed onto solid catalysts because of the importance of such reactions in organic synthesis and the chemical process industry. The most common kinetic treatments of these reactions involve the Langmuir adsorption isotherm; this topic is covered adequately by most physical chemistry texts and is not further developed here. Heterogeneous catalysis in natural water chemistry would not seem to be of general importance, although examples of its occurrence can be cited. Theoretical treatment of this phenomenon has apparently not been highly developed. Morgan and Stumm (42) and Hem (58) have discussed one example of heterogeneous catalysis in natural waters — the oxygenation of aqueous manganese(II). These workers found that manganese dioxide (MnO_2) formed initially as a product. Manganese(II) oxidation results in solid, nonstoichiometric products whose average degrees of oxidation range from $MnO_{1.3}$ to $MnO_{1.9}$ in the pH range 9–10, and these effects have been interpreted as resulting from the sorption or exchange of Mn^{2+} ions onto incipiently formed manganese dioxide (42).

The oxygenation of aqueous manganese(II) is an example of auto-catalysis, i.e., catalysis of the reaction by a product. In this case the maximum rate occurs after some finite reaction time rather than initially, since the concentration of the catalyst (a reaction product) increases with time. Thus, a plot of the quantity reacted versus time produces an S-shaped curve. First-order autocatalytic reactions follow the rate expression

$$dx/dt = k_1(a-x) + k_2 x(a-x) \qquad (80)$$

the integrated form of which is

$$(k_1 + k_2 a)t = 2.303 \log \frac{a(k_2 x + k_1)}{k_1(a-x)} \qquad (81)$$

Figure 8(a) shows the increase in oxidation products of manganese(II) versus time for several pH values (42). The results suggested an auto-catalytic model for Mn(II) oxidation, and Fig. 8(b) shows that Eq. (81) fits the data reasonably well. Autocatalytic equations can also be used to describe the process of sludge digestion, BOD progression where a seeding lag is present, and the general curve for population growth of microorganisms.

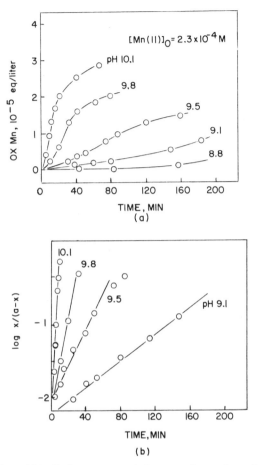

Fig. 8. (a) Oxidation of Mn(II) in aqueous solution at various values of pH. (b) Data of (a) plotted according to autocatalytic model. [Both (a) and (b) reprinted from Ref. *138* by courtesy of the American Chemical Society.]

C. Nucleation and Precipitation

A third type of heterogeneous reaction important in natural water chemistry is the precipitation of sparingly soluble salts from solution. The precipitation of metal sulfides in anoxic water and of calcium carbonate and apatite $[Ca_5(PO_4)_3X$, where $X =$ hydroxide or fluoride] in the ocean, in hard water lakes, and in water softening plants are examples of important precipitation reactions in natural waters. Any precipitation process can be considered as consisting of two steps: nucleation and crystal growth. In the absence of any nuclei, precipitation in super-saturated solutions is often exceedingly slow. This metastable state arises because there is a critical size for nuclei. Below the critical size, the nuclei are unstable and tend to dissolve completely while above it they will continue to grow. This critical size, n^*, is of the order 10–100 atoms or molecules per nucleus(59). The phenomenon of critical size arises from the fact that surface molecules or ions are in a state of higher potential energy than are interior species because the surface ions form fewer and weaker bonds (the crystal lattice is discontinuous at the surface) than interior ions. This excess energy is not compensated by sufficient increase in entropy (i.e., surface and interior ions in crystals are restricted in movement or ordered nearly to the same extent). Thus in nuclei where $n < n^*$, surface effects predominate and ΔG for nucleus growth is positive. For larger nuclei ($n > n^*$) the difference in chemical potential for the two phases (dissolved and crystalline) overrides the surface effects and ΔG becomes negative (assuming the solution is supersaturated).

Nucleation can also take place on other particles suspended in solution; this is known as heteronucleation. Heteronuclei laboratory precipitations arise from such sources as airborne dust, the walls of the vessel, or the presence of less soluble substances in solution. Heteronuclei tend to reduce the period of induction and allow nucleation to occur at lower con-centrations (i.e., lesser degree of supersaturation) than in a homogeneous solution, particularly when the heteronucleus surface matches well with the crystal. The wide variety and often high concentrations of solid particles suspended in natural waters suggest that precipitation in nature probably occurs typically by a heteronucleation mechanism.

Strictly speaking, nucleation is not a chemical reaction of definite order, but Nielsen(59) has found that

$$J = k_n C^{n^*} \tag{82}$$

where J is the rate of nucleation, C is concentration, k_n is the nucleation rate constant, and n^* is critical nucleus size, is a good approximation of the kinetics. Since n^* is a fairly large exponent, it is apparent that the rate of nucleation is highly dependent on concentration.

The rate of crystal growth during precipitation is influenced by two main phenomena: diffusion and surface nucleation. Diffusion of ions from the bulk solution to the crystal surface is generally the controlling mecha-nism for high rates of precipitation. Several mechanisms of surface nucleation can be distinguished. We can think of crystal growth occurring as the sequential addition of ions to a simple cube to build up a new layer. When the layer is completed, growth stops because ions from solution don't find a place of sufficient attraction (where several bonds can be formed) on the completed surface. In order to initiate a new layer, several ions must meet simultaneously in a two-dimensional surface nucleus establishing a sufficient number of bonds. Two limiting cases can occur: (1) the time between two consecutive nucleations on the crystal surface is much larger than the time it takes for a layer to be completely covered; (2) surface nucleation is so fast that each crystal layer is the result of intergrowth of many individual surface nuclei. The former mechanism is called mononuclear layer growth, the latter polynuclear layer growth. In addition to these mechanisms, Nielsen(59) discussed a third mechanism, called dislocation controlled growth, which may be important at low saturation ratios where surface nucleation becomes very slow. Saturation ratio, S, is defined as concentration/solubility, or c/s; some crystals are known to grow rapidly at values of S less than two. Dislocation growth is thought to occur by a spiral step at the surface, winding itself around a screw dislocation. Such dislocations can arise spontaneously, by growth around a foreign particle, by growth on a heteronucleus with unexact matching of lattices, or by other "erratic" events(59).

The kinetics of crystal growth by the above mechanisms is treated mathematically by Nielsen. Development of the kinetic expressions is complicated and beyond the scope of this chapter, but Nielsen describes methods of treating empirical data to distinguish among the possible mechanisms.

Compared to most other inorganic reactions (e.g., acid–base, complexation), precipitation reactions are slow processes, and because of the nucleation step, they are often characterized by long induction periods. In natural waters where S would seldom be large, kinetic control of precipitation is undoubtedly common. In fact, hard waters, including the oceans, are often supersaturated with respect to calcium carbonate. For example, Hawley(60) found the epilimnion of Lake Mendota, Wisconsin, was supersaturated about a hundredfold during certain times of the year. When pure calcium carbonate was added to lake water, equilibration took place rapidly, indicating that the lack of nuclei was the cause of supersaturation. Stumm(18) calculated the equilibrium concentration of phosphate in contact with hydroxyapatite to be 0.03 mg P/liter. As Lee(61) states however, many hard-water eutrophic lakes have inorganic phosphate concentrations 10 to 50 times this value, presumably because the kinetics of nucleation of hydroxyapatite are so slow.

Stumm and Leckie(62) have recently studied the kinetics of apatite formation in laboratory systems containing suspended calcite and have interpreted their data on the kinetics of phosphate loss from solution (Fig. 9) as indicating epitaxial growth of apatite crystals on calcite particles. Thus the calcite crystals act as heteronuclei, but rather than immediate growth of apatite on the calcite, Stumm and Leckie found a long lag period followed a rapid initial uptake. They interpreted this as (rapid) chemisorption of phosphate accompanied by formation of amorphous calcium phosphate nuclei followed by a slow transformation of

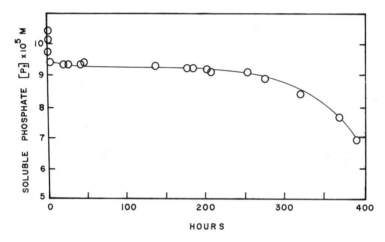

Fig. 9. Rate of soluble phosphate uptake from solution onto calcite at pH 8.25 and 25°C. [Reprinted from Ref. *62* by courtesy of Pergamon Press.]

the nuclei to crystalline apatite. Finally crystal growth of apatite occurred and the rate of phosphate uptake from solution increased (after about 275 hr). Fluoride concentrations as low as $5 \times 10^{-5} M$ were found to increase the rate of apatite formation; since fluoride can substitute for hydroxide in apatite, adding fluoride in effect increases the degree of supersaturation. The long induction time in apatite precipitation suggests that phosphate removal by lime addition in waste water treatment plants is not controlled by apatite equilibrium, but more probably by calcium phosphate solubility.

The reverse reaction of precipitation is dissolution; this of course occurs (on a net basis) only when a solution in contact with a solid substance is unsaturated with respect to that substance. Since nucleation is not involved in dissolution, the rate determining factor usually is diffusion (*59*). A complicating factor in this mechanism is the armoring of particles by other materials, which prevents solubilization of the particle even though the water is undersaturated. For example, Chave (*63*) reported that calcium carbonate particles in the sea are often coated with organic substances (apparently derived from microorganisms) which inhibits their dissolution in calcium carbonate unsaturated sea water. On the other hand, Hawley(60) did not find any evidence for organic coatings on Lake Mendota (Wisconsin) calcium carbonate.

IV. Biochemical Reaction Kinetics and Dynamics

Biological reactions are important in controlling the concentrations of many elements and compounds in natural waters. Distributions of organic compounds such as dissolved amino acids, vitamins, and sugars are controlled largely by rates of excretion and uptake by microorganisms. Nutrient elements like nitrogen, phosphorus, and trace metals are controlled to varying extents by biologically mediated reactions, and many other elements and compounds are also affected indirectly.

Biological reactions are mediated by enzymes, which are highly specific catalysts. All enzymes have a basic protein structure. The detailed mechanisms of enzyme action are beyond the scope of this discussion and in many cases are poorly understood, but evidently enzymes act to decrease the activation energy for a reaction so that it can proceed rapidly at physiological pH values and temperatures. Enzymes can be quite specific in their catalytic actions; for example, the enzyme urease catalyzes the hydrolysis of urea, $(NH_2)_2CO$, to ammonia and carbon dioxide, but does not affect the hydrolysis of substituted ureas, e.g., methyl urea, $(NH_2)(CH_3NH)CO$. Evidently the active sites on enzymes like urease have geometrical configurations such that only compounds of specific stereochemical configurations can interact with them. Other enzymes are less specific; for example, the enzyme lipase hydrolyzes a wide variety of neutral fats.

A. The Michaelis–Menten Equation

The kinetics of enzyme catalysis has long been under development. Henri(64) first theorized in 1903 that the mechanism of enzyme catalysis is through formation of an enzyme-substrate complex. He, and Michaelis and Menten(65) in 1913, assumed that the complex was in equilibrium with free enzyme and substrate and derived a rate equation based on this approach. Briggs and Haldane(66) showed in 1925 that a theoretically more accurate steady-state approach could be used to derive the same equation. The basic equation describing enzyme kinetics is usually called the Michaelis–Menten equation, regardless of the approach used to derive it or the fact that Henri's derivation preceded theirs by 10 years. The basic Michaelis–Menten equation can be derived by the steady-state approach as follows. For a single-substrate single-product enzyme-catalyzed reaction, the simplest mechanism is given by

$$E + S \underset{k_2}{\overset{}{\rightleftharpoons}} ES \underset{k_4}{\overset{k_3}{\rightleftharpoons}} E + P \tag{83}$$

where E represents the enzyme, S the substrate, P the product, and ES the "active complex." Initially P will be low and the reverse reaction $(E + P \xrightarrow{k_4} ES)$ can be neglected. Rate equations for the above two-step sequence are:

$$-d[S]/dt = d[P]/dt = k_1[E][S] - k_2[ES] = k_3[ES] \tag{84}$$

and

$$d[ES]/dt = k_1[E][S] - k_2[ES] - k_3[ES] \tag{85}$$

It is assumed that the overall reaction is controlled by k_3, so that the overall rate is proportional to $[ES]$. If E is present in only catalytic quantities compared to S, after an initial transient period $[ES]$ will reach steady state (i.e., $d[ES]/dt = 0$). Then Eq. (85) simplifies to

$$[ES] = k_1[E][S]/(k_2 + k_3) \tag{86}$$

The conservative equation for enzyme concentration is

$$E_t = [E] + [ES] \tag{87}$$

If the value $[E]$ in Eq. (87) is substituted in Eq. (86), we obtain

$$[ES] = k_1[E_t][S]/(k_2 + k_3 + k_1[S]) \tag{88}$$

which, when substituted into Eq. (84), gives the rate of product formation:

$$-d[S]/dt = d[P]/dt = k_1 k_3[E_t][S]/(k_2 + k_3 + k_1[S]) \tag{89}$$

The Michaelis–Menten equation is an abbreviated version of Eq. (89). It is not practical to measure k_1, k_2, and k_3 separately; the combination $(k_2 + k_3)/k_1$ is defined as the Michaelis constant, K_M. The term $k_3[E_t]$ represents the rate of reaction when all the enzyme is present as the "active complex," i.e., $[E_t] = [ES]$. For a given enzyme concentration this represents the maximum rate possible for the reaction; hence we can define $V_{max} = k_3[E_t]$. Substituting K_M and V_{max} in Eq. (89) and defining v as the rate of reaction yields

$$v = V_{max}[S]/(K_M + [S])$$ (90)

which is the usual form of the Michaelis–Menten equation.

Equation (90) indicates that $v = V_{max}$ when $[S] \gg K_M$. In this case the reaction follows zero-order kinetics. On the other hand, when $K_M \gg [S]$, first-order kinetics is followed, and when K_M and $[S]$ are not too different, description of the rate requires the complete equation. The rate of enzymatic reaction as a function of substrate concentration is shown in Fig. 10(a) for reactions fitting the Michaelis–Menten model. The above mathematical expressions are consistent with the known properties of enzyme-catalyzed reactions. At high substrate concentrations the enzyme-active sites are saturated and reaction rate is independent of substrate concentration (i.e., zero-order); at low substrate concentrations the concentration of "active complex" is proportional to the concentration of substrate, and the reaction is first order (with respect to S).

From Eq. (90) it is apparent that the Michaelis constant, K_M, has units of concentration. When $K_M = [S]$, $v = 0.5 V_{max}$. The usual method of determining K_M and V_{max} is to plot rate versus substrate concentration data according to the reciprocal of Eq. (90), i.e. [Fig. 10(b)]:

$$1/v = 1/V_{max} + K_M/V_{max}[S]$$ (91)

In order to avoid complexities resulting from concentration changes, short incubations are employed and initial velocities (v_0) determined. A plot of $1/v$ versus $1/[S]$ (or $1/v_0$ versus $1/[S]_0$) yields a straight line with a slope of K_M/V_{max} and a y intercept of $1/V_{max}$ if the Michaelis–Menten equation holds. Such a plot is called a Lineweaver–Burk plot, and Eq. (91) is referred to as the Lineweaver–Burk equation. Several other variations of Eq. (90) are used to linearize enzyme kinetic data and determine K_M and V_{max} graphically(67). For example, multiplying Eq. (91) by $[S]$ results in

$$[S]/v = [S]/V_{max} + K_M/V_{max}$$ (92)

and a plot of $[S]/v$ versus $[S]$ gives a straight line with a slope of $1/V_{max}$ and intercept K_M/V_{max} [Fig. 10(c)]. Experimental anomalies occurring at high values of $[S]$ will lie out to the right on this graph rather than near the intercept as they would in a standard Lineweaver–Burk plot. Thus, this type of plot poses fewer problems in extrapolating the linear portions of the data to obtain K_M and V_{max}. If Eq. (91) is multiplied by $V_{max}(v)$ and rearranged,

$$v = -K_M(v/[S]) + V_{max}$$ (93)

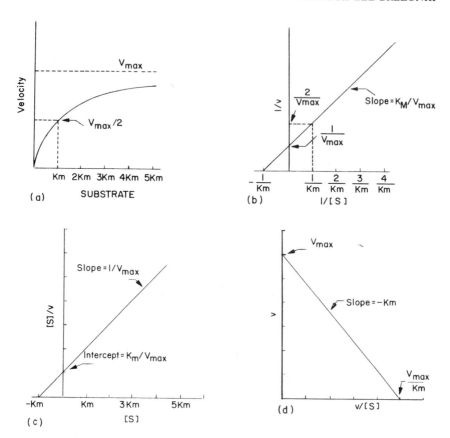

Fig. 10. (a) Theoretical curve of velocity versus substrate concentration for an enzyme catalyzed reaction according to the Michaelis–Menten model [see Eq. (90)]. (b) Lineweaver–Burk reciprocal plot of velocity versus substrate concentration for enzymatic reactions according to Eq. (91). (c) Graphical analysis of enzyme kinetic data according to Eq. (92). (d) Graphical analysis of enzyme kinetic data according to Eq. (93).

and a plot (Fig. 10(d)) of v versus $v/[S]$ yields a straight line with slope $-K_M$ and y intercept V_{max}. Finally, if Eq. (91) is multiplied by $V_{max}[S]$ and rearranged,

$$[S] = V_{max}[S]/v - K_M \qquad (94)$$

and a plot of $[S]$ versus $[S]/v$ yields a straight line with a slope of V_{max} and y intercept of $-K_M$. It should be noted that Eqs. (92) and (94) are essentially equivalent in terms of estimating the kinetic constants; the same variables are plotted in both equations, but the ordinate and abscissa are reversed. Estimates of V_{max} and K_M from the above linear transformations of the Michaelis–Menten equation have been compared by Dowd and Riggs(68). Although the Lineweaver–Burk plot is by far the most commonly used transformation, these authors found it gave by far the least reliable results compared to plots of $[S]/v$ versus $[S]$ and plots of v versus $v/[S]$. Differences among the transformations arise in part because $[S]$ and especially v are subject to experimental error; taking the reciprocal of v gives undue emphasis to the smallest values of v, and these are the values likely to have the largest percentage error. When the error in v was small, plots of $[S]/v$ versus $[S]$ were slightly superior to plots of v versus $v/[S]$, but the reverse was found when the error of v was large. Plots of Eq. (93) have the further advantage of warning when data deviate from the theoretical relationship since this form tends to magnify departures from linearity. Ironically, Dowd and Riggs found that Lineweaver–Burk plots often give a deceptively good straight-line fit with unreliable points, but lines fitted to Lineweaver–Burk plots give the least reliable estimates of K_M and V_{max}.

The Michaelis constant is a specific property of an enzyme, the value of which generally indicates the sensitivity of the enzyme for its substrate. Thus, large values of K_M imply that high concentrations of substrate are required to reach V_{max}. However, K_M is not necessarily a direct measure of the affinity of the enzyme for its substrate, since the turnover of ES in Eq. (83) may not be the rate-limiting step and there may be more than one isomer of the enzyme-substrate complex involved in the reaction. It should be noted that the Michaelis–Menten equation is valid whether or not k_3 is rate limiting.

Most enzyme reactions are at least bimolecular, and the above treatment of enzyme kinetics presents a simplified view of a complex subject. If the concentration of all but one of the reactants can be held constant, the reaction can be made apparently first order, and we can determine apparent first-order rate constants, which are products of the true higher order constants and the concentrations of the other reactants.

The Michaelis–Menten theory has been refined and further developed by many workers(69–74). Detailed kinetic descriptions have been developed for complex enzyme reactions involving three or four reactants and products in a variety of mechanisms [see Refs. (74, 75) for examples], but application of such formulations to reactions in natural waters has not yet been accomplished. The use of enzyme kinetic treatments in natural water studies by and large has not progressed beyond the simple Michaelis–Menten formulation.

B. Inhibition of Enzyme-Catalyzed Reactions

Inhibition of enzyme-mediated reactions can be accomplished by a number of mechanisms. For example, a substance can combine with the active site of an enzyme but be unable to undergo reaction. The classic example of this is inhibition of succinic dehydrogenase by malonate. Succinate ($^-$OOC—CH$_2$—CH$_2$—COO$^-$) can undergo dehydrogenation to yield fumarate ($^-$OOC—CH=CH—COO$^-$), whereas the structure of malonate ($^-$OOC—CH$_2$—COO$^-$) precludes this reaction. This type of inhibitor (e.g., malonate) is called a dead-end inhibitor, and its net effect is to decrease the amount of enzyme available for reaction with succinate substrate. Products can also inhibit by recombining with the enzyme form from which they were dissociated. Some enzymes can use alternate substrates and produce alternate products. If present in the reaction medium, alternate substrates or products can act as inhibitors, and these result in complicated kinetic patterns.

More recently it has been found that substances can react with enzymes at other sites besides the active or reaction site. Such substances, called allosteric (second-site) modifiers, alter enzyme conformation and can act either as inhibitors or activators. In some cases substrate itself can act as an allosteric modifier, either as activator or inhibitor; Mahler and Cordes(79) treat the kinetics of this complicated situation in detail.

Velocity patterns caused by inhibitors can be accommodated and described by expansions of the Michaelis–Menten equation. Three general inhibition patterns are observed from reciprocal plots of $1/v$ versus $1/[S]$ at various inhibitor concentrations. The plots are usually straight lines but the slope, intercept, or both may vary. If only the slope varies (Fig. 11(a), inhibition is competitive and $1/V_{max}$ is unaffected by inhibitor(76). This corresponds to the case where the inhibitor combines reversibly with the same enzyme form as the substrate. (More complicated interpretations are also possible.) In competitive inhibition, raising substrate concentration overcomes the inhibitor so that at substrate saturation ($1/[S] \to 0$) no inhibition occurs (V_{max} is unaffected). The Michaelis–Menten expression for competitive inhibition is given by

$$v = \frac{V_{max}K_{ic}[S]}{K_M K_{ic} + K_M[I] + K_{ic}[S]} \tag{95}$$

and the equivalent Lineweaver–Burk expression is

$$\frac{1}{v} = \frac{K_M}{V_{max}}\left[1 + \frac{[I]}{K_{ic}}\right]\frac{1}{[S]} + \frac{1}{V_{max}} \tag{96}$$

where $[I]$ is inhibitor concentration and K_{ic} is the constant for competitive inhibition. If only the intercept varies [Fig. 11(b)], inhibition is termed uncompetitive and V_{max} decreases with increasing inhibitor con-

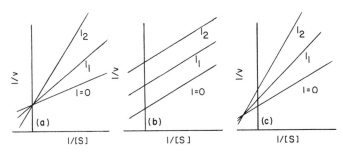

Fig. 11. (a) Lineweaver–Burk pattern for competitive inhibition. (b) Lineweaver–Burk pattern for uncompetitive inhibition. (c) Lineweaver–Burk pattern for simple linear noncompetitive inhibition. I_1 and I_2 refer to different concentrations of an inhibitor; I_2 represents a higher concentration of inhibitor than I_1 in each case.

centration. This corresponds to the case where inhibitor combines irreversibly with a form of the enzyme separated (by an irreversible step) from the form with which the substrate combines. Thus, a fraction of the enzyme is made unavailable for reaction with S, and this is unaffected by saturation with S. The Michaelis–Menten and Lineweaver–Burk expressions for this case are, respectively:

$$v = \frac{V_{\max}[S]K_{inc}}{K_M K_{inc} + [S]K_{inc} + [S][I]}$$

(97)

and

$$\frac{1}{v} = \left[\frac{K_M}{V_{\max}}\right]\frac{1}{[S]} + \frac{1}{V_{\max}}\left[1 + \frac{[I]}{K_{inc}}\right]$$

(98)

where K_{inc} is the constant for uncompetitive inhibition. This case has also been called noncompetitive in some older treatments (77,78).

In many cases inhibition affects both slope and intercept of Lineweaver–Burk plots. This type is called noncompetitive inhibition(76,79), or mixed inhibition in older treatments (78), and results when an inhibitor combines with several forms of an enzyme, one form yielding competitive, the other form yielding uncompetitive inhibition with the variable substrate. More complicated theoretical interpretations can also be made. Linear noncompetitive inhibition is the most common type, with both slopes and intercepts being linear functions of $[I]$ [Fig. 11(c)]. Reciprocal plots always intersect at a point to the left of the $1/v$ axis in this case. The Lineweaver–Burk equation for linear noncompetitive inhibition is

$$\frac{1}{v} = \frac{K_M}{V_{\max}}\left[1 + \frac{[I]}{K_{ic}}\right]\frac{1}{[S]} + \frac{1}{V_{\max}}\left[1 + \frac{[I]}{K_{inc}}\right]$$

(99)

Many other more complex inhibition patterns are possible. Plots of slope or intercept versus [I] may be parabolic or hyperbolic rather than linear. In noncompetitive inhibition, slope can be one function of [I] and intercept a different function. Knowledge of the nature of the inhibition patterns is extremely useful in delineating the mechanism of an enzyme reaction. For further treatment of this subject the reader is referred to specific texts and articles(76,78,79).

C. *Application of Enzyme Kinetics to Environmental Studies*

Enzyme kinetic theories have been applied to several problems in environmental studies. Fruh and Lee(53,54) developed a model based on enzyme kinetics for sorption of cesium on stratified mica (see Sec. III.A). Various graphical analyses (reciprocal plots of sorption rate versus cesium concentration at different inhibitor concentrations) were developed to distinguish among several types of inhibition that could result from the presence of other cations in solution.

Hartmann and Laubenberger(80) applied Michaelis–Menten kinetics and inhibition patterns to toxicity measurements in activated sludge systems. They reported that O_2 uptake rates (a measure of microbial activity) of activated sludge versus chemical oxygen demand (a measure of substrate concentration) followed Michaelis–Menten kinetics. Heavy metal (e.g., copper and chromium) additions to activated sludge resulted in oxygen uptake rates that fit Lineweaver–Burk plots for noncompetitive inhibition [Fig. 12(a)]. Furthermore, inhibition was a linear function of inhibitor concentration, as indicated by plots of slope and intercept versus metal concentration in Fig. 12(b)(81). These results have potential ramifications for further studies on the action of toxicants in mixed microbial systems. However, there is danger in attaching too much significance on the microbial level to the inhibition patterns observed, since inhibition in a mixed system could be the result of a toxicant's action on many different enzyme systems(81). Furthermore, it is not always possible to duplicate the Michaelis–Menten response of respiratory activity to increased food supply. Batch-fed activated sludge from bench-scale units in the author's laboratory did not respond to increases in exogenous food (milk solids)(82). Apparently, organisms grown in this way have sufficient food storage supplies so that changes in exogenous food have little effect on respiratory activity, at least over periods of a few hours. Perhaps if batch-fed organisms were starved for long periods of time (e.g., 24 hr or more), much of their food reservoir would be exhausted and then varying additions of substrate would yield a hyperbolic (Michaelis–Menten) response in respiratory activity.

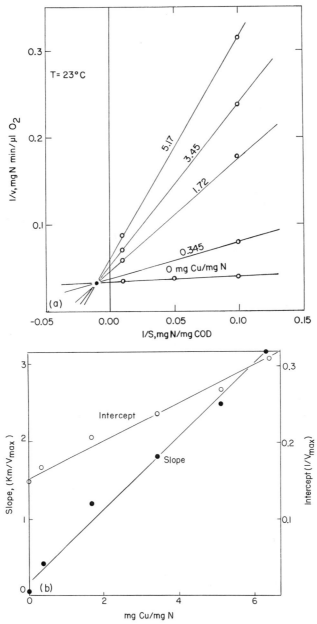

Fig. 12. (a) Lineweaver–Burk plot showing the effect of copper on activity (per milligram of N as a measure of biomass) of activated sludge as a function of substrate (COD) concentrations. [Reprinted from Ref. *80* by courtesy of the American Society of Civil Engineers.] (b). Plot of inhibitor (copper) concentration (per unit of biomass) versus slope and intercept in (a).

However, this situation would hardly correspond to the normal environment of the organisms in a waste treatment facility. On the other hand, activated sludge organisms from continuously fed laboratory units have exhibited Michaelis–Menten response to slug increases in food (milk solids) supply.

Michaelis–Menten kinetics has recently been applied to the dynamics of several organic solutes in natural waters. Parsons and Strickland(83) measured the uptake of carbon 14-labeled organic solutes by microorganisms to evaluate the "relative heterotrophic potential" in seawater and estimated the substrate concentration in water by use of Michaelis–Menten formulations. The method was further developed and refined by Wright and Hobbie(84) and by Vaccaro and Jannasch(85,86), who also discussed some limitations of the method as applied to mixed cultures in natural waters. The method relies on a Lineweaver–Burk plot of uptake rate versus substrate concentration. An equation similar to Eq. (92) was derived(84) whereby v is replaced by a term giving the rate of substrate uptake from isotope tracer data and $[S]$ is replaced by $[S+A]$, S being the original substrate concentration and A the amount of substrate added. The resulting equation is

$$\frac{C\mu t}{c} = \frac{K}{V} + \frac{[S+A]}{V} \tag{100}$$

where C is the counts per minute from $1 \mu Ci$ of ^{14}C in the counting assembly, c is the radioactivity (counts/min) of the filtered organisms after incubation, μ is the microcuries of ^{14}C added, t is length of incubation, K is a constant similar to K_M, and V is analogous to V_{max}. A plot of $C\mu t/c$ versus $[S+A]$ should yield a straight line with a slope of $1/V$. If the ordinate is constructed at $[S]$, i.e., $[A]=0$, the value $[K+S]$ can be calculated. If $K \ll S$, the amount of substrate initially present is approximated. According to Wright and Hobbie(84), V is the best estimate for natural uptake rates under conditions where uptake is independent of substrate concentration during the experimental time. Later studies(87) implied that this method cannot be used to estimate $[S]$ as originally intended, but only $[K+S]$. For trace organic solutes it is difficult to determine either $[K]$ or $[S]$ independently, but the value of $[K+S]$ sets an upper limit on both. Figure 13 shows a plot of data on glycollic acid uptake rates in a small Massachusetts lake(87) according to Eq. (100). The data fit the kinetic scheme reasonably well in this case and indicate a value for V of 8.1×10^{-4} mg/liter per hr and a $[K+S]$ value of 0.06 mg/liter. However,

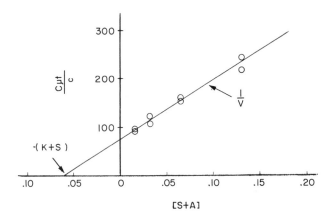

Fig. 13. Uptake of glycollic acid in Gravel Pond, Massachusetts, August 11, 1967, at 3 m depth, $V = 8.1 \times 10^{-4}$ mg/liter per hour, $(K + S) = 0.060$ mg/liter. See text for meaning of symbols. [Reprinted from Ref. *87* by courtesy of the author.]

this study(*87*), earlier ones(*84–86*), and more recent studies(*88,89*) also reported experiments in which data did not fit the modified Lineweaver–Burk plot. There are some questions concerning the theoretical validity of this method as a means to study assimilation of dissolved organic substrates by mixed populations in natural waters. Uptake of organic solutes may at times be diffusion controlled in natural waters, resulting in rates that are incompatible with Michaelis–Menten formulations. Diffusion control apparently caused some of the lack of fit mentioned above. Even if the reaction is controlled by enzyme kinetics, a plot of data from natural (mixed) populations according to Eq. (100) may result in a non-linear function since kinetic constants for uptake may differ for each species(*86*). Another, perhaps obvious cause for the lack of fit is the failure to take respiration of ^{14}C-labeled substrates to CO_2 into account in the above studies. Hobbie and Crawford(*90*) found that from 8% (for arginine) to 60% (for aspartic acid) of the labeled material taken up by microorganisms was immediately respired. Thus only measuring the amount of label in the particulate phase after incubation would significantly underestimate total uptake of the labeled compound. If the per cent of labeled compound respired versus per cent assimilated is a function of concentration, failure to account for respiration could cause the lack of fit reported in the above studies.

Dugdale(*91*) recently derived a simplified model to investigate nutrient limitation in the sea. The model assumed steady-state conditions and ignored regeneration of nutrients from detritus. Uptake of major nutrients was assumed to follow Michaelis–Menten kinetics. Some data taken from the literature substantiated this assumption. In a more recent paper, MacIsaac and Dugdale(92) presented evidence from ^{15}N tracer experiments that uptake of ammonia and nitrate by natural marine plankton samples follows the Michaelis–Menten equation. Addition of increasing amounts of nitrogen as nitrate or ammonia to samples from nutrient-depleted waters of the eastern Pacific resulted in increased uptake rates according to the typical hyperbolic relation of the Michaelis–Menten equation (Fig. 14). Calculated K_t values (transport constants analogous to

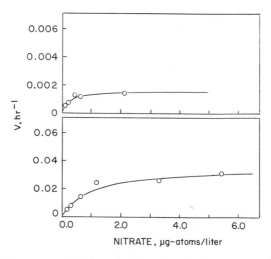

Fig. 14. Rate of nitrate assimilation as a function of nitrate concentration in marine surface waters. Upper graph illustrates a low nutrient, low productivity zone; lower graph represents a productive, nutrient-enriched area. [Reprinted from Ref. 92 by courtesy of Pergamon Press.]

Michaelis constants) ranged from less than 0.2 μg-atom/liter to 4.2 μg-atom/liter for nitrate and from 0.1 to 1.3 μg-atom/liter for ammonia. Low transport constants correlated with oligotrophic (nutrient-depleted) areas, while the higher values for K_t correlated with eutrophic (nutrient-enriched) waters. It was suggested that phytoplankton of oligotrophic populations are adapted to low ambient concentrations and can assimilate nutrients more rapidly under these conditions than would phytoplankton from nutrient-enriched regions.

Two other methods have been reported to evaluate half saturation constants for uptake of ammonia and nitrate. Eppley et al. (93) used laboratory cultures (presumably axenic) of marine algae grown in an enriched seawater medium and determined rates of ammonia and nitrate uptake as functions of concentration by direct measurement of NH_3 and NO_3^- concentration initially and after 15 to 120 min of incubation. Ammonia or nitrate (1–20 μM) was added to N depleted media at the start of an experiment, and kinetic constants were evaluated using Eq. (94). Half saturation constants (K_s) for 16 marine species varied from 0.1 to 9.3 μM for nitrate, and values varied approximately in proportion to cell size and inversely with specific growth rate. Eppley et al. (93) considered the K_s values important characteristics of organisms living in nitrogen-limited environments. They were able to use the values to predict competitive advantage of one species over another by calculating specific growth rates as functions of nitrate and ammonia levels for species with known growth responses to temperature, light, and photoperiod.

Thomas (94) calculated K_s values for growth rates of mixed natural populations of marine phytoplankton in the eastern tropical Pacific by direct measurement of phytoplankton growth as determined by *in vivo* chlorophyll fluorescence increases. Sea water samples were enriched with a complete mixture of non-nitrogenous nutrients and varying concentrations of nitrate or ammonia and were incubated in natural light aboard ship. However, Michaelis–Menten (hyperbolic) plots were obtained in only two out of four experiments. The K_s for ammonia was estimated at 1.5 μM, and K_s for nitrate was 0.75 μM for these two experiments. While this technique is attractive in that near natural (in situ) conditions are used and difficult ^{15}N measurements are not required, it should be pointed out that the K_s values obtained are for algal growth rather than for short term nitrogen uptake. Thus the K_t values reported by MacIsaac and Dugdale (92) and the K_s values of Epply et al. (93) may not be directly comparable with the values reported by Thomas. For strict comparability of the values we would have to assume that nitrogen uptake is directly proportional to growth and that nitrogen concentration is the single limiting variable for growth under the experimental conditions. Nevertheless, it is interesting to note that the values obtained by the three methods are at least of the same magnitude. The three studies also graphically reveal the ability of marine phytoplankton to efficiently assimilate nitrogenous nutrients at very low levels; whether fresh water species are equally efficient remains to be determined.

The kinetics of acetylene reduction by nitrogen fixing organisms also fit the Michaelis–Menten equation. Acetylene is a competitive inhibitor of nitrogen for all nitrogen-fixing organisms, and the nitrogen-fixing enzyme reduces it to ethylene. This fact supplies the basis for a recently proposed (95) assay procedure for nitrogen fixation in natural waters, i.e., gas chromatographic determination of the amount of ethylene produced from acetylene in an incubated sample is an indirect measure of the rate of nitrogen fixation in the sample. Since acetylene reduction is enzymatic, it would be expected to follow Michaelis–Menten-type kinetics, and it does (Fig. 15). Furthermore, inhibition of acetylene reduction by molecular

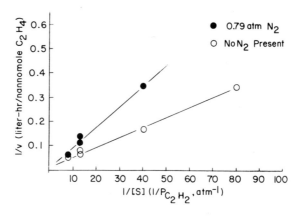

Fig. 15.   Lineweaver–Burk plot of acetylene reduction to ethylene by a natural population of blue-green algae (primarily *Aphanizomenon* sp.). Note that the effect of atmospheric nitrogen on acetylene reduction follows a pattern of competitive inhibition (P. L. Brezonik, unpublished data).

nitrogen is competitive (as expected since the converse is known to be true); this relationship may be useful in determining whether or not acetylene reduction is related to the presence of nitrogen–fixing organisms in a sample.

D. Diffusion-Controlled Kinetics for Biological Assimilation

An enzyme-catalyzed reaction is not always the rate-limiting step in a biological process. In order for most biological reactions to occur, the substrate must be transported through a cell wall to the location of the enzyme system. Frequently, organisms grow in a mat or within a gelatinous sheath and the substrate concentration within the mat or sheath becomes depleted. Transport of substrate through the cell membrane or diffusion of substrate through the mat, sheath, or bulk solution (in the case of quiescent systems) may in fact be the rate-limiting step. Substrate transport across cell membranes can be active (i.e., linked to metabolic activity) or passive (i.e., controlled by simple diffusion). The former process can overcome concentration gradients, requires an expenditure of energy, and follows Michaelis–Menten kinetics. In passive transport or simple diffusion, movement of substrate follows a concentration gradient; i.e., substrate diffuses from regions of high concentration into regions of low concentration. The diffusion equation relates flux across a membrane to the product of a diffusion coefficient and the concentration difference across the membrane:

$$\phi = D(C_0 - C_i) \tag{101}$$

where ϕ is the flux in mg/cm^2 per hr, D is a coefficient in cm/hr, and C_0 and C_i are external and internal concentrations (mg/cm^3), respectively. Equation (101) can be expressed in terms of the rate of change in external concentration:

$$dC_0/dt = -K_D(C_0 - C_i) \tag{102}$$

If substrate is metabolized as soon as it enters the cell, C_i remains constant (in fact, it may approach zero). Equation (102) then becomes a simple first-order expression. In these circumstances, K_D is interpreted as a mass transfer coefficient rather than as a reaction rate constant.

Appropriate kinetic plots of substrate utilization data should thus indicate whether the uptake rate is diffusion controlled or reaction (enzyme) controlled. If the reaction is zero order, enzymatic control is indicated and the substrate is saturating ($v = V_{max}$). If a reciprocal plot of $1/v$ versus $1/S$ is linear, enzymatic control is also indicated. The limiting enzymatic step could be either reaction within the cell or active transport across the cell membrane; which of these two would control would not be readily distinguishable. If diffusion controls the rate of substrate utilization, a semilog plot of substrate concentration in the bulk solution versus time will be linear (indicating first-order kinetics). Another means of determining the importance of diffusion is to determine the uptake rate at varying conditions of turbulence or mixing. If an increase in the mixing rate results in an increase in substrate utilization, mass transport (diffusion) presumably is the limiting factor.

A number of recent studies have indicated the possible importance of diffusion-controlled kinetics. Wright and Hobbie(*84*) reported evidence for diffusion-controlled uptake of glucose and acetate in natural populations of several Swedish lakes. Swilley et al.(*96*) discussed the significance of mass transport in various biological oxidation processes, including the BOD procedure. A significant increase was found in the rate of BOD exertion in a stirred bottle compared to a quiescent bottle. Gulevich et al. (97) used a rotating disk covered with a biological slime to determine the effect of the external velocity field on the rate of nutrient uptake; rate of glucose uptake was found to be influenced by the velocity field in the laminar flow region, implying control of the overall transformation by diffusion. According to Mueller et al. (98), oxygen diffusion through the floc matrix controls the rate of oxygen utilization by a *Zoolglea ramigera* strain (a common floc former in activated sludge) at dissolved oxygen concentrations below a range of critical values, 0.6–2.5 mg/liter (depending on the size and activity of the floc). Similarly, Baillod and Boyle (99) reported that mass transfer limitations affected glucose uptake rates by *Z. ramigera*. Critical glucose concentrations were 2.2 mg/liter for unblended floc and 1 mg/liter for blended floc; above these concentrations uptake was zero order, implying that diffusion through the floc matrix was not rate limiting. The glucose uptake rate differences between flocculated and blended cultures were ascribed to diffusional resistance to glucose transfer afforded by the floc matrix itself rather than to any stagnant liquid film resistance.

E. Biological Dynamics in Natural Waters

It is somewhat misleading to speak of kinetics in describing the turn-over of organic substrates and inorganic nutrients in natural waters. The term kinetics generally refers to specific chemical reactions. Nutrient and organic substrate assimilation are generally effected by a variety of organisms, whose assimilation rate constants and mechanisms of uptake may vary considerably. Furthermore, the nutrient or substrate may be simultaneously involved in several different reactions. Rigorous kinetic formulations under such circumstances are difficult if not impossible to derive. Because of these factors, the general term "dynamics" is frequently used to describe the turnover of biogenic substances.

However, biological reaction data often fit simple kinetic expressions that belie the complexity of the mechanism. For example, growth curves of microorganisms in culture are frequently described by simple first-order kinetics:

$$dM/dt = kM; \qquad M = e^{kt} \tag{103}$$

where M is the mass of microorganisms. Obviously this implies nothing about the mechanism of growth. Nutrient assimilation rates have also been described by first-order kinetics in some cases without implying anything about mechanisms. Reactions like these are usually termed "apparent first order."

The relationship between changes in microbial mass and concentration of a growth-limiting substrate is given by

$$dM/dt = -adS/dt - bM \qquad (104)$$

where S represents the growth-limiting substrate, a is a yield constant, and b is a decay or die-off constant(100). Under proper circumstances (for example, in a young, healthy culture in the log-growth phase), the second term in Eq. (104) can be neglected and the increase in biomass is directly proportional to changes in substrate concentration:

$$M = M_0 + a(S_0 - S) \qquad (105)$$

where M_0 and S_0 represent initial values and M and S are the values after some time. The change in concentration of a growth-limiting substrate with time was shown by Monod(101) to be approximated by:

$$dS/dt = -kMS/(K_s + S) \qquad (106)$$

which is analogous to the Michaelis–Menten expression for enzyme kinetics. K_s in Eq. (106) represents an empirical half-velocity or saturation constant. First-order growth kinetics for microorganisms [Eq. (103)] is seen to derive from the Monod expression as a special case; when $[S] \gg K_s$, Eq. (106) simplifies to $dS/dt = -kM$. Substituting this expression into Eq. (104) and assuming no die-off results in the first-order expression for microbial growth [Eq. (103)]. Thus, microbial growth is first order when the growth-limiting substrate is present in saturating concentrations, or more generally when none of the substances required for growth are limiting. The kinetics of substrate uptake in pure and mixed cultures has been further discussed by Mateles and Chian(102).

Another familiar example of apparent first-order kinetics describing a complex biological phenomenon is the BOD formulation. The rate of BOD (biochemical oxygen demand) exertion is most commonly fit to the first-order equation; $d(\text{BOD})/dt = -k(\text{BOD})$. The integrated expression is commonly written as

$$y = L(1 - 10^{-kt}) \qquad (107)$$

where y is the BOD exerted at any time and L is the ultimate or total BOD.

In the past it was common to determine only 5-day BOD values, but the fact that these values by themselves mean little, especially in comparing different wastes, has become appreciated in recent years. Of far more value for predictive and comparative purposes are determinations of k and L. While L can be considered as the amount of BOD present initially, its value is usually not known, nor is it easily determined. This presents difficulties in solving the BOD equation for k, the first-order rate constant.

The most straightforward method of estimating L would be to plot measured y values versus time, which yields a hyperbolic curve from which L can be extrapolated. Unless the data are complete for large portions of the curve and show little statistical scatter from the theoretical curve, it is difficult to extrapolate the asymptotically approached value for L with any degree of confidence. Hence this method is seldom used. Linear graphical solutions for k require a semilog plot of $(L - y)$ versus t. A variety of techniques have been developed to determine L from measured y values(103–109). The most accurate method is that of Theriault and Reed and Theriault(103,104), but it involves tedious least squares fit of a curve to the data by trial and error. The present possibilities for computer-assisted solutions now make this less of a problem. A considerably simpler least squares method is the method of moments(107). This method is widely used and is almost as accurate as the Reed–Theriault procedure. Curves of k versus $\Sigma y/\Sigma ty$ and of $nL/\Sigma y$ versus k were calculated(107) for various time sequence observations of y (i.e., n determinations of BOD according to a specified sequence, e.g., y measured at 1, 2, 3, 4, 5, 10 days; $n = 6$ in this case). Calculation of these curves requires extensive work, but once they are available, the procedure of solving for k and L is simple. Figure 16 shows the curves of k versus $\Sigma y/\Sigma ty$ for four time sequences. From measured y values, $\Sigma y/\Sigma ty$ is computed and k is read from the appropriate curve. Figure 17 is a plot of $nL/\Sigma y$ versus k for the same time sequences. From the previously determined k, a value for $nL/\Sigma y$ is read off the appropriate curve, and L is determined as this value times $(\Sigma y)/n$. Table 2 presents an example of k

TABLE 2
Sample Calculation of k and L for BOD by Method of Moments[a]

t (days)	0	1	2	3	4	5	6	7	
y_i (BOD, ppm)	0	80	110	150	162	175	190	200	$\Sigma y_i = 1067$
$y_i t_i$		0	80	220	450	648	875	1140	1400 $\Sigma y_i t_i = 4813$

[a]$\Sigma y/\Sigma yt = 0.2217$. From Fig. 16, series III, with $\Sigma y/\Sigma yt = 0.2217$, read $k = 0.185$. From Fig. 17, series III, with $k = 0.185$, read $nL/\Sigma y = 1.35$. Thus, $L = 1.35 \times \Sigma y/n = 1.35(1067)/7 = 206$ ppm.

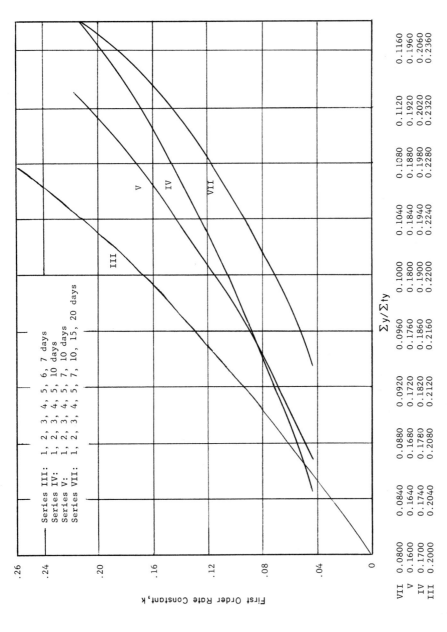

Fig. 16. Method of moments for BOD evaluation. Plot of k versus $\Sigma y/\Sigma ty$ for four time series of BOD measurements. Values below abscissa represent scales for the term $\Sigma y/\Sigma ty$ for the different time series. [Reprinted from Ref. *107* by courtesy of the Water Pollution Control Federation.]

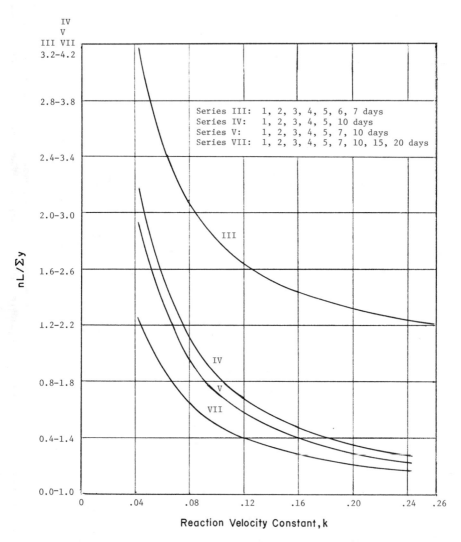

Fig. 17. Method of moments for BOD evaluation. Plot of $nL/\Sigma y$ versus k for four time series of BOD measurements. Reprinted from Ref. *107* by courtesy of the Water Pollution Control Federation.

and L computation by this procedure. A simple graphical method has also been developed by Thomas(*108*). Plots of $(t/y)^{1/3}$ versus t are linear for BOD data and they yield an intercept (A) and slope (B) whose relationships to k and L are:

$$k = 2.61B/A; \qquad L = 1/6A^2B \qquad\qquad (108)$$

Table 3 and Fig. 18 illustrate this method for computation of k and L from BOD measurements. The advantage of this method is that any time sequence of BOD measurements can be used, whereas the method of moments requires sequences for which appropriate curves are available.

TABLE 3

Sample Calculation of k and L for BOD by Graphical Method[a]

t (days)[b]	0	1	2	4	6	8
y (BOD, ppm)	0	32	57	84	106	111
$(t/y)^{1/3}$	—	0.315	0.327	0.362	0.384	0.416

[a]From Fig. 18, $A = 0.30$ and $B = 0.0145$; substituting these values into Eq. (108): $k = 2.61\ B/A = 0.13$ day^{-1}; $L = 1/6\ A^2B = 129$ ppm.

[b]See Fig. 18 for a plot of $(t/y)^{1/3}$ versus t for these data.

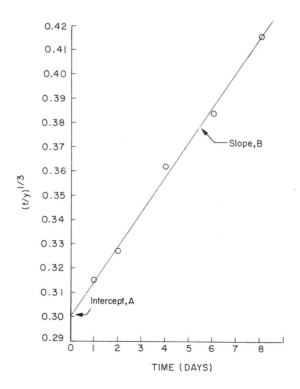

Fig. 18.   Graphical analysis of BOD data according to method of Thomas (108). See Table 3 for computation of k and L from this graph.

It should be noted that for any of the above methods to yield accurate k and L values, the kinetics must be approximated by the first-order expression; the better the approximation, the more accurate the calculated values.

There is no theoretical basis for the fit of BOD rates to first-order kinetics; in fact, the fit is sometimes poor, and several workers have attempted to fit BOD data to other kinetic expressions. Thomas(106) feels second-order kinetics apply equally well to BOD rates. Buswell et al.(110) and others have concluded that oxygen uptake is related to bacterial numbers growth rather than to the amount of oxidizable substrate. Busch(111) concluded from previous studies that BOD exertion is a two-stage process and proposed a short-term BOD test(112,113) using only the first stage of the reaction, in which soluble substrate is converted to cell material and storage products. Numerous workers(110,111,114) have found a plateau in oxygen consumption after short periods of incubation (8 hr to 2 days) corresponding to completion of the first stage of bacterial decomposition. However, the 5-day BOD test recommended by *Standard Methods*(115) is still the most widely accepted test. It should be pointed out that the BOD test is basically empirical and suffers from many theoretical deficiencies. Sophisticated and extended efforts to apply other kinetic expressions to BOD data may not be justified so long as the first-order expression provides an empirical fit of the data.

While many biological phenomena fit first-order kinetic expressions, even these descriptions can become complex in the case of sequential reactions. For example, renewal of inorganic nitrogen from organic nitrogen in natural waters is a three-step process:

$$\text{organic N} \xrightarrow{\text{(A)}} NH_4^+ \xrightarrow[-\frac{3}{2}O_2\ 2H^+ + H_2O]{\text{(B)}} NO_2^- \xrightarrow[1/2O_2]{\text{(C)}} NO_3^- \qquad (109)$$

Reaction (A) is termed ammonification and reactions (B) and (C) represent the two-step process of nitrification. The latter two reactions are mediated by different organisms: *Nitrosomonas* is the best-known genus of bacteria involved in ammonia oxidation, and *Nitrobacter* is the most common nitrite oxidizer. Sequential kinetics required to describe the overall process were discussed earlier (Sec. II.F.); Fig. 19 illustrates the buildup and decay of ammonia, nitrite, and nitrate during nitrification(116). First-order kinetics with $k_1 = 2k_2$ are assumed; nitrate and nitrite are initially absent and all nitrogen is present initially as ammonia.

Rate formulations for biological processes in nature are complicated by the kinetics of organism growth and decay. Stratton and McCarty(100) derived a rate equation for the two-step nitrification process based on the

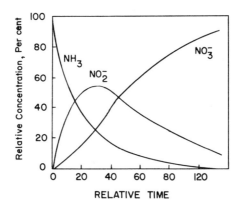

Fig. 19.   Theoretical plot of nitrification time course with $k_{NO_2^-} = 2k_{NO_3^-}$; all nitrogen present initially as ammonia. First-order kinetics is assumed. [Reprinted from Ref. *116* by courtesy of McGraw-Hill, Inc.]

Monod expression for microbial growth. From the stoichiometry of nitrification and the equation for rate of substrate utilization, the rate of oxygen consumption resulting from nitrification was expressed as

$$d[O_2]/dt = 3.43[kMC/(Ks+C)]_{NH_3} + 1.14[kMC/(K_s+C)]_{NO_2^-}$$
$$(110)$$

where subscripts NH_3 and NO_2^- refer to rate parameters for ammonia and nitrite oxidation, respectively; k is a substrate utilization constant (milligrams per day per milligrams of organisms); M is the concentration of organisms; C is the concentration of ammonia or nitrite; and K_s is the half-velocity constant for ammonia or nitrite. The kinetic parameters were evaluated with computer formulations of the equations using data from laboratory batch tests on nitrification. According to the authors, the obtained values approximate results from the natural environment and the derived method can be used to predict the effects of initial nitrogen concentration, initial nitrifying organism concentration, and temperature on the in situ rate of nitrification (and thus on the rate of oxygen consumption resulting from nitrification). The validity of this approach was questioned by Wezernak and Gannon(*117*), who contend that in situ studies are necessary to properly evaluate the nitrification kinetic parameters.

    In recent years it has become a popular endeavor to model biological activities in ecosystems mathematically by use of differential equations. For example the dynamics of algal populations can be described mathematically as a function of light, temperature and limiting nutrient concentration (which affect growth rate) and zooplankton concentration, sinking velocity and natural mortality rates (which affect loss rates). Other equations can be written to describe the dynamics of zooplankton, nutrient concentration, etc., and the set of simultaneous differential equations can be solved by numerical methods on a computer. Most of the models developed to date utilize linear equations (i.e. rates are considered to be additive functions of first-order processes), which are much simpler to solve numerically than are nonlinear equations, but may not realistically reflect ecosystem dynamics. The subject of mathematical modeling of plankton productivity and of whole ecosystems is interesting and important, but further discussion of this topic is beyond the scope of this chapter. However, the interested reader can find further information on modeling theory, techniques and applications in several recent texts and articles (118–123) (See also Chap. 5).

Often the methods used to measure biological dynamics preclude determination of the kinetic pattern that fits the reaction, and zero-order kinetics (constant rate with time) is tacitly assumed. For example, measurement of primary production involves incubating samples in bottles for several hours and measuring the oxygen evolved or $^{14}CO_2$ assimilated. It is doubtful that the photosynthetic rate would remain constant over the period of incubation; the result represents an average rate over the length of incubation, which is sufficient for most purposes. Similar problems arise in measuring other biological reactions such as nutrient assimilation in natural waters. Development of isotope tracer methods (e.g., ^{32}P for phosphate assimilation, ^{15}N for nitrogen cycle reactions, of ^{14}C-tagged compounds for carbon cycle reactions) has enabled the measurement of many biological reaction rates in the range of nanograms per liter hour in natural waters(124–128). But even with these advances, the incubation time required to build up a measurable activity of the tracer in the reaction products may be on the order of several hours, thus preventing detailed time studies. In long-term tracer experiments on element circulation through ecosystems (e.g., use of ^{32}P to trace the movement of phosphorus through the various abiotic and biotic components of lake, pond, or laboratory carboy ecosystems), exponential (first-order) tracer movement is generally assumed. However, in the absence of kinetic data on the natural population, it is probably best to assume steady-state conditions (and zero-order kinetics) in nutrient and substrate assimilation rate studies and to approach these conditions by minimizing incubation time. The importance of minimizing the incubation time can hardly be overemphasized. Growth characteristics and species composition of the biota change rapidly when natural water samples are placed in bottles; nutrient levels in bottles may decrease as a result of rapid assimilation or sorption onto the bottle walls or may increase as a result of bacterial growth on the bottle walls. Furthermore, mixing and turbulence within a bottle differ from natural conditions.

The factors that affect the rates of biological reactions are too complex for any detail in this discussion. For reactions which follow first-order kinetics, substrate concentration obviously is an important factor. The concentration of organisms which mediate the reaction and organism viability (growth and death rates) are of fundamental importance in the dynamics of biological reactions. Population dynamics for microorganisms may be affected by a wide range of environmental factors, including a variety of habitat variables (e.g., light intensity, temperature), resource variables (macro- and micronutrients), and organism variables. An example of the last factor is the predation of the reaction-mediating organisms by other organisms in higher trophic levels. Temperature has a marked effect on biological reaction rates; a 10°C rise in temperature may double the rate. The ratio of reaction rates at temperatures 10°C apart is symbolized by Q_{10}, i.e., $Q_{10} = k_t/k_{t-10}$. Many workers have used the Arrhenius (i.e., an exponential) relationship Eq. (50) to express biological rates as a function of temperature. However, this expression is usually valid only over narrow temperature ranges, and it does not account for decreasing rates when the organism's temperature optimum is exceeded. The fact that different organisms with different temperature optima conduct the same reaction leads to complex temperature effects on biological rates. Temperature increases within the range of an organism's thermal preference will in general lead to reaction rate increases as predicted by chemical reaction rate theory, but larger temperature shifts may cause a change in species composition to organisms having different metabolic rates and different responses to further temperature shifts. For example, Zanoni[129] has reported on the effect of temperature on the rate of BOD exertion. He found that the first-order rate constant was not a simple Arrhenius function of temperature over the range 2–32°C, but that two distinct ranges, 2–15°C and 15–32°C, with different Arrhenius slopes, were necessary to fit the data. Also it is well known that microbial methane formation rates exhibit two temperature maxima, 35 and 55°C, the former being optimum for mesophilic methanogens and the latter being optimum for thermophilic strains. Without further elaborating on this complex situation, it can be said that our knowledge regarding the dynamics of biological reactions in natural waters is to a large extent empirical, and few unifying principles are available for theoretical treatments. More detailed discussions of these phenomena can be found in texts on microbial ecology[118,130–132].

V. Analytical Implications

A. Effect of Analytical Capabilities on Kinetic Studies

Kinetic and dynamic studies obviously deal with changing concentrations of reactants and products in essentially unstable situations. This poses certain requirements for analytical procedures; for example, the time for measurement of reactant or product concentration must be small compared to the length of time for complete reaction. Alternately, the reactant or product being measured can be rapidly changed into a nonreacting form or the reaction can be quenched by a variety of methods at appropriate time intervals. If a reactant or product is itself colored, the decrease or increase in color can be monitored continuously and instantaneously by spectrophotometric methods. In other reactions instantaneous concentrations in the medium can be continuously monitored by nondestructive electrometric methods. Several examples are applicable to studies of reactions in natural systems. The kinetics of oxygen uptake by activated sludge organisms has been studied by continuous monitoring with dissolved oxygen electrodes (98,133,134). Recently specific ion electrodes have been used in kinetic studies; Srinivasan and Rechnitz (135) used a fluoride electrode to study the kinetics of fluoride complexation with iron and aluminum. Electrodes are now available to measure the activities of such ions as fluoride, sulfide, nitrate, calcium, ammonium, sodium, and others; applications to kinetic studies involving these ions are potentially great (see Chaps. 27 and 32).

When continuous analysis is used to monitor reaction kinetics, cognizance must be taken of instrument response time in interpreting instrument output. Response time is determined by various physical and electronic features of an instrument. In optical instruments (e.g., spectrophotometers) response time depends primarily on electronic signal dampers (e.g., capacitors) since light transmission itself is essentially instantaneous. Response times of some electrometric instruments (e.g., conductivity meters) depend primarily on electronic factors, while in others (e.g., pH meters) response time also is affected by the time required for equilibration of the electrode membrane. Response times for pH and other ion electrodes range from a few seconds to as much as a minute or more (for specific ion electrodes measuring concentrations near their lower limit of response). In dissolved oxygen electrodes, response time depends on the rate of oxygen diffusion through the semipermeable polyethylene membrane, and response times for various of these electrodes are on the order of a minute or less.

AutoAnalyzer procedures are becoming increasingly popular for routine water analysis and have also been used to continuously monitor reactions in kinetic studies(99). The output signal from these instruments can differ significantly from the input signal; even when the transport time through the instrument is taken into account, mixing and diffusion afforded by long tubing lines and mixing coils tend to dampen out short-term changes. The response characteristics of any instrument can be determined by observing its response to a known input signal. Figure 20

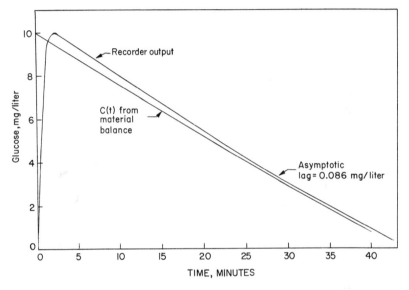

Fig. 20. AutoAnalyzer response to linear input signal of glucose. [Reprinted from Ref. *99* by courtesy of the American Chemical Society.]

shows the response characteristics of an AutoAnalyzer system measuring glucose by the glucose oxidase procedure as determined by Baillod and Boyle(99) using a stirred reactor to simulate a linear decrease in glucose with time. Details of the method are given in their paper, but it is pertinent to note that the output function consistently lagged the input function. The differences between theoretical (input) concentration and measured response were especially large initially; the concentration by which the output lagged the input asymptotically approached a value of about 0.1 mg/liter. Instrument response time was calculated from the asymptotic concentration by which output function lagged input function and the known rate of change of glucose concentration in the reactor according to the procedure of Mueller et al.(*134*). In this case the response time was found to be approximately 0.35 min. It can be seen from Fig. 20 that the effect of instrumental lag was rather small after long sampling times (these were the times of greatest interest to the authors), but at short sampling times response characteristics, if not taken into account, could cause misleading interpretations of the data.

Many reactions of interest in natural water studies are not amenable to continuous monitoring, and discrete samples must be taken for analysis. A variety of methods can be used to stop reactions in samples. If the reaction is biological, inhibitors or poisons like trichloroacetic acid, mercuric ion, formaldehyde, or heat can be used. Many reactions are pH dependent, and addition of acid or base will stop the reaction. Oxygenation of manganese(II) is an example(42); the reaction is greatly retarded at pH values less than 9. Often the first reagent in the analytical scheme changes the reactant being measured into a nonreactive form essentially immediately.

In reactions with k values greater than several seconds, conventional methods such as described above are adequate for kinetic studies. Most reactions whose rates are of interest in environmental studies belong in this category. Reactions with k values less than 1 sec or so are too rapid to be followed by most conventional analytical methods. Often such reactions are simply considered to be "instantaneous," and their kinetics are considered no further. This point of view is satisfactory if one is interested only in the rate of reaction, since the time scale of greatest interest in environmental studies is that in which we can personally comprehend changes (e.g.. fractions of minutes or greater). However, a major purpose of kinetic studies is to determine reaction mechanisms, and it is desirable to know the mechanisms of even "instantaneous" reactions in order to control conditions for optimum reaction. A variety of techniques have been developed to study rapid reaction kinetics; these include relaxation methods and rapid flow reactor systems. Discussion of these techniques is beyond the scope of this chapter, and the interested reader is referred to specialized texts(1,37,136).

Practically every kinetic study on reactions in natural waters is unique and presents its own problems. However, analytical capabilities have frequently limited the conditions under which a reaction could be studied and the ease with which the study could be conducted. Kinetic studies of many important reactions have been delayed by lack of suitable analytical procedures; probably many studies still await development of reliable methodologies. The oxidation of manganese(II) in water is a case in point. Morgan and Stumm(137) found it necessary to develop and evaluate methods to differentiate the several oxidation states of manganese before they could study the kinetics of manganese oxidation in dilute aqueous solutions(42,138,139). There is still much room for improvement of analytical methods, especially for Mn(II) and Mn(III), if the latter actually exists in water. Presumably the kinetics of manganese reactions in natural waters will become better known as analytical capabilities are improved. Other examples of analytical problems and their solutions can be found in many kinetic studies discussed earlier.

B. Effects of Reaction Kinetics on Analyses

The implications of kinetics and dynamics on analyses of constituents in natural waters should be obvious from the preceding discussions. For example, it is highly desirable to have rapid reaction rates in titrimetric procedures. Fortunately, most ion reactions (acid–base, precipitation, complex formation) are rapid, but some redox reactions and reactions involving organic compounds are slow and limit the possible use of a reaction for titrimetric analysis. Several possible alternatives are available to speed up slow reactions, including the use of higher temperatures. Many reactions occur appreciably only at elevated temperatures, and this also provides a means of quenching a reaction at any desired point. For many reactions, catalysts can be added to the medium. Catalysis by acids and bases is often used to bring a reaction of analytical importance to rapid conclusion. Metal ions are often useful as catalysts; for example, mercury(II) and copper(II) are commonly used to speed the decomposition of organic matter in Kjeldahl digestions, and silver sulfate enhances the oxidation of straight chain aliphatic compounds in the chemical oxygen demand (COD) test. Excess quantities of reagents can be added to the reaction mixture to promote rapid and complete conversion of the sought-for substance into an analytically measurable form. Addition of excess reagents followed by back titration can sometimes be used in titrimetry. A change of solvents may also speed up a slow reaction. Other relatively slow reactions can be used in titrimetry if an end point detection technique can be employed which does not require observations at or very near the equivalence point (140). Conductance and amperometric titrations are examples.

For ease and accuracy in analysis, it is preferable for analytical reactions to yield stable products. Reactions with unstable products require close control of reaction time for all samples and standards. Automated procedures, such as AutoAnalyzer methods in which the reaction time is identical for each sample and standard, are attractive in such cases. Because of the possibility of product breakdown or of incomplete product formation, the analyst should heed directions in analytical manuals regarding both the time allowed for reaction before measurements are made and the time period during which the product is stable.

For the most part, the kinetics of analytically important reactions have been determined by analytical chemists or are their proper domain. However, the practicing water chemist or environmental engineer cannot always assume that a method will yield reliable data just because it appears in the literature. The presence of interfering substances in particular water samples may obviate the use of an analytical procedure which would be acceptable elsewhere. Analytical interferences are not always well specified; the analyst should keep this fact in mind and evaluate the possibilities of positive and negative interferences before adapting a new procedure. Proper evaluation of an analytical method involves such techniques as standard curve determinations, recovery studies, and determination of matrix effects and interference levels. Details of these procedures can be found in most quantitive analysis texts [e.g.,(*141,142*)] and are not repeated here. The importance of kinetics and reaction rates in chemical analysis has been further discussed by Laitinen(*140*).

C. Kinetic Analyses

The kinetic principles discussed in earlier sections can be applied to practical problems in analysis, and a growing body of such studies are being reported in the literature. In some reactions, the desired component has equilibrium chemical properties that render it unmeasurable or indistinguishable from other reactants; analysis by reaction rate methods is particularly useful in such cases. The advantages of kinetic methods include great sensitivity (if catalytic or inhibition reactions can be used), high resolution for mixtures of chemically similar substances, and adaptability to automated measurement. Disadvantages include relatively poor accuracy and lack of suitability for continuous monitoring of instantaneous concentrations of substances(*143*).

The term kinetic analysis can be applied to a rather diverse group of analytical procedures, which in general determine concentrations of substances by measuring the rate of a reaction rather than the amount of product formed at equilibrium. In some procedures the reaction time course is followed by continuous monitoring of the concentration changes of reactant or product. In other cases reaction rate is determined implicitly by measuring the reactant consumed or product formed in a specified reaction time. Since reaction rates depend on reactant concentrations, the latter can be estimated from previously described kinetic equations. By using a working curve of reaction rate versus known concentrations, reaction rates of unknown samples can be translated into concentration values even when the reaction rate equation is unknown. In order that reaction rate depend only on the concentration of the sought-for substance, it is necessary to treat all standards and samples identically (i.e., the same amounts of all other reactants must be added to each sample and standard).

The field of kinetic analysis is too broad for any more than cursory coverage here, and the interested reader is referred to several review articles (143–145), monographs (146, 147), and other general papers (148, 149) for further details. Recent and potential applications of kinetic analysis in determinations of natural water constituents are emphasized in the remainder of this section.

The selectivity of kinetic analysis is illustrated by the reports of Pausch and Margerum (150) and Margerum et al. (151), who developed differential kinetic analyses for mixtures of alkaline earth, lanthanide, transition, and group III metals. Exchange reactions between Pb(II) and alkaline earth complexes with CyDTA (trans-1,2-diaminocyclohexane-N,N,N′,N′-tetraacetate) proceed at sufficiently different velocities to permit determination of Mg, Ca, Sr, and Ba in mixtures (150). The ultraviolet absorbance of $PbCyDTA^{-2}$ at 260 mμ is used to monitor the reaction. Margerum et al. (151) further reported on a similar procedure to differentiate 30 metal ions by means of exchange rates of the metal-CyDTA complex with acid, copper, or lead. Rate constants vary by a factor of 10^{12} for the metals, and qualitative analysis is achieved by determination of absorbance time scans of various durations (milliseconds to minutes) and at various pH values. A metal or group of metals is observed at a specified combination of pH and time scan. Quantitative determinations at the $10^{-6} M$ level with $\pm 10\%$ accuracy were reported. The authors used a stopped flow reaction system with oscilloscope and Polaroid film recording of the absorbance time scans.

Rates of catalyzed homogeneous reactions are frequently proportional to the concentration of catalyst and therefore can be used for determinations of catalyst concentration. Major advantages of analyses based on catalytic reactions are their great sensitivity and (usually) high selectivity. Catalysts such as transition metal ions are often effective in the parts per billion range and can be determined at these levels by measuring their effect on rates of reactions between much more highly concentrated (hence more easily measured) reactants. For example, Fishman and Skougstad (152) developed a method for low level (0.1–8.0 μg/liter) vanadium analyses in natural waters based on the catalytic effect of vanadium on the rate of gallic acid oxidation by persulfate. Fernandez et al. (153) proposed a kinetic approach to manganese analysis based on oxidation of the leuco-base of malachite green by periodate at pH 4 with Mn acting as catalyst. The method has a reported sensitivity of 0.1 μg of Mn/liter but has the disadvantage of requiring 8 hr to complete the analysis. Janjic et al.(154) recently reported a more rapid method for manganese determination based on its catalysis of alizarin S oxidation by hydrogen peroxide. The loss in alizarin S color at 335 nm is followed for 10 min following reaction start. The procedure has a reported sensitivity of 0.3 µg of Mn/liter and is highly selective for manganese. Wilson(155) based an analytical method for molybdenum on its catalysis of iodide oxidation by perborate. The method uses a continuous flow system and recording photometer to measure the iodine formed by the reaction. Sensitivity of the reported method is $10^{-7} M$ molybdate, but Wilson(155) suggested that a detection limit of $2 \times 10^{-9} M$ would be possible if certain modifications are made. Others(156–158) have reported analytical methods for molybdate based on its catalysis of iodide oxidation by peroxide. Selenium can also be determined at low levels using a catalytic method(159). The reduction of methylene blue by sodium sulfide is catalyzed by selenium; as little as 0.1 µg of selenium in 10 ml of water can be measured. The procedure is very simple and requires no instrumentation since the time required for complete decolorization of methylene blue is measured.

The *Standard Methods*(160) procedure for iodide is based on its ability to catalyze the reduction of ceric ion by arsenious acid. The effect is not linear, but a working curve can be made. It is difficult to read the loss of ceric ion (which absorbs at 410 nm) directly since the color fades while being read in the photometer. The reaction is therefore stopped after a specific time interval (15 min) and the amount of ceric ion remaining is determined by addition of ferrous ammonium sulfate to form an equivalent amount of ferric ions:

$$Ce(IV) + Fe(II) \longrightarrow Ce(III) + Fe(III) \qquad (111)$$

Ferric ion forms a relatively stable complex with thiocyanate, the intensity of which is read at 525 nm. Several other catalytic methods have been published for iodide based on the cerate-arsenite reaction. Rodriguez and Pardue(161) studied the kinetics of the reaction in sulfuric acid and proposed an analytical method(162) sensitive to $10^{-8} M$ iodide. This procedure is much simpler to conduct than the *Standard Methods* procedure since it follows the disappearance of Ce(IV) directly with a recording photometer. Initial rates can be obtained from recorded per cent T versus time curves. A specially designed, highly stable photometer was used to measure the small absorbance changes accompanying slow reaction rates at low (10^{-8}–$10^{-7} M$) iodide concentrations. The authors' system(161) is capable of reading 0.03% T differences, and total transmittance differences measured were less than 3% T. However, the procedure can also be used with commercially available spectrophoto-meters if lower sensitivities can be tolerated. For example, Fig. 21(a) illustrates the $\%T$−time response for iodide concentrations of 10^{-7}–10^{-6} M measured with a commercial recording spectrophotometer using a recorder range expansion of 5×. Figure 21 shows the resulting

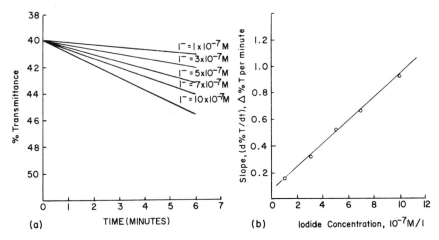

Fig. 21. (a) Rate of Ce(IV) reduction by As(III) at various concentrations of catalyst (I^-) according to procedure in Ref. *161*. (b) Plot of slopes from (a) versus iodide concentration. Both (a) and (b) unpublished data by P. L. Brezonik using a Beckman DBG spectro-photometer.

standard curve of reaction rate slope ($d\% \ T/dt$) versus iodide con-centration. A 10^{-7} M concentration of iodide is equivalent to 12.6 μg/liter, which indicates that the method has sufficient sensitivity for most routine water analyses. The procedure can also be used to determine iodate, which itself does not catalyze the reaction but is reduced to catalytic iodide by As(III). Silver(I) and mercury(II) can be determined at levels as low as 2.5×10^{-8} M by using the fact that stoichiometric amounts of these ions completely inhibit the catalytic activity of iodide. Another variation of the ceric-arsenite reaction for iodide was reported by Mitchell (*163*), who used silver nitrate to stop the reaction at a specified time and measured the amount of Ce(IV) remaining. Fishman and Robinson(*164*) review several papers from the eastern European literature which describe iodide analyses based on the above reaction.

Present methods(*165*) for sulfide and sulfate in natural waters leave much to be desired, especially for low level analyses, and it is interesting to note that kinetic methods have recently been reported for these ions. Babko and Maksimenko(*166*) determined traces of sulfide by its catalytic effect on silver ion reduction by iron(II). Hems et al.(*167*) described a method for sulfate analysis based on its catalysis of the zirconium methyl thymol blue method.

The selectivity and sensitivity of enzyme-catalyzed reactions make them attractive candidates for kinetic analyses. Because enzymes are catalysts and affect only the rate of a reaction but not the equilibrium, enzyme activity must be measured by a kinetic method or by a direct titration of the active site(*168*). In order to determine enzyme activity by kinetic analysis, all substrates must be present in saturating concentrations. In these circumstances reaction rate follows zero-order kinetics at a given level of enzyme, and reaction rate is directly proportional to enzyme concentration. Enzymes are used to determine substrate concentrations under conditions where all other substrates are present at saturation levels and the substrate being analyzed limits the rate of reaction. Biochemical literature is replete with kinetic methods to determine enzyme activities and/or substrate concentrations.

A number of enzymatic analyses have direct application to environmental studies. Several sensitive, highly selective methods for glucose use the enzyme glucose oxidase, which forms an equivalent amount of hydrogen peroxide from glucose. In one method(*169, 170*) peroxidase then catalyzes the oxidation by H_2O_2 of a reduced colorless chromogen to a colored oxidized form. The procedure has been adapted to the AutoAnalyzer and used in glucose assimilation studies with activated sludge(*99, 170*). In another variation of the method, hydrogen peroxide oxidizes iodide to iodine, which is measured directly by a spectrophotometer(*171*) or is used to oxidize *o*-tolidine to a colored product(*172*).

Enzymatic analyses for ATP in natural populations of marine organisms(*173, 174*) and activated sludge(*175, 176*) have been developed from the fact that ATP is required in the firefly light reaction. Extracts containing luciferin and luciferase from firefly lanterns emit light in direct proportion to the amount of ATP added. Procedures based on this reaction are extremely sensitive and can detect as little as 10^{-11} g (10 pg) of ATP in a sample(*173*). In the procedure developed by Patterson et al. (*176*), a liquid scintillation spectrometer is used to monitor the emission of light when ATP is added to firefly lantern extracts. The change in light emission with time generally follows an exponential decay after a short lag (Fig. 22(a)); standard curves of the light emission rate at a specified reaction time versus ATP concentration show a linear response over a wide range of ATP levels (Fig. 22(b)). ATP analyses have much promise as measures of viable microorganism concentrations in natural samples and metabolic activity in biological waste treatment.

The fact that enzyme reactions can be inhibited by specific substances

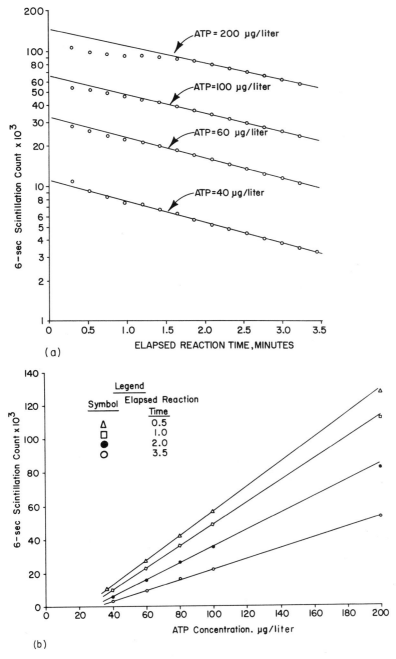

Fig. 22. (a) Luminescence decay of ATP-firefly lantern extract reaction with time. (b) Standard curves for light emission versus ATP concentration at several elapsed reaction times. [Reprinted from Ref. *176* by courtesy of the American Chemical Society.]

can be used advantageously for analytical purposes. An enzymatic technique for organophosphorus pesticides is based on their inhibition of cholinesterase, which catalyzes the hydrolysis of carboxylic acid esters. The rate of production of carbon ester hydrolysis products is inversely related to the inhibitor (pesticide) concentration. Various detection methods for hydrolysis products have been used, depending on the nature of the carbon ester substrate. These include simple esters which produce acid products (hence a pH change upon hydrolysis)(177), esters which hydrolyze to form colored species from colorless ones, and nonfluorescent esters which hydrolyze to fluorescent species(178). The cholinesterase procedure is not specific for a single pesticide but measures organophosphate pesticides as a group. If analyses of individual pesticides are desired by this method, thin layer or column chromatography can be used for prior separation. Guilbault and Sadar(179) recently reported an enzymatic technique for chlorinated hydrocarbon pesticides (2,4-D, DDT, aldrin, lindane, heptachlor) and the carbamate sevin based on their inhibition of lipase activity. A fluorimetric technique is used to monitor lipase activity using 4-methyl umbelliferone heptanoate as the substrate. The rate of production of highly fluorescent 4-methyl umbelliferone from the nonfluorescent ester is measured with a recording fluorimeter. The method is sensitive at parts per million levels but suffers from interferences by organophosphorus compounds. Again, mixtures of pesticides must be separated by chromatography in order to determine concentrations of individual pesticides.

The activation of certain enzymes by heavy metals and their inhibition by other metals can be used for metal determinations at ppm and sub-ppm levels. Kratochvil et al.(180) described a procedure for magnesium, manganese, and cobalt based on their activation of isocitrate dehydrogenase. The production of reduced nicotinamide adenine dinucleotide phosphate ($NADPH_2$) by the enzyme is monitored by ultraviolet spectrophotometry. Advantages of the method are high sensitivity and the small volume of sample required. In the absence of interferences, $10^{-6} M$ manganese in 0.1 ml of manganese solution can be determined with $\pm 5\%$. However, many metals inhibit the enzyme at low levels, and care must be taken to prevent these effects or to take them into account.

Kinetic methods have recently been devised for low levels of nutrients in natural waters. Crouch and Malmstadt(181) investigated the mechanism of the molybdenum blue method for phosphate and developed an automatic reaction rate method based on the initial rate (first 20–30 sec) of molybdenum blue formation from phosphate, molybdate, and ascorbic acid(182). In a later paper(183), a fast reaction rate system for phosphate analysis using an automated stopped flow system for determination of

12-molybdophosphoric acid from phosphate and Mo(VI) was reported. This system uses a measurement time of 0.3 sec and a total analysis time of 0.7 sec. For high precision and accuracy, an average of 10 results can be read out in about 10 sec. In addition to the obvious advantages of speed, these rapid reaction rate methods also eliminate interference from slower reacting substances. This was demonstrated(*184*) by comparing results of the reaction rate method and the manual procedure in analyzing mixtures of orthophosphate and ATP. The reaction rate method gave identical results for orthophosphate in the presence or absence of ATP, but slow hydrolysis of ATP under the acidic reaction conditions caused interferences in the manual method.

A reaction rate method for ammonia using the Berthelot reaction has been developed by Weichselbaum et al.(*185*). The reaction involves formation of a blue indophenol from ammonia, hypochlorite, and phenol with sodium nitroprusside as catalyst:

$$NH_3 + OCl^- \rightleftharpoons NH_2Cl + OH^-$$

(112)

indophenol blue

The reaction is sensitive and widely used in clinical analysis but suffers from a complex mechanism. The reaction products are unstable; completing and side reactions cause imprecise and inaccurate results with manual equilibrium methods based on the reaction. However, good results are obtained on natural water samples with an automated (AutoAnalyzer) version of the procedure(*186*). The kinetic method of Weichselbaum et al.(*185*) uses a continuous reaction rate system that directly measures the maximum value of the reaction rate, which was found to be directly proportional to the initial ammonia in the sample. An instrument was developed(*187*) that reads $dAbs/dt$ (the derivative of absorbance time curves) and prints out the peak, which corresponds to maximum rate. Analysis time is 2.5 min/sample, and a $\pm 2\%$ reproducibility is claimed. The procedure was developed for the higher levels (50–1000 mg N/liter) found in clinical analysis but could probably be modified for analyses at the parts per billion level in natural water samples.

In summary, analysis by kinetic or reaction rate methods, while no panacea, offers some advantages for particular reactions and substances. These methods are particularly useful for reactions that yield unstable products and for low level determinations of substances that act as catalysts when only small sample volumes are available. The sensitivities and selectivities obtainable with catalytic reactions and the adaptability of kinetic methods to automated measurement techniques ensure a growing number of applications of these methods in the analysis of natural water constituents.

REFERENCES

1. W. J. Moore, *Physical Chemistry*, Prentice-Hall, Englewood Cliffs, N.J., 1962.
2. S. Glasstone and D. Lewis, *Elements of Physical Chemistry*, 2nd ed., Van Nostrand, Princeton, N.J., 1960.
3. S. H. Maron and C. F. Prutton, *Principles of Physical Chemistry*, Macmillan, New York, 1958.
4. F. Daniels and R. A. Alberty, *Physical Chemistry*, Wiley, New York, 1961.
5. H. Freiser and Q. Fernando, *Ionic Equilibria in Analytical Chemistry*, Wiley, New York, 1963.
6. J. N. Butler, *Ionic Equilibrium. A Mathematical Approach*, Addison-Wesley, Reading, Mass., 1964.
7. K. Denbigh, *The Principles of Chemical Equilibrium*, 2nd ed., Cambridge Univ. Press, Cambridge, England 1966.
8. J. E. Ricci, *Hydrogen Ion Concentration. New Concepts in a Systematic Treatment*, Princeton Univ. Press, Princeton, N.J., 1952.
9. T. S. Lee and L. G. Sillen, *Chemical Equilibrium in Analytical Chemistry*, Wiley (Interscience), New York, 1959.
10. R. M. Garrels and C. L. Christ, *Solutions, Minerals, and Equilibria*, Harper and Row, New York, 1965.
11. W. Stumm, ed., *Equilibrium Concepts in Natural Water Systems, Advan. Chem. Ser. 67*, American Chemical Society, Washington, D.C., 1967.
12. S. D. Faust and J. V. Hunter, eds., *Principles and Applications of Water Chemistry*, Wiley, New York, 1967.
13. W. Stumm and J. J. Morgan, *Aquatic Chemistry*, Wiley, New York, 1970.
14. W. J. Weber, Jr., and W. Stumm, *J. Am. Water Works Assoc.*, **55**, 1553 (1963).
15. L. G. Sillen, in *Oceanography* (M. Sears, ed.), *Publ. No. 67*, American Association for the Advancement of Science, Washington, D.C., 1961, p. 549.
16. L. G. Sillen, *Science*, **156**, 1189 (1967).
17. L. G. Sillen, in *Equilibrium Concepts in Natural Water Systems* (W. Stumm, ed.), *Advan. Chem. Ser. 67*, American Chemical Society, Washington, D.C., 1967, pp. 45, 57.
18. W. Stumm, in *Proc. Symp. Environ. Measurement* (J. S. Nader and E. C. Tsivoglou, eds.), *U.S. Public Health Serv. Publ. No. 999–WP–15*, U.S. Dept. Health, Education, and Welfare, Washington, D.C., 1964, p. 299.
19. J. R. Kramer, in *Greak Lakes Res. Div., Univ. Mich., Publ. No. 11*, 1964, p. 43.
20. R. M. Garrels and M. E. Thompson, *Am. J. Sci.*, **260**, 57 (1962).
21. R. M. Garrels and F. T. MacKenzie, in *Equilibrium Concepts in Natural Water Systems* (W. Stumm, ed.), *Advan. Chem. Ser. 67*, American Chemical Society, Washington, D.C., 1967, p. 222.

22. E. S. Gould, *Mechanism and Structure in Organic Chemistry*, Holt, New York, 1959.
23. A. A. Frost and R. G. Pearson, *Kinetics and Mechanism*, 2nd ed., Wiley, New York, 1961.
24. F. A. Cotton and G. Wilkinson, *Advanced Inorganic Chemistry*, 2nd ed., Wiley (Interscience), New York, 1966, pp. 158–190.
25. Chemical Rubber Company, *Handbook of Chemistry and Physics*, 48th ed., Cleveland, 1967, p. B–7.
26. J. N. Pitts, Jr., A. U. Khan, E. B. Smith, and R. P. Wayne, *Environ. Sci. Technol.*, **3**, 241 (1969).
27. N. L. Clesciri and G. F. Lee, *Intern. J. Air Water Pollution*, **9**, 743 (1965).
28. N. L. Clesciri and G. F. Lee, *Intern. J. Air Water Pollution*, **9**, 723 (1965).
29. G. F. Lee and J. C. Morris, *Intern. J. Air Water Pollution*, **6**, 419 (1962).
30. G. F. Lee, in *Principles and Applications of Water Chemistry* (S. D. Faust and J. V. Hunter, eds.), Wiley, New York, 1967, pp. 54–72.
31. J. C. Morris, in *Principles and Applications of Water Chemistry* (S. D. Faust and J. V. Hunter, eds.), Wiley, New York, 1967, p. 23.
32. J. D. Cline and F. A. Richards, *Environ. Sci. Technol.*, **3**, 838 (1969).
33. K. Y. Chen and J. C. Morris, *Proc. Internat. Water Pollution Res. Conf.*, *5th*, San Francisco, 1970, Pergamon Press (in press).
34. M. C. Rand and S. B. Gale, in *Principles and Applications of Water Chemistry* (S. D. Faust and J. V. Hunter, eds.), Wiley, New York, 1967, p. 380.
35. S. Glasstone, K. J. Laidler, and H. Eyring, *The Theory of Rate Processes*, McGraw-Hill, New York, 1941.
36. E. S. Amis, *Kinetics of Chemical Change in Solutions*, Macmillan, New York, 1949.
37. K. J. Laidler, *Chemical Kinetics*, McGraw-Hill, New York, 1950.
38. H. Eyring, *J. Chem. Phys.*, **3**, 107 (1935).
39. W. Stöber, in *Equilibrium Concepts in Natural Water Systems* (W. Stumm, ed.), *Advan. Chem. Ser. 67*, American Chemical Society, Washington, D.C., 1967, pp. 161–182.
40. D. E. Carritt and S. Goodgal, *Deep-Sea Res.*, **1**, 24 (1954).
41. R. Haque, F. T. Lindstrom, V. H. Freed, and R. Sexton, *Environ. Sci. Technol.*, **2**, 207 (1968).
42. J. J. Morgan and W. Stumm, *Proc. Intern. Water Pollution Res. Conf.*, *2nd*, *Tokyo*, *1964*, p. 103.
43. W. J. Weber, Jr., in *Principles and Applications of Water Chemistry* (S. D. Faust and J. V. Hunter, eds.), Wiley, New York, 1967, p. 89.
44. C. H. Wayman, in *Principles and Applications of Water Chemistry* (S. D. Faust and J. V. Hunter, eds.), Wiley, New York, 1967, p. 127.
45. W. Gardner and G. F. Lee, *Intern J. Air Water Pollution*, **9**, 553 (1965).
46. W. J. Weber, Jr., and R. R. Rumer, Jr., *Water Resources Res.*, **1**, 361 (1965).
47. W. J. Weber, Jr., and J. C. Morris, *J. Sanit. Eng. Div. Am. Soc. Civil Engrs.*, **89**, SA2, 31 (1963).
48. J. Crank, *The Mathematics of Diffusion*, Oxford Univ. Press, London, 1956.
49. F. J. Edeskuty and N. R. Amundson, *Ind. Eng. Chem.*, **44**, 1698 (1952).
50. F. Helfferich, *J. Phys. Chem.*, **66**, 39 (1962).
51. S. Masamune and J. M. Smith, *J. Am. Inst. Chem. Engrs.*, **10**, 246 (1964).
52. D. D. Eley, *Trans. Faraday Soc.*, **49**, 643 (1949).
53. E. G. Fruh, Ph.D. thesis, Univ. Wisconsin, Madison, 1965.
54. E. G. Fruh and G. F. Lee, in *Principles and Applications of Water Chemistry* (S. D. Faust and J. V. Hunter, eds.), Wiley, New York, 1967, pp. 168–214.

55. J. T. Cookson, Jr., and W. J. North, *Eviron. Sci. Technol.*, **1**, 46 (1967).

56. J. T. Cookson, Jr., *J. Am. Water Works Assoc.*, **61**, 52 (1969).

57. W. Stumm and G. F. Lee, *Ind. Eng. Chem.*, **53**, 143 (1961).

58. J. D. Hem, *U.S. Geol. Surv. Water Supply Paper No. 1667-A*, Washington, D.C., 1963.

59. A. E. Nielsen, *Kinetics of Precipitation*, Pergamon, New York, 1964.

60. J. E.Hawley, M.S. thesis, Univ. Wisconsin, Madison, 1967.

61. G. F. Lee, *Factors Affecting the Transfer of Materials Between Water and Sediments, Lit. Rev. No. 1*, Water Resources Center, Univ. Wisconsin, Madison, 1970.

62. W. Stumm and J. O. Leckie, *Proc. Internat. Water Pollution Res. Conf., 5th*, San Francisco, 1970, Pergamon Press (in press).

63. K. E. Chave, *Science*, **148**, 1723 (1965).

64. V. C. Henri, *Lois Generales de l'Action des Diastases*, Hermann, Paris, 1903.

65. L. Michaelis and M. L. Menten, *Biochem. Z.*, **49**, 333 (1913).

66. G. E. Briggs and J. B. S. Haldane, *Biochem. J.*, **19**, 338 (1925).

67. V. R. Williams and H. B. Williams, *Basic Physical Chemistry for the Life Sciences*, Freeman, San Francisco, 1967.

68. J. E. Dowd and D. S. Riggs, *J. Biol. Chem.*, **240**, 863 (1965).

69. J. M. Reiner, *Behavior of Enzymes Systems*, Burgess, Minneapolis, 1959.

70. H. L. Segal, in *The Enzymes* (P. D. Boyer, H. Lardy, and K. Myrbäck, eds.), Vol. I, 2nd ed., Academic, New York, 1959, p. 1.

71. J. Z. Hearon, S. A. Bernhard, S. L. Friess, D. J. Botts, and M. F. Morales, in *The Enzymes* (P. D. Boyer, H. Lardy, and K. Myrbäck, eds.), Vol. I, 2nd ed., Academic, New York, 1959, p. 49.

72. R. A. Alberty, in *The Enzymes* (P. D. Boyer, H. Lardy, and K. Myrbäck, eds.), Vol. I, 2nd ed., Academic, New York, 1959, p. 143.

73. E. L. King and C. Altman, *J. Phys. Chem.*, **60**, 1375 (1956).

74. W. W. Cleland, *Biochim. Biophys. Acta*, **67**, 104 (1963).

75. W. W. Cleland, *Biochim. Biophys. Acta*, **67**, 188 (1963).

76. W. W. Cleland, *Biochim. Biophys. Acta*, **67**, 173 (1963).

77. E. S. West and W. R. Todd, *Textbook of Biochemistry*, 3rd ed., Macmillan, New York, 1961.

78. M. Dixon and E. C. Webb, *Enzymes*, Academic, New York, 1959.

79. H. R. Mahler and E. H. Cordes, *Biological Chemistry*, Harper and Row, New York, 1966.

80. L. Hartmann and G. Laubenberger, *J. Sanit. Eng. Div. Am. Soc. Civil Engrs.*, **94**, SA2, 247 (1968).

81. J. W. Patterson and P. L. Brezonik, *J. Sanit. Eng. Div. Am. Soc. Civil Engrs.*, **95**, 775 (1969).

82. J. W. Patterson, P. L. Brezonik, and H. D. Putnam, *Proc. Ind. Waste Conf., 24th*, Purdue University, Lafayette, Ind., 1969.

83. T. R. Parsons and J. D. H. Strickland, *Deep-Sea Res.*, **8**, 211 (1962).

84. R. T. Wright and J. E. Hobbie, *Limnol. Oceanog.*, **10**, 22 (1965).

85. R. F. Vaccaro and H. W. Jannasch, *Limnol. Oceanog.*, **11**, 596 (1966).

86. R. F. Vaccaro and H. W. Jannasch, *Limnol. Oceanog.*, **12**, 540 (1967).

87. R. T. Wright, *Proc. Symp. Organic Matter in Natural Waters*, Univ. Alaska, College, 1968.

88. R. D. Hamilton and J. E. Preslan, *Limnol. Oceanog.*, **15**, 395 (1970).

89. R. F. Vaccaro, S. E. Hicks, W. H. Jannasch, and F. G. Carey, *Limnol. Oceanog.* **13**, 356 (1968).

90. J. E. Hobbie and C. C. Crawford, *Limnol. Oceanog.*, **14**, 528 (1969).

91. R. C. Dugdale, *Limnol. Oceanog.*, **12**, 685 (1967).

92. J. J. MacIsaac and R. C. Dugdale, *Deep-Sea Res.*, **16**, 45 (1969).

93. R. W. Eppley, J. N. Rogers, and J. J. McCarthy, *Limnol. Oceanog.*, **14**, 912 (1969).

94. W. H. Thomas, *Limnol. Oceanog.*, **15**, 386 (1970).

95. W. D. P. Stewart, G. P. Fitzgerald, and R. H. Burris, *Proc. Natl. Acad. Sci. U.S.*, **58**, 2071 (1967).

96. E. L. Swilley, J. D. Bryant, and A. W. Busch, *Proc. Ind. Waste Conf., 19th, Purdue University, Lafayette, Ind., 1964*, p. 821.

97. W. Gulevich, C. E. Renn, and J. C. Liebman, *Environ. Sci. Technol.*, **2**, 113 (1968).

98. J. A. Mueller, W. C. Boyle, and E. N. Lightfoot, *Proc. Ind. Waste Conf., 21st, Purdue University, Lafayette, Ind., 1966*, p. 964.

99. C. R. Baillod and W. C. Boyle, *Environ. Sci. Technol.*, **3**, 1205 (1969).

100. F. E. Stratton and P. L. McCarty, *Environ. Sci. Technol.*, **1**, 405 (1967).

101. J. Monod, *Recherches sur la Croissance des Cultures Bacteriennes*, Herman, Paris, 1942.

102. R. I. Mateles and S. K. Chian, *Environ. Sci. Technol.*, **3**, 569 (1969).

103. E. J. Theriault, *The Oxygen Demand of Polluted Waters, U.S. Public Health Serv. Bull. No. 173*, U.S. Dept. Health, Education, and Welfare, Washington, D.C., 1927.

104. L. J. Reed and E. J. Theriault, *J. Phys. Chem.*, **35**, 673, 950 (1931).

105. G. M. Fair, *Sewage Works J.*, **8**, 430 (1936).

106. H. A. Thomas, Jr., *Sewage Works J.*, **9**, 425 (1937).

107. E. W. Moore, H. A. Thomas, Jr., and W. B. Snow, *Sewage Ind. Wastes*, **22**, 1343 (1950).

108. H. A. Thomas, Jr., *Water Sewage Works*, **97**, 123 (1950).

109. A. F. Gaudy, Jr., K. Komolrit, R. H. Follett, D. F. Kincannon, and D. E. Modesitt, *Methods for Evaluating the First Order Constants k, and L_0 for BOD Exertion*, Center for Water Research, Oklahoma State University, Stillwater, 1968.

110. A. M. Buswell, H. F. Mueller, and I. Van Meter, *Sewage Ind. Wastes*, **26**, 276 (1954).

111. A. W. Busch, *Sewage Ind. Wastes*, **30**, 1336 (1958).

112. A. W. Busch, *Water Sewage Works*, **108**, 255 (1961).

113. A. W. Busch, L. Grady, T. Shivaji Rao, and E. L. Swilley, *J. Water Pollution Control Federation*, **34**, 354 (1962).

114. S. R. Hoover, L. Jasewicz, and N. Porges, *Sewage Ind. Wastes*, **25**, 1163 (1953).

115. *Standard Methods for the Examination of Water and Wastewater*, 12th ed., American Public Health Association, New York, 1965, p. 415.

116. C. N. Sawyer and P. L. McCarty, *Chemistry for Sanitary Engineers*, 2nd ed., McGraw-Hill, New York, 1967.

117. C. T. Wezernak and J. J. Gannon, *Proc. Ind. Waste Conf., 24th, Purdue University, Lafayette, Ind., 1969*, in press.

118. K. E. F. Watt, *Ecology and Resource Management*, McGraw-Hill, New York, 1968.

119. E. C. Pielou, *An Introduction to Mathematical Ecology*, Wiley, New York, 1969.

120. G. L. Atkins, *Multicompartment Models for Biological Systems*, Methuen, London, 1969.

121. B. C. Patten, *Intern. Revue Ges. Hydrobiol.*, **53**, 357 (1968).

122. *Proc. Conf. Modeling the Eutrophication Process*, (H. D. Putnam, ed.), Department of Environmental Engineering, Univ. Florida, 1970.

123. R. V. Thomann, D. J. O'Connor, and D. M. DiToro, *Proc. Internat. Water Pollution Res. Conf., 5th*, San Francisco, 1970, Pergamon Press (in press).

124. E. Steeman-Nielsen, *Nature*, **167**, 684 (1951).

125. F. H. Rigler, *Limnol. Oceanog.*, **9**, 511 (1964).

126. F. R. Hayes, J. A. McCarter, M. L. Cameron, and D. A. Livingstone, *Ecology*, **40**, 202 (1952).

127. J. C. Neess, R. C. Dugdale, V. A. Dugdale, and J. J. Goering, *Limnol. Oceanog.*, **7**, 163 (1962).

128. P. L. Brezonik, Ph.D. Thesis, Univ. Wisconsin, Madison, 1968.

129. A. E. Zanoni, *Water Res.*, **1**, 543 (1967).

130. R. W. Eppley and J. D. H. Strickland, in *Advances in Microbiology of the Sea*, (M. R. Droop and E. J. F. Wood, eds.), Vol. 1, Academic, New York, 1968, p. 23.

131. E. J. F. Wood, *Marine Microbia Ecology*, Reinhold, New York, 1965.

132. T. D. Brock, *Principles of Microbial Ecology*, Prentice Hall, Englewood Cliffs, N.J., 1966.

133. J. A. Mueller, W. C. Boyle, and E. N. Lightfoot, *Appl. Microbiol.*, **15**, 672 (1967).

134. J. A. Mueller, W. C. Boyle, and E. N. Lightfoot, *Appl. Microbiol.*, **15**, 674 (1967).

135. K. Srinivasan and G. A. Rechnitz, *Anal. Chem.*, **40**, 1955 (1968).

136. C. H. Bamford and C. F. H. Tipper, eds., *Comprehensive Chemical Kinetics*, *The Practice of Kinetics*, Vol. I, Elsevier, New York, 1969.

137. J. J. Morgan and W. Stumm, *J. Am. Water Works Assoc.*, **57**, 107 (1965).

138. J. J. Morgan and W. Stumm, paper presented at Division of Water and Waste Chemistry, 146th Meeting, American Chemical Society, New York, September, 1963.

139. J. J. Morgan, in *Principles and Applications of Water Chemistry* (S. D. Faust and J. V. Hunter, eds.), Wiley, New York, 1967, p. 561.

140. H. A. Laitinen, *Chemical Analysis*, McGraw-Hill, New York, 1960.

141. W. J. Blaedel and V. W. Meloche, *Elementary Quantitative Analysis*, 2nd ed., Harper and Row, New York, 1963.

142. R. A. Day. Jr., and A. L. Underwood, *Quantitative Analysis*, 2nd ed., Prentice Hall, Englewood Cliffs, N.J., 1967.

143. G. A. Rechnitz, *Anal. Chem.*, **38**, 513R (1966).

144. G. A. Rechnitz, *Anal. Chem.*, **40**, 455R (1968).

145. G. G. Guilbault, *Anal. Chem.*, **42**, 334R (1970).

146. K. B. Yatsimirskii, *Kinetic Methods of Analysis*, Pergamon, London, 1966.

147. H. B. Mark and G. A. Rechnitz, *Kinetics in Analytical Chemistry*, Wiley (Interscience), New York, 1968.

148. H. V. Malmstadt and S. R. Crouch, *J. Chem. Educ.*, **43**, 340 (1966).

149. H. L. Pardue, *Record Chem. Prog.*, **27**, 151 (1966).

150. J. B. Pausch and D. W. Margerum, *Anal. Chem.*, **41**, 226 (1969).

151. D. W. Margerum, J. B. Pausch, G. A. Nyssen, and G. F. Smith, *Anal. Chem.*, **41**, 233 (1969).

152. M. S. Fishman and M. W. Skougstad, *Anal. Chem.*, **36**, 1643 (1964).

153. A. A. Fernandez, C. Sobel, and S. L. Jacobs, *Anal. Chem.*, **35**, 1721 (1963).

154. T. J. Janjic, G. A. Milovanovic, and M. B. Celap, *Anal. Chem.*, **42**, 27 (1970).

155. A. M. Wilson, *Anal. Chem.*, **38**, 1784 (1966).

156. T. P. Hadjiionannou, *Anal. Chim. Acta*, **35**, 351 (1966).

157. G. Svehla and L. Erdey, *Microchem. J.*, **7**, 206, 221 (1963).

158. K. B. Yatsimirskii, *Z. Anal. Khim.*, **11**, 327 (1956).

159. P. W. West and T. V. Ramakrishna, *Anal. Chem.*, **40**, 966 (1968).

160. *Standard Methods for the Examination of Water and Wastewater*, 12th ed., American Public Health Assoc., New York, 1965, p. 152.

161. P. A. Rodriguez and H. L. Pardue, *Anal. Chem.*, **41**, 1369 (1969).

162. P. A. Rodriguez and H. L. Pardue, *Anal. Chem.*, **41**, 1376 (1969).

163. C. G. Mitchell, *U.S. Geol. Surv. Water Supply Paper No. 1822*, 1966, p. 77.

164. M. J. Fishman and B. P. Robinson, *Anal. Chem.*, **41**, 323R (1969).

165. *Standard Methods for the Examination of Water and Wastewater*, 12th ed., American Public Health Assoc., New York, 1965, p. 287.

166. A. Babko and T. Maksimenko, *Zh. Anal. Khim.*, **22**, 570 (1967).

167. R. Hems, G. Kirkbright, and T. West, *Talanta*, **16**, 789 (1969).

168. G. G. Guilbault, *Anal. Chem.*, **40**, 459R (1968).

169. J. B. Hill and G. Kessler, *J. Lab. Clin. Med.*, **57**, 970 (1961).

170. J. S. Jeris and R. R. Cardenas, Jr., *Appl. Microbiol.*, **14**, 857 (1966).

171. A. G. Ware and E. P. Marbach, *Clin. Chem.*, **11**, 792 (1965).

172. R. Thompson, *Clin. Chim. Acta*, **13**, 133 (1966).

173. O. Holm-Hansen and C. R. Booth, *Limnol. Oceanog.*, **11**, 510 (1966).

174. R. D. Hamilton and O. Holm-Hansen, *Limnol. Oceanog.*, **12**, 319 (1967).

175. J. W. Patterson, Ph.D. thesis, University of Florida, Gainesville, 1970.

176. J. W. Patterson, P. L. Brezonik, and H. D. Putnam, *Environ. Sci. Technol.*, **4**, 569 (1970).

177. Technicon Corp., *Bull. No. P7*, Chauncey, N.Y.

178. J. Epstein and M. M. Demek, *Anal. Chem.*, **39**, 1136 (1967).

179. G. G. Guilbault and M. H. Sadar, *Anal. Chem.*, **41**, 366 (1969).

180. B. Kratochvil, S. L. Boyer, and G. P. Hicks, *Anal. Chem.*, **39**, 45 (1967).

181. S. R. Crouch and H. V. Malmstadt, *Anal. Chem.*, **39**, 1084 (1967).

182. S. R. Crouch and H. V. Malmstadt, *Anal. Chem.*, **39**, 1090 (1967).

183. A. C. Javier, S. R. Crouch, and H. V. Malmstadt, *Anal. Chem.*, **41**, 239 (1969).

184. S. R. Crouch and H. V. Malmstadt, *Anal. Chem.*, **40**, 1901 (1968).

185. T. E. Weichselbaum, J. C. Hagerty, and H. B. Mark, Jr., *Anal. Chem.*, **41**, 848 (1969).

186. R. A. Yorton and P. L. Brezonik, unpublished data, 1969.

187. T. E. Weichselbaum, W. H. Plumpe, Jr., R. E. Adams, J. C. Hagerty, and H. B. Mark, Jr., *Anal. Chem.*, **41**, 725 (1969).

Chapter **17** **Effect of Structure and Physical Characteristics of Water on Water Chemistry**

R. A. Horne†

ARTHUR D. LITTLE, INC.
CAMBRIDGE, MASSACHUSETTS

I. Introduction

A. The Anomalous Nature of Water

If we were to try to characterize the total chemistry of the human environment on this planet in as sweeping, yet as exact, terms as possible, I do not think we could do better than to describe it as the chemistry

†Present address: JBF Scientific Corp., Burlington, Massachusetts.

of aqueous electrolytic solutions. The unique feature of our planet compared to its sisters in the solar system is the abundance of water on its surface. The chemistry of the hydrosphere and biosphere is almost entirely water chemistry, and water continues to play an important role even in the chemistry of the lithosphere and atmosphere.

Now the full impact of our generalization may not be immediately obvious. Not only is the Earth's watery mantle extraordinary among planets, but water is extraordinary, one can even say unique, among chemical substances. Thus the chemistry of man and his environment is remarkable chemistry indeed.

Water is anomalous in *all* of its physical–chemical properties. Table 1 compares some of the properties of water with those of a "normal" liquid of comparable boiling point. Particularly noteworthy are the high values of the thermodynamic properties of water, for these, along with

TABLE 1

Comparison of the Physical Properties of Liquid
Water and *n*-Heptane[a]

Physical properties	Water $(H_2O)_n$	Normal heptane (C_7H_{16})
Molecular weight	$(18)_n$	100
Dipole moment	1.84×10^{-18}	$> 0.2 \times 10^{-18}$
Dielectric constant	80	1.97
Density	1.0	0.73
Boiling point, °C	100	98.4
Melting point, °C	0	−97
Specific heat	1.0	0.5
Heat of evaporation, cal/g	540	76
Melting heat, cal/g	79	34
Surface tension at 20°C, dyne/cm	73	25
Viscosity at 20°C, poise	0.01	0.005

[a]From Ref. *1*.

the very high surface tension and viscosity, give us the first hint of the disinclination of water molecules to be separated one from another. The very high dielectric constant of water is also of great interest to us here, for this property accounts in part for the excellent solvent capabilities of water. An even more striking comparison illustrating the anomalous nature of water can be made between water and the closely related compounds H_2S, H_2Se, and H_2Te (Fig. 1); as the molecular weights of the members of this family decrease, the melting and boiling points decrease

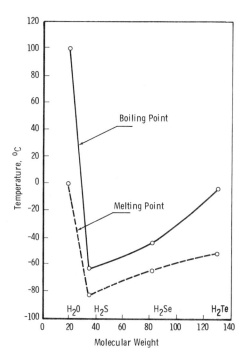

Fig. 1. Transition temperatures of the group VI-A hydrides. [Reprinted from Ref. *18* by courtesy of Academic Press.]

in a regular way, yet H_2O, representing the lowest molecular weight in the series, has melting and boiling points much higher than those of the other substances. It cannot be overemphasized that water is easily the strangest, the most studied, and the least understood substance known to man.

B. The Importance of Water Structure in Water Chemistry

Similarly, the importance of water chemistry cannot be exaggerated, not only for its central role in the chemistry of the total ecology of this planet, but also because the chemistry of life is aqueous solution chemistry. The life processes, even the form of the various biosubstances, are determined by the physical chemistry and structural properties of the circumambient water. So essential is water in living things that we can say, confidently, that where there is no water there can be no life, and anywhere in this universe where there is ample liquid water there may be life.

What makes liquid water so different, so complex? The answer is deceptively simple. It is the ability of the molecules in liquid water to associate with one another and thereby form complex — if you like, polymeric — structures. Customarily we think of gases and liquids as being distinguished from solids by their lack of structure or repeating organization, but in liquid water there is a great level of local order.

Every species in water, whether electrolyte, polyelectrolyte, or non-electrolyte, hydrophilic or hydrophobic, microscopic or macroscopic, alters the local water structure in its immediate neighborhood, and the interaction of each species in water with any other species must take place via the mediacy of this water structure. Every chemical reaction in water, every interaction, even of the most tenuous nature, entails alterations of the water structure.

II. The Structure of Pure Liquid Water

A. The Water Molecule and the Structure of Ice-I_h

Our knowledge of the structure of liquid water is very imperfect and has been a subject of intense controversy for the last 30 or more years. But at least the area of confusion is delimited by two extremes of clarity: The structure of the monomeric molecule in the vapor and the structure of the lattice of ice-I_h (the familiar 1-atm form of the solid) are well known and can therefore serve as convenient points of departure.

There is very little intermolecular association of water in the vapor state. Gaseous water is for the most part monomeric, with a few dimers and perhaps some very rare trimers.

The electronic cloud of the water molecule is like an abbreviated, distorted jack in appearance (Fig. 2). The H—O—H bond angle is 104°31′(2). The O—H distance is 0.96 Å in the vapor and somewhat greater in the solid, 0.99 Å. The two pairs of unshared electrons form distributions which project upward toward the opposite corners of the opposite face of the containing cube from the direction of the proton-containing clouds (Fig. 2). This separation and directionality of negative electrification is responsible for the molecule's large dielectric constant (see Table 1) and, by attracting the positive partial charges of the protons on nearby water molecules, for the tendency of water molecules to associate.

In ice-I_h a water molecule associates with its four closest neighbors in a tetrahedral configuration (Fig. 3) to form a tridymite-like arrangement of the O centers consisting of a layered network of puckered, hexagonal rings (Fig. 4). Less certainty surrounds the positioning of the H centers

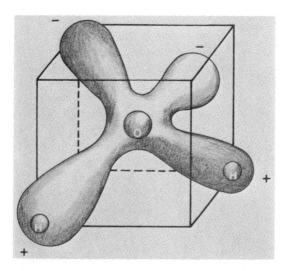

Fig. 2. Electronic cloud of the water molecule. [Reprinted from Ref. *24* by courtesy of John Wiley & Sons (Interscience).]

in ice. Pauling(*3*) proposed that the H of the hydrogen bond is not midway between the oxygen atoms but rather prefers two sites in the O—H—O bond, 0.96 Å from one O atom and 1.81 Å from the other. Neutron diffraction studies on D_2O ice appear to have confirmed this view(*4*). The formation of an H bond enhances the charge separation of the molecules involved, thereby facilitating further H bond formation with neighboring waters and cooperatively extending the bonding over a bigger volume.

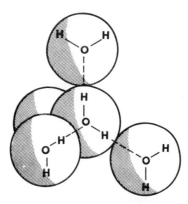

Fig. 3. Tetrahedral arrangement of the water molecules in ice. [Reprinted from Ref. *24* by courtesy of John Wiley & Sons (Interscience).]

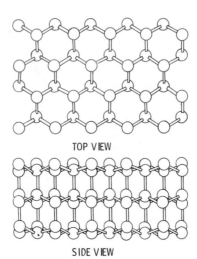

TOP VIEW

SIDE VIEW

Fig. 4. The ice–I_h lattice. [Reprinted from Ref. *59* by courtesy of the American Institute of Physics.]

B. A Brief History of Theories of Water Structure

While modern scientists appear to be only just beginning to appreciate the central role of water in the scheme of things, the ancients had a clearer view of the priorities. The first scientist, Thales of Miletus (fl. ca. 546 B.C.), based his entire cosmological system on water. All substances, he maintained, were only different forms of water. Although subsequently replaced by the Empedoclean–Aristotelian theory of four elemental substances, the idea persisted. Physicists are not inclined to advertise Newton's alchemical leanings, so it is not widely known that the great man subscribed to some form of the Thalesian view. The notion of the primacy of water (or of its constituent, hydrogen) did not really die until the 19th century when highly accurate analyses laid Prout's hypothesis to rest. Perhaps we shall see an idea which once dominated physics and chemistry come to reign in biology.

By the end of the 19th century it had become abundantly clear that water is not a simple liquid, and scientists such as Armstrong(5) and Vernon(16) were proposing that it contains complexes or aggregates of water molecules. The subsequent early theories of water structure tried to account for the anomalous properties of the liquid, such as the density maximum at 4°C, in terms of the association of water molecules to form mixtures of polymers — "dihydrol" $(H_2O)_2$, "trihydrol" $(H_2O)_3$, etc. These early theories of water structure were reviewed by Chadwell in 1927(7). In the light of the cooperative nature of the hydrogen-bonding process (see above), more recent theories have discounted the presence of significant concentrations of small polymers in liquid water. Wicke(8), however, has suggested that the inadequacies of present theories may compel us to reconsider a "third state" consisting of small aggregates.

The modern theories of the structure of liquid water, in one way or another, take as their point of departure the classic paper of Bernal and Fowler(9) published in the first volume of the *Journal of Chemical Physics*. They proposed that liquid water consists of two structural forms: ice or tridymite-like water I of density 0.91 g/cm³, and quartz-like water II of 1.08 g/cm³ density. Below 4°C, water I is the dominant form, but above this temperature it is replaced by water II and the transition gives rise to the density maximum. This model has one conspicuous fault which its critics were not slow to belabor: How can a substance composed of only two specific crystallographic forms be so fluid at room temperature? Only by such extensive rupture and/or violent bending of hydrogen bonds that the crystallographic forms become blurred into meaninglessness. But the deficiencies of this theory have been almost as fruitful as its successes in provoking fresh ideas concerning the structure of liquid water.

C. Continuum versus Mixture Models

Bernal and Fowler(9) proposed with great caution that the two structural forms in liquid water "are not distinct and pass continuously into one another." The liquid is homogeneous and there is no "mixture of volumes with different structures." Such water models are called "continuum" or uniformist theories, and in insisting that there is no microregion in water structurally different, on the average, from any other arbitrarily chosen volume element, they differ from the so-called "mixture" models which treat the liquid as a mixture of at least two water states.

Over the past 10 years mixture models appear to have enjoyed the greater favor in the United States, perhaps reaching the zenith of their popularity in Nemethy and Scheraga's(10) quantitative analysis of the Frank–Wen(11) flickering cluster mixture model (described below). However, in recent years objections have been made which, at least by implication, seem to support continuum models(12–14). These cirticisms serve to remind us of the very inadequate state of our knowledge, but they have not, I must say, shaken my faith in the fundamental qualitative correctness of the Frank–Wen theory. This model is used throughout the remainder of this paper. The theory envisions liquid water as consisting of constantly forming, constantly dissolving (i.e., "flickering") clusters of

H-bonded water molecules immersed in more or less "free" or monomeric water (Fig. 5). The clusters are constantly forming and "melting"
with the random microscopic thermal fluctuations in the liquid. Their half-
life is of the order of 10^{-10}–10^{-11} sec — "short enough to correspond to the

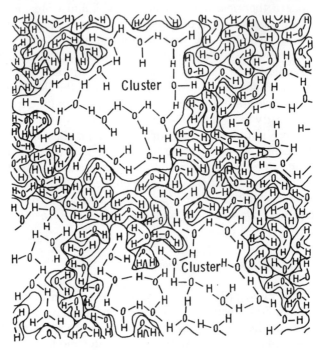

Fig. 5.   Frank–Wen flickering cluster model of liquid water.

dielectric relaxation time of water, but long enough [10^2–10^3 times a molecular vibration period] to constitute a meaningful existence of the cluster"
(*11*). Because of the cooperative nature of H bond formation, small
clusters are disfavored. Nemethy and Scheraga(*10*) have calculated that
at room temperature the average cluster contains about 30 water molecules, whereas other theorists have postulated much larger clusters of
hundreds and even thousands of water molecules(*15–17*).

Quite apart from the difficulties in the quantitative application of the
Frank–Wen cluster theory, two qualitative ambiguities have caused a
great deal of needless confusion. Frank and Wen(*11*), in their description
of the clusters, were always careful to place terms such as "ice-like-ness"
and "icebergs" in qualifying quotation marks. Similarly, Nemethy and
Scheraga(*10*) cautiously declared that "the model is not dependent on
the postulation of a particular semicrystalline structure — the term 'ice-

like' [does not] necessarily imply that the clusters have the tridymite arrangement of ordinary ice — rather irregular arrangements of the molecules in the clusters are not excluded by the model." Yet despite these admonitions, some of the critics of the model have clearly not rid themselves of the notion that the structure of the clusters is specifically that of ice-I_h. But in fact, there may be good reason for believing that the structure of the cluster is not ice-like(18), and at the present time we must conclude that the details of structure of the clusters are unknown. The second difficulty arises from the description of the nonclustered water as "free." Again, despite the qualifying quotation marks, some writers have interpreted this to mean that the monomeric water in the liquid phase enjoys the same freedom to execute translational and rotational movements as do water molecules in the gas phase. On the contrary, although it need not be specifically bound to its neighbors in the ordinary sense, a a water molecule is nevertheless "aware" of their presence. It might be "free" enough to move, but this motion is not without restraint. I am ready to agree that the intermolecular forces run a range from the very tenuous to the quite strong(13,14), but in subscribing to a mixture theory I do not admit that this spectrum is a continuous one(14), i.e., that all energy states are equally preferred by the water molecules.

I do not wish to conclude this discussion of water models without at least mentioning some of the other types of water structure theory that are being put forward. Marchi and Eyring(19) have attempted to apply the theory of significant structures to liquid water. There are a number of variations on the theory that liquid water represents a broken ice structure in which the hydrogen bonds bend and stretch but do not break and in which the interstices in the lattice can be occupied by one or more "free" water molecules(20).

A particularly fascinating model was proposed by Pauling(21) and developed by Frank and Quist(22); it pictures liquid water as consisting of clathrate cages occupied by other water molecules, analogous to the clathrates of certain inert gases and tetraalkylammonium ions.

For more detailed reviews of theories of water structure, the reader is referred to Kavanau(23), Samoilov(20), Wicke(8), and Horne(18,24).

III. The Effect of Temperature and Pressure on the Structure of Pure Water

A. Temperature and Higher Structural Transitions

Opinion now appears to be unanimous that as temperature increases the structure of liquid water, whatever it may be, is loosened. But agree-

ment does not extend very far beyond this entirely qualitative conclusion. With respect to this topic, the current controversies center around
two questions. Quantitative attempts, based on the different theories, to
estimate the extent of structure (i.e., the extent of hydrogen bonding) as
a function of temperature give different answers (Fig. 6), while attempts
based on different types of experimental measurement give values for the
percentage of broken hydrogen bonds in the liquid at 0°C ranging from
2.5 to 71.5(*14*). It is easy to overestimate the significance of these

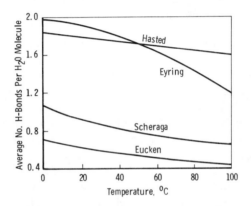

Fig. 6. Extent of water structure as a function of temperature based on different theoretical
models. [Reprinted from Ref. *8* by courtesy of Verlag Chemie GmbH.]

difficulties and let them obscure the important fact that liquid water still
has a lot of structure even at the boiling point, in other words, that
temperature is a relatively mild structure-breaking influence.

The second controversy centers on the possible existence of inflections
or "kinks" in the temperature dependence of water properties in the
neighborhood of certain temperatures, such as 30–40°C, possibly arising
from higher structural transitions in liquid water. Drost-Hansen(*25*) has
accumulated from the literature a wealth of evidence for these kinks, but
Falk and Kell(*26*) argue, unconvincingly, that this evidence should be
dismissed as experimental error. In many instances the phenomenon in
question is indeed small, yet even in the absence of unassailable experimental evidence, I am inclined to believe both in the reality and the very
great importance of these higher structural transitions, and I would consider it most tragic if negative thinking were to prevail and attempts to
unravel these mysteries were abandoned.

B. The Effect of Pressure

The Frank–Wen clusters are less dense than the surrounding un-
structured water (Fig. 5). By the application of the principle of Le
Chatelier we expect the application of hydrostatic pressure to favor the
less voluminous form, that is to say, pressure should destroy the structured regions in liquid water. A water property that is a good indicator of
the extent of structure present is the macroscopic viscosity. For example,
the relative viscosity of water decreases with increasing temperature,
the decrease closely paralleling the decrease in the size of the clusters
(Fig. 7). The effect of pressure on the structure of water is most extraordinary (Fig. 8). First, the relative viscosity, $\eta_P/\eta_{1\,atm}$ versus P, goes
through a minimum. At still higher pressures, $\eta_P/\eta_{1\,atm}$ increases with
increasing pressure, as expected of a "normal" liquid. The implications of
this unique behavior are interesting: Pressure, in contrast to temperature
(see above), appears to be a very powerful structure-breaking influence,
so powerful, in fact, that by the time that pressures of 1500 kg/cm² are
reached all of the structure in liquid water is destroyed and it behaves
like a "normal," unassociated fluid (27).

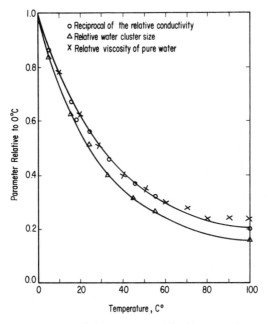

Fig. 7. Viscosity as a measure of the structure of liquid water: the temperature dependence of the relative viscosity, electrical conductance, and water cluster size. [Reprinted
from Ref. 60 by courtesy of the American Chemical Society.]

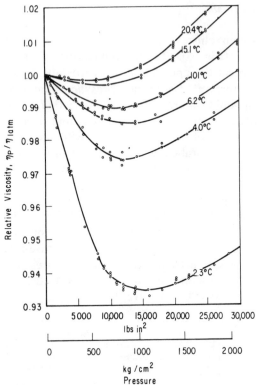

Fig. 8. Pressure dependence of the relative viscosity of pure water. [Reprinted from Ref. 27 by courtesy of the American Chemical Society.]

In condensed phase systems the volume changes and hence the pressure effects are commonly modest, yet they are important in achieving a detailed understanding of the chemistry of natural waters, for high temperature-high pressure environments are encountered in certain geothermal waters, and the ocean depths are distinguished by the particularly complex and little-understood combination of low temperature (2–5°C) and high pressure (up to about 1000 kg/cm²).

IV. Aqueous Electrolyte Solutions

A. The Primary Hydration of Ions and the Short-Range Perturbation of Water Structure

Hitherto we have confined our discussion to pure water, but natural waters are hardly ever pure and, thanks to human activity, are very rapidly becoming increasingly less so. The most conspicuous component of the Earth's hydrosphere, the ocean, is an aqueous solution roughly $0.5\,M$ NaCl–$0.05\,M$ MgSO$_4$ with a trace of just about everything else imaginable from gold to DNA. Figure 9 is an attempt to classify the substances found in natural waters in general and seawater in particular Although this table is very crude, it contains a number of valuable qualitative generalizations and we refer to it again and again throughout

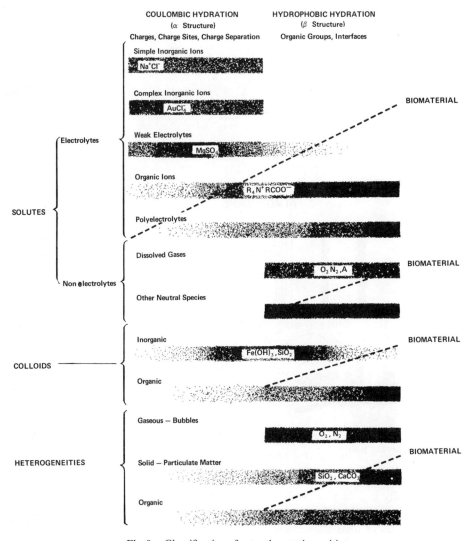

Fig. 9. Classification of natural water impurities.

this discussion. I might say a word here about the origins of the substances in, say, seawater. These substances fall into two categories:
inorganic constituents and organic constituents that are described in
Fig. 9 as biomaterial. The inorganic material comes largely from the
Earth's lithosphere; the cations from the weathering and leaching of
surface rock, the anions, usually via the atmosphere, from plutonic
effusions(24). The organic compounds have the Earth's biosphere,
both terrestrial and marine, as their origin, and they range from complex
systems such as living creatures down through protein fragments and
amino acids to the relatively simple end products of the decay processes.
Strictly speaking, some of the inorganic substances such as O_2, SiO_2,
$CaCO_3$, HCO_3^-, etc., also pass through the biosphere in the course of
their chemical history (Fig. 10).

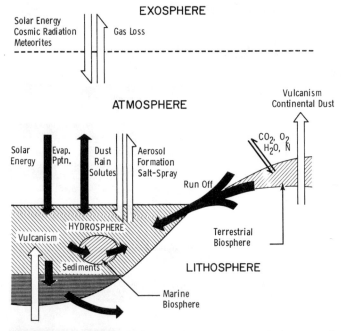

Fig. 10.   Chemical exchange between the exosphere, atmosphere, hydrosphere, lithosphere,
and biosphere of Earth.

When an electrolyte, that is, a material capable of dissociating into
electrically conductive ions, such as NaCl, is added to water, the structure and properties of water are profoundly altered. In this section we deal
with the more drastic of these changes, while in the next section we turn
our attention to more subtle, long-range perturbations of water structure.

The dissolution of an electrolytic solute in water is accompanied by appreciable temperature changes. In some cases, such as H_2SO_4, these changes are so great as to constitute a hazard. This means that strong chemical bonds are being broken and new bonds, between the solute and the solvent, are being formed. Secondly, the addition of an electrolyte changes the colligative properties of water and markedly increases the electrical conductivity. The Arrhenius theory of electrolytic dissociation has become so familiar to the modern chemist that he may have lost sight of how remarkable the coexistence of oppositively charged ions, without recombination, really is. Here again we see that there must be very powerful forces at work in aqueous solutions capable of vitiating the strong coulombic attraction between unlike charged ions. The solvent molecules bound by the electrolyte must form a hydration atmosphere capable of "screening," i.e., attenuating, the charge densities, the ion's mutual interactions. Finally, we observe that the addition of an electrolyte to water is commonly accompanied by an overall volume decrease. The polar water molecules are so strongly attracted by the charge field of an ion that they crowd tightly toward it; thus, the density of the water immediately adjacent to an ion is greater than the density of water at some great distance from the ion ("electrostriction"). The very tightly bound, electrostricted water immediately adjacent to an ion is sometimes referred to as the primary hydration sphere of the ion. The experimental determination of "hydration number" tends to yield the number of water molecules in the primary hydration sphere as a lower limit. For many electrolytes these waters are so strongly held that they cling to the solute even when it is crystallized out of solution, for example, Epsom salt ($MgSO_4 \cdot 7H_2O$).

B. The Viscosity B Coefficient and More Long-Range Perturbations of Water Structure

Earlier I pointed out that the macroscopic viscosity often provides a valuable insight into what is happening to the structure of water. The effect of the addition of various electrolytes on the viscosity of water (Fig. 11) is very interesting. Some solutes, especially 2:2 electrolytes such as $MgSO_4$, make the water more viscous, presumably by increasing the extent of water structure, whereas other solutes, such as CsCl, actually make the water less viscous and their presence appears to break up the structure of liquid water(28). These structure-making and structure-breaking phenomena are evident even at modest electrolyte concentrations (Fig. 11); hence the perturbation of the water structure must extend out from the ions into the solution for a considerable distance.

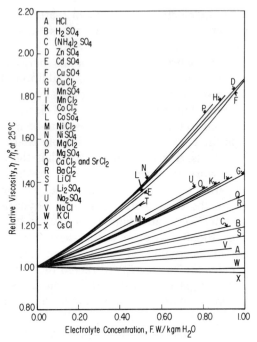

Fig. 11.   Effect of electrolyte addition on the viscosity of water.

The viscosity of a solution is related to the viscosity of the pure solvent, η_0, by the empirical relation of Jones and Dole (29):

$$\eta = \eta_0(1 + A\sqrt{c} + Bc) \tag{1}$$

where c is the concentration of solute. Of particular interest to us here is the B coefficient, inasmuch as it provides a quantitative measure of the effects of individual ions on water structure. A positive value of B indicates that the solute is a structure maker; a negative value indicates a structure breaker (28).

In order to account for these long-range effects of electrolytes on water structure, Frank and Wen (11) proposed a two-zone model for the hydration atmosphere of ions in aqueous solution: an inner, structure-enhanced region of tightly bound, electrostricted water molecules and an outer region of broken water structure in which the coulombic field of the ion is strong enough to disrupt the "normal" structure of bulk water yet insufficiently potent to reorient the water molecules into some new configuration. Water is only about 55.5 M in water, so as the concentration of electrolyte is increased, the supply of water molecules available for hydration soon becomes exhausted and the hydration atmospheres of the ions begin to overlap. Just like the viscosity itself, although, for reasons still unclear, not always covariant with it, the Arrhenius activation energy of transport processes such as diffusion, electrical conduction, and viscosity also is highly sensitive to structural changes in the liquid and provides a useful tool for examining such change. As the outer structure-broken zones overlap, transport processes are facilitated and the Arrhenius activation energy of viscous flow, $E_{a,\text{vis.}}$, decreases (Fig. 12), but then as the inner structure-enhanced regions begin to overlap, the energy required for the movement of species increases$(30,31)$. The minimum in the plot of $E_{a,\text{vis.}}$ versus electrolyte concentration (Fig. 12) corresponds to the onset of overlap of the structure-enhanced regions and the number of water molecules in the ion's total hydration atmosphere; that is, all the waters in the structure-enhanced region can be estimated from the electrolyte concentration at which the minimum occurs(31). The results (Table 2) are remarkable indeed. The number of water molecules in the *total* hydration atmosphere of Na^+, for example, is very nearly the same as the average number of water molecules in the Frank–Wen cluster in pure bulk water at the same temperature. In other words, a hydrated ion

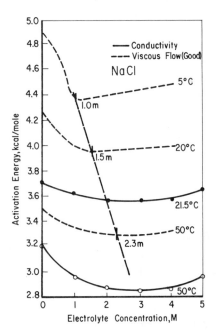

Fig. 12. Effect of electrolyte concentration on the Arrhenius activation energies of viscous flow and electrical conductivity of aqueous NaCl solutions. [Reprinted from Ref. *31* by courtesy of *Electrochimica Acta*, Oxford, England.]

TABLE 2
The Hydration of Some Selected Ions in Dilute Aqueous Solution

Ion	Temp, °C	Primary hydration number[a], 1 atm	Total hydration number[b], 1 atm	Average no. H₂O molecules in Frank–Wen cluster[c]	Total hydration number[a], 5000 kg/cm²
Li⁺	5	4	60	57	—
	20	4	56	38	1
	50	4	32	20	—
Na⁺	5	4	52	57	—
	20	4	34	38	1
	50	4	21	20	—
K⁺	25	1(?)	20	—	0(?)
Rb⁺	25	1	18	—	0
Cs⁺	25	1	16	—	0
F⁻	25	>1	—	—	1
Cl⁻	25	1	—	—	0
Br⁻	25	1	—	—	0
I⁻	25	0	—	—	0

[a]From Ref. *33*.
[b]From Ref. *31*.
[c]From Ref. *10*.

can be described as an ion-containing Frank–Wen cluster. The structure-enhanced region can be further divided into two zones: the dense, tightly bound, innermost electrostricted water and a Frank–Wen cluster-like envelope of relatively low density (Fig. 13).

Different experimental techniques yield different hydration numbers; for example, the values reported in the literature for Na⁺ range from 4 to 70(*32*), and the reason is now clear—the different methods measure water molecules bound with different strengths. Some methods measure only the most tightly bound of the primary hydration shell, while other techniques count the more loosely bound waters as well. In order to circumvent this difficulty, the Soviet school(*20*) prefers to define hydration in terms of residence time of water molecules near an ion relative to the residence time of a water molecule at some location a great distance from the ion.

C. Transport Processes in Aqueous Solution

The movements of ions and molecules in aqueous solution are highly dependent on water structure and can, in turn, yield valuable insight into

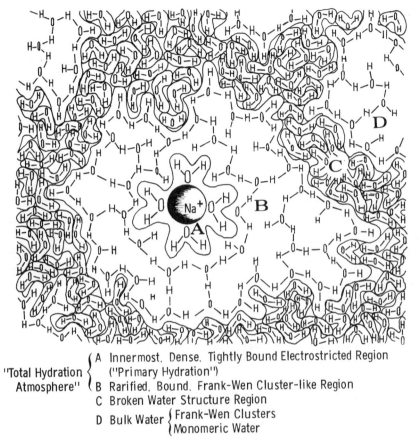

"Total Hydration Atmosphere" {
A Innermost, Dense, Tightly Bound Electrostricted Region ("Primary Hydration")
B Rarified, Bound, Frank-Wen Cluster-like Region
C Broken Water Structure Region
D Bulk Water { Frank-Wen Clusters / Monomeric Water

Fig. 13. Cationic coulombic hydration: the local water structure near Na^+ ion at 25°C and 1 atm.

the nature and extent of that structure. We have already seen this in the instance of the viscosity. Collie et al. (*34*) observed only a single dielectric relaxation time for water at a given temperature, indicating that there is only one rotationally orientable molecular species in the liquid — presumably the monomer. In more recent years, nuclear magnetic resonance (NMR) has provided a powerful means for examining the rotational movements of water molecules, especially those in the hydration atmosphere of ions [see the review of Wicke(*8*) cited earlier], and has added further support to the conclusions drawn earlier from the viscosity effects, namely, some ions such as K^+, Cs^+, and halide anions increase the fluidity of water and facilitate rotation, whereas others such as Na^+ and Mg^{2+} hinder the rotation of water molecules.

At 25°C, dielectric relaxation, electrical conductivity, viscous flow, and diffusion in aqueous solution all have similar Arrhenius activation energies (3–5 kcal/mole), and from this coincidence the conclusion was drawn that they all have the same rate-determining step in their mechanisms (35–37). But more detailed examination of the concentration, temperature, and pressure dependencies of the activation energies of the several processes has indicated that they go not by the same but by distinctly different, if sometimes closely related, mechanisms (33).

Our knowledge concerning ionic movement in solution in both the absence (diffusion) and presence (electrical conductivity) of an electrical field is in a sorry state. One proposed mechanism (the Debye–Hückel theory) has carried the entire weight of the whole modern theory of electrical conduction in aqueous electrolytic solution since the first decade of this century and is based on the extravagantly oversimplified model of a charged sphere moving with nonaccelerated velocity through a viscous continuum. The theoretical implications of the second, and far more realistic, mechanism — that of the formation of a "hole" or vacancy in the solvent followed by the jump of a nearby ion to fill the hole (36) — have not been pursued with the vigor they warrant.

D. The Effect of Temperature and Pressure on Hydration

Earlier we discussed the effect of temperature and pressure on the structured regions in pure bulk water. Now let us examine the influence of these variables on the local water structure in the hydration atmosphere of ions in solution.

Walden's rule, derivable from Stokes' law and based on the viscous continuum model, states that the product of the electrical conductance and viscosity is a constant independent of temperature. This rule works much better than we have any reason to expect. Even for so complex a solution as seawater it holds remarkably well; in going from 0 to 20°C the conductance of seawater changes by 61% and the viscosity by 49%, but their product changes by only 5% (38). Apparently the so-called hydrodynamic radius of a hydrated ion in solution is not highly temperature dependent; that is, temperature has little effect on the hydration envelope of an ion, at least on that fraction of its total hydration atmosphere that accompanies the ion when it jumps.

Walden's rule fails, however, under the condition of variable hydrostatic pressure (38,39). As the pressure increases, aqueous electrolytic solutions, even of strong electrolytes, at first become more conductive than they should be on the basis of the volume and viscosity changes.

The effective radii of the ions appear to decrease. The application of pressure destroys the local water structure in the hydration atmospheres of ions(18,39,40). At pressures greater than about 2000 kg/cm², the hydration atmospheres of the ions are reduced to a single water molecule or are completely stripped away (Table 2), the conductance decreases exactly as one would expect from the volume and viscosity changes, and the conduction correlates nicely with the crystal radii of the ions (Figs. 14 and 15). Kay and Evans(41) have argued that, as a consequence of electrostriction, pressure should stabilize rather than destroy ionic hydration. Not only is this contrary to observation(42) but it is not expected on the basis of the detailed model of ionic hydration (Fig. 13). As pressure is increased the higher specific volume cluster-like region (B in Fig. 13) first "melts." Now the dense, electrostricted region (A in Fig. 13) is highly incompressible; hence, as the pressure continues to increase, a point will finally be reached where the density of the bulk water (D) exceeds that of the electrostricted water and the primary hydration envelope will then be destroyed.

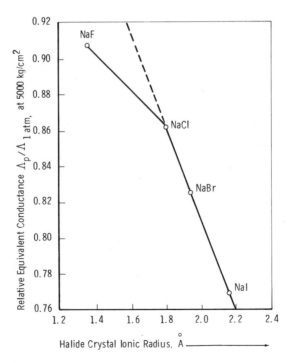

Fig. 14.   Dependence of conductance under pressure on crystal radius – anions. [Reprinted from Ref. 61.]

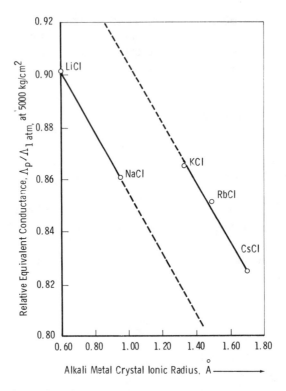

Fig. 15.   Dependence of conductance under pressure on crystal radius — cations. [Reprinted from Ref. *61.*]

Figure 16 contrasts schematically the effects of temperature and pressure. As the temperature increases the structure of the bulk water decreases, but the local water structure in the hydration atmospheres of ions remains intact; however, the application of pressure destroys both the bulk and the local water structure.

V. Interfacial Water Structure and Hydrophobic Hydration

A. *Water Structure near the Air–Water and Solid–Water Interfaces*

Hitherto we have confined our attention only to electrolytes, to the upper left-hand corner of Fig. 9. But this represents only a fraction of the materials found in natural waters, albeit a very important fraction; we now consider the other categories of substances, and in so doing we con-

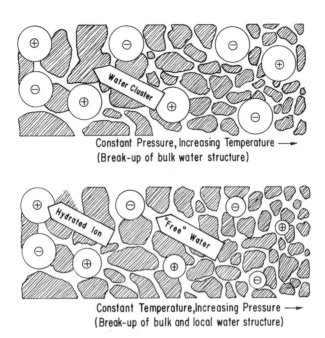

Constant Pressure, Increasing Temperature ⟶
(Break-up of bulk water structure)

Constant Temperature, Increasing Pressure ⟶
(Break-up of bulk and local water structure)

Fig. 16. Comparison of the effects of temperature and pressure on water structure. [Reprinted from Ref. 62 by courtesy of Water Resources Research, Washington, D.C.]

front a new type of water structure, different from that of the Frank–Wen clusters and the hydration atmospheres of ions.

The water molecules at the surface of the liquid must be bound together very tightly. A relatively high temperature is required to provide the energy necessary to enable molecules to escape across the interface (Fig. 1), and the heat of evaporation is large (Table 1). The surface tension of water is even more startling — it is much greater than that of other liquids (Table 1). Much of the more important chemistry of natural water systems takes place at interfaces — the atmosphere/hydrosphere, lithosphere/hydrosphere, and biosphere/hydrosphere interactions — and as our concern with the chemistry of our total environment intensifies, there almost certainly will be a growing appreciation of the profound significance of the extraordinary interfacial properties of water. The average molar surface entropy of water is about 14 J/°C less than for nonpolar liquids. indicating that the molecules near the surface are highly ordered(43).

What is the nature of this order and how deeply into the bulk phase of the liquid does it extend? Fletcher(44) has estimated that a dipolar water molecule at the surface gains some 10^{-12} ergs by orienting itself with its negative (oxygen) vertex extended outward to the low dielectric constant phase. Claussen(*45*) pictures the surface layer as composed of planes of puckered hexagons reminiscent of the ice–I_h structure (Fig. 4). But while we cannot yet say with any certainty what the structure of the interfacial water is, we can say what it is not. For reasons which will become clearer subsequently, it almost certainly is not the same as the structure of ice–I_h or of the Frank–Wen clusters in the bulk phase. With respect to the latter, the temperature dependence of the viscosity (a bulk structure property) of pure water is entirely different from the temperature dependence of the surface tension (a surface structure property) while the differences between these same two properties for a "normal" liquid such as *n*-octane is very much less pronounced (Fig. 17).

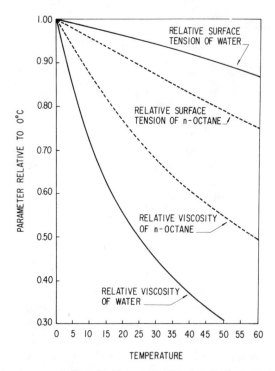

Fig. 17. Difference between bulk and surface water structure illustrated by the temperature dependence of the viscosity and surface tension. [Reprinted from Ref. *50* by courtesy of *Electrochimica Acta*, Oxford, England.]

In Fig. 18 I have tried to represent schematically, with no attempt to be quantitative, the enhancement of water structure as the liquid-air interface is approached from the liquid side.

Another interesting property of the interfacial water structure is its

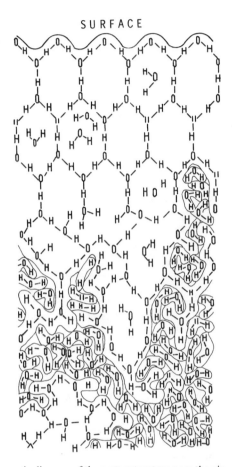

Fig. 18. Schematic diagram of the water structure near the air − water interface.

ability to selectively exclude electrolyte [see Ref. *46* for a review of the effects of solutes on surface tension]. The Onsager–Samaras (*47*) theory of surface tension attributes this exclusion to electrostatic image forces − the ion is repulsed by the low dielectric constant air phase and is pushed deeper into the high dielectric constant water phase. Cation exclusion from the solid–aqueous solution interfaces causes water molecules to assume a preferred orientation falling off rapidly in an exponential manner so that the structured region is no more than 13 molecules (or about 23 Å) thick. This is in agreement with what has hitherto been the majority opinion. However, work on water-permeated porous solids and capillaries in recent years has resulted in a growing body of evidence that the structured region may be very extensive indeed, extending into the bulk solution for hundreds and perhaps even thousands of angstroms [see Ref. *24*, Chap. 12].

The properties of this interfacial water are most remarkable—it does not freeze until about −40°C, its specific volume increases with decreasing temperature below 0°C (*48*), and its viscosity is 10–15 times that of "normal" water (*49*). This new type of water structure also seems to be more stable with respect to both temperature and pressure than the Frank–Wen clusters (*50*).

To summarize, we are proposing that liquid water contains at least two different structural forms in addition to the "free" or monomeric water: the α form, which represents the substance of the Frank–Wen clusters, and the β form, which becomes important near a second phase or even a hydrophobic solute (Table 3).

TABLE 3

Properties of the Two Structural Forms of Liquid Water

α Form	β Form
Present in large amounts of bulk water in the form of Frank–Wen clusters	Present in small amounts in bulk water
Surrounds ions and charge sites	Envelops interfaces (liquid/water, gas/water, and solid/water), hydrophobic materials, and hydrophobic portions of molecules
Structure unknown, may be random; probably is not ice-I_h	Structure unknown, possibly Pauling clathrate type; probably is not ice-I_h
Relatively temperature stable; gradually destroyed in going from 0 to 100°C; gives rise to the density maximum at 4°C	Somewhat greater thermal stability; gives rise to anomalies in water properties near 40°C
Completely destroyed by 1000 kg/cm²	Still evident at 5000 kg/cm²

In a finely porous medium such as a gel or an ion exchange membrane all of the internal solution water can be in the "abnormal" β form, and even in a soil or other coarsely porous medium an appreciable fraction of the interstitial water will be "abnormal." For example, cation exclusion studies on 80- to 120-mesh silica (*51*) indicated that at least 0.5% of the interstitial water is sufficiently structured to exclude alkali metal cations, and this in turn indicates that the layer of interfacial perturbation of water structure surrounding the solid particle is at least 1000 Å thick. Many of the substances in natural waters (Fig. 9) are present as a second phase

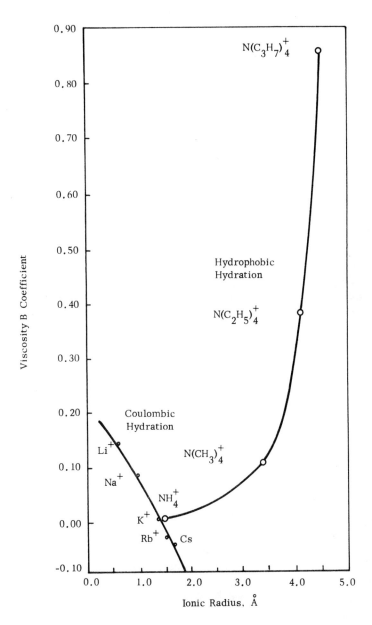

Fig. 19. Comparison of the size dependence of the viscosity *B* coefficient for coulombic and hydrophobic cations in aqueous solution.

and are therefore defined by interfaces. Not only is the chemistry of our environment largely water chemistry, but, perhaps less well realized, a very sizable and important fraction of that chemistry is the chemistry of "abnormal" interfacial water.

Before concluding this brief discussion of interfacial water structure I would like to interject two further remarks. Hitherto we have confined our attention to the interaction of water molecules with each other as a consequence of the presence of a surface. The water molecules may also interact with the surface itself and be strongly chemisorbed. Silica may have as many as three such layers of immobilized chemisorbed water (52). Then too, the structure of the solid phase may influence the water structure. Ice and silicate minerals have very nearly the same crystal lattice paramaters—the O—O distances are 4.52 and 4.51 Å, respectively—a coincidence that has led to the proposal that an ice-like layer of water is present on the surface of such solids (53). Weiss (54) has pointed out more recently that the microdimensions of the surface water layer on silicates correspond nicely to the structural configuration in the gas clathrate hydrates. Secondly, it should be mentioned that if the solid surface is charged, as is frequently the case, and the water contains an ionic electrolyte, the interfacial water structure is very much complicated by the formation of an electrical double layer.

B. Hydrophobic Hydration and the Hydration Sheath of Biopolymers

The formation of the β structural form in liquid water represents the response to the presence of a foreign substance. The intruder can be a second gaseous, solid, or liquid phase, but it need not be a macroregion. It can be any size, a colloid-sized heterogeneity even a single molecule or ion. Thus, while species characterized by an envelope of β structure (lower right of Fig. 9) are, in general, not as soluble and hence are found in smaller concentrations than coulombically hydrated solutes (Fig. 9), they represent enormous variety. The response of the water molecules to an intruder, be it an ion or molecule or large heterogeneity, is to "join hands," surround it, if possible exclude it, and at least try to minimize its volume, thereby minimizing the perturbation and masking its presence. This cooperative effort, not from attraction to one another but out of distaste for the alien, is not a feeble process. The so-called hydrophobic bond is so strong that it can dominate what we ordinarily consider as strong coulombic interactions. For example, rather than the expected maximum charge separation, hydrophobic bonding may be sufficiently strong to segregate the charge sites on a biopolymer all in one volume and the hydrophobic segments all in another. For those readers interested in pursuing the topic of hydrophobic bonding further, I recommend the important series of papers by Nemethy and Scheraga published in 1962 (*10,55,56*).

The most uncluttered view of hydrophobic hydration is provided by the dissolution of chemically inert gases in water. When an ionic electrolyte is added to water, the energetics of the process are dominated by the transference of charge from one dielectric constant medium to another, and for more dilute solutions it is often possible to calculate the thermodynamic changes accompanying this dissolution from the Born equation. In the case of uncharged species, such as inert gases, N_2, O_2, A (but not CO_2), and hydrocarbons, this coulombic effect is absent and the observed thermodynamic properties derive entirely from structural changes in the solvent. The dissolution of an inert gas in a "normal" liquid loosens the solvent intermolecular forces and results in an increase in the partial molal entropy, but, in strong contrast, the dissolution of these same gases in water gives rise to a pronounced *negative* entropy change, that is to say "... when a rare gas atom or a nonpolar molecule dissolves in water ... it modifies the water structure in the direction of greater 'crystallinity' ... the water, so to speak, builds a microscopic iceberg around it" (*57*).

In this connection a particularly interesting series of solutes are the salts of the tetraalkylammonium cations, R_4N^+, for as the size of R increases from H up through C_nH_{2n+1} the dominant character of the species undergoes a gradual transformation from coulombic to hydrophobic, and the nature of its hydration envelope correspondingly changes gradually from α to β structure. If we look at the size dependence of the viscosity B coefficients of these substances (Fig. 9), we see that not only are they very powerful water structure makers, but they represent a family entirely distinct from, say, the alkali metal cations, although the two families do have a common member, NH_4^+. Further information concerning these cations can be found in the review paper of Wicke (*8*). Recent electrical conductivity studies (*58*) on solutions of these salts have shown that the local β structure surrounding these ions tends to be stable with respect to hydrostatic pressure, in contrast to the α structure surrounding "normal" ions, and that for the larger members of the series hydrophobic bonding is sufficiently strong to overcome coulombic repulsion so that cation–cation interactions become significant.

As a consequence of interactions with the terrestrial and marine biospheres, natural waters, both fresh and seawater, contain highly varied and variable quantities of biomaterial (Fig. 9) ranging all the way from intact living organisms down through biopolymers to the smaller molecules that represent the end products of excretion and the decay processes.

These biosubstances, like everything else, are enveloped in a hydration crust that can determine not only the spatial configuration but the chemistry of biopolymers. Countless examples could be cited here. Among the most familiar are the coagulation of proteins when the hydration sheath is disrupted by temperature or pressure and denaturation by water structure-altering solutes (the Hoffmeister or lyotropic series). The importance of water structure to the life processes can hardly be overemphasized. The preservation of large protobiological molecules by their protective hydration sheath was one of the crucial factors that enabled life to first appear in the hydrosphere of this planet.

Biomolecules have as their framework a hydrocarbon skeleton, so we might expect their hydration crust to be largely composed of the hydrophobic β structure. But in addition, they have many charge sites and partial charge separations, and in the immediate neighborhood of these centers of coulombic field the hydration crust is expected to assume an α structure character (Fig. 20).

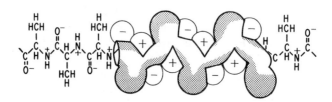

Fig. 20. Hydration of a polyfunctional macromolecule with α structure surrounding charge sites and β structure enveloping hydrophobic segments.

VI. Water Structure and Chemistry

In summary, at the risk of belaboring the point, I repeat the truism that the chemistry of our total environment is almost entirely water chemistry, that is to say, the chemistry of aqueous solutions. There are three further points that bear repeating in order to assure that they have been given the emphasis merited by their importance.

When dealing with the chemistry of natural waters we are *never* concerned with the species A. The actual reactant is always $A(H_2O)_n$. This has been half-realized in the case of the proton, which it is now fasionable to represent in the textbooks as H_3O^+ ("hydronium ion") rather than H^+ (although $H_9O_4^+$ could be more realistic in the light of modern findings).

And we are never, not even in the case of "simple" metathesis, concerned with a reaction of the type

$$A + B \rightleftharpoons C \tag{2}$$

but rather with a series of water reactions such as

$$A(H_2O)_n \rightleftharpoons A(H_2O)_{n-1} + H_2O \tag{3}$$

$$B(H_2O)_m \rightleftharpoons B(H_2O)_{m-1} + H_2O \tag{4}$$

$$A(H_2O)_{n-1} + B(H_2O)_{m-1} \rightleftharpoons C(H_2O)_{m+n-2} \tag{5}$$

$$C(H_2O)_{m+n-2} \pm (l-m-n+2)H_2O \rightleftharpoons C(H_2O)_l \tag{6}$$

Although very rapid (see Fig. 21), the transference of a water molecule in or out of the hydration atmosphere of a species is not instantaneous, and there are undoubtedly a great many hydrologically important reactions for which it is the rate-determining step. For all its importance (the future existence of the human race may be contingent on the rate of dissolution of CO_2 in seawater), the kinetics of hydrological reactions has only just begun to receive long-deserved attention.

The "in" crowd today is always seeking where the action is, and this brings me to our third point to remember. The "action" in environmental

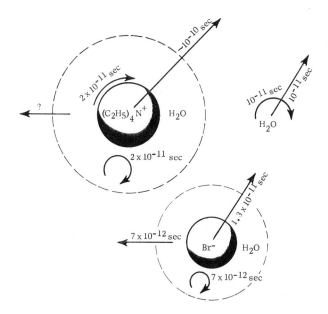

Fig. 21. Half-lives of the rotational and translational movements of water molecules in the hydration atmospheres of ions. [Reprinted from Ref. 8 by courtesy of Verlag Chemie GmbH.]

chemistry is at interfaces, the boundaries where the lithosphere, atmosphere, and biosphere interact with the hydrosphere. I predict confidently that these are the areas into which the imaginative and creative environmental scientists will move. This scientific vanguard will have to come to grips with the most complex aspect of water chemistry — the nature of the interfacial water structure.

REFERENCES

1. G. Dietrich, *General Oceanography*, Wiley, New York, 1963.
2. B. T. Darling and D. M. Dennison, *Phys. Rev.*, **57**, 128 (1940).
3. L. Pauling, *J. Am. Chem. Soc.*, **57**, 2680 (1935).
4. E. D. Wollan, W. L. Davidson, and C. G. Shull, *Phys. Rev.*, **75**, 1348 (1949).
5. H. E. Armstrong, *J. Chem. Soc. (London)*, **53**, 131 (1888).
6. H. M. Vernon, *Phil. Mag.*, **31**, 387 (1891).
7. H. M. Chadwell, *Chem. Rev.*, **4**, 375 (1927).
8. E. Wicke, *Angew. Chem.*, **5**, 106 (1966).
9. J. D. Bernal and R. H. Fowler, *J. Chem. Phys.*, **1**, 515 (1933).
10. G. Nemethy and H. A. Scheraga, *J. Chem. Phys.*, **36**, 3382 (1962).
11. H. S. Frank and W. Y. Wen, *Discussions Faraday Soc.*, **24**, 133 (1957).
12. K. L. Mysels, *J. Am. Chem. Soc.*, **86**, 3503 (1964).
13. T. F. Wall and D. F. Hornig, *J. Chem. Phys.*, **43**, 2079 (1965).
14. M. Falk and T. A. Ford, *Can. J. Chem.*, **44**. 1699 (1966).
15. G. W. Stewart, *Phys. Rev.*, **37**, 9 (1931).
16. O. Nomoto, *J. Phys. Soc. Japan*, **11**, 827 (1956).
17. W. Luck, *Ber. Bunsenges Ges. Physik. Chem.*, **69**, 626 (1965).
18. R. A. Horne, *Surv. Progr. Chem.*, **4**, 1968.
19. R. P. Marchi and H. Eyring, *J. Phys. Chem.*, **68**, 221 (1964).
20. O. Y. Samoilov, *Structure of Aqueous Electrolyte Solutions and the Hydration of Ions*, Consultants Bureau, New York, 1965.
21. L. Pauling, *Science*, **134**, 15 (1961).
22. H. S. Frank and A. S. Quist, *J. Chem. Phys.*, **34**, 604 (1961).
23. J. L. Kavanau, *Water and Solute-Water Interactions*, Holden-Day, San Francisco, 1964.
24. R. A. Horne, *Marine Chemistry*, Wiley (Interscience), New York, 1969.
25. W. Drost-Hansen, in *Equilibrium Concepts in Natural Water Systems*, *Advan. Chem. Ser. No. 67* (W. Stumm, ed.), American Chemical Society, Washington, D.C., 1967.
26. M. Falk and G. S. Kell, *Science*, **154**, 1013 (1966).
27. R. A. Horne and D. S. Johnson, *J. Phys. Chem.*, **70**, 2182 (1966).
28. R. W. Gurney, *Ionic Processes in Solution*, McGraw-Hill, New York, 1955.
29. G. Jones and M. Dole, *J. Am. Chem. Soc.*, **51**, 2950 (1929).
30. W. Good, *Electrochim. Acta*, **9**, 203 (1964).
31. R. A. Horne and J. D. Birkett, *Electrochim. Acta*, **12**, 1153 (1967).
32. A. J. Rutgers and Y. Hendrikx, *Trans. Faraday Soc.*, **58**, 2184 (1962).
33. R. A. Horne, R. A. Courant, and D. S. Johnson, *Electrochim. Acta*, **11**, 987 (1966).
34. C. M. Collie, J. B. Hasted, and D. M. Ritson, *Proc. Phys. Soc. (London)*, **60**, 145 (1948).
35. J. B. Hasted, R. M. Ritson, and C. N. Collie, *J. Chem. Phys.*, **16**, 1 (1948).

36. S. Glasstone, K. L. Kaidler, and H. Eyring, *The Theory of Rate Processes*, McGraw-Hill, New York, 1941.

37. J. H. Wang, C. U. Robinson, and I. S. Edelman, *J. Am. Chem. Soc.*, **75**, 466 (1953).

38. R. A. Horne and R. A. Courant, *J. Geophys. Res.*, **69**, 1971 (1964).

39. R. A. Horne, *Nature*, **200**, 418 (1963).

40. W. A. Zisman, *Phys. Rev.*, **39**, 151 (1932).

41. R. L. Kay and D. F. Evans, *J. Phys. Chem.*, **70**, 2325 (1966).

42. R. R. Sood and R. A. Stager, *Science*, **154**, 388 (1966).

43. R. J. Good, *J. Phys. Chem.*, **61**, 810 (1957).

44. N. H. Fletcher, *Phil. Mag.*, **7**, 255 (1962).

45. W. F. Claussen, *Science*, **156**, 1226 (1967).

46. W. Drost-Hansen, *Ind. Eng. Chem.*, **57** (4), 18 (1965).

47. L. Onsager and N. Samaras, *J. Chem. Phys.*, **2**, 528 (1934).

48. J. S. Schufle and M. Venuglopalan, *J. Geophys. Res.*, **72**, 327 (1967).

49. B. V. Derjaguin, N. N. Fedjakin, and M. V. Talayev, *Dokl. Akad. Nauk SSSR*, **167**, 376 (1966).

50. R. A. Horne, A. F. Day, R. P. Young, and N. T. Yu, *Electrochim. Acta*, 1968.

51. R. A. Horne and R. P. Young, *Tech. Rept. No. 32*, Arthur D. Little, Inc., December 15, 1967, Office of Naval Research Contract Nonr–4424(00); see also R. A. Horne, *J. Phys. Chem.*, **70**, 1335 (1966).

52. A. A. Antoniou, *J. Phys. Chem.*, **68**, 2756 (1964).

53. H. H. Macey, *Trans. Brit. Ceramic Soc.*, **41**, 73 (1942).

54. A. Weiss, *Kolloid Z. Z. Polymer*, **211**, 94 (1966).

55. G. Nemethy and H. A. Scheraga, *J. Chem. Phys.*, **36**, 3401 (1962).

56. G. Nemethy and H. A. Scheraga, *J. Chem. Phys.*, **36**, 1773 (1962).

57. H. S. Frank and M. W. Evans, *J. Chem. Phys.*, **13**, 507 (1945).

58. R. A. Horne and R. P. Young, *J. Phys. Chem.*, 1968.

59. C. M. Davis, Jr., and T. A. Litovitz, *J. Chem. Phys.*, **42**, 2563 (1965).

60. R. A. Horne and R. A. Courant, *J. Phys. Chem.*, **68**, 1258 (1964).

61. R. A. Horne, *Tech. Rept. No. 3*, Arthur D. Little, Inc., September 30, 1964, Office of Naval Research Contract Nonr–4424(00).

62. R. A. Horne, *Water Resources Res.*, **1**, 263 (1965).

Chapter **18** **The Use of Standards in the Quality Control of Treated Effluents and Natural Waters**

Alan H. Molof
SCHOOL OF ENGINEERING AND SCIENCE
NEW YORK UNIVERSITY
BRONX, NEW YORK

I. Introduction

Water is one of our basic natural resources. Since the total amount of water on earth is essentially constant, any waste of this resource will reduce the small readily available supply. Thus, the concept of control was introduced and adopted.

One degree of water quality degradation can be called "pollution" — impairment which does not create an actual hazard to public health but which does adversely and unreasonably affect the usage of the water. "Contamination" is impairment to a degree which creates an actual hazard to public health, i.e., poisoning or disease transmission. In the following discussion, the waters are considered polluted but not contaminated.

II. Historical Development

In the past there was a period when water quality was only affected by natural systems such as volcanoes, floods, and forest fires due to lightning. Then, as time progressed, man began to degrade the quality of the water. Prior to the development of cities, small groups of people did not significantly affect the water. Degradation became more severe when increased population and urban settlements resulted in the problems of domestic waste water and storm water. Man's agricultural activities also began to affect natural systems; plowing and disturbing the earth created erosion conditions, and fertilizers and insecticides were swept into the groundwaters and rivers. Industrial wastes have greatly contributed to pollution of rivers since the Industrial Revolution. Suspended material, chemicals, both organic and inorganic, thermal wastes, and radioactive materials are now being dumped into rivers.

Thus nature, man, agriculture, and industry all contribute to the degradation of our water quality. The importance of these pollution sources is apparent when we consider the needs of water users. Each use dictates the water quality need. The highest priority use is man's need for drinking water. Other uses related directly to man are fishing and water recreation. Commercial uses include irrigation and industrial process and cooling water. In addition, nature has its own requirements for maintaining an ecological balance.

Thus we have a conflict between the users of water and the polluters of water. However, all user groups are also polluters. For example, people consume food for survival and excrete wastes which pollute water. This consumption–excretion cycle has created situations in certain river basins where one town discharges its waste to the river downstream, while another town uses the river as its source of drinking water. Thus, in a broad sense, the waste water of one becomes the drinking water of another. Waters used for fishing and swimming are also affected by domestic waste water.

Industry needs quality process waters. Cooling waters must not be excessively polluted. However, their discharges to rivers can be a very serious pollution problem. Agriculture requires large volumes of water for irrigation, often this creates problems of salinity buildup. In coastal areas a severe problem is caused by seawater intrusion into aquifers previously occupied by fresh groundwater which has been removed for domestic and industrial use.

Projected water requirements for the year 2000 have been made for the various users(1). The requirements are divided into gross withdrawal, consumptive use, and amount returned (Table 1). The difference between

TABLE 1

Water Use Requirements for the Year 2000[a]

(Billions of gallons per day)

Use	Gross withdrawal	Consumptive use	Amount returned
Irrigation	184.5	126.3	58.2
Municipal	42.2	5.5	36.7
Manufacturing	229.2	20.8	208.4
Mining	3.4	0.7	2.7
Power cooling	429.4	2.9	426.5
Total	888.7	156.2	732.5

[a]From Ref. 1.

gross withdrawal and consumptive use represents the amount returned to the river with a deterioration in water quality.

III. Organized Concern on Pollution Policy

To assure the rights of all the users of water, the government must take a role in reconciling the conflict between disposal and use. The rules for use have consisted of the Law of Riparian Rights and the Appropriation Doctrine. Initially the Law of Riparian Rights stated that the owner of land bordering on a natural stream is entitled to receive the full natural flow of the stream without change in the quality or quantity. Protection is provided against diversion of water except for domestic use upstream and flood control. This concept has been modified to provide for actual use of water for beneficial purposes such as irrigation, industry, and sewage effluent dilution.

The Appropriation Doctrine uses a "first come, first served" principle without regard to land bordering the stream. That is, the first to appropriate the water for beneficial use has priority of use of the water and his right' extends to any other water use. His needs, up to the limit of his appropriation, must be satisfied prior to downstream use. However, a person may not be able to divert water if it is in short supply, if there is not a favorable appropriation in point of time. The appropriate right is to a specific amount of water, but more water cannot be taken than can be put to beneficial use.

Due to increasing demands for water and to increased pollution by users, there is now a need to establish policy concerning quality objectives. This policy should be a reflection of the needs of various segments of the population, including the conservation groups, such as the Isaac Walton League; recreation groups, such as sport fishermen; potable

water groups, such as the American Water Works Association; and industrial groups with varying opinions on what policy should guide our water
resources. At one extreme, some people want our rivers restored to their
condition before the advent of man, and at the other extreme some feel
that a river should be used to dilute effluents as long as no nuisance
such as odor exists. Others feel aesthetic considerations should be a
factor. But in the end, policy should reflect everybody's interest in preservation not only for the present but, more importantly, for the future.

The development of the government role in protecting water resources
must come from all three branches: legislative, executive, and judicial.
The idea of cooperation of all agencies at the Federal level must also
apply to smaller governmental units (the Ohio River Valley Water Sanitation Commission is one example).

The policy will of necessity be guided by a yardstick that measures
human values versus dollars. Therefore, there is a great need for more
accurate representation of the dollar value of human benefits. Only in
this way can dollars be used realistically as one of the guidelines in setting
water resource policies.

IV. Implementation of Pollution Policy by the Use of Standards and Criteria

Once a policy has been set, methods of implementation will follow. One
such method is the setting of criteria and standards. Before further
discussion, it is necessary to understand the critical differences between
the meaning of the words "criterion" and "standard." The following is
based on a comprehensive report on water quality criteria by McKee
and Wolf(2):

A "standard" is a definite rule, principle, or measure established by
authority. Since it is established by authority, it is official or quasi-legal.
However, because something is termed a standard does not mean that it
is rationally based on the best scientific knowledge and engineering
practice. Use of a standard tends to eliminate improvement and sustain
inflexibility.

The term "criterion" designates a scientific requirement on which
a judgement or decision to support a particular use may be based. A
criterion should be capable of quantitative evaluation by existing analytical tests and also be capable of definite resolution. In contrast to
standards criteria have no connotation of authority. When data are
gathered to be used as a yardstick of water quality, "criterion" is the
proper term.

Other terms referred to in the report(2) include "requirement" and "objective." "Requirement" is used to describe an administrative decision by a regulatory body to fulfill a given mission. It does not necessarily have scientific justification. Requirements for discharge of a treated effluent are used by New York State. An "objective" is an aim or goal toward which to strive. It is not as rigid and authoritative as a standard and does not have the enforcement element of requirements.

No matter which category is considered, all must be subject to review and change.

Criteria or standards can be applied in two areas — the effluent, and the water receiving or being affected by the effluent. A stream standard refers to rivers, lakes, estuaries, oceans, or groundwater, while an effluent standard encompasses effluents from municipal, industrial, and agricultural pursuits (see Chap. 19). In New York State both types of standards are used. For plants discharging to the Hudson River, secondary treatment will be required. In addition, classifications and standards have been set for the Hudson River on the basis of best use, i.e., drinking, fishing, recreation, agriculture, and process use.

The parameters used in establishing these stream standards can be illustrated by those utilized by the State of New York(3). These are listed in the Appendix. It is evident that specific chemical standards are mainly concerned with drinking water requirements, while a few apply to fishing waters. With the growth of knowledge about the ecological balance, the need for new and revised specific chemical standards will be evident for all water use categories. An obvious need is for a mercury standard.

Stream standards reflect the assimilative capacity of the water body receiving effluent; their purpose is to preserve the stream at a certain minimum quality. The degree of treatment of waste water will vary depending on the desired condition of the water body, thereby utilizing the assimilative capacity of the stream most economically. However, it should be noted that this type of standard is generally difficult to administer.

Effluent standards, which apply to the material being discharged to the receiving water, do not consider the most economic use of the assimilative capacity of streams. However, they are simple and easy to administer. In general, these standards either restrict the quantity of pollutants in the effluent or set the desired degree of treatment. Use of a concentration standard (e.g., milligrams per liter) for a substance can be misleading since dilution can mask the true polluting level. A more appropriate method is to limit the total amount of pollutant (e.g., pounds) which can be discharged in a given time, thus including both concentration and quantity of flow.

According to stream standards, any effluent can be discharged which does not cause the receiving water to fall below its best usage state. As discharges to the river increase, a point is reached at which all the river's capacity is being used. Then, if more treatment plants are to be established which will discharge effluents to the river, some type of agreement must be made as to effluent levels for all those discharging into the river. The greater the number of effluents, the tighter the effluent standard must be. There is also a tendency toward setting stricter effluent standards because the receiving waters are not in a satisfactory state.

The basis for setting these standards involves the best scientific techniques and knowledge. But in such a complex field as environmental pollution, this scientific information and/or ability is often lacking. It is of course dangerous to set a standard that cannot be reached or to set a standard based on fallacious information. Thus, standards depend upon the state of scientific development, no matter how deficient the scientific data. Professional responsibility requires that the available information be considered in reaching a decision.

Complicating our use of standards is the interaction of the parts that comprise the whole in a river. The interaction of chemicals, microorganisms, plants, fish, etc., with their environment is called ecology. This is a highly complicated scientific field; we are finding more and more that stress at any one point in the environment is transmitted throughout the entire natural biological cycle that exists in a river.

We do know that progress doesn't stop because the answers are not known. Thus, stream and effluent standards are set based on our best thinking at present and the latest scientific knowledge and engineering. On this basis, as technology improves and better methods are developed for economically delivering high quality effluents, a specific river standard can be revised upward. There is probably more chance of upgrading treatment in the near future than of learning the story of river ecology. However, both areas must be investigated, since we are definitely committed as a nation to protecting our environment.

V. Differing Viewpoints on Standards

Economists have a different outlook on how standards should be used than do the engineers(4). For example, an economist has stated that standards are too rigid and are only valuable as minimums. In addition, they require conditions or actions that restrict better and less costly solutions. To replace such legally established rules, the idea of effluent charges has been suggested. These charges would be in proportion to damage to the river and would result in industry limiting pollution to maximize profit. However, there is not much agreement with this idea by others concerned. For example, the conservationist feels that such a method would permit the rich polluters to pay the price and continue to pollute. Industry feels that it is too costly and requires too much information. Government administrators consider this method too uncertain in its results, as it leaves too much to chance.

Other thoughts on the use of economic aids to standard enforcement include interim effluent price systems as penalty until standards are met. The funds could be used for monitoring and surveillance of standards. This type of pricing standard and rigorous enforcement constitute some type of incentive in themselves. There is also the concept of pricing the value of the river and selling for waste disposal that capacity of the river above the minimum standards.

Another view states (5) that standards

1. should not confuse people,
2. should not be unworkable statistical methods,
3. should not be changed often,
4. should not create a credibility gap, and
5. should be set as minimum but controlled as maximum.

Standards allow industries or municipalities to dump a certain amount of pollutants into water bodies. However, standards should not be set at minimum levels lest water quality be destroyed for future generations. Standards should also prevent a polluter from profiting at the expense of neighboring water users. Another view supports flexibility in standards, compensating, for example, for differences in summer and winter and dry versus wet years.

Water quality standards could be accepted as goals and be related to effluent standards. By the use of stream monitoring it would be possible to tell if the public is getting the water quality represented by the standards. But of what quality or how clean should water be? One view says it is up to the people and their willingness to pay for clean water, above and beyond the cost that engineers assess as minimum to attain a less dirty water. Perhaps the cost differential is not appreciable. It is important for the engineers to remember that man is the final authority on what he wants. Water quality enhancement should not be considered a present financial liability, but rather protection of a future asset.

VI. Government Action

A national commitment toward clean water is expressed in the Federal Water Act of 1965 (P.L. 89–234), which provided for the establishment of water quality standards by the states with approval of these standards by the Secretary of the Interior. The legislation stated that the water quality standards shall "... enhance the quality of water."

The Department of the Interior's nondegradation clause has caused perhaps more problems than any other single item. This clause infers that no water should be polluted more than it is at present. An industrialist pointed out that this concept would economically harm underdeveloped states with relatively clean water and favors heavily industrialized states with relatively polluted water. Does this indicate that a National Water Plan should be adopted?

Before the formation of this act, the states had the responsibility for river water quality. However, for rivers that involve many states the political aspects become significant. Certain solutions to this problem have been accomplished, as evidenced by the eight state groups plus the Federal Government comprising the Ohio River Valley Water Sanitation Commission (Orsanco)(6), the four states and the District of Columbia comprising the Interstate Commission on the Potomac River Basin, established in 1940(7), and the four states plus the Federal Government comprising the Delaware River Basin Commission (DRBC), established in 1961(8).

The DRBC has been most active in controlling the water quality of the Delaware River. In fact, it has adopted the most advanced concepts for achieving clean rivers. In 1967 the DRBC enacted river standards, and in 1968 effluent standards were adopted based on division of the estuary into four parts and their respective assimilative capacities. These zones were later further subdivided into industrial and municipal treatment plant allotments saving 10% reserve for future growth.

In 1969, tentative maximum allowable organic discharge quantities were set for 91 treatment plants. A total of 74% of these plants accepted the assigned shares; at the end of 1969, 16 appeals were pending. The nontidal portion of the river is not included and has a separate effluent requirement of 85% removal of biochemical oxygen demand (BOD).

Orsanco was formed in 1948 to prescribe how pollution should be controlled in the Ohio River Basin. The commission can establish standards for river water quality and for treatment. In addition, it is empowered to secure compliance with the standards it sets. The Federal Water Quality Act of 1965, involving individual states, has conflicted with this type of interstate concept. For example, a difference in state schedules for implementing the new standards has caused problems. Also, the establishment of different standards applicable to the same stretch of river bordering two states calls for the development of regional compatibility.

Recent changes have been made by existing interstate groups. One such group is the New England Interstate Water Pollution Control Commission, which was established in 1947 and includes Connecticut, Maine, Massachusetts, New Hampshire, New York, Rhode Island, and Vermont(9). A revaluation in 1968 led to the signing of a charter and to changes which included the establishment of a regional water quality surveillance network and provision for enforcement authority for water pollution abatement in interstate waters.

As of 1970, federally approved standards have not yet been established for every state due to disagreements between the two levels of government, state and federal, on the interpretation of the language of the statute. In fact, there are states forming groups to deal collectively with the Department of the Interior. One of the major disagreements concerns the necessity for secondary rather than primary treatment regardless of the receiving water conditions. In such a situation effluent standards rather than receiving water standards are used. However, most states do have approved standards for their waters.

There has been one instance in which the Secretary of the Interior, Walter J. Hickel, has established Federal standards for a state under the Water Quality Act of 1965. In November 1969, the Secretary published proposed regulations in the Federal Register for the State of Iowa. He called for secondary treatment of all waste water from Iowa being discharged to the Mississippi and Missouri Rivers as of December 31, 1973. The action was taken since that state did not comply in setting standards within the six-month period allotted, and by law the Federal standards then took effect.

VII. Response to Setting of Standards

Once the government standards have been set to protect the water resource for uses such as potable water, swimming, fishing, process cooling, or irrigation, the response by the various polluters becomes critically important.

Municipalities must build sewers to collect and transmit the waste water to a treatment plant before discharge to the river. Attention must be given to the differences between domestic waste water and storm waters, since one sewer often carries both. When the runoff due to rain is excessive, the flow is allowed to partially bypass the treatment plant, which is unable to handle these excessive flows. Much effort has been spent on solving this combined sewer problem. One suggestion involves storing excess flows until the large runoff stops followed by pumping this stored runoff to the plant. Another suggestion involves separating domestic and storm flow by using pressurized sewer lines inside the larger storm sewers. At the end of the sewer line is the treatment plant, which can remove 0–100% of the organic pollutants in waste water. These treatment plants may also have to handle industrial waste waters.

It should be cautioned that design and construction of a wastewater treatment plant is only the first part of an extensive pollution abatement program. The plant must be operated to meet the design objectives; therefore sufficient funds must be budgeted for proper operation, maintenance, and personnel or the whole purpose of the design will be defeated and a large amount of money will be wasted.

To ensure that the plant will be able to perform in accordance with the approved design objectives, it should be the responsibility of the design engineer to supervise the start-up until the plant is operating as designed. A percentage of the design engineer's fee should be withheld until the plant is transferred to the plant operations group.

In measuring plant performance, monitoring of the treated effluent will be the key test. Therefore, an effluent standard should be set for every newly constructed plant to ensure that it is functioning as represented to the taxpayers and government who supply the funds. Some penalty must be provided for nonperformance. Once the plant is handed over by the design engineer, it then becomes the responsibility of the operating group. Effluent monitoring must continue to ensure that the treatment plant is fulfilling its mission.

For the municipality to construct the facilities to meet the stream standard, some form of incentive must be present since many dollars are involved. This incentive must come from enlightened officials with the backing of the citizen groups interested in protecting water resources. For example, the people in New York State were allowed by the state government to vote on a billion dollar bond issue for upgrading waste water treatment in the state. In addition to local pressure, there is the interest of the Federal Government in acting cohesively in its legislative, executive, and judicial branches. The Federal Government must set guidelines for action and provide financial help. Such aid is of vast importance.

The other major source of pollution, industrial waste water, is really a different aspect of the problem. Industry is usually not in the business of treating waste water and therefore will have little reason to devote money to this area without tremendous pressure from all levels of government plus the general population. This type of pressure can be successful, as evidenced by past improvement through process research in which water is more heavily reused and certain polluting species have been eliminated. However, only constant pressure by all will force those in industry and agriculture to clean their waters. Perhaps the recent announcement that Eli Lilly and Co. has pledged to provide "maximum — not just acceptable — control of wastes" is a first bright light in the coming dawn.

It must be stressed that the mere setting of standards without a corresponding compliance by municipalities and industries will not lead to any improvement in our water resources. Water quality improvement is the responsibility of all, individually and jointly. Action by all levels of government and the population is a prerequisite to water quality improvement. The DRBC has led the way in making stream standards effective by creating effluent standards to meet its goals. This dual approach is the logical way, since streams standards will not ensure cleaner water unless effluent standards are also simultaneously set.

VIII. Methods to Meet Stricter Effluent and Natural Water Standards

Even if we can cleanse our rivers and natural water bodies of today's pollution and provide sufficient water for all our uses, what about the future? Population will increase, chemical waste waters will increase in complexity as well as volume, technological advances will add to the problem, and the temperature of our rivers will be still higher. The future will require new standards, e.g., standards for viruses, insecticides, toxicity levels, and specific organic chemicals. Therefore, in all aspects of water usage, a new care must be exercised. A greater volume of waste water effluents will become drinking water influents. Effluent quality must be improved.

Reuse of water today is becoming a necessity. However, new technology must be developed to enable economical and realistic reuse of waste water for potable water production. At present, waste water is reused for purposes other than potable water — for irrigation, process water, and recreation lakes. An example is the South Lake Tahoe Water Reclamation, which is using effluent to create a 165-acre recreational lake, called the Indian Creek Reservoir, with a storage capacity of over a billion gallons. The waters are being used for irrigation and are also stocked with rainbow trout and have been approved for all water contact sports by the California State Department of Health. Since the water reclamation plant discharge is reused, effluent standards have been set by both the state and county. These are shown in Table 2.

The plant, at a capacity of 2.5 mgd (million gallons/day) and without

TABLE 2
South Lake Tahoe Utilities District Effluent Characteristics[a]

| | Effluent standards | | | |
| | State of California Maximum, % of time[c] | | | |
Effluent quality parameters	50	80	100	Alpine County, Calif.
BOD, mg/liter	3	5	10	5
COD, mg/liter	20	25	50	30
Suspended solids, mg/liter	1	2	4	2
Turbidity, JU[b]	3	5	10	5
pH	6.5–9.0			6.5–8.5
MBAS[b], mg/liter	0.3			0.5
Coliform, MPN[b]/100 ml	Median less than two. Maximum number of consecutive samples greater than 50, two			Adequately disinfected

[a]From Ref. 10.
[b]JU = Jackson candle or turbidity units; MBAS = methylene blue active substances; MPN = most probable number.
[c]E.g., 50% of the time, BOD is a maximum of 3 mg/liter.

ammonia stripping, was completed in July 1965. The 7.5-mgd plant expansion completed in the fall of 1968 included ammonia stripping, solids handling facilities with sludge incineration and lime recalcining, and effluent disposal facilities to the new recreation lake about 26 miles away from the treatment plant site.

The Lake Tahoe tertiary waste water treatment plant, with a capacity of 7.5 mgd, consists of primary sedimentation, secondary activated sludge, and tertiary chemical coagulation with lime for phosphate removal, stripping ammonia at the high pH, neutralizing with carbon dioxide, and filtering followed by activated carbon adsorption. Disinfection is the final step.

Advances in municipal waste water treatment are being made in the chemical-physical treatment area as well as in biological treatment methods other than activated sludge, such as fixed film disk processes. The removal of settleable, supracolloidal, and colloidal solids in waste water can be achieved by present technology using chemical coagulation. Removal of the soluble organic fraction, however, has not been economically feasible.

Adsorption on activated carbon is the major treatment method for soluble organic removal, but it has not been consistent and efficient in removing all organic compounds in waste water. In addition, activated sludge, the most used secondary treatment process, has been described as efficient in reducing settleable, supracolloidal, and colloidal solids, but inefficient in reducing soluble organic matter(*11*). Further work by Ciaccio(*12*) on the properties of the organic matter left in activated sludge effluents has resulted in the isolation of several fractions after freeze concentration of the effluent. A nondialyzable fraction composed of proteinaceous matter was subjected to gel permeation chromatography and indicates the presence of molecular weights in the following ranges: >156,000, 100,000, 60–70,000 and ≤3,000. The insolubility of about 30% of this nondialyzable fraction intimates the presence of polymers with molecular weights substantially greater than 156,000.

Recently, nonbiological treatments for removing soluble organic substances have been described. Although at a pilot plant stage, the results and economics are most promising. A physical-chemical pilot plant operated by the FMC Corporation as part of a Federal Water Quality Administration (FWQA) contract was described recently(*12*). The process used is chemical coagulation, with ferric chloride, of a primary effluent, followed by filtration and activated carbon adsorption. Another pilot plant described as chemical-physical (Z-M Process) was operated by EcoloTech Research, Inc., as part of a New York State Department of Health contract(*14*). Laboratory studies used as a basis for this pilot plant have recently been published(*15*). The Process (now a part of Envirotech, Inc.) consisted of chemical coagulation (with lime) of raw waste water, raising of pH by lime, detention time to produce hydrolysis, and adsorption by activated carbon. The FWQA has also reported on pilot plant work on nonbiological treatment processes(*16*).

The impact of this type of newly developed process on standards, criteria, requirements, and objectives can be substantial. If it is now possible to economically remove substantially all organic substances from waste water, we can now set an objective of reuse and provide the means for achieving this goal of high quality water. This objective can then be implemented with standards set on treated water. This, in turn, will relate to standards set on the receiving water.

An excellent discussion on standards for high quality water was recently made by Coulter(*5*), who stated:

> Government's assertion of the right to establish standards of quality higher than those required to protect health, safety, and commerce has a rather revolutionary effect. It amounts to an assertion that quality of water is a property right. Furthermore, it implies that quality is owned by the public and that government is the legitimate custodian of this rather unique property right ... It is inherent in the concept of quality as a public property, that discharges may be made as a matter of public necessity and convenience. Everyone must discharge some waste material, but to discharge any more polluting substance than that which is necessary, represents a trespass ... Standards should be set as minimum permissible quality levels, but, at the same time, every source of pollution should be controlled to the maximum feasible extent.

IX. Enforcement of Standards

The drive toward clean water must build on the best of these various views on effluent versus receiving water standards and the many ramifications of each that are possible. That standards are required is an acknowledged fact. The enforcement of these standards, which is essential for clean water, is the big question of the future.

The implementation of any standard will depend primarily on the capacity to detect and monitor any and every specific pollutant that violates the standard (see Chap. 26). An increase in analytical ability is becoming more pronounced with the introduction of new developments in sensors and instruments such as specific ion electrodes and low level organic detection by COD, TOC, and TCOD analyzers (see Chap. 27). With better data-handling facilities, these monitoring tools will be the backbone of the enforcement of the standards.

Thus, by defining our objective, we can set requirements for river and effluent quality levels and establish standards on both river and effluent quality. Meeting both these standards, with the aid of monitoring instrumentation and enforcement actions, will be our best approach to the goal of clean water resources for any and all purposes.

REFERENCES

1. *Waste Management and Control, Publ. 1400*, National Academy of Sciences-National Research Council, Washington, D.C., 1966.
2. J. E. McKee and H. W. Wolf, *Water Quality Criteria*, 2nd ed., *Publ. No. 3-A*, State Water Quality Control Board, Sacramento, Calif. 1963.
3. *Classifications and Standards Governing the Quality and Purity of Waters of New York State*, Parts 700–703, Title 6, New York State Department of Environmental Conservation, Albany, 1970.
4. James J. Flannery, *Am. Soc. Civil Engrs. Conf. Proc., Is Water Quality Enhancement Feasible*, p. 57, American Society Civil Engineers, New York, 1970.
5. James B. Coulter, *Am. Soc. Civil. Engr. Conf. Proc. Is Water Quality Enhancement Feasible*, p. 97, American Society Civil Engineers, New York 1970.
6. *Ohio River Valley Water Sanitation Commission Annual Report*, Cincinnati, Ohio, 1969.
7. *Interstate Commission on the Potomac River Basin Annual Report*, Washington, D.C., 1968.
8. *Delaware River Basin Commission Annual Report*, Trenton, N.J., 1969.
9. *New England Interstate Water Pollution Control Commission Annual Report*, Boston, Mass., 1969.
10. H. E. Moyer, *Public Works*, **12**, 82 (1968).
11. D. A. Rickert and J. V. Hunter, *J. Water Pollution Control Federation*, **39**, 1475 (1967).
12. L. L. Ciaccio, personal communication, 1970.
13. W. J. Weber, C. B. Hopkins, and R. Bloom, *J. Water Pollution Control Federation*, **42**, 83 (1970).

14. M. M. Zuckerman and A. H. Molof, *5th Intern. Conf. Water Pollution Res. Proc.*, San Francisco, July, 1970.

15. M. M. Zuckerman and A. H. Molof, *J. Water Pollution Control Federation*, **42**, 437 (1970).

16. J. J. Convery, *New York Water Pollution Control Assoc. Annual Meeting*, New York, January 1970.

Appendix

Classifications and Standards Governing The Quality and Purity of Waters of New York State(*3*).

Part 700
Tests or Analytical Determinations

(Statutory authority: Public Health Law, art. 6.)

Sec.
700.1 Collection of samples

Sec.
700.2 Tests or analytical determinations

Section 700.1 Collection of samples. In making any tests or analytical determinations of classified water to determine compliance or non-compliance of sewage, industrial wastes or other wastes discharges with established standards, samples shall be collected in such manner and at such locations as are approved by the Water Pollution Control Board as being representative of the receiving waters after opportunity for reasonable dilution and mixture with the wastes discharged thereto.

700.2 Tests or analytical determinations. Tests or analytical determinations to determine compliance or non-compliance with standards shall be made in accordance with methods and procedures approved by the Water Pollution Control Board.

Part 701
Classifications and Standards of Quality and Purity

(Statutory authority: Public Health Law, art. 6)

Sec.
701.1 Definitions

701.2 Conditions applying to all classifications and standards

Sec.
701.3 Classes and standards for fresh surface waters

701.4 Classes and standards for tidal salt waters

Section 701.1 Definitions. The several terms, words or phrases hereinafter mentioned shall be construed as follows:

(a) *Best usage* of waters as specified for each class shall be those uses as determined by the board in accordance with the considerations prescribed by section 109 of the Public Health Law.

(b) *Approved treatment* as applying to water supplies means treatment accepted as satisfactory by the authorities responsible for exercising supervision over the sanitary quality of water supplies.

(c) *Source of water supply for drinking, culinary or food processing purposes* shall mean any source, either public or private, the waters from which are used for domestic consumption or used in connection with the processing of milk, beverages, foods or for other purposes which require finished water meeting U.S. Public Health Service drinking water standards.

(d) *Fishing* shall include the propagation of fish and other aquatic life.

(e) *Agricultural* shall include use of waters for stock watering, irrigation and other firm purposes but not as source of water supply for drinking, culinary or food processing purposes.

(f) *Tidal salt waters* shall mean all tidal waters which are so designated by the board and which generally shall have a chloride ion content in excess of 250 parts per million.

701.2 Conditions applying to all classifications and standards. (a) In any case where the waters into which sewage, industrial wastes or other wastes effluents discharge are assigned a different classification than the waters into which such receiving waters flow, the standards applicable to the waters which receive such sewage or wastes effluents shall be supplemented by the following:

> "The quality of any waters receiving sewage, industrial wastes or other wastes discharges shall be such that no impairment of the best usage of waters in any other class shall occur by reason of such sewage, industrial wastes or other wastes discharges."

(b) Natural waters may on occasion have characteristics outside of the limits established by the standards. The standards adopted herein relate to the condition of waters as affected by the discharge of sewage, industrial wastes or other wastes.

701.3   Classes and standards for fresh surface waters.

CLASS AA

Best usage of waters. Source of water supply for drinking, culinary or food processing purposes and any other usages.

Conditions related to best usage. The waters, if subjected to approved disinfection treatment, with additional treatment if necessary to remove naturally present impurities, meet or will meet U.S. Public Health Service drinking water standards and are or will be considered safe and satisfactory for drinking water purposes.

Quality Standards for Class AA Waters

Items	*Specifications*
1. Floating solids; settleable solids; oil; sludge deposits; tastes or odor producing substances	None attributable to sewage, industrial wastes or other wastes.
2. Sewage or wastes effluents	None which are not effectively disinfected.
3. pH	Range between 6.5 and 8.5.
4. Dissolved oxygen	For trout waters, not less than 5.0 parts per million; for non-trout waters, not less than 4.0 parts per million.
5. Toxic wastes, deleterious substances, colored or other wastes or heated liquids	None alone or in combination with other substances or wastes in sufficient amounts or at such temperatures as to be injurious to fish life, make the waters unsafe or unsuitable as a source of water supply for drinking, culinary or food processing purposes or impair the waters for any other best usage as determined for the specific waters which are assigned to this class.

Note 1: In determining the safety or suitability of waters in this class for use as a source of
water supply for drinking, culinary or food processing purposes after approved
treatment, the Water Pollution Control Board will be guided by the standards
specified in the latest edition of *Public Health Service Drinking Water Standards*
published by the United States Public Health Service.

Note 2: With reference to certain toxic substances as affecting fish life, the establishment
of any single numerical standard for waters of New York State would be too
restrictive. There are many waters, which because of poor buffering capacity and
composition will require special study to determine safe concentrations of toxic
substances. However, based on non-trout waters of approximately median alkalinity (80 p.p.m.) or above for the State, in which groups most of the waters near
industrial areas in this State will fall, and without considering increased or decreased toxicity from possible combinations, the following may be considered as
safe stream concentrations for certain substances to comply with the above
standard for this type of water. Waters of lower alkalinity must be specially
considered since the toxic effect of most pollutants will be greatly increased.

Ammonia or Ammonium compounds	Not greater than 2.0 parts per million (NH_3) at pH of 8.0 or above
Cyanide	Not greater than 0.1 part per million (CN)
Ferro- or Ferricyanide	Not greater than 0.4 parts per million ($Fe(CN)_6$)
Copper	Not greater than 0.2 parts per million (Cu)
Zinc	Not greater than 0.3 parts per million (Zn)
Cadmium	Not greater than 0.3 parts per million (Cd)

CLASS A

Best usage of waters. Source of water supply for drinking, culinary or
food processing purposes and any other usages.

Conditions related to best usage. The waters, if subjected to approved
treatment equal to coagulation, sedimentation, filtration and disinfection,
with additional treatment if necessary to reduce naturally present impurities, meet or will meet U.S. Public Health Service drinking water
standards and are or will be considered safe and satisfactory for drinking
water purposes.

Quality Standards for Class A Waters

Items	*Specifications*
1. Floating solids; settleable solids; sludge deposits	None which are readily visible and attributable to sewage, industrial wastes or other wastes or which

deleteriously increase the amounts of these constituents in receiving waters after opportunity for reasonable dilution and mixture with the wastes discharged thereto.

2. Sewage or waste effluents

None which are not effectively disinfected.

3. Odor producing substances contained in sewage, industrial wastes or other wastes

The waters after opportunity for reasonable dilution and mixture with the wastes discharged thereto shall not have an increased threshold odor number greater than 8, due to such added wastes.

4. Phenolic compounds

Not greater than 5 parts per billion (Phenol).

5. pH

Range between 6.5 and 8.5.

6. Dissolved oxygen

For trout waters, not less than 5.0 parts per million; for non-trout waters, not less than 4.0 parts per million.

7. Toxic wastes, oil, deleterious substances, colored or other wastes or heated liquids

None alone or in combination with other substances or wastes in sufficient amounts or at such temperatures as to be injurious to fish life, make the waters unsafe or unsuitable as a source of water supply for drinking, culinary or food processing purposes or impair the waters for any other best usage as determined for the specific waters which are assigned to this class.

Note: Refer to notes 1 and 2 under class AA, which are also applicable to class A standards.

CLASS B

Best usage of waters. Bathing and any other usages except as source of water supply for drinking, culinary or food processing purposes.

Quality Standards for Class B Waters

Items	*Specifications*
1. Floating solids; settleable solids; sludge deposits.	None which are readily visible and attributable to sewage, industrial wastes or other wastes or which deleteriously increase the amounts of these constituents in receiving waters after opportunity for reasonable dilution and mixture with the wastes discharged thereto.
2. Sewage or wastes effluents	None which are not effectively disinfected.
3. pH	Range between 6.5 and 8.5.
4. Dissolved oxygen	For trout waters, not less than 5.0 parts per million; for non-trout waters, not less than 4.0 parts per million.
5. Toxic wastes, oil, deleterious substances, colored or other wastes or heated liquids	None alone or in combination with other substances or wastes in sufficient amounts or at such temperatures as to be injurious to fish life, make the waters unsafe or unsuitable for bathing or impair the waters for any other best usage as determined for the specific waters which are assigned to this class.

Note: Refer to note 2 under class AA, which is also applicable to class B standard.

CLASS C

Best usage of waters. Fishing and any other usages except for bathing as source of water supply for drinking, culinary or food processing purposes.

Quality Standards for Class C Waters

Items	*Specifications*
1. Floating solids; settleable solids; sludge deposits	None which are readily visible and attributable to sewage, industrial wastes or other wastes or which deleteriously increase the amounts of these constituents in receiving waters after opportunity for reasonable dilution and mixture with the wastes discharged thereto.
2. pH	Range between 6.5 and 8.5.
3. Dissolved oxygen	For trout waters, not less than 5.0 parts per million; for non-trout waters, not less than 4.0 parts per million.
4. Toxic wastes, oil, deleterious substances, colored or other wastes, or heated liquids	None alone or in combination with other substances or wastes in sufficient amounts or at such temperatures as to be injurious to fish life or impair the waters for any other best usage as determined for the specific waters which are assigned to this class.

Note: Refer to note 2 under class AA, which is also applicable to class C standards.

CLASS D

Best usage of waters. Agricultural or source of industrial cooling or process water supply and any other usage except for fishing, bathing or as source of water supply for drinking, culinary or food processing purposes.

Conditions related to best usage. The waters will be suitable for fish survival; the waters without treatment and except for natural impurities which may be present will be satisfactory for agricultural usages or for industrial process cooling water; and with special treatment as may be needed under each particular circumstance, will be satisfactory for other industrial processes.

Quality Standards for Class D Waters

Items	*Specifications*
1. Floating solids; settleable solids; sludge deposits.	None which are readily visible and attributable to sewage, industrial wastes or other wastes or which deleteriously increase the amounts of these constituents in receiving waters after opportunity for reasonable dilution and mixture with the wastes discharged thereto.
2. pH.	Range between 6.0 and 9.5.
3. Dissolved oxygen	Not less than 3.0 parts per million.
4. Toxic wastes, oil, deleterious substances, colored or other wastes, or heated liquids.	None alone or in combination with other substances or wastes in sufficient amounts or at such temperatures as to prevent fish survival or impair the waters for agricultural purposes or any other best usage as determined for the specific waters which are assigned to this class.

Note: Refer to note 2 under class AA, which is also applicable to class D standards.

Historical Note
Sec. amd., filed May 26, 1967 to be eff. May 26, 1967. Class E and F deleted.

Chapter 19 Design of Measurement Systems for Water Analysis

Herbert E. Allen and K. H. Mancy
DEPARTMENT OF ENVIRONMENTAL AND INDUSTRIAL HEALTH
SCHOOL OF PUBLIC HEALTH
UNIVERSITY OF MICHIGAN
ANN ARBOR, MICHIGAN

I. Introduction

Water pollution characterization is deceptively simple. The availability of a number of monographs and manuals listing water analysis recipes inadvertently gives the impression to the general worker in the field that water pollution characterization is an easy task. As a result, water analyses in the majority of treatment plants in the United States are usually assigned to personnel with minimum qualifications. This is in contrast to the practice in Europe, where the majority of treatment plants have qualified chemists in charge of water analysis. This misconception can also be seen in the lack of appropriations from federal, state, and local authorities for the development and improvement of water quality characterization techniques.

Water pollution control, as a specialized technology, has its own analytical needs. Any pollution control measure will be limited by the reliability and applicability of its methods of pollution characterization. Thus the water analyst is the "eyes and ears" of any control measure, for without valid, meaningful analyses our efforts to control pollution will be fruitless.

Once water pollution characterization consisted of analysis for a limited number of parameters using titrimetric or gravimetric procedures. Now, with the obvious increase in the magnitude and diversity of man-made pollution, there is an ever-increasing need for more elaborate information on water quality, which frequently requires the use of advanced instrumental and monitoring procedures.

The application of principles of analytical chemistry to natural and waste water systems requires subtle correlation between theory and experience, an insight into the nature and mode of action of interferences, and the ability to interpret analytical results in correlation with pertinent field observations. Because of the complexity of the system under investigation, water pollution characterization is one of the most challenging tasks likely to confront the analytical chemist.

II. Objectives

Definition of the purpose and objectives of analysis is the first step in the design of any measurement system; this includes the definition of particular problems to which solutions are sought. The analyses may be performed on surface or ground waters, municipal or industrial water supplies, irrigation waters, or municipal or industrial waste water. The analyses may be required to:

(1) Determine the suitability of a water for its intended use and to establish the degree of treatment necessary prior to its use.
(2) Estimate the possible detrimental effects of a waste effluent on the quality of the receiving water for subsequent downstream use.
(3) Evaluate necessary treatment requirements in view of water reuse.
(4) Determine the quantities of valuable by-products which could be recovered from a waste effluent.

(5) Evaluate and optimize industrial processes on a continuous or batch basis.

(6) Provide background information on the present quality of streams and lakes which can be used to demonstrate future changes in their quality.

Natural waters can be classified according to intended uses, including: public water supply, industrial water supply, fish or aquatic life propagation, water supply for stock and wildlife, irrigation, navigation, recreation, hydroelectric power generation, and disposal of sewage and industrial wastes. The Water Quality Act of 1965 required the states to adopt water quality standards for their interstate and coastal waters and a plan for implementing and enforcing the standards adopted. The standards and plans accepted by the Secretary of the Interior became the water quality standards applicable to the states' interstate and coastal waters.

The water quality criteria and standards relate to specific desired uses for the waters and therefore are based on specific chemical requirements necessary for these uses. The parameters important for these uses are discussed in Sec. III, "Water Quality Parameters."

Water quality standards may be based on the quality to be maintained in a receiving water, or they may relate to the quality of the effluent. Standards relating to the receiving water are termed "stream standards" while those relating to the waste discharge are termed "effluent standards." Both types of standards have their strengths and weaknesses (see Chap. 18).

Stream standards, which have been developed to comply with the provisions of the Water Quality Act, permit the full capacity of a stream to be utilized in the assimilation of wastes. In establishing stream standards the quality of the receiving stream is maintained at a level above that required for the desired usage. This permits maximum utilization of the stream's capacity to assimilate wastes and also permits materials discharged into different streams to be subjected to different degrees of treatment prior to discharge. In constrast, if effluent standards were used, all discharges of the same type to streams designated for the same use would be required to receive identical levels of treatment. The level of treatment would be consistent with the capability of advanced technology. The effect of the effluent on the receiving water would differ with each stream and there would be no assurance that the concentration of pollutant in the stream would not exceed a level critical to the desired use.

It is readily apparent that both systems of standards have flaws. It is likely that future standards will incorporate features of both stream and effluent standards. At present such "combined" standards are coming into effect for waste discharge of phosphate. In the Lake Erie basin a goal of 95% phosphate removal from industrial and municipal discharges has been established(1); the 95% reduction constitutes an effluent standard whose value was based upon desired stream standards and the capabilities of waste treatment technology.

The chemical analyses associated with the control and treatment of industrial and municipal waste waters which are discharged to streams, lakes, or oceans are dependent on the type of water quality standards to which they must conform. The analyses and data interpretation required to ensure that discharge of an effluent will not violate stream standards are usually more complicated than those to ensure compliance with effluent standards. In the former case the analysis is directed not only to characterization of the quality of the waste effluent, but also to determination of its effect on the aquatic ecosystem of the receiving water, while in the latter case only the quality of the waste is of concern.

In the determination of the deleterious effects of a waste effluent upon a receiving water it has been common practice to treat the waste water as an entity by analyzing for nonspecific parameters. Prior to analysis the waste water is diluted to the same extent that it would be diluted in the receiving water. The diluted sample is then analyzed for such nonspecific parameters as taste, odor, color, and toxicity to fish, depending upon the intended use. Such analyses may provide preliminary information on the ability of a stream to assimilate a waste water, but it is highly doubtful if significant conclusions can be drawn from these data.

Industrial wastes are often discharged to municipal sewers, after which they are treated by conventional waste treatment methods. The industrial wastes must be analyzed to ensure compliance with effluent standards established by the municipality. Effluent standards are established to protect the treatment facility from components of industrial wastes which might interfere with the operation of the waste treatment facility. It is essential to control the introduction of toxic materials which might inhibit essential biological waste treatment processes. Analyses of these effluents are also conducted for the assessment of charges, which are usually related to the strength and quantity of a particular waste component.

Chemical analysis of industrial waste waters, for example, may be performed for in-plant operations for one or more of the following reasons(2):

(1) Estimation of material balances for processes to permit evaluation of unit efficiencies and to relate material losses to production operations.
(2) Evaluation of continuing conformance with limits set for performance efficiency of certain unit processes.

(3) Determination of sources and temporal distributions of waste loads for segregation of flows, relative to strength and type, for separate treatment.

(4) Provision for immediate recognition of malfunctions, accidents, spills, or other process disturbances.

(5) Determination of the type and degree of treatment required for recovery of certain substances from waste effluents.

(6) Evaluation of conformance with standards set for effluent quality and/or stream quality.

(7) Provision for control of treatment and discharge of waste effluents according to present standards and/or according to variations in the conditions of the receiving water.

(8) Provision of a current record of costs associated with discharge of waste effluents to municipal sewers when such costs are at least partially based on the chemical characteristics of the waste.

To achieve some of the objectives of the industrial waste water analyses listed above only infrequent analyses are necessary, while for others, continuous monitoring and control are required.

An important objective of many surveys of lakes and streams is to provide a base of information to be used to evaluate future changes in water quality. Such data have been used by Beeton[3] and others in demonstrating the eutrophication of the Great Lakes. These data are important sources for other investigators. The U.S. Geological Survey has compiled data on the quality and quantity of water of surface waters of the United States[4] and has publications on the quality of surface water for irrigation[5] and of municipal water supply for industrial use[6]. Other water quality data have been published by the Public Health Service and the Federal Water Pollution Control Administration[7,8].

III. Water Quality Parameters

A. Introduction

After the objectives of the water quality study have been established, the next step in the design of the measurement system is to decide on those water quality characteristics to be measured. The analyst familiar with water quality characterization will often select the parameters to be measured based on experience and intuition. For certain uses water quality criteria have been established which include quantitative as well as qualitative information[9]. In most cases these guidelines should be followed. The analyst should familiarize himself with the significance of important parameters and the factors which determine the concentrations and distributions of the constituents to be measured. Although hundreds of different water quality parameters exist, in actual practice the number of analyses which are applied to an individual water sample is always quite small. As an example of the large number of parameters which can be considered, the Federal Water Pollution Control Administration's STORET system of data collection, processing, analysis, and reporting includes 424 parameters in six major groups (Table 1)(*10*).

It is convenient to group water quality parameters according to common usage (Tables 2, 3, 5, and 11). The mineral constituents of Table 2 include

TABLE 1
Number of STORET Listings for Water Analysis[a]

Parameters by groups	Example	Number of parameters in group
General physical and chemical	Alkalinity, COD, iron turbidity, zirconium	149
Physical observations	Algae, foam, oil	12
Radionuclides	Gross alpha and beta, strontium-90	141
Microbiological	Coliform by MPH and MF, total plate count	18
Organic materials Carbon adsorption data	Chloroform and alcohol extractables	12
Natural organics	Chlorophyll, tannins	4
Synthetic organics	ABS, phenols	2
Halogenated hydrocarbons	Aldrin, heptachlor, toxaphene	62
Phosphorated hydrocarbons	Malthion, parathion	10
Miscellaneous pesticides	Silvex	8
Treatment-related observations	Available chlorine	6

[a]From Ref. *10*.

TABLE 2
Mineral Constituents of Natural Waters

Major constituents	Minor constituents
Calcium	Iron
Magnesium	Manganese
Sodium	Aluminum
Potassium	Silica
Carbonate	Phosphate
Bicarbonate	Fluoride
Sulfate	Nitrate
Chloride	

the major constituents of most natural waters and are routinely included in many water analyses.

The presence of most of the major constituents is due to the phase equilibria of the natural water with various mineral species. Equilibrium concepts have been applied to natural water systems by many authors including Sillen(11), Garrels and Christ(12), and Stumm(13). Kramer(14) has developed a model predicting the chemical composition of the Great Lakes which involves the equilibrium of calcite, dolomite, apatite, kaolinite, gibbsite, sodium and potassium feldspars, and atmospheric carbon dioxide. A relatively good agreement exists between the actual and predicted values for calcium, magnesium, sodium, potassium, alkalinity, silica, and pH. The model predicts all the major constituents with the exception of sulfate and chloride. Although high sulfate concentrations may arise from gypsum or sodium sulfate deposits, most waters are very undersaturated with respect to these minerals. Chloride probably undergoes no simple phase equilibrium, and the concentrations of both sulfate and chloride may be controlled by atmospheric and pollutional input and surface runoff rather than by phase equilibria. All of the minor constituents of Table 2, with the exception of fluoride and nitrate, have been included in Stumm's(13) equilibrium model.

Equilibrium models may be useful in explaining existing water quality if their application is not abused. Most natural waters must be considered dynamic rather than static systems and may not have achieved thermodynamic equilibrium. Thermodynamics cannot predict whether a certain substance is present, but rather if a system containing the substance is stable. In addition, thermodynamic calculations do not predict the time necessary to achieve equilibrium in the system.

The presence of trace elements (see Table 3) is extremely important

TABLE 3
Abundance of Trace Elements in Natural Water

Common elements	Scarce elements	Rare elements
Aluminum	Arsenic	Antimony
Barium	Cadmium	Beryllium
Boron	Chromium	Bismuth
Copper	Lead	Cobalt
Iron	Lithium	Mercury
Manganese	Molybdenum	Selenium
Strontium	Nickel	Vanadium
Zinc	Tin	
	Bromide	
	Iodide	

for many water uses. Certain of these elements may impair the use of the water for domestic water supply, industrial water supply, irrigation, fish and aquatic life, and for watering of stock and wildlife. Some trace metals are essential nutrients and are required in low concentration by aquatic life. Green algae require the trace metals cobalt, copper, iron, manganese, molybdenum, vanadium, and zinc in addition to other elements(15). Although in trace quantities a metal may be an essential micronutrient, in high concentrations it may be inhibitory or toxic to biological systems. Copper is required as an essential element in many plant enzymes, and the electron transport system in chloroplasts includes a copper protein (16). Copper has not yet been found to be a limiting nutrient factor, but it may be inhibiting to algal growth when its concentration is high(17) and copper sulfate is used for the control of undesirable plankton blooms in reservoirs, lakes, and streams. The controlling role of metal ions in the ecological balance and productivity of fresh waters has been emphasized by several authors(17–21). For example, molybdenum, considered an essential micronutrient for nitrogen fixation and the formation of the plant enzyme nitrogen reductase, controls the primary productivity of Castle Lake, California(18).

The Federal Water Pollution Control Administration has monitored dissolved trace metals at 130 sampling points as a part of its Water Quality Surveillance Program. The data from over 1500 filtered surface samples which were collected from October, 1962 to October, 1967 are summarized in Table 4(8). Few data exceeded limits established by Water Quality Criteria(9) for public water supplies. Manganese exceeded the desirable limit of 50 μg/liter in 74 samples, of which 58 were from the

TABLE 4

Summary of Trace Elements in Waters of the United States[a,b]

Element	No. of positive occurrences	Frequency of detection, %	Observed positive values, μg/liter		
			Min.	Max.	Mean
Zinc	1207	76.5	2	1183	64
Cadmium	40	2.5	1	120	9.5
Arsenic	87	5.5	5	336	64
Boron	1546	98.0	1	5000	101
Phosphorus	747	47.4	2	5040	120
Iron	1192	75.6	1	4600	52
Molybdenum	516	32.7	2	1500	68
Manganese	810	51.4	0.3	3230	58
Aluminum[c]	456	31.2	1	2760	74
Beryllium	85	5.4	0.01	1.22	0.19
Copper	1173	74.4	1	280	15
Silver	104	6.6	0.1	38	2.6
Nickel	256	16.2	1	130	19
Cobalt	44	2.8	1	48	17
Lead	305	19.3	2	140	23
Chromium	386	24.5	1	112	9.7
Vanadium	54	3.4	2	300	40
Barium	1568	99.4	2	340	43
Strontium	1571	99.6	3	5000	217

[a]From Ref. 8.

[b]1577 samples (Oct. 1, 1962–Sept. 30, 1967).

[c]1464 aluminum analyses.

Ohio River basin and 12 were from the Lake Erie basin. Arsenic exceeded the 50 μg/liter limit in 41 samples and lead exceeded 50 μg/liter in 27 samples. Iron exceeded the limit of 0.3 ppm in 25 samples. Cadmium and hexavalent chromium exceeded the limits of 10 and 50 μg/liter in six and four samples, respectively. No samples were analyzed in which the criteria for zinc (5 mg/liter) or copper (1 mg/liter) were exceeded.

B. Nonspecific Parameters

Some of the most important and most frequently used tests in the analysis of water are the nonspecific tests listed in Table 5. These tests often measure a property of a group of substances such as alkalinity, where the capacity of the water to neutralize hydrogen ions is measured, or a physical parameter such as density, or a physiological property such

TABLE 5
Tests Used for Nonspecific Characterization of Water Properties

Physical properties	Chemical properties	Physiological properties
Filterable residue	Hardness	Taste
Salinity	Alkalinity and acidity	Odor
Density	Biochemical oxygen demand	Color
Electrical conductance	Chemical oxygen demand	Suspended matter
	Total carbon	Turbidity
	Chlorine demand	

as odor. Many of these tests are used to determine the suitability of natural waters for industrial or municipal use and to determine the type and degree of treatment needed to render them acceptable.

1. SALINITY

The quantity of salts dissolved in seawater is usually expressed as salinity, S. In fresh water it is called the filterable residue. Seawater evaporated to dryness and heated to a red heat to remove traces of water loses variable amounts of hydrogen chloride, carbon dioxide, and bromide from its salts. These losses cannot be compensated with sufficient accuracy by applying a correction. Since 1902 salinity has been defined as the total amount of solid material in grams contained in 1 kg of seawater when all the bromide and iodide have been replaced by the equivalent amount of chloride, all the carbonate has been converted to oxide, and the organic material has been completely oxidized(22).

In practice the constancy of proportions of ions present in seawater is used to permit chloride to be a measure of the salinity by means of the Knundsen equation:

$$S^o/_{oo} = 1.8050\ Cl^o/_{oo} + 0.030 \tag{1}$$

This equation is a definition and it should not be assumed that the salinity determined in this manner will in all cases be the same as that determined gravimetrically. The chlorinity, $Cl^o/_{oo}$, of a sample is defined as 0.3285234 times the weight of silver ions precipitated as silver halides from 1 kg of seawater. Chlorinity includes all halogen ions with the exception of fluoride.

Recently a new definition of salinity was proposed (23). This expression,

$$S^o/_{oo} = 1.80655 \; Cl^o/_{oo} \tag{2}$$

does not include a constant such as that required in the original Knundsen equation to account for the results obtained with water from the Baltic Sea which were used for the low concentrations. The new expression gives the same results as the Knundsen equation at a salinity of 35 $^o/_{oo}$ and differs by only $0.026^o/_{oo}$ at salinities of 32 and $38^o/_{oo}$.

The new definition of salinity is based on the work of Cox et al. (24), who determined the following polynomial expression relating conductivity and chlorinity:

$$Cl^o/_{oo} = -0.4980 + 15.66367 \; R_{15} + 7.08993 \; (R_{15})^2$$
$$- 5.91110 \; (R_{15})^3 + 3.31363 \; (R_{15})^4 - 0.73240 \; (R_{15})^5 \tag{3}$$

This expression is based on the R_{15}, the ratio of the electrical conductivity of a sample of seawater to that of seawater of salinity $35^o/_{oo}$, both being at 15°C and 1 atm pressure.

2. Total and Dissolved Residues

Total residue includes both suspended and dissolved materials in water. Filtration of the sample will separate the dissolved compounds and salts from the heterogenous suspension of dispersed solid material. The dissolved material is termed filterable residue or dissolved solids, while the dispersed solid matter is the nonfilterable residue or suspended solids.

Total residue is determined by evaporating a sample of water and bringing it to constant weight at 103–105 or 179–181°C (25). Residues dried at 103–105°C may retain occluded water as well as water of hydration. Because of the difficulty in removal of occluded water at this lower temperature, attainment of constant weight may be slow. Residues dried at the higher temperature will lose mechanically occluded water and may lose some organic matter and inorganic salts by volatilization. Samples which have been dried at 179–181°C in general yield residue values which conform more closely to the sum of the individually determined mineral salts than do the total residue values determined at the lower temperature.

Filterable residue or dissolved solids is determined by evaporating to dryness a volume of water which has been filtered through a membrane filter, filter paper, porous-bottom crucible, diatomaceous filter candle, or Gooch crucible. The residue may be dried to constant weight at either 103–105 or 179–181°C. Nonfilterable residue or suspended solids may be determined directly by filtering an appropriate volume of sample through a preweighed filter and drying the filter at 103–105 or 179–181°C. The increase in weight is due to the nonfilterable residue. The nonfilterable residue may also be determined as the difference between the total residue and the filterable residue.

Dissolved solids is one of the most important criteria for many uses of water including municipal water supply, industrial water supply, irrigation water, and water for stock and wildlife and fish and aquatic life. The limit of 500 mg/liter for domestic water supply (26) is based primarily on taste thresholds. Limiting concentrations of dissolved solids for various industrial applications are summarized in Table 6. The concentration of dissolved material present in boilers must be kept low to prevent corrosion and the deposition of scale. The higher the pressure of the boiler, the lower the concentration of dissolved solids which can be tolerated (Table 7).

TABLE 6

Limiting Concentrations of Dissolved Solids for Industrial Waters[a]

Industrial use	Limiting concentration, mg/liter
Brewing, light beer	500
Brewing, dark beer	1000
Brewing and distilling, general	500–1500
Canning and freezing	850
Carbonated beverages	850
Confectionery	50–100
Dairy washwaters	850
Food equipment washing	850
Food processing, general	850
Ice manufacture	170–1300
Plastics, clear	200
Paper manufacture	
Fine papers	200
Groundwood papers	500
Kraft paper, bleached	300
Kraft paper, unbleached	500
Soda and sulfate pulp	250
High grade paper products	80
Lower grade products	150–200
Rayon manufacture	100–200

[a]From Ref. 27.

TABLE 7
Water Quality Tolerances of Boiler Water[a]

Operating pressure, psi	Total dissolved solids, mg/liter	Total alkalinity, mg/liter	Suspended solids, mg/liter
0–300	3500	700	300
301–450	3000	600	250
451–600	2500	500	150
601–750	2000	400	100
751–900	1500	300	60
901–1000	1250	250	40
1001–1500	1000	200	20
1501–2000	750	150	10
2001 and higher	500	100	5

[a]From Ref. 28.

Table 8 lists the classifications of irrigation waters based on their dissolved solids concentration. It is considered that 1000 mg/liter is a concentration which approaches the limit for the best crop growth (27).

TABLE 8
Classification of Irrigation Waters
Based on Dissolved Solids[a]

Dissolved solids concentration, mg/liter	Classification
175 or less	Excellent
175–525	Good
525–1400	Permissible
1400–2100	Doubtful
2100 or more	Injurious

[a]From Ref. 27

3. DENSITY MEASUREMENTS

Density measurements are necessary in waters of high salinity to convert volume measurements to a weight basis. By definition, concentrations expressed in terms of parts per million represent the weight of dissolved matter in one million times the weight of solution (i.e., milligrams of solute per kilogram of solution). It is common practice to measure the quantity of sample used for an analysis volumetrically and to express results as a weight per unit volume of solution (e.g., milligrams of solute per liter of solution). Although parts per million and milligrams per liter are identical only when the density of the sample is precisely 1, for many samples little error is introduced by assuming unit density in converting concentrations from a volume to a weight basis.

The United States Geological Survey corrects for density only if the concentration of dissolved solids is greater than 7000 ppm(29). The density of a solution is increased by less than 1 g/liter or 0.1% due to the presence of 1000 ppm dissolved salts. The Geological Survey's choice of 7000 ppm dissolved solids as the dividing line for making density corrections permits an error of no more than 0.7% to be propagated into the computation of parts per million from milligrams per liter. Often the measurement of dissolved solids is not included in analytical schemes for the analysis of water. In these cases a specific conductance of 10,000 μmhos/cm can be used to separate those samples for which density corrections should be applied(29). This conductivity criterion should be applied carefully since un-ionized material will affect the density but not the specific conductance of the solution.

4. ELECTRICAL CONDUCTANCE

Electrical conductance is used as a measure of the total concentration of ionic species in a water sample. The conductance, L, of a solution can be represented by the equation:

$$L = k \sum_{i=1}^{n} C_i \lambda_i Z_i \qquad (4)$$

where k is the cell constant which is dependent on the geometry of the conductance cell, C is the molar concentration, λ is the equivalent conductance, and Z is the ionic charge. Conductivity may bear no simple relationship to the concentration of dissolved solids, and an equality of conductance between two samples does not necessarily imply an equality of total dissolved solids. In practice it has been found that the conductance at 25°C, multiplied by a factor between 0.55 and 0.70, can be used to predict the concentration of dissolved solids unless the water is of an unusual composition(25).

Conductivity measurements are usually included in automated multiparameter monitoring systems to monitor changes in the quality of surface water or the strength of a given waste water. Due to the constancy of the proportions of ions present in seawater, conductivity methods have been extensively used for determination of salinity (30).

Although it is possible to use direct current measurements for conductivity determinations, in practice most determinations are based on alternating current measurements using either conventional electrode or inductive methods, depending on how the current is generated in the solution. In the more common system, two or more electrodes are placed in the solution and an ac current is passed between them. This system suffers from the possibilities of polarization and electrode fouling. Both problems have been eliminated by the development of transformer bridge salinometers using an inductive seawater loop in place of electrode cells. Such an inductive system is fundamentally sounder than the more conventional electrode system and will gain wider acceptance in fresh water as well as oceanographic studies (30).

The effect of temperature on electrical conductance is quite complex. The temperature coefficient varies with both ionic composition and the temperature of the sample. The temperature coefficient of the conductivity of seawater is 3% per degree at 0°C and only 2% per degree at 25°C. The conductivity of seawater at 30°C is almost double that at 0°C (30). Temperature corrections for conductivity measurements of fresh waters have been reported by Smith (31).

5. HARDNESS

Hardness is caused by metallic ions which are capable of precipitating soap. Most hardness in water is due to the presence of calcium and magnesium, but soap may be precipitated by other polyvalent ions such as strontium, iron, manganese, aluminum, and zinc. Hard waters are undesirable for domestic water supplies because large amounts of soap are required to produce lather in them. The formation of scale at elevated temperatures makes hard waters objectionable for industrial boiler

TABLE 9
Classification of the Hardness of Water[a]

Hardness range (ppm of calcium carbonate)	Hardness description
0–60	Soft
61–120	Moderately hard
121–180	Hard
More than 180	Very hard

[a] From Ref. 32.

water supplies. However, hard waters reduce the toxic effects of heavy metals on fish.

The description of a water as hard or soft depends upon the hardness to which one has become accustomed. A water supply which seems hard to an Easterner might seem soft to many Westerners. The degrees of hardness have been classified by Dufor and Becker(*32*) (Table 9).

6. ALKALINITY AND ACIDITY

Alkalinity is the capacity to neutralize acids and acidity is the ability to neutralize bases. Although alkalinity of water usually is due to the presence of bicarbonate, carbonate, and hydroxide ions, care should be taken in assigning the alkalinity of a sample only to these three substances. Other weak acids such as borate, silicate, phosphate, sulfide, and humate will contribute to the alkalinity. Various waste waters may contain appreciable concentrations of acidity- or alkalinity-producing constituents other than carbonate species, hydrogen ions, and hydroxyl ions.

7. ORGANIC CARBON

The nonspecific measurement of organic carbon by such tests as biochemical oxygen demand (BOD), chemical oxygen demand (COD), or total organic carbon (TOC) is widely used as a measure of the organic pollutional load of a water or waste water. Although a relationship may be empirically established among BOD, COD, and TOC measurements for a given waste, no basic correlation among these measurements is possible since the fundamental basis of each test is different.

a. BOD

The BOD procedure is a bioassay which measures the oxygen consumed by bacteria and other organisms utilizing the organic matter present in the sample. The curve of BOD plotted against time has an initially steep slope which gradually decreases until a plateau is reached when all the decomposible organics have been utilized. This relation is described by the equation:

$$y = L(1 - 10^{-k_1 t}) \tag{5}$$

where y is the BOD at time t, L is the ultimate oxygen demand, and k_1 is the rate constant. The rate constant is affected by such factors as substrate concentration, inorganic nutrient concentration, pH, temperature, toxic compounds, and species of microorganisms present. For domestic sewage and many industrial wastes the 5-day BOD is 70–80% of the ultimate BOD, but the actual rate constant must be determined by experiment (see Chap. 15).

b. COD

Because the time required for completion of a BOD analysis, which may be 5 days, is too great for the test to have application in many situations, the analysis of COD, which requires 3 hr or less, is often preferred. The COD procedure is based on the fact that most organic compounds can be oxidized to carbon dioxide and water by strong oxidants under acid conditions.

In the determination of COD a standard dichromate solution is refluxed with the sample containing organic matter until the reaction is complete. The excess of dichromate is measured by titration with standardized ferrous ammonium sulfate(25). Substances which reduce dichromate will interfere with the COD reaction and yield high results. Chloride interference by the reaction:

$$6Cl^- + Cr_2O_7^{2-} + 14H^+ \rightarrow 2Cr^{3+} + 3Cl_2 + 7H_2O$$

can be eliminated by the addition of mercuric sulfate prior to refluxing. This procedure ties up the chloride as a soluble mercuric chloride complex.

Most samples contain a high percentage of material which can be chemically, but not biochemically, oxidized; such samples will have COD values greater than their BOD values. For example, textile, paper mill, and other wastes containing high concentrations of cellulose have a high COD, but a low BOD. On the other hand, distillery and refinery wastes have a high BOD and low COD, unless the COD procedure incorporates silver sulfate to catalyze the oxidization of straight-chain alcohols and acids.

Advantages of the COD test include speed of analysis, which facilitates plant control. COD oxidation conditions can be better standardized than BOD conditions, resulting in improved test precision for the COD test.

COD results are not applicable for estimating BOD except as a result of experimental evidence by both methods on a sample of a given type. The results of COD analyses of wastes cannot be used to predict oxygen depletion in receiving waters. Certain compounds, including saturated hydrocarbons, pyridine, and toluene, are not susceptible to oxidation in the COD procedure. Because of the oxidation of chloride to chlorine, precision may be low for saline waters.

c. TOC

Total carbon analysis is carried out by an instrumental method providing results in a matter of minutes and requiring much less than 1 ml of sample. The procedure involves the introduction of a micro sample into a catalytic combustion tube maintained at 950°C which vaporizes the water. In a carrier stream of pure oxygen the organic matter is converted to CO_2, H_2O, and N_2. The water is condensed and the gas stream is passed through a continuous flow sample cell in a nondispersive infrared CO_2 analyzer(33). The amount of CO_2 recorded is proportional to the carbon content of the sample. The procedure does not distinguish between organic and inorganic carbon in the initial sample, but acidification of the sample and sparging with an inert gas prior to introduction of the sample removes most of the inorganic carbon as CO_2. The sample can then be analyzed for total organic carbon (TOC).

In another instrumental procedure for the analysis of organic carbon the sample is carried by CO_2 through a platinum catalytic combustion furnace which oxidizes the organic material to CO and H_2O(34). Following removal of water and passage through a second platinum-catalyzed oxidation the CO concentration is measured by an infrared analyzer.

Total carbon analysis is an especially valuable tool in process control where even the analysis of COD may be too slow to be useful. The total carbon procedure responds to all types of organic substances and does not respond to elements other than carbon. The total organic carbon methods are more reproducible than either BOD or COD methods and are applicable to the measurement of carbon in large numbers of samples (see Chap. 27).

8. CHLORINE DEMAND

The chlorine demand of water is the difference between the amount of chlorine added to a water sample and the amount of free, combined, or total chlorine remaining at the end of a specified contact period. The chlorine demand is due to the oxidation of ferrous, manganous, nitrite, sulfite, and sulfide ions, as well as the reaction with ammonia and organic amino compounds to form chloramines. Chlorine reacts with phenolic compounds to produce chlorophenols which have high odor intensities, but a higher chlorine concentration often completely oxidizes these compounds.

The chlorine demand of a water, sewage, or industrial wastewater is used in design specifications to ensure sufficient chlorination to produce a chlorine residual. Since the chlorine demand of a sample depends on the contact time, pH, and temperature, it is essential that these data be recorded with the test results (see Chap. 22).

9. TASTE AND ODOR

The physiological properties of taste and odor are measured by the sensory response of an analyst. Although the taste and odor sense organs are extremely sensitive, they are not precise. Standard procedures involve the use of a panel of not less than five persons to evaluate the sample (25). The tests for taste and odor are based on subjective comparison since no absolute base or unit exists for these measurements.

Waters free of taste and odor are particularly desirable for municipal water supplies and for food processing industries such as dairying, distilling, and brewing. Tastes and odors can be imparted to fish and shellfish by organic compounds of natural or pollutional origin.

Although taste and odor responses are closely related, certain nonvolatile substances which do not cause odor can produce taste sensations. Among the taste-producing substances not evoking odor responses are iron, manganese, sodium, potassium, copper, and zinc salts (25). The tolerance to dissolved salts is an individual reaction, and taste tolerances can be developed, but some dissolved gases and minerals are generally considered essential to make water palatable.

Odors often can be related to the presence of certain biological forms in the water, with the most frequent offenders being algae. Other causes of tastes and odors in water supplies include the decay of organic material, deoxygenation of rivers and lakes, and the chlorination of water supplies which have been polluted by organic compounds. Industries such as pulp and paper; explosives; petroleum, gasoline, and rubber; wood distillation; coke and coal tar; gas; tanneries, meat packing, and glue; chemicals and dyes; and milk products, canneries, beet sugar, distillation, and other food products industries discharge potentially odoriferous compounds (27).

The chemicals responsible for tastes and odors include halogens, sulfides, ammonia, turpentine, phenols and cresols, picrates, various hydrocarbons and unsaturated organic compounds, mercaptans, tar and tar oils, detergents, and pesticides (27). The various types of odors have been classified by chemical type (25), as shown in Table 10.

Odor intensity is expressed by the threshold odor number (25, 35), which is the ratio to which the sample must be diluted with odor-free water to be just perceptible to the individual. It has been recommended that odor tests be run at 25 and 60°C (25) or 40 and 60°C (35). Samples for taste tests should be at 40°C. Since taste and odor are dependent on temperature, the sampling and test temperatures should be reported.

TABLE 10
Classification of Odors by Chemical Types[a]

Odor class	Perception level of odor characteristics				Chemical types included	Examples
	Sweetness	Pungency	Smokiness	Rottenness		
Estery	High	Medium	Low to medium	Medium	Esters, ethers, lower ketones	Lacquer, solvents, most fruits, many flowers
Alcoholic	High	Low to medium	Low to medium	Medium	Phenols, cresols, alcohols, hydrocarbons	Creosote, tars, smokes, liquor, rose and spicy flowers, spices and herbs
Carbonyl	Medium	Medium	Low to medium	Medium	Aldehydes, higher ketones	Rancid fats, butter, violets, grasses, and vegetables
Acidic	Medium	High	Low to medium	Medium	Acid anhydrides, organic acids, sulfur dioxide	Vinegar, perspiration, rancid oil, resins, garbage
Halide	High	Medium to high	Medium to high	Low to high	Quinones, oxides and ozone, halides, nitrogen compounds	Insecticides and weed killers, musty and moldy odors, medicinal odors, earth, peat
Sulfury	Medium	Medium	High	High	Selenium compounds, arsenicals, mercaptans, sulfides	Skunks, bears, foxes, rotting fish and meat, cabbage, onion, sewage
Unsaturated	High	Medium	Medium	High	Acetylene derivatives, butadiene, isoprene	Paint thinners, varnish, kerosine, turpentine, essential oils, cucumber
Basic	High	Medium	Low to medium	High	Vinyl monomers, amines, alkaloids, ammonia	Fecal odors, manure, fish and shellfish, stale flowers, lilac, lily

[a]From Ref. 35.

10. COLOR AND TURBIDITY

Color, like taste and odor, is a property measured by a physiologic response, but there are methods for the measurement of color which are not subjective. It is necessary to distinguish between the true color of water caused by substances in true solution and the apparent color which is caused by the effects of suspended and colloidal material.

a. Color

Color in water may result from the presence of material of a vegetable origin, such as tannins, humic acids, peat material, plankton, or weeds. Metal ions, including iron, manganese, copper, and chromium, may also impart color. Color is objectionable for domestic water supplies and for use in many industries, including food and beverage processing, pulp and paper manufacturing, and textile production.

Color is determined by comparison of a sample with colored solutions of known concentration or with special colored glass disks. Standard platinum-cobalt color solutions are used for visual comparison. One color unit is produced by 1 mg of platinum per liter, in the form of chloro-platinate ion, and approximately 0.5 mg of cobalt per liter. For field use, samples are compared to colored glass disks which have been calibrated to correspond to the platinum–cobalt scale (25).

Visual comparison techniques depend on the response of the individual viewing the sample. Certain types of industrial wastes may produce colors which cannot be matched well using the standard platinum–cobalt scale.

Tri-stimulus colorimetric techniques permit a more accurate determination of water color (26, 36). The light transmission characteristics of a filtered sample are measured at certain wavelengths using a spectrophotometer or colorimeter. The results, expressed by the dominant wavelength, luminance, and purity, approximately describe the visual response of an individual. The dominant wavelength, which is expressed in nanometers, is a measure of the hue (red, yellow, green, etc.). The degree of brightness is designated by luminance and the saturation (pastel, pale, etc.) by purity. Both luminance and purity are reported in units of per cent.

b. Turbidity

Turbidity is a measure of the light-scattering characteristics of water caused by the presence of colloidal and particulate matter suspended in water. The weight of the material filtered from a sample is used as a measure of the total suspended solids. The standard for turbidity is a silica suspension in which 1 mg/liter equals one unit of turbidity. Because the turbidity of a given weight of material is dependent on its state of division, the particle size of standards must be controlled(35).

Turbidity is undesirable in domestic water supplies, for industrial applications where the product is destined for human consumption, and for other industrial applications such as pulp and paper manufacture and boiler feed water. The turbidity of natural waters is an important factor in the control of productivity. Turbidity interferes with the penetration of light and can therefore limit photosynthesis and primary productivity.

The standard instrument for measuring turbidity is the Jackson turbidimeter. The test is based on measuring the length of the light path through which the flame of the standard candle just becomes indistinguishable from the background(35).

The Jackson turbidimeter can be used to measure samples with turbidities between 25 and 1000 turbidity units. Waters with turbidity in excess of 1000 units are diluted prior to measurements. For waters with low turbidity, light scattering or nephelometric techniques are commonly used(35, 37). Both the theoretical and practical aspects of turbidity measurement with the Jackson turbidimeter and more sophisticated instruments have been discussed by Black and Hannah(37).

C. Specific Parameters

The specific chemical analysis to be performed in a water pollution study will depend on the types of materials discharged and on the desired uses of the receiving water. Some of the more frequently measured specific parameters in pollution studies are listed in Table 11.

1. DISSOLVED OXYGEN

Dissolved oxygen is often the most important parameter in studies of pollution. Severe depletion of the oxygen concentration may indicate an excessive load of organic wastes which are exerting a high biochemical oxygen demand in the receiving water. Low dissolved oxygen will have an adverse effect on the fish and other aquatic life of a body of water.

An extensive portion of the sanitary engineering and limnological literature has been devoted to the significance of dissolved oxygen in the environment and to its measurement. Mancy and Jaffe(38), in a comprehensive review of methods for the determination of dissolved oxygen, discuss 23 titrimetric methods in addition to colorimetric, gas exchange, gas chromatographic, gasometric, radiometric, conductometric, coulometric, voltammetric, and membrane electrode methods.

The concentration of dissolved oxygen in water at equilibrium with the

TABLE 11

Tests Used for the Measurement of Water Pollution

Nutrient demand	Specific nutrients	Nuisances	Toxicity
Dissolved oxygen Biochemical oxygen demand	Nitrogen: Ammonia	Sulfide Sulfite	Cyanide Heavy metals
Chemical oxygen demand	Nitrate	Grease and oil	Pesticides
Total carbon	Nitrite	Detergents	
	Organic nitrogen	Phenols	
	Phosphorus:		
	Orthophosphate		
	Polyphosphate		
	Organic phosphorus		

atmosphere is a function of temperature, pressure, and the presence of other solutes. The solubility of oxygen decreases from 14.6 mg/liter at 0°C to 8.4 mg/liter at 25°C in pure water. At 0°C increasing the chloride concentration from 0 to 20 g/liter causes a 3.3 mg/liter decrease in the solubility of oxygen, from 14.6 to 11.3 mg/liter, while at 25°C the decrease from (8.4 to 6.7 mg/liter) is only 1.7 mg/liter(25). The effect of hydrostatic pressure on the solubility of oxygen is slight. For example, at 300°K (27°C) the solubility of dissolved oxygen is decreased by less than 0.1% at 1000 m below the surface. At equilibrium the concentration of dissolved oxygen is proportional to the partial pressure of oxygen in the gas phase. The solubility of oxygen in water follows Henry's law up to 4 atm of oxygen(38).

Uncontaminated surface waters are usually nearly saturated with oxygen. A diurnal variation of dissolved oxygen is caused by the photosynthetic activity of phytoplankton, and during blooms rapid changes in dissolved oxygen may take place. For example, during a plankton bloom in western Lake Erie in August, 1969, the dissolved oxygen increased from 142 to 195% saturation in 2.4 hr(39). Dissolved oxygen reaches a maximum in the late afternoon, but at night in the absence of light, algal respiration reduces the level of dissolved oxygen. The lowest oxygen concentrations usually occur just before daylight. Primary production, especially the occurrence of algal blooms, is dependent on the availability of nutrient elements as well as on such physical factors as light and temperature.

Deep lakes and reservoirs may develop thermal stratification during the summer months. The upper layer, or epilimnion, extends to about 40 ft, depending on the wind and other mixing action. A zone of sharp decrease in water temperature, the thermocline, characterizes the middle layer, known as the metalimnion. The hypolimnion is the lower layer and in most deep lakes it remains relatively cold throughout the period of thermal stratification. During the period of stratification, relatively little oxygen from the epilimnion or atmosphere is contributed to the hypolimnetic water, and oxygen depletion may develop due to the exertion of an oxygen demand by substances in the sediment or water. The central basin of Lake Erie, for example, appears to undergo yearly oxygen depletion with as much as 1390 sq miles containing less than 1.0 ppm dissolved oxygen(40). Depletion of oxygen to extremely low levels has been shown by Mortimer(41, 42) to cause the release of substances from the sediment. In the case of Lake Erie, iron, manganese, phosphate, and ammonia concentrations are at least 20-fold higher in the deoxygenated hypolimnion than at the same depth after the period of thermal stratification(43). The water, which is low in dissolved oxygen content, is uninhabitable by fish and is a serious potential threat to the survival of aerobic benthic organisms. High chlorine demands are associated with hypolimnetic waters low in oxygen, due to the presence of appreciable concentrations of reduced iron and manganese, ammonia, and other materials capable of reacting with chlorine. At one water plant in Cleveland, Ohio, which occasionally draws hypolimnetic water, the chlorine demand of the raw water increases three- to fivefold during low oxygen conditions. Under these conditions, not only is treatment more expensive, but concurrent taste and odor problems may greatly reduce consumer acceptance of the finished water(44).

The analyses for biochemical oxygen demand, chemical oxygen demand, and total carbon are used extensively for the detection and quantitation of organic pollution and the determination of its effects on the oxygen budget of the receiving water. The decomposition of organic matter by microorganisms utilizes oxygen and produces the characteristic variation in streams called the "dissolved oxygen sag curve." The oxygen deficit was related to the BOD by the equation proposed in 1925 by

Streeter and Phelps(45):

$$D = \frac{K_1 L_A}{K_2 - K_1} (e^{-K_1 t} - e^{-K_2 t}) + D_A e^{-K_2 t} \tag{6}$$

where

D is the dissolved oxygen deficit (mg/liter)
t is the time of flow (days)
L_A is the ultimate carbonaceous BOD at the upstream end of the reach (mg/liter)
D_A is the dissolved oxygen deficit at the upstream end of the reach (mg/liter)
K_1 is the BOD rate coefficient determined in the laboratory (per day)
K_2 is the reaeration coefficient (per day)

The equation is based on a number of assumptions of uniformity of conditions within the reach of the stream being studied. Since the required uniformity is unlikely for great lengths, the stream is usually modeled in successive short reaches.

2. NUTRIENTS

Nutrients discharged to a watercourse promote biological responses which may interfere with some desired uses of the water. Plankton blooms and excessive growths of attached algae such as Cladophera result from the overfertilization of lakes and rivers. The unaesthetic appearance and odor of plankton growths make waters unfit for recreation and also for other purposes such as municipal and industrial water supply.

Although many elements and compounds are required algal nutrients, most attention has been focused on nitrogen and phosphorus because the quantity of each in waters has been shown to limit algal populations. Both are common pollutants arising from municipal and industrial wastes and agricultural runoff.

The increase in the concentration of nutrients in a body of water is called eutrophication and is one of the most important pollution problems. Eutrophication has affected Lake Erie(3, 46) and Lake Washington(47) as well as other important bodies of water. The problem of eutrophication has been reviewed by Fruh(48) and Stewart and Rohlich(49) and was the subject of a 1967 conference sponsored by the National Academy of Science(50) (see Chap. 4).

a. Nitrogenous Compounds

In addition to dissolved nitrogen molecules, ammonia, nitrate, nitrite, and organic nitrogen compounds are the most abundant forms of nitrogen in natural water. Certain species of blue-green algae are among the few aquatic organisms capable of utilizing molecular nitrogen. During periods of low nitrate concentration, blue-green algae capable of nitrogen fixation may become the dominant species and undesirable blue-green blooms may occur(*51*).

Of the three inorganic combined nitrogen forms, ammonia is by far the most abundant form in municipal waste effluents receiving only primary treatment, but where aerobic biological treatment is used there is oxidation to nitrite or nitrate.

Table 12 shows the relative importance of various nitrogen and phosphorus sources as nutrient supplies for the surface waters of Wisconsin (*52*).

Nitrogen in groundwater is predominant in the form of nitrate, which averages 1.2 mg/liter nitrogen(*51*). Nitrate is not significantly adsorbed onto soil, but ammonia can undergo significant reactions with soil constituents including physical adsorption, chemical sorption, and ion exchange reactions(*53*). Fertilizers contribute heavily to the input of nitrogen to groundwater. Nitrogen fertilizer usage increased by almost 250% between 1953 and 1963, and fertilizer manufacture was responsible for the consumption of 76% of commercially utilized nitrogen

TABLE 12

Estimated Contributions of Nitrogen and Phosphorus to Wisconsin Surface Water[a]

Source	Nitrogen, % of total	Phosphorus, % of total
Municipal treatment facilities	24.5	55.7
Private sewage systems	5.9	2.2
Industrial wastes	1.8	0.8
Rural sources:		
Manured lands	9.9	21.5
Other cropland	0.7	3.1
Forest land	0.5	0.3
Pasture, woodlot, and other lands	0.7	2.9
Groundwater	42.0	2.3
Urban runoff	5.5	10.0
Precipitation on water areas	8.5	1.2

[a]From Ref. *52*.

in 1957(*54*). Municipal waste effluents, contributing almost one quarter of the total nitrogen, are the second largest nitrogen source after groundwater.

Many of the transformations of nitrogenous compounds are bacterially mediated and the relationships between forms of nitrogen are frequently expressed by the diagram known as the nitrogen cycle. Nitrate, ammonia, and atmospheric nitrogen can be converted to plant and bacterial protein. Animals are dependent upon plants as nitrogen sources, and excess nitrogen is excreted and converted to ammonia. Ammonia is oxidized to nitrite in the nitrification reaction by autotrophic bacteria such as *Nitrosomonas*, which uses the energy released in the reaction for growth. Nitrification of nitrite to nitrate is carried out by *Nitrobacter*, while denitrification of nitrite to form nitrogen gas can be accomplished by various heterotrophic bacteria.

The presence of large quantities of ammonia or nitrite in surface waters may indicate sewage pollution. Both are usually present in extremely low concentrations and most inorganic nitrogen is present as nitrate. Although nitrate concentrations are usually greater than 0.1 mg/liter nitrogen, during the summer nitrate concentration of lakes may decrease to almost undetectable levels(*55*). During this period, the growth of blue-green algae which are capable of nitrogen fixation is favored(*51*).

Organic nitrogen compounds may arise from the residues of plants and animals, their waste products, or wastes from industries including food processing, pharmaceutical, plastics, and textiles. While some of these compounds, such as urea and protein arising from muscle tissue, are readily biodegradable, the fibrous protein of hair, skin, and hoofs and many of the synthetic organic compounds are resistant to biological degradation.

(1) Ammonia. The most widely used method for the analysis of ammonia is the nesslerization reaction. The test is based on the formation of a yellowish brown colloidal dispersion on the addition to the sample of Nessler's reagent, a strongly alkaline solution of potassium mercuric iodide(*25*):

$$2K_2HgI_4 + NH_3 + 3KOH \longrightarrow I-Hg-O-Hg-NH_2 + 2H_2O + 7KI \tag{7}$$

Samples may be pretreated by a flocculation step using zinc sulfate and alkali, which removes some ions (e.g., Ca^{2+}, Mg^{2+}, Fe^{3+}, S^{2-}) which form precipitates with Nessler's reagent. Precipitation of calcium and magnesium can also be prevented by the addition of EDTA or Rochelle salt to the sample. Although direct nesslerization is most often preferred, due to its speed and ease, ammonia must be separated from some samples by distillation to remove turbidity, natural color, and certain organics which produce an off-color with Nessler's reagent.

Samples containing in excess of 1 mg/liter of ammonia may be analyzed by a titrimetric procedure. The sample is maintained at pH 7.2–7.4 by a phosphate buffer and the ammonia is distilled into a boric acid solution which is back-titrated using a standard strong acid.

A sensitive colorimetric procedure is gaining wide acceptance for the analysis of samples containing low levels of ammonia and for the automated colorimetric analysis of samples. In the phenol-hypochlorite method, chloramine is formed by the reaction of ammonia and hypochlorite. The chloramine reacts with phenol and base to form an indophenol compound by the reaction (56):

$$NH_4^+ + OCl^- \longrightarrow NH_2Cl \xrightarrow{\text{phenol}} OC_6H_4NCl \xrightarrow{\text{phenol}} \tag{8}$$

$$OC_6H_4NC_6H_4OH \xrightarrow{OH^-} OC_6H_4NC_6H_4O^-$$

The sensitivity of the method can be enhanced by use of nitroprusside, which catalyzes the reaction. An automated procedure for the analysis of ammonia in concentrations less than 0.1 mg/liter has been developed (57).

(2)   Organic Nitrogen Compounds. Organic nitrogen is converted to ammonia by the Kjeldahl method. Organic matter in the sample is destroyed by oxidation with sulfuric acid. The oxidation temperature is slightly above the boiling point of sulfuric acid (340°C), which is achieved by the addition of potassium sulfate. A catalyst of mercuric sulfate is included to ensure complete destruction of the organic matter. After the digestion is complete the pH is adjusted and the ammonia distilled. Analysis of the distillate may be by any ammonia method suitable for the expected concentration. If ammonia was not removed from the sample prior to digestion, the results should be reported as "total Kjeldahl nitrogen." Organic nitrogen may be determined as the difference between total Kjeldahl nitrogen and ammonia or by analysis for Kjeldahl nitrogen of samples from which the ammonia has been removed by distillation.

(3)   Nitrites. The standard method (25) for the analysis of nitrite in water is based on the diazotization of sulfanilic acid by nitrite in an acidic solution followed by coupling of the diazonium compound with α-naphthylamine hydrochloride to form an intense red azo dye which has an absorbence maximum at 520 nm. This method is often referred to as the Griess–Ilosvay method (58). Sulfanilamide can be used instead of sulfanilic

acid and the α-naphthylamine hydrochloride can be replaced by N-(1-naphthyl) ethylenediamine hydrochloride, according to Shinn(59) and Bendschneider and Robinson(60). This procedure has a more rapid color development and is less sensitive to pH and the amount of reagent used than is the Griess–Ilosvay method.

An excellent review of 52 spectrophotometric methods for nitrite has been published by Sawicki et al.(61). The authors critically evaluated the sensitivity, color stability, conformity to Beer's law, simplicity, and precision of the methods.

(4) Nitrates. Nitrate has been determined by colorimetric procedures based on the nitration or oxidation of organic compounds, ultraviolet spectrophotometry of the nitrate ion, and reduction of nitrate to nitrite or ammonia. Phenoldisulfonic acid reacts with nitrate ion in acidic solution to produce a nitro derivative. When the solution is made alkaline, the compound rearranges to produce a yellow compound with a maximum absorbence at 410 nm. Small amounts of chloride do not interfere, but nitrite should be removed by treatment with sodium azide(35).

Nitrate is capable of oxidizing a variety of reagents in acid solution to form colored products. Oxidizable reagents include brucine, strychnidine, diphenylamine, resorcinol, and diphenylbenzidine(62). Brucine has been the most widely used of these compounds, but the color intensity varies with time, making it necessary to develop the color of samples and standards simultaneously. A modification(63) has rendered the method applicable to the analysis of seawater and brackish water. Conditions are controlled so that Beer's law is followed and concentrations below 1 mg/liter nitrate nitrogen can be determined.

Nitrate ion in an acid medium exhibits strong absorption of ultraviolet radiation with an absorbence maximum at 206 nm(64). Although chloride is not an interference in the method, dissolved organic compounds, nitrite, and chromium interfere.

Nitrate can be reduced to ammonia using (a) aluminum foil in alkaline solution, (b) a zinc–copper couple in acetic acid solution, (c) Devarda's alloy, or (d) alkaline ferrous sulfate(65). The ammonia produced may be separated by steam distillation and estimated in the distillate by nesslerization or other suitable ammonia analysis.

Nitrate may also be determined by reduction of nitrate to nitrite, which is then determined by the Griess–Ilsovay method. Hydrazine in a buffered solution containing catalytic quantities of cupric ion has been used for the reduction of nitrate to nitrite(66). This method requires 24 hr for the reduction and is sensitive to the presence of particulate matter and to hydrogen sulfide which deactivates the copper catalyst.

A number of solid reductants have been used for the reduction of nitrate to nitrite. Zinc powder in acid solution has been applied to both natural waters and waste waters(67). Extremely simple methods of reduction have been developed from oceanographic use. The reductants are used in columns through which the samples are poured. Both amalgamated cadmium(68) and cadmium–copper(69) have been used and produce nitrite yields in excess of 90%.

Automated procedures have been developed for the determination of nitrate based on homogeneous or heterogeneous reduction of nitrate to nitrite. Kamphake et al.(70) have developed a hydrazine reduction procedure in which the temperature is controlled to provide equivalent absorbences for equimolar nitrate and nitrite standards. Correction for the nitrite contribution to the measured concentration is thereby simplified. Small columns containing a copper–cadmium couple have been used in oceanographic analysis by Bernhard and Macchi(71), while Brewer and Riley(72) have used coarse cadmium filings to reduce nitrate to nitrite.

b. Phosphorus Compounds

(1) Sources. Phosphorus may be present in aquatic systems as orthophosphate (PO_4^{-3}); condensed linear phosphates such as pyrophosphate ($P_2O_7^{-4}$) and tripolyphosphate ($P_3O_{10}^{-5}$); and condensed ring phosphates such as trimetaphosphate ($P_3O_9^{-3}$) and tetrametaphosphate ($P_4O_{12}^{-4}$). In addition to the inorganic phosphates, a significant fraction of the phosphorus in natural and waste waters may be present in organic phosphorous compounds of biological origin. These compounds include organic orthophosphates such as sugar phosphates (glucose-6-phosphate, fructose-6-phosphate), deoxyribonucleic acids, phospholipids (glycerophosphate, phosphatidic acids, phosphatidyl choline, lecithin), inositol phosphates (inositol hexaphosphate, inositol monophosphate), phosphoamides, creatine phosphate, and phosphoproteins and organic condensed phosphates such as adenosine-5-triphosphate and coenzyme A(53).

A large proportion of the phosphorus in surface waters originates from municipal waste effluents. A large proportion of this phosphorus is of detergent origin. A survey of nutrient sources for Wisconsin water(52), Table 12, disclosed that over half the phosphorus originated from municipal wastes and over one-fifth originated from the manuring of frozen land. Municipal wastes, runoff from manured land, and urban runoff contributed 87.2% of the phosphorus of Wisconsin surface water. In a report on the water quality of Lake Erie(2), municipal wastes were reported to be the source of 72% of the phosphorus entering the lake. Detergents were reported to be responsible for 66% of the total phosphorus in municipal wastes; hence at least 47.5% of the phosphorus entering the lake originates from detergents.

(2) Orthophosphate. Phosphate is generally determined by a colorimetric procedure involving the complexation of phosphate with molybdate in an acid medium and reduction of the complex to an intensely colored molybdenum-blue compound. The reaction between orthophosphate and acidified molybdate has generally been assumed to be specific for the ortho form of phosphate. Strickland and Parsons(73) point out that the chemical analysis will include not only the inorganic orthophosphate ion, but also organic phosphates such as those easily hydrolyzable sugar phosphates which form inorganic orthophosphate by the acidic condition prevailing during the analysis (approximately 5 min). They have called the results of this analysis on filtered water samples "soluble reactive phosphate." Rigler has reported(74) that in some lake waters substantial portions of the soluble reactive phosphate are not inorganic orthophosphate but are results of the hydrolysis of organic phosphorus compounds.

An extensive review of the measurement of orthophosphate has been conducted by Olsen(75), who discussed the terminology of phosphorus data reporting, differentiation of phosphorous forms, and analytical methods for phosphorus analyses. Both Jones(76) in his comparison of methods for the determination of phosphate in seawater, and Strickland and Parsons(73) in their excellent manual of seawater analysis recommend the phosphorus procedure of Murphy and Riley(77) which uses ascorbic acid as a reducing agent and introduces potassium antimonyl tartrate to speed up the reaction. Strickland and Parsons(73) have characterized this method as "so superior to other methods in terms of the rapidity and ease of analysis that it probably represents the ultimate in sea-going techniques."

Inorganic polyphosphates and organic ortho- and polyphosphates are first hydrolyzed or digested to convert the phosphorus to orthophosphate and then analyzed by the same colorimetric procedures used for inorganic orthophosphate. The digestion can be accomplished by boiling the sample with acid(25), autoclaving the sample with acid(78), or autoclaving or heating the sample with potassium persulfate(79). This last method is gaining in popularity due to its simplicity and the completeness of digestion of orgainc matter. A differentiation between ortho- and polyphosphate can be achieved by the oxidation of organic matter by high intensity ultraviolet radiation(80). This procedure, which has been extended by Grasshoff(81) to the automated analysis of phosphate, utilizes a high intensity mercury vapor lamp capable of radiating light below 2500 Å. Inorganic and organic polyphosphates are not converted to orthophosphate by this procedure, but organic orthophosphates are converted to inorganic orthophosphate.

Olsen(75) has extensively reviewed the methods for differentiation of phosphorus forms in natural water. Differentiation of the phosphorus components of natural waters is commonly based on size and chemical reactivity. Samples may or may not be filtered; the filter most commonly used is a 0.45-μm membrane filter. The samples may be hydrolyzed or the reactive phosphorus may be determined.

c. Sulfides

Sulfide is a frequent water pollutant which may originate from sewage or industrial wastes of tanneries, paper mills, textile mills, chemical plants, and gas-manufacturing works. Sulfide is also commonly present in anaerobic hypolimnetic waters of lakes and estuaries. Hydrogen sulfide is objectionable due to its toxicity to fish and because of its odor. Waters with high concentrations of hydrogen sulfide are therefore undesirable for drinking water supplies or for use in the food and beverage industries. Since hydrogen sulfide blackens lead paint and copper and brass objects and causes corrosion of concrete, its presence in industrial waste waters is deleterious.

Total sulfides may be determined in the range 0.02–20 mg/liter by a sensitive colorimetric procedure(25). Hydrogen sulfide is allowed to react, under suitable conditions, with p-aminodimethylaniline and ferric chloride to produce methylene blue. The test measures free sulfides as well as acid-soluble sulfides bound to metals such as iron, manganese, and lead. Mercury and copper sulfides are too insoluble to react.

Sulfides in industrial waste effluents can be measured by a titrimetric procedure(25). Sulfide is volatilized from an acidic sample and collected in a zinc acetate solution. After the addition of excess acidified standard iodine solution, the solution is back-titrated with standard thiosulfate solution.

The methods of sulfide analysis described above measure total sulfides, i.e., free sulfide, acid-soluble sulfides, polysulfides, and sulfanes. Free sulfide alone may be measured with an ion-selective electrode(82). The sulfide electrode responds only to S^{2-} in the sample, but the measurement of both S^{2-} and pH permits the concentrations of H_2S, HS^-, and S_T (analytical concentration of free sulfide) to be calculated by the following equilibrium relationships:

$$\log\,[S_T] = \log\,[S^{2-}] + \log\left[\frac{[H^+]^2}{K_1K_2} + \frac{H^+}{K_1} + 1\right] \qquad (9)$$

$$\log [H_2S] = \log [S^{2-}] + \log \frac{[H^+]^2}{K_1 K_2} \tag{10}$$

$$\log [HS^-] = \log [S^{2-}] + \log \frac{[H^+]}{K_1} \tag{11}$$

where K_1 and K_2 are the acidity constants for the weak acid H_2S.

d. Sulfites

Since sulfites are used for the preparation of cellulose from woods, they are commonly found in the waste waters from pulp and paper mills. Sulfites may be analyzed by an iodometric titration procedure subsequent to precipitation and filtration of sulfides as zinc sulfide. This test is subject to interference from organic matter and other reduced materials such as ferrous iron(25).

e. Grease and Oil

Grease and oil have been determined gravimetrically after solvent extraction of the sample and evaporation of the solvent(25). In this procedure the more volatile oils may be lost during the evaporation procedure, and high molecular weight materials may not be extracted with the same efficiency as lower molecular weight material. For purposes of identification of greases or oils present, gas chromatography may be used for separation of components of the mixture and infrared spectroscopy for the identification of their structure(83). One of the greatest problems in the analysis of grease and oils is obtaining a representative sample. These materials are soluble to a very limited extent, and oils may occur as extremely thin films on the surface of a body of water. Not only is the sampling of these systems difficult, but the evaluation of the data in terms of the amount of oil in the environment and the thickness of the film involved is almost impossible.

f. Detergents

Detergents are classified as anionic, cationic, and non-ionic based on their ionization properties in water. Anionic detergents are the most widely used and contribute proportionally more to water pollution problems than do the other detergent types. Cationic detergents are used in products requiring a germicidal capacity, and non-ionic surfactants are used in "controlled suds" detergents.

Anionic surfactants of both the alkyl benzene sulfonate (ABS) and linear alkylate sulfonate (LAS) types can be determined in the methylene blue method and are therefore called methylene blue-active substances (MBAS). In this test a blue-colored salt which is soluble in chloroform is formed when methylene blue reacts with anionic sulfactants. A number of organic and inorganic compounds interfere in the test(25). Positive interferences are due to organic sulfates, sulfonates, and phenols which complex methylene blue, and inorganic chlorides, nitrates, thiocyanates, and cyanates which form ion pairs with methylene blue. Organic compounds, especially amines, also react with methylene blue. This competition results in low values for surfactant content. Confirmation of the presence of surfactants may be made by an infrared procedure.

g. Phenols

Phenols are important pollutants since their presence may produce bad tastes in fish and at higher levels may be toxic to aquatic life. The presence of traces of phenol in domestic water supplies is undesirable due to the objectionable taste resulting from the action of chlorine on phenols. For this reason the standard for phenol in drinking water is 1 μg/liter(26).

Phenolic wastes arise from the distillation of wood, from spent and crude gas liquors, coke oven effluents, gas works, and many chemical industries. The generally used colorimetric procedures do not distinguish between various phenols present in a sample. The standard method for phenol determination, the 4-aminoantipyrine reaction, is applicable to phenols not substituted in the *para* position. Phenols react with 4-aminoantipyrine in basic solution in the presence of ferricyanide to form a chloroform-extractable antipyrine dye. The intensity of the color is a function of the structure of the phenol being analyzed. For example, the relative absorbences of the cresols are o-cresol, 74%; m-cresol, 69%; and p-cresol, 3%, compared to the absorbence of phenol(84).

Gas chromatography has been utilized in the resolution and analysis of complex mixtures of phenolic compounds. Baker and Malo(85) have reviewed the technique of direct aqueous injection of samples. Many phenols could be analyzed at the 1 mg/liter level using flame ionization detection, but separation of the m- and p-cresols and monochlorophenols and certain dichlorophenols was not possible.

h. Cyanides

Cyanides may be found in significant quantities in the effluents from coke ovens, electroplating and metal cleaning, and steel plants. Cyanides are extremely toxic and are responsible for many kills. Toxicity to fish may result from cyanide concentrations as low as 0.05 mg/liter of HCN(27).

Cyanide may be present as "free cyanides," which includes HCN and its conjugate base, CN^-. Stable metal cyanide complexes, such as $K_4Fe(CN)_6$ and $K_3Co(CN)_6$, may be found in waste effluents. Total cyanide is determined by breaking down complexed cyanide and distilling the liberated HCN. The decomposition may be accomplished by the Serfass method(86), in which the sample is distilled with mercuric chloride, magnesium chloride, and sulfuric acid.

The cyanide which is distilled into a solution of base may be determined by the Liebig test(25). This test, which can detect 0.1 mg/liter of cyanide if a 500-ml sample is distilled, is based on the titration of cyanides with silver nitrate to form the soluble cyanide complex, $Ag(CN)_2^-$. After all the cyanide has been complexed and a small excess of silver is present, the solution immediately turns from yellow to salmon due to the reaction of the silver with the indicator p-dimethylaminobenzalrhodanine.

Cyanide can also be determined by colorimetric analysis. The sample is halogenated by reaction with chloramine-T, which converts cyanide to cyanogen chloride, CNCl. After the reaction is complete, a pyridine-pyrazolone reagent is added which forms a blue dye with the CNCl. The absorbence of the aqueous solution may be measured at 620 nm, or the solution may be extracted with n-butyl alcohol and the absorbence read at 630 nm. Extraction improves the detection limit from 0.5 μg of cyanide for aqueous samples to 0.1 μg for extracted samples and increases the precision of the analysis.

i. Pesticides

Pesticides are present in water in extremely low concentrations, generally less than 0.1 μg/liter. Gas chromatography is the usual method employed in the analysis of pesticides in water as well as in sediments and organisms. The electron capture and microcoulometric detectors employed for pesticide analysis are quite sensitive. Direct aqueous measurement is not feasible due to the extremely low concentration of pesticides which may be present and due to the presence of interfering substances.

Liquid–liquid extraction using 15% ethyl ether in a hexane solvent is the preferred technique for the concentration of pesticides(87). After the extraction, the organic phase may be cleaned up by column adsorption chromatography or thin layer chromatography. Column chromatography on silica gel or Florisil columns may be performed using a variety of solvents for the elution of the pesticides. Gas chromatographic conditions (columns, temperature, etc.) suitable for the separation of pesticides are summarized by Faust and Suffet(88) (see Chap. 23).

IV. Selection of Analytical Methods

A. Introduction

After the objectives of the program have been established and the parameters to be measured have been chosen, a suitable method of analysis must be selected. There is no prescribed procedure which is applicable to all situations—the best method for any given situation must be based upon consideration of many factors. Some of the more important factors are summarized in Table 13. The relative importance assumed by each of these factors depends on the objectives of the program. The decision on the best method of analysis ultimately rests with the analyst, with the advice of biologists, sanitary engineers, and other personnel in the program.

TABLE 13
Some Important Factors in the
Selection of Analytical Methods

Required sensitivity
Accuracy of method
Interferences present
Number of samples to be analyzed
Necessity of field or in situ analyses
Rapidity of analysis
Availability of required instruments
Availability of skilled laboratory personnel
Required use of standard or referee methods

In selecting an analytical method the chemist should carefully distinguish between specific and nonspecific methods of analysis and between methods which measure intensive properties and those which measure extensive properties. A number of nonspecific tests used in water quality characterization are listed in Table 5 and have been discussed earlier in this chapter. In the analysis of organic substances the analyst must often choose between specific and nonspecific methods. For example, waste effluents may be analyzed for total carbon (nonspecific) or for specific organic compounds. Even if the analyst chooses the specific analysis, he must distinguish between methods which measure a class of compounds (e.g., the aminoantipyrine method for phenols) and those for specific compounds such as gas chromatographic analysis for individual phenols.

B. Intensive and Extensive Measurements

Physical and chemical parameters for water quality characterization can be categorized conveniently as intensive or extensive measurements. The use of these categories should not be considered in terms of rigorous thermodynamic entities but rather in terms of conceptual quantitative properties of the system under investigation. It is for reasons of convenience, which will become apparent later, that we should differentiate between intensive and extensive water quality parameters.

In textbook terminology extensive properties are additive in the sense that the total value of a system is the sum of the individual values for each of its constituent parts. Conversely, intensive properties are not additive and can be specified for any system without reference to the size of that system. This is illustrated in Table 14.

TABLE 14
Intensive and Extensive Parameters

System designation	Intensive parameter	Extensive parameter	Work done by the system
Gravitational	Height (h) × gravitational acceleration (g), cm × cm/sec^{2-}	Mass (m) g	$mg(h_2 - h_1)$, ergs
Thermal	Temperature (T), °C	Heat capacity (C_p), cal/deg	$C_p(T_2 - T_1)$, cal
Electrical	Potential (E), V	Charge (q), coulombs	$q(E_2 - E_1)$, joules
Chemical	Chemical potential (μ), cal/mole	Number of moles (n)	$n(\mu_2 - \mu_1)$, cal

The chemical potential or the molar free energy change $(\partial G/\partial n)_{T,P}$ is further defined:

$$\mu = \mu^\circ = RT \ln a \qquad (12)$$

where a is the activity given in a molar scale and μ° is the chemical potential at a reference state. The activity can be related empirically to concentration C by the equation:

$$a = \gamma C \qquad (13)$$

where γ is the activity coefficient. Accordingly, the activity, a, is an intensive parameter and is a measure of the difference between the chemical potential in the actual and in the reference state.

In defining a chemical system it is important to distinguish between intensive parameters based on chemical potential measurements and extensive parameters based on counting the number of moles of a given substance. This can easily be illustrated in comparing data from potentiometric measurements of pH, pX, or pM (where H, X, and M refer to hydrogen ions, anions, and cations, respectively) with those from titrimetric determinations of acidity, anions, or cations. In the former case measurement is based on potential determinations which are essentially intensive parameters, while in the latter case measurement is based on stoichiometric calculation. Results of analysis of either type may not agree particularly if interferences are present which may cause the activity coefficient to deviate from unity.

Similarly, in the case of the voltammetric membrane electrode systems, such as the galvanic cell oxygen analyzer(38), the measured parameter is essentially an intensive factor, since the diffusion current is solely dependent on the difference in the chemical potentials of molecular oxygen across the membrane. Accordingly, values derived from measurements with the galvanic cell oxygen analyzer do not have to be equal to values obtained by titration methods for dissolved oxygen, such as the Winkler test. In the former case, the activity of molecular oxygen is the parameter measured, while in the latter case, the total number of oxygen molecules present in the test volume is measured. For the majority of natural and waste waters it is unlikely that the two kinds of measurements will give exactly the same results, although in many applications the differences may be negligible for all practical purposes.

C. Sensitivity and Limits of Detection

Sensitivity and the limit of detection are often the most important considerations in the selection of an analytical method. Sensitivity is the ability to distinguish between small differences in concentration and is measured by the slope of the curve relating the measured signal and the concentration of the species being measured. The concentration of many constituents is near or below the limit of detection of many methods of analysis, thus forcing the analyst to select the most sensitive methods for analyzing a specific constituent or to increase the sensitivity of the method or to increase the concentration of the desired constituent by physical or chemical means.

The limit of detection of an analytical method is the lowest concentration whose signal can be distinguished from the blank signal. This value depends on the sensitivity of the method, as well as on the signal-to-noise ratio required to discern the response due to a sample. Advances in electronics have brought about the design of instruments with greater inherent stability and, therefore, lower limits of detection. Use of an on-line digital computer in fast-sweep derivative polarography has permitted the resolution of closely spaced peaks and has extended the analytical sensitivity of the technique by more than an order of magnitude (89).

The analyst should utilize methods which are sufficiently sensitive that analyses will not be performed near the limit of detection for the method. Typical detection limits for electrochemical methods of analysis are listed in Table 15, and detection limits for instrumental methods are listed in Table 16. As the concentration of the desired constituent approaches the detection limit of the method, the error increases greatly. The error in colorimetric analyses using the standard series method becomes constant when the concentration is greater than approximately 15 times the detection limit (91). There is, however, no need to employ methods of high sensitivity where methods of lower sensitivity would suffice.

TABLE 15

Detection Limits of Electroanalytical Methods of Analysis[a,b]

Method	Detection limit
ac Polarography, chronopotentiometry, potentiometry	10^{-4}–10^{-5} M
Classical polarography, coulometry at controlled potential, precision null-point potentiometry	10^{-5}–10^{-6} M
Derivative polarography, square wave polarography, linear sweep voltammetry	10^{-6}–10^{-7} M
Pulse polarography, amperometry with rotating electrodes	10^{-7}–10^{-8} M
Anodic stripping with hanging mercury drop electrodes	10^{-8}–10^{-9} M
Anodic stripping with thin film electrodes or solid electrodes	10^{-9}–10^{-10} M

[a]From Ref. 90.
[b]For sensitivity of nonelectroanalytical methods, see Table 16.

TABLE 16

Detection Limits of Instrumental Methods of Analysis[a]

Method	Principle	Detection limit
Molecular absorption spectrophotometry	Absorption of radiation by dissolved molecules	10^{-5}–10^{-6} M
Molecular fluorescence spectrophotometry	Re-emission of radiation absorbed by dissolved molecules	10^{-7}–10^{-8} M
Atomic absorption spectrophotometry	Absorption of radiation by free atoms	10^{-6}–10^{-7} M
Atomic fluorescence spectrophotometry	Re-emission of radiation absorbed by free atoms	10^{-7}–$10^{-8}M$
Optical and X-ray spectroscopy	Spectral emission analysis by flame, arc, or X-ray excitation	10^{-5}–10^{-6} M
Neutron activation analysis	Nuclear activation by thermal neutrons	10^{-9}–10^{-10} M

[a]For sensitivity of electroanalytical methods, see Table 15.

Most analytical procedures require the running of a blank to take into account the amount of the desired constituent contributed by the reagents. The minimum concentration which can be detected in a sample must be due to a statistically greater response of the sample than of the blank(92). Therefore, although a reagent blank giving a small response is desirable, a zero response is not necessarily a sign of absence of the desired constituent(91). Such a result may indicate the presence of an interfering substance which has converted the desired constituent to an unreactive form.

The sensitivity of many analytical procedures can be increased by variation in the conditions of analysis or by improvement in instrument or experimental design. In colorimetric analysis sensitivity can be increased by using longer cells. The sensitivity of activation analysis can be increased by irradiation in a higher neutron flux or by irradiating the sample for a longer time. Increasing deposition time for the plating step or increasing the rate of voltage scan in the stripping step will increase the sensitivity of anodic stripping voltammetry(93).

Perhaps the most common method of improving the sensitivity of analyses is through preconcentration of the desired constituent by physical or chemical processes. Inherent in some concentration procedures is a separation of the desired constituent from many possible interferences. Care must be taken in all concentration procedures that the desired constituent is not lost by volatilization, adsorption, ion exchange, or some other process. Ion exchange, adsorption on activated carbon, coprecipitation, extraction, evaporation, freeze drying, and electrodeposition have been extensively used for the concentration of trace constituents. Many of these techniques, as well as limitations of concentration methods, have been reviewed by Mizuike(94) and Minczewski(95) (see Chap. 11).

D. Accuracy

The accuracy required is dictated by the objectives of the program, and the analyst must select a method capable of providing the required accuracy. Accuracy can often be improved by changes in technique or modification of equipment. Replicate determinations reduce the possibility of reporting an erroneously high or low result, but only at the expense of reducing the number of samples which can be analyzed. Analysis of the sample by a second, independent method can also be used to increase or ensure accuracy.

The method of standard additions is used to test the recovery of known amounts of the desired constituent which have been added to the sample. If the expected amount is recovered, more confidence may be placed in the accuracy of the analysis (25).

E. Interferences

1. TYPES

In selecting an analytical method the possibility of interfering substances in the sample must be considered. In some cases the presence of interfering substances will rule out use of a particular method, but generally the analyst can cope with interferences.

Interferences may be either specific or nonspecific. For example, in a colorimetric analysis turbidity or color in the sample would be a nonspecific interference, while a specific interference would be one that reacts in the same manner as the desired constituent. The interference of arsenic in phosphate analyses by the formation of arsenomolybdic acid rather than phosphomolybdic acid is an example of a specific interference.

Some substances can interfere with the sensor system. The oxygen electrode system, when used for in situ analyses, often becomes less sensitive after being in a stream for a period of time due to the accumulation on the membrane of a film of suspended solids which decreases the permeability of the membrane to dissolved oxygen. The presence of hydrogen sulfide also causes an interference in the determination of dissolved oxygen using a membrane electrode. The hydrogen sulfide permeates the membrane and reacts with the electrode, changing its electrochemical characteristics.

Certain substances are interferences because they react with the species under test in an undesired manner. For example, in the presence of aluminum, fluoride is bound according to the reaction

$$6F^- + Al^{3+} \rightarrow [AlF_6]^{3-} \tag{14}$$

In this case, aluminum would be an interference in the measurement of total fluoride by a specific ion electrode.

2. REMOVAL AND COMPENSATION OF INTERFERENCES

Interferences may be removed from the sample or their effect compensated for in a number of ways. Commonly, interfering substances are removed from the sample by physical or chemical separation techniques.

Often, running an appropriate blank allows the investigator to compensate for the effects of interferences. In colorimetric analyses the effect of turbidity or background color may be compensated for by running a sample with distilled water instead of a color-producing reagent and subtracting the absorbence of this blank from that of the sample prepared in the conventional manner.

Samples may contain two or more colored substances whose absorbence spectra overlap. For a two-component system, the absorbence measured at the wavelength of maximum absorption of the desired constituent is due not only to the absorbence of the desired constituent but also to that of the interfering substance. By measuring the absorbence of the sample system at two predetermined wavelengths the concentration of both components can be determined if the molar absorptivities of both components are known at these wavelengths, because the absorbences are additive(96). This principle is commonly used in the measurement of chlorophyll pigments(97). Chlorophyll pigments a, b, and c have absorbence maxima at 665, 645, and 630 nm, respectively. After extraction of the pigments by acetone the absorbence of the solution is measured in 10-cm cells. The equations for determining the concentrations of the chlorophylls are:

$$C \text{ (chlorophyll } a) = 11.6A_{665} - 1.31A_{645} - 0.14A_{630} \tag{15}$$
$$C \text{ (chlorophyll } b) = 20.7A_{645} - 4.34A_{665} - 4.42A_{630} \tag{16}$$
$$C \text{ (chlorophyll } c) = 55A_{630} - 4.64A_{665} - 16.3A_{645} \tag{17}$$

where C and A represent the concentration and absorbence, respectively, for the indicated chlorophyll types.

Ion-selective electrodes have gained wide acceptance for analysis of water and waste water. The response of these electrodes is given by the empirical equation(98):

$$E = \text{constant} + \frac{2.3RT}{zF} \log\left[a + \sum_{i=1}^{n} K_i a_i^{z/z_i}\right] \tag{18}$$

where E is the electrode potential, $2.3RT/F$ is the Nernst factor (59.16 mV at 25°C), z is the charge of the ion being measured, a is the activity of the ion being measured, n is the number of interfering ions, K_i is the selectivity coefficient for the ith interfering ion, a_i is the activity of the ith interfering ion, and z_i is the charge of the ith interfering ion. The equation holds well for low levels of interferences and permits the analyst to calculate whether an ion-sensitive electrode can be used. Selectivity constants for several commonly used electrodes are given in Table 17 (see Chap. 32).

TABLE 17
Selectivity Coefficients for Ion-Selective Electrodes[a]

Ion measured	Selectivity constants
Calcium	Zn^{2+} 3.2, Fe^{2+} 0.80, Pb^{2+} 0.63, Cu^{2+} 0.27, Mg^{2+} 0.01, Na^+ 0.0016
Hardness (divalent cation)	Zn^{2+} 3.5, Fe^{2+} 3.5, Cu^{2+} 3.1, Ni^{2+} 1.35, Ca^{2+} 1.0, Mg^{2+} 1.0, Ba^{2+} 0.94, Na^+ 0.01
Nitrate	I^- 20, Br^- 0.9, S^{2-} 0.57, NO_2^- 0.06, CN^- 0.02, HCO_3^- 0.02, Cl^- 0.006, F^- 0.009, SO_4^{2-}, 0.0006
Chloride	I^- 17, NO_3^- 4.2, Br^- 1.6, OH^- 1.0, HCO_3^- 0.19, SO_4^{2-} 0.14, F^- 0.10

[a]From Ref. 98.

F. Modes of Analysis

1. AUTOMATED ANALYSIS

Many water pollution programs require the analysis of large numbers of samples. The development of automated techniques has made feasible the analysis of these samples using reasonable numbers of laboratory personnel. Besides increasing the number of samples which can be analyzed, automation increases the precision attainable because very reproducible conditions are maintained. The automation of analyses often reduces the number of technicians required in the laboratory, but additional skilled personnel may be required to deal with more sophisticated equipment and to handle and interpret large amounts of data. In the automated laboratory, it is difficult to deal with samples with widely varying compositions since the instrumental techniques employed are usually designed to analyze samples with relatively constant composition where the concentration of the desired constituent does not vary greatly (see Chap. 27).

2. FIELD AND IN SITU ANALYSIS

Many programs require the in situ or field analysis of samples because samples may not be stable or because the quantity of samples required or the speed of data acquisition necessary make it impossible to transport samples to the laboratory. In such cases, the analyst may require relatively simple equipment, such as portable pH, dissolved oxygen, or conductivity meters.

Some programs require analysis for large numbers of parameters in the field. In such cases a mobile laboratory may be necessary, equipped with instrumentation for continuous monitoring of water quality parameters and analysis of discrete samples. Electrical power may have to be generated if the mobile facility is to be used at locations remote from electrical transmission lines. In general, the amount of electrical power available will be less than in permanent installations and this means there must be careful planning of the equipment to be used.

Continuous and in situ methods are being used more extensively in water quality measurements as applicable techniques become available. Continuous analysis is essential if the highest degree of control of operating conditions is to be maintained. To reduce the possibility of contamination of the sample, in situ analysis is desirable. In situ analysis also may provide more accurate data since the physical and chemical form of the desired constituent will not be altered as a result of removing the sample from its natural environment. Many methods of continuous and in situ analysis are included in the review by Blaedel and Laessig (99) (see Chap. 26).

G. Analysis Time

The speed required to obtain analytical data is an important consideration in selection of a method of analysis. Many tests take a long time to complete, which renders their data unusable in the control of treatment facilities or for other purposes. The biochemical oxygen demand test, which requires 5 days for its completion, is not applicable to a situation where the current status must be known. In such a case the analysis of chemical oxygen demand, which takes 2 hr, or total organic carbon, which can be done in minutes, would be better choices. For specific wastes the results of these tests can be related to the biochemical oxygen demand if this information is also needed.

H. Practical Considerations

The availability of required instrumentation is often a deciding factor in the selection of an analytical procedure. If instrumentation for one procedure is available and that for another would have to be purchased, the former procedure will generally be the one selected.

In selecting an analytical method, it must also be determined if there are sufficient laboratory personnel to complete the required analysis and if they possess the necessary skills.

Standard methods of analysis often must be used to satisfy legal requirements and to ensure comparability of data obtained by different laboratories. Compilations of standard or referee methods are available in *Standard Methods*(25), ASTM's manual(35), and in the U.S. Geological Survey's procedures(29).

V. Conclusion

The proper design of measurement systems is the single most important step in any water quality control program. It is essential in defining the type and magnitude of the water quality problem as well as in determining the effectiveness of the control program.

The design of any water quality measurement system is based on three questions: (a) Why is the measurement needed? (b) What are the parameters to look for? and (c) How are these parameters to be measured? In this chapter the authors have discussed the answers to these questions and their applications to various natural and waste waters.

Despite the fact that discussions in this chapter have been entirely concerned with aquatic systems, the basic concepts in the design of aquatic measurement systems are applicable to air and soil media. It is becoming more apparent, as time goes on, that air, water, and soil quality problems must not be dealt with separately but rather as interrelated parts of unified environmental quality models. In order to define quality characteristics of an environmental segment, the measurement program should be designed to determine physicochemical and biochemical transformations across environmental boundaries of the atmosphere, hydrosphere, lithosphere, and biosphere. Attempts should be made to coordinate and interrelate air, water, and soil quality monitoring systems where measurement data are fed into a single computer modeling program.

Insofar as methodology and techniques of measurement are concerned, an overview of classical and advanced analytical procedures has been presented. Emphasis has been placed on the utility and need for in situ and automated measurement systems. Environmental quality control programs will tend to rely more heavily on remote in situ and automatic monitoring systems. There is a pressing need for reliable sensor systems to detect various environmental parameters. This seems to be the most serious limiting step in all present air and water quality monitoring programs.

Acknowledgment

We wish to acknowledge the assistance of Deena Allen in editing the manuscript.

REFERENCES

1. *Pollution of Lake Erie, Lake Ontario and the International Section of the St. Lawrence River*, Vol. I, *Summary*, International Joint Commission, Washington, D.C., 1969.
2. K. H. Mancy and W. J. Weber, Jr., in *Treatise on Analytical Chemistry* Vol. 2, Part III (I. M. Kolthoff and P. J. Elving, eds.), Wiley (Interscience), New York (in press).
3. A. M. Beeton, *Limnol. Oceanog.*, **10**, 240 (1965).
4. Annual records of chemical quality, suspended sediment and water temperature, published in six volumes annually in the U.S. Geological Survey Water Supply Paper Series; for example, for the water year 1964, *Water Supply Papers 1964–1969*.
5. S. K. Love, *Quality of Surface Waters for Irrigation, Western States, 1963, Water Supply Paper 1952*, U.S. Geological Survey, Washington, D.C., 1967.
6. E. W. Lohr and S. K. Love, *The Industrial Utility of Public Water Supplies in the United States, 1952*, Parts I and II, *Water Supply Papers 1922 and 1300*, U.S. Geological Survey, Washington, D.C., 1954.
7. Annual compilation of data from the Federal Water Pollution Control Administration Water Quality Surveillance Program; for example, *Water Pollution Surveillance System Annual Compilation of Data, October 1962–September 1963, Publ. 663* (revised 1963 ed.), U.S. Public Health Service, Washington, D.C.
8. J. F. Kopp and R. C. Kroner, *Trace Metals in Waters of the United States*, Federal Water Pollution Control Administration, Cincinnati.
9. *Water Quality Criteria: Report of the National Advisory Committee to the Secretary of the Interior*, Washington, D.C., April 1, 1968.
10. R. C. Kroner, in *Advances in Automated Analysis, Technicon International Congress 1969*, Vol. II, Mediad, New York, 1970, p. 129.
11. L. G. Sillen, in *Oceanography* (M. Sears, ed.), *Publ. No. 67*, A.A.A.S., Washington, D.C., 1961, p. 549.
12. R. M. Garrels and C. L. Christ, *Solutions, Minerals and Equilibria*, Harper and Row, New York, 1965.
13. W. Stumm, in *Symposium: Environmental Measurements, Publ. 999-WP-15*, U.S. Public Health Service, Washington, D.C., 1964, p. 229.
14. J. R. Kramer, in *Equilibrium Concepts in Natural Water Systems* (R. F. Gould, ed.), *Advances in Chemistry Series, No. 67*, Washington, D.C., 1967, p. 243.
15. H. J. M. Bowen, *Trace Elements in Biochemistry*, Academic, New York, 1966.
16. S. Katoh, *Nature*, **186**, 533 (1960).
17. C. R. Goldman, in *Primary Productivity in Aquatic Environments, Mem. Ist. Ital. Idrobiol.*, 18 Suppl. (C. R. Goldman, ed.), University of California Press, Berkeley, Calif., 1965, p. 121.
18. C. R. Goldman, in *Chemical Environment in the Aquatic Habitat* (H. L. Golterman and R. S. Clymo, eds.), N. V. Noord-Hollandsche Uitgevers Maatschappij, Amsterdam, 1967, p. 229.
19. C. R. Goldman, *Verh. Intern. Ver. Limnol.*, **15**, 365 (1964).
20. J. Shapiro, in *Chemical Environment in the Aquatic Habitat*, (H. L. Golterman and R. S. Clymo, eds.), N. V. Noord-Hollandsche Uitgevers Maatschappij, Amsterdam, 1967, p. 202.

21. J. Shapiro, in *Chemical Environment in the Aquatic Habitat*, (H. L. Golterman and R. S. Clymo, eds.), N. V. Noord-Hollandsche Uitgevers Maatschappij, Amsterdam, 1967, p. 219.
22. C. Forch, M. Knudsen, and S. P. Sorensen, *Kgl. Danske Videnskab. Selskab Skr., Raekke, Naturv., Mat.*, **6**, Afd. XII, 1, 151 (1902).
23. W. S. Wooster, A. J. Lee, and G. Dietrich, *Limnol. Oceanog.*, **14**, 437 (1969).
24. R. A. Cox, F. Culkin, and J. P. Riley, *Deep-Sea Res.*, **14**, 203 (1967).
25. *Standard Methods for the Examination of Water and Wastewater*, 12th ed., American Public Health Association, New York, 1965.
26. *Public Health Service Drinking Water Standards*, *Publ. 956*, U.S. Public Health Service, Washington, D.C., 1962.
27. J. E. McKee and H. W. Wolf, *Water Quality Criteria*, *Publ. No. 3-A*, 2nd ed., State Water Quality Control Board, Sacramento, 1963.
28. Betz Laboratories, Inc., *Betz Handbook of Industrial Water Conditioning*, 6th ed., Philadelphia, 1962, p. 425.
29. F. H. Rainwater and L. L. Thatcher, *Methods for Collection and Analysis of Water Samples*, *Water Supply Paper 1454*, U.S. Geological Survey, Washington, D.C., 1960.
30. A. Cox, in *Chemical Oceanography* (J. P. Riley and G. Skirrow, eds.), Academic, New York, 1965, Chap. 3.
31. S. H. Smith, *Limnol. Oceanog.*, **7**, 330 (1962).
32. C. N. Durfor and E. Becker, *Public Water Supplies of the 100 Largest Cities in the United States, 1962*, *Water Supply Paper 1812*, U.S. Geological Survey, Washington, D.C., 1964.
33. C. E. Van Hall, J. Safranko, and V. A. Stenger, *Anal. Chem.*, **35**, 315 (1963).
34. V. A. Stenger and C. E. Van Hall, *ibid.*, **39**, 206 (1967).
35. *1970 Annual Book of ASTM Standards*, Part 23, *Water; Atmospheric Analysis*, American Society for Testing Materials, Philadelphia, 1970.
36. M. G. Mellon, *Analytical Absorption Spectroscopy*, Wiley, New York, 1950, Chap. 9.
37. A. P. Black and S. A. Hannah, *J. Am. Water Works Assoc.*, **57**, 901 (1965).
38. K. H. Mancy and T. Jaffee, *Analysis of Dissolved Oxygen in Natural and Waste Waters* *Publ. 999-WP-37*, U.S. Public Health Service, Cincinnati, 1966.
39. J. R. Kramer, H. E. Allen, G. W. Baulne, and N. M. Burns, *Lake Erie Time Study (LETS)*, Canada Centre for Inland Waters Paper No. 4, Burlington, Ontario, 1970.
40. J. F. Carr, *Proc. 5th Conf. Great Lakes Res.*, *Publ. No. 9*, Great Lakes Res. Div., University of Michigan, 1962, p. 1.
41. C. H. Mortimer, *J. Ecol.*, **29**, 280 (1941).
42. *Ibid.*, **30**, 147 (1942).
43. *Environ. Sci. Tech.*, **1**, 212 (1967).
44. C. Potos, *Proc. 11th Conf. Great Lakes Res.*, 571 (1968).
45. H. W. Streeter and E. B. Phelps, *A Study of the Pollution and Natural Purification of the Ohio River*, *Public Health Bull. 146*, Washington, D.C., 1925.
46. A. M. Beeton, *Trans. Am. Fisheries Soc.*, **90**, 153 (1961).
47. W. T. Edmondson, G. C. Anderson, and D. P. Peterson, *Limnol. Oceanog.*, **1**, 47 (1956).
48. E. G. Fruh, *J. Water Pollution Control Federation*, **39**, 1449 (1967).
49. K. M. Stewart and G. A. Rohlich, *Eutrophication—A Review*, *Publ. No. 34*, California State Water Resources Control Board, 1967.
50. *Eutrophication: Causes, Consequences, Correctives*, National Academy of Science, Washington, D.C., 1969.
51. R. E. Ogawa and J. F. Carr, *Limnol. Oceanog.*, **14**, 342 (1969).

52. R. B. Corey, A. D. Hasler, G. F. Lee, F. H. Schraufnagel, and T. L. Wirth, *Excessive Water Fertilization, Report to the Water Subcommittee*, Natural Resources Committee of State Agencies, Madison, Wis., 1967.

53. Committee Report, *J. Am. Water Works Assoc.*, **62**, 127 (1970).

54. Task Group 2610-P Report, *ibid.*, **59**, 344 (1967).

55. H. E. Allen and J. F. Carr, Division of Water, Air, and Waste Chemistry, 157th Meeting, American Chemical Society, Minneapolis, Minn., April, 1969.

56. A. C. Docherty, paper presented at Technicon International Symposium, New York, 1964.

57. H. E. Allen and L. Myers, unpublished work, 1969.

58. B. F. Rider and M. G. Mellon, *Ind. Eng. Chem., Anal. Ed.*, **18**, 96 (1946).

59. M. C. Shinn, *ibid.*, **13**, 33 (1941).

60. K. Bendschneider and R. J. Robinson, *J. Marine Res.*, **11**, 1 (1952).

61. E. Sawicki, T. W. Stanley, J. Pfaff, and A. D. Amico, *Talanta*, **10**, 641 (1963).

62. D. F. Martin, *Marine Chemistry*, Vol. I, Dekker, New York, Chap. 13.

63. D. Jenkins and L. L. Medsker, *Anal. Chem.*, **36**, 610 (1964).

64. R. Navone, *J. Am. Water Works Assoc.*, **59**, 1193 (1967).

65. L. Klein, *River Pollution: I. Chemical Analysis*, Butterworths, London, 1959, Chap. 5.

66. J. B. Mullin and J. P. Riley, *Anal. Chim. Acta*, **12**, 464 (1955).

67. G. P. Edwards, J. P. Pfafflin, L. H. Schwartz, and P. M. Lauren, *J. Water Pollution Control Federation*, **34**, 1112 (1962).

68. A. W. Morris and J. P. Riley, *Anal. Chim. Acta*, **29**, 272 (1963).

69. E. D. Wood, F. A. J. Armstrong, and F. A. Richards, *J. Marine Biol. Assoc. U.K.*, **47**, 47 (1967).

70. L. J. Kamphake, S. A. Hannah, and J. M. Cohen, *Water Res.*, **1**, 205 (1967).

71. M. Bernhard and G. Macchi, in *Automation in Analytical Chemistry, Technicon Symposia 1965*, Mediad, New York, 1966, p. 255.

72. P. G. Brewer and J. P. Riley, *Deep-Sea Res.*, **12**, 765 (1965).

73. J. D. H. Strickland and T. R. Parsons, *A Manual of Sea Water Analysis, Bull. No. 167*, Fisheries Research Board, Canada, 1968.

74. F. H. Rigler, *Limnol. Oceanog.*, **13**, 7 (1968).

75. S. Olsen, in *Chemical Environment in the Aquatic Habitat* (H. L. Golterman and R. S. Clymo, eds.), N. V. Noord-Hollandsche Uitgevers Maatschappij, Amsterdam, 1967, p. 63.

76. P. W. Jones, *J. Marine Biol. Assoc. U.K.*, **46**, 19 (1966).

77. J. Murphy and J. P. Riley, *Anal. Chim. Acta*, **27**, 31 (1962).

78. H. W. Harvey, *J. Marine Biol. Assoc. U.K.*, **27**, 337 (1948).

79. D. W. Menzel and D. Corwin, *Limnol. Oceanog.*, **10**, 280 (1965).

80. F. A. J. Armstrong, P. M. Williams, and J. D. H. Strickland, *Nature*, **211**, 481 (1966).

81. K. Grasshoff, in *Automation in Analytical Chemistry, Technicon Symposia 1966*, Vol. I, Mediad, New York, 1967, p. 573.

82. Y. Demirijian and K. H. Mancy, Symposium on the Applications of Selective Ion Electrodes, Pittsburgh Conference on Analytical Chemistry and Applied Spectroscopy, Cleveland, Ohio, April, 1969.

83. F. K. Kawahara, *Environ. Sci. Technol.*, **3**, 150 (1969).

84. M. Dannis, *Sewage Ind. Wastes*, **23**, 1516 (1951).

85. R. A. Baker and B. A. Malo, *Environ. Sci. Technol.*, **1**, 997 (1967).

86. E. J. Serfass, R. B. Freeman, B. F. Dodge, and W. Zabban, *Plating*, **39**, 267 (1952).

87. FWPCA Method for Chlorinated Hydrocarbon Pesticides in Water and Wastewater, Federal Water Pollution Control Administration, Cincinnati, April, 1969.

88. S. D. Faust and I. H. Suffet, in *Microorganic Matter in Water, Spec. Tech. Publ. 448*, American Society for Testing and Materials, Philadelphia, 1969, p. 24.
89. S. P. Perone, J. E. Harrar, F. B. Stephens, and R. E. Anderson, *Anal. Chem.*, **40**, 899 (1968).
90. H. A. Laitinen, in *Trace Characterization: Chemical and Physical* (W. W. Meinke and B. F. Scribner, eds.), *Monograph 100*, National Bureau of Standards, Washington, D.C., 1967, p. 75.
91. E. B. Sandell, *Colorimetric Determination of Traces of Metals*, Wiley (Interscience), New York, 1959, p. 82.
92. G. H. Morrison and R. K. Skogerboe, in *Trace Analysis: Physical Methods* (G. H. Morrison, ed.), (Interscience), New York, 1965, p. 1.
93. H. E. Allen, W. R. Matson, and K. H. Mancy, *J. Water Pollution Control Federation*, **42**, 573 (1970).
94. A. Mizuike, in *Trace Analysis: Physical Methods* (G. H. Morrison, ed.), Wiley (Interscience), New York, 1965, p. 103.
95. J. Minczewski, in *Trace Characterization: Chemical and Physical* (W. W. Meinke and B. F. Scribner, eds.), *Monograph 100*, National Bureau of Standards, Washington, D.C., 1967, p. 385.
96. H. H. Willard, L. L. Merritt, Jr., and J. A. Dean, *Instrumental Methods of Analysis*, 4th ed., Van Nostrand, Princeton, N.J., 1965, p. 94.
97. T. R. Parsons and J. D. H. Strickland, *J. Marine Res.*, **21**, 155 (1963).
98. J. W. Ross, Jr., in *Ion-Selective Electrodes* (R. A. Durst, ed.), *Spec. Publ. 314*, National Bureau of Standards, Washington, D.C., 1969, p. 57.
99. W. J. Blaedel and R. H. Laessig, in *Advances in Analytical Chemistry and Instrumentation* (C. N. Reilley and F. W. McLafferty, eds.), Vol. 5, Wiley (Interscience), New York, 1966, p. 69.

Chapter **20** **Organic Analytical Chemistry in Aqueous Systems**

Joseph V. Hunter and David A. Rickert*
DEPARTMENT OF ENVIRONMENTAL SCIENCES
RUTGERS–THE STATE UNIVERSITY
NEW BRUNSWICK, NEW JERSEY

*Present address: Urban Water Program, Water Resources Division, U.S. Geological Service, Washington, D.C.

I. Introduction

Almost all natural waters contain measurable concentrations of organic matter. The organic materials arise through the natural processes of biological synthesis and degradation, and through the various activities of man. Whatever the source, there has been increasing attention paid to the problems associated with these materials, resulting in increased interest in the methods employed in their identification and estimation.

In certain instances, the concern is about the presence of specific organic compounds which have entered the aquatic environment and which are believed to be actual or potential causes of problems. For example, there has been concern about water foaming caused by the presence of alkyl benzene sulfonates(1), and about the potential toxic, accumulative effects of herbicides and insecticides(2).

Although the concentrations of these compounds are low, and considerable potential interferences are present, the analytical problem is simplified in that the nature and properties of the compounds are known. This knowledge greatly simplifies the selection of suitable concentration, separation, and estimation procedures. Though considerable effort may be required in optimizing the procedures, there is an excellent chance of ultimate success. This is witnessed by the abundance of information on the presence and distribution of ABS and pesticides in the aquatic environment.

Another area of major interest is the nature of the organic materials associated with obvious problems of a less defined nature. Typical of these problems are color(3) and taste and odor(4). There are fairly well-established techniques available to delineate the nature and intensity of such problems. However, if a more detailed investigation is desired to delineate what types of organic materials contribute to these problems, major analytical difficulties are encountered.

There is also concern over the presence of organic substances in the environment which are not presently associated with known problems. Such concern has reflected itself in a recommendation for organic matter limits in potable waters(5), and also in the examination for carcinogenic properties of organic materials recovered from natural waters(6). In both cases, for optimum information, the analyst would be required to identify and estimate all the organic compounds in a dilute aqueous solution and/or suspension. This would be an impossible task, since little is known about the nature or total number of the different molecular species which exist in natural waters.

There are a number of approaches to this problem. One is to guess from previous experience and data what organic compounds might be present, and to then examine the mixture for these. Another is to divide the mixture into different physical and chemical groupings and then apply the first method to the subgroupings(7). In many cases, classes of materials, rather than the individual compounds, are detected and estimated. Typical examples of such classes are organic acids, hexoses, pentoses, and amino acids.

The aquatic chemist is thus faced with problems arising from the physical and chemical complexity of the aquatic environment, the low concentrations of organic materials present, and the difficulty sometimes encountered in interpreting the results of organic analysis. Despite these difficulties, a fair amount of work has been done, revealing considerable information on the molecular organic composition of the aquatic environment.

The search for new information, however, continues to intensify, and it is the purpose of this chapter to note and discuss the tools now available to aid the analyst in this search.

II. Preliminary Fractionation

A. Particulate Fractionation

Organic materials in the aquatic environment occur as substances soluble in water and also as colloidal and larger-sized particulates. In both the soluble and particulate forms there are many thousands of organic compounds, as substantiated by the comprehensive reviews of Vallentyne (8) and Lee(9). Accompanying the organic materials in both phases are numerous inorganic substances.

Because of the chemical complexity of the aquatic environment, procedures which physically fractionate and define constituent organic materials greatly simplify the analysis. In attempting to detect and estimate a specific organic compound, these procedures may be used to establish the distribution of the compound among the liquid and solid phases.

Such procedures should keep losses of organic materials at a minimum and should cause no change in the structural makeup of the organic compounds. This also implies that the procedures should be rapid and performed in a manner that negates biological alteration of the compounds. As further analyses are usually made on the separated materials, these procedures are essentially preparative in nature and should be capable of dealing with relatively large volumes.

Most separation procedures are designed to remove colloidal and larger-sized particulates from samples, leaving bulk residuals consisting of the soluble materials. The methods generally used are sedimentation and filtration or a combination of the two. In addition to separating particulate from soluble materials, several procedures also provide for stepwise size fractionation of the particulate materials (Table 1).

TABLE 1
Size Classification for Solids
Fractionated from Waste
Waters[a]

Fraction	Size
Settleable solids	$> 100 \, \mu m$
Supracolloidal solids	$1–100 \, \mu m$
Colloidal solids	$1 \, nm–1 \, \mu m$
Soluble solids	$< 1 \, nm$

[a]From Refs. *11, 13*.

The usual preliminary step is sedimentation under normal gravity for periods of approximately 1 hr. This procedure removes gross particulates, down to $100 \, \mu m$ in size, and has been employed in many waste water analytical systems because it is simple, rapid, and easily controlled (*10–13*).

After preliminary sedimentation, centrifugation or combinations of centrifugation and filtration have been employed to remove particulates smaller than $100 \, \mu m$. Centrifugation has been employed by a number of investigators to remove particulates down to $1 \, \mu m$ in size.

Removal of particulates smaller than $1 \, \mu m$ to obtain a "soluble" fraction has been accomplished by candle filtration (*12*), cellulose membrane filtration (*11*), and high speed centrifugation (*13*). Colloid scientists generally place the soluble-colloidal boundary at 1 nm. The degree to which this boundary is approached by a separation has been investigated for several procedures(*14*). In all other procedures the terms soluble and colloidal are defined entirely by the technique of separation, and their relationship to absolute size limits is unknown.

It must also be noted that filtration and centrifugation remove particulates by different mechanisms, and it is therefore doubtful that they represent equivalent procedures. A marked difference between these procedures was indicated in studies of the electromobility of sewage particulates(*15*). It was found that when the particulates were divided into size fractions by cellulose acetate membrane filtration, there was no trend in the mean electromobility values. However, using high speed centrifugation, the electromobility values increased as the particle size decreased.

During separation by filtration or centrifugation the particulates are concentrated on a surface and must be removed for further analysis. For this reason, losses of materials occur during this separation process(*16*). The situation is complicated by the fact that such losses may be selective for hydrophobic materials such as fats and oils.

Another method of removing particulates from water is chemical coagulation using iron or aluminum salts. Although such techniques obviously do remove particulates down to the colloidal range, there is evidence that the chemical nature of the removed organic materials is different from that of physically removed organic material(*7*) (see Chap. 12).

B. Soluble Fractionation

Upon removal of particulate matter from a water, a very dilute solution of soluble materials is left behind. In the past, the soluble materials have been analyzed as a group, sometimes after preliminary concentration(*11, 17,18*). It is probable, considering the complexity of the aquatic environment, that a better approach would be the further separation of soluble organic materials into various fractions based on their molecular weight–size relationships. Unlike the separation and segregation of particulates, little has been done in this area. Two methods which show great promise, however, are gel (molecular sieve) filtration and hanging curtain electrophoresis. In the former, fractionation occurs primarily as a function of size and shape; in the latter, primarily on the basis of shape and charge-to-mass ratio. Since the techniques work on different principles, the nature of the fractions obtained should be different.

Gel filtration was recently used to divide organic matter in natural waters into as many as 10 fractions. The apparent molecular weight limits of the fractions ranged from <700 to $>200,000$, and the apparent molecular radii from 10^{-8} to 10^{-6} cm. All fractionation was carried out after removal of gross particulates and preliminary concentration. Loss of organic matter on the molecular sieve columns was observed (19).

It appears that hanging curtain electrophoresis has yet to be used for fractionation of organic materials in natural waters. The technique has been applied to the study of organic matter in soils (20), and its use for similar studies with aqueous systems seems imminent.

Thus, there are techniques available for the segregation of aquatic organic matter into many particulate and soluble fractions. It is difficult to specify in advance how many procedures should be used in an investigation, for this is a function of the purpose and extent of the given study. Even though separation procedures may be time-consuming and may lead to losses of organic matter, they greatly simplify analytic interpretation and provide additional insight into the distribution of organic materials.

III. Concentration and Separation

A. Concentration Procedures

As noted in the previous section, physical segregation procedures usually result in concentrated particulate suspensions and a dilute solution of soluble organic materials. Further study of the soluble materials usually requires preliminary concentration. The two general methods of concentration involve the removal of water from the liquid phase. In one method the water is removed in vapor form, in the other as a solid.

Water vapor is usually removed from the liquid aqueous system under reduced pressure at temperatures substantially lower than the boiling point of water. This has been done for both natural waters (21) and waste waters (7). A variation of this general method is the removal of water vapor from the solid (frozen) aqueous system, again under reduced pressure. This process, called lyophilization, has been employed for waste water samples (7).

Both of these methods result in the loss of volatile organic materials such as low molecular weight (1–3 carbon atoms) alcohols, ketones, and acids. Due to the principles involved, there is less chance of codistillation losses during the lyophilization process. In addition, lyophilization uses very low temperatures, yielding less chance for reactions to occur during the concentration procedure, and it results in fine powders which are simple to store and composite.

Concentration through the removal of water as ice crystals has also been investigated(22,23). It was noted in one of these studies that there was selective loss of an organic material (dye) to the solid phase, which did not occur with inorganic materials. This suggests that this method may also be selective, perhaps toward the loss of organic materials that tend to concentrate on surfaces.

B. Separation Procedures

Many physicochemical processes have been used to separate organic materials from the aquatic environment and from each other. From one standpoint, these processes also have concentration functions, as they frequently result in a more concentrated solution of organic materials or facilitate such concentrations. These processes are highly selective in nature and therefore do not yield similar results. Due to the place such processes occupy in the analytical picture, this may be a major advantage.

Typical of such procedures are partition and, of course, adsorption and partition in the form of their respective chromatographic procedures. Due to their selectivity, these processes are of great value when a specific organic compound is to be detected and estimated in the aquatic environment.

Organic materials have been removed from the aquatic environment on activated carbon adsorption columns(24,25). Studies using this technique have resulted in suggested standards for organic matter in potable water(5). Once adsorbed, the organic matter must be removed for study, and extraction of the organic materials with different solvents may give different results(24). The selectivity of adsorption on activated carbon has also been demonstrated(26).

Partition processes have been used in isolating colored materials from natural waters(21) and in separation of phenols(27), pesticides(28), and organic acids(29). Partition processes have also been used in the division of solvent soluble organic materials into classes, such as acids, bases, neutrals, and amphoterics, through pH adjustments of the aqueous phase (24). The selectivity aspects of solvent extraction of organic matter from the aquatic environment are obvious. However, note must be taken of the possibility of reactions occurring between solvent and extracted organic materials, or among the extracted organic materials themselves. This is especially true in the pH classification system noted, and during the evaporation of solvents to concentrate the organic materials.

Distillation has been employed in the removal of volatile (low molecular weight) organic acids and phenols(30) before determination. Due to its limited nature, however, this process is not widely employed in the separation of aqueous organic materials.

Precipitation has been used to remove humic materials through pH adjustment(31,32), or high molecular polysaccharides through water-soluble organic solvent addition(33). This is also a very limited type of separation process and is used only in certain specific areas.

Chromatographic procedures have been widely used to separate organic materials found in the aquatic environment. The major advantage of chromatographic procedures lies in their ability to separate organic materials that have very similar properties. Thus, they are frequently employed after preliminary separations, especially following partition. Specific applications of chromatographic procedures are discussed in a later section.

It can be seen that there are many methods available for the separation and concentration of organic materials in aqueous systems. These methods can be used in the analysis of specific organic compounds or in general separation of organic materials of unknown composition into groups according to chemical and/or physical behavioral characteristics. Especially in the latter case it is important to note that concentration and separation yield selective losses. Furthermore, such processing may cause reactions which yield organic materials that are by-products of the procedure and do not represent the materials present in the initial sample. Despite these reservations, present knowledge renders such procedures essential in any attempt to elucidate the organic composition of the aquatic environment (see Chap. 11).

IV. Basic Structural Considerations in Qualitative Organic Analysis

Identification of an organic compound requires a knowledge of the identity and structural peculiarities of the compound's atomic makeup. Structural peculiarities occur in organic molecules due to the directive nature of the covalent bond. The elucidation of organic structure involves the determination of the spatial geometry and atomic order of molecules. This effort constitutes a major portion of the work in organic analytical chemistry.

The determination of the molecular formula is thus only the first step in qualitative organic analysis. Even for simple hydrocarbons of the general formula C_nH_{2n+2}, the formula C_4H_{10} could represent

$$CH_3-CH_2-CH_2-CH_3 \quad \text{or} \quad CH_3-CH-CH_3 \atop | \atop CH_3$$

For a hydrocarbon of the formula $C_{20}H_{42}$, there would be 366,319 such "structural" isomers (34).

For unsaturated hydrocarbons, the formula C_4H_6 could represent

$$CH_2\!\!=\!\!CH\!\!-\!\!CH\!\!=\!\!CH_2 \quad \text{or} \quad H\!\!-\!\!C\!\!\equiv\!\!C\!\!-\!\!CH_2\!\!-\!\!CH_3$$

and the formula C_6H_{12} could be

$$CH_3(CH_2)_3CH\!\!=\!\!CH_2 \quad \text{or} \quad CH_2(CH_2)_4CH_2$$

When another element such as oxygen is added, the picture may become even more complex. For example, C_3H_8O could represent three isomers,

$$CH_3\!\!-\!\!CH_2\!\!-\!\!OCH_3, \quad CH_3\!\!-\!\!CH_2\!\!-\!\!CH_2\!\!-\!\!OH, \quad \text{or} \quad CH_3\!\!-\!\!CHOH\!\!-\!\!CH_3$$

while C_2H_4O could be

$$\begin{array}{c} H_2C\!\!-\!\!CH_2 \\ \diagdown\!\diagup \\ O \end{array} \quad \text{or} \quad CH_3\!\!-\!\!CH\!\!=\!\!O$$

Such differentiations are of utmost importance, since configuration determines to a considerable degree the properties exhibited by the molecule.

Another form of isomerization is the reversible migration of an atom between two or more sites in an organic molecule. The classic example of this "tautomerism" is that of the keto-enol type:

$$\begin{array}{ccc} \overset{O}{\underset{\parallel}{}} & & \overset{OH}{\underset{|}{}} \\ -CH_2\!\!-\!\!C\!\!- & \longleftrightarrow & -C\!\!=\!\!C\!\!- \\ \text{keto} & & \text{enol} \end{array} \quad (1)$$

Depending on the nature of the neighboring groups, one form or the other may be highly favored. The general form of such tautomers is

$$\begin{array}{ccc} (A)\!\!-\!\!C\!\!- & & (A)\!\!=\!\!C\!\!- \\ |\quad\parallel & \longleftrightarrow & | \\ H\ (B) & & (B)H \end{array} \quad (2)$$

where (A) usually is carbon or nitrogen, and (B), oxygen or sulfur.

In ethylene, all the atoms lie in a single plane and there is restricted rotation about the carbon–carbon double bond. As a consequence, "geometric" isomerism occurs in disubstituted ethylenes of the formula

CHX=CHX. The two general forms are

	and	
Cis		*Trans*

where *cis* isomers have the hydrogens on the same side, and *trans* isomers have them on opposite sides.

The number of such isomers that have the same basic structure increases with the degree of unsaturation. Geometric isomers also occur, but to a much lesser degree, in the chemistry of compounds having carbon–nitrogen and nitrogen–nitrogen double bonds.

Consideration of isomerism leads to one of the most important concepts of organic chemistry, that of the "functional group." A functional group may be defined as "a combination of elements within a molecule — which behave and react as a unit"(*35*). This concept is of great importance, because the chemical and physical behavior of organic molecules is largely determined by the functional groups they contain. Thus, each isomer of the formula C_3H_8O contains a different radical, and the resultant properties of the three compounds are vastly different.

A molecule frequently contains more than one functional group. Here, however, the properties of the compounds are not simply the sum of the group properties, but are determined to a considerable extent by neighboring group effects. Such effects result from steric factors, inductance, resonance, tautomerization, etc. In certain cases, these effects can lead to a completely new behavioral pattern. For example, if a hydroxy (OH) and a carbonyl (C=O) group combine, the resultant carboxylic group (COOH) has a set of properties so different that it is classified as a distinct functional group. In most cases, however, the result is more subtle and merely leads to deviations from the general rules of behavior. A list of some of the more familiar functional groups is found in Table 2.

No single method will reveal the basic formula and structure of an organic molecule. There are, however, a large number of physical and chemical techniques that can be used to obtain parts of the required information. These techniques are not equivalent from the standpoint of theory, in practice, or even in the types of information they reveal. Some are exploratory, others confirmatory. Which and how many are used is usually decided through a compromise between practicability and theoretical considerations. These techniques are employed by the aquatic

TABLE 2
Common Functional Groups[a]

Elements	Group	Description
C, H	—CH$_3$	Methyl
	—CH$_2$—	Methylene
	—CH=CH—	Ethylenic
	—C≡C—	Acetylenic
	—ϕ	Phenyl
	—CH$_2$—ϕ	Benzyl
C, H, O	$\overset{\text{O}}{\overset{\|}{-\text{CH}}}$	Carbonyl (aldehyde)
	$\overset{\text{O}}{\overset{\|}{-\text{C}-}}$	Carbonyl (ketone)
	$\overset{\text{O}}{\overset{\|}{-\text{COH}}}$	Carboxyl
	$\overset{\text{O}}{\overset{\|}{-\text{COR}}}$	Ester
	$\overset{\text{O O}}{\overset{\| \|}{-\text{COC}-}}$	Anhydride
	(Aliphatic)$\overset{\text{O}}{\overset{\|}{-\text{CO}-}}$	Acyl
	(Aromatic)$\overset{\text{O}}{\overset{\|}{-\text{CO}-}}$	Aroyl
	—OH	Hydroxy
	—OOH	Peroxy
	—OCH$_3$	Methoxy
	—OC$_2$H$_5$	Ethoxy
	—HC—CH$_2$ (epoxide ring with O)	Epoxy
C, H, N	—NH$_2$	Amine (primary)
	＞N—H	Amine (secondary)
	＞N—	Amine (tertiary)
	=NH	Imino
	—C≡N	Nitrile
	—$\overset{+}{\text{N}}$≡C$^-$	Isonitrile
	—N=N—	Azo
	—NHNH—	Hydrazo
	N (pyridine ring)	Pyridyl

TABLE 2 – *continued*

Elements	Group	Description
	$RN^{+}\langle\bigcirc\rangle$	Pyridinium
C, H, S	—SH	Thiol
	$\overset{S}{\underset{\parallel}{—CH}}$	Thial
	$\overset{S}{\underset{\parallel}{—C—}}$	Thione
	—S—	Thioether
	—S—S—	Dithio
	$\overset{S}{\underset{\parallel}{—C—SH}}$	Dithionic acid
C, H, P	$\rangle P—$	Phosphine
	$\rangle P\langle$	Phosphorane
C, H, O, N	—O—C≡N	Cyanate
	—N=C=O	Isocyanate
	—N=O	Nitroso
	$—NO_2$	Nitro
	$—ONO_2$	Nitrate
	—ONO	Nitrite
	$\rangle C=NOH$	Oxime
	$\overset{O}{\underset{\parallel}{—CNH_2}}$	Amide
	$\overset{O\quad O}{\underset{\parallel\quad\parallel}{—CHNC—}}$	Imide
	$\rangle SO_2$	Sulfonyl
	$\rangle SO$	Sulfinyl
	SO_4^{2-}	Sulfate
	$\overset{O}{RSO—}$	Sulfonic
	$\overset{O}{SO_3^{2-}}$	Sulfite
	$\overset{O}{\underset{\parallel}{—CSH}}$	Thiolic acid

TABLE 2 — *continued*

Elements	Group	Description
	$\overset{\displaystyle S}{\underset{\displaystyle\parallel}{}}$ —COH	Thionic acid
C, H, O, P	$\overset{O}{\underset{O}{R-PO-}}$	Phosphonic
	$\overset{O}{\underset{H}{RPO-}}$	Phosphinic
	PO_4^{3-}	Phosphate
	PO_3^{3-}	Phosphite
C, H, N, S	—S—C≡N	Thiocyanate
	—N=C=S	Isothiocyanate
C, H, O, N, S	—SO$_2$NH$_2$	Sulfonamide
	—SONH$_2$	Sulfinamide

*Reprinted from Ref. 35 by courtesy of Wiley, New York.

chemist to assign structures to natural organic materials, and also for tentative and confirmatory identification of "known compounds" which have entered natural waters. Excellent descriptions of the former use are found in studies attempting to elucidate the structure of humic substances in soils(36) and in natural waters(37). The latter use is exemplified by the work on organic pesticides(28).

Some of the methods are fairly specific for certain functional groups and can be used to detect their presence in mixtures. These methods thus reveal the general nature of the compounds present in a given system.

Any division of this broad topic into sections would of necessity be arbitrary. For purposes of organization, the decision was made to divide the subject into sections discussing basic structural considerations, physical identification methods, and chemical identification methods. The remainder of this section concerns the basic structural considerations.

The first step in qualitative organic analysis is the determination of the molecular formula, that is, the identity and atomic proportion of each element in the molecule. Molecular formulas are calculated from molecular weight and per cent composition data. After designation of the formula it is possible, by functional group equivalent weight determinations, to ascribe numbers of certain functional groups to the molecule. Once this has been accomplished, the final task is to correctly assign the remaining structural constituents to the molecule. These assignments are made through the use of the many physical and chemical tests available for this purpose.

The described procedure is of value mainly in working with unknown materials isolated from the environment. In cases where the identity of a material is suspected (e.g., a pesticide), the analyst should proceed directly to those tests required for confirmation.

A. Molecular Weight

One of the most important characteristics of an organic compound is its molecular weight. This property is useful for confirming the identity of "known" organic compounds and for ascribing structure to those of unknown character. In conjunction with equivalent weight it can be used to ascertain the number of certain functional groups in a molecule. When combined with per cent composition data, it gives the molecular formula.

Molecular weight determinations are generally applicable only to pure compounds, although certain of the techniques do have other uses. Accuracy is determined by the purity of the compound and how closely it approximates ideal behavior. As the techniques differ depending upon the magnitude of the molecular weight to be determined, methods for low and high molecular weight compounds are discussed separately.

1. LOW MOLECULAR WEIGHT METHODS

a. Vapor Density

If an organic compound is volatile, it is possible to determine its molecular weight by vapor density measurements. This is done by measuring the volume of air displaced when a weighted sample is volatilized (38). The process is usually carried out at a temperature 20°C above the boiling point of the compound, since the ideal gas law does not apply to vapors near their condensation points. The molecular weight of the gas is then calculated by use of the ideal gas law. Thus

$$\text{molecular weight} = WRT/PV \qquad (3)$$

where W is the weight of sample (g), P is the pressure (atm), T is the absolute temperatures of experiments (°K), V is the volume of air displaced (ml), and R is the gas constant of 82.05 (ml-atm/°K/mole).

Even for volatile materials, the determination is limited by compliance of the compound's behavior with the ideal gas law. Such compliance is questionable for highly associated organic liquids. Other gas equations, such as those of Van der Waal or Berthelot, are not applicable unless the identity of the compound is known, so that the appropriate constants can be calculated. Since the molecular weight of a known compound can be easily calculated, these equations are of little value in this area.

If a compound exists as a dimer or trimer in the vapor state, the molecular weight will be erroneous by a factor of 2 or 3. Thus, those volatile organic acids that occur as dimers in the vapor state will give a molecular weight that is double that of the actual parent compound.

b. Vapor Pressure Lowering

It has long been recognized that when a substance is dissolved in a liquid, the vapor pressure of the liquid is decreased. This observation was placed on a quantitative basis by Roault, who proposed the equation (39)

$$(P^0 - P)/P^0 = m_2/(m_1 + m_2) \tag{4}$$

where P^0 is the vapor pressure of solvent, P is the vapor pressure of solution, m_1 is the number of moles of solvent, and m_2 is the number of moles of solute.

As the number of moles equals weight/molecular weight, Eq. (5) can be derived from Eq. (4). Thus, the determination of molecular weight can be obtained from the following equation:

$$(P^0 - P)/P^0 = (W_2/M_2)/(W_1/M_1 + W_2/M_2) \tag{5}$$

where W_1 is the weight of solvent, W_2 is the weight of solute, M_1 is the molecular weight of solvent, and M_2 is the molecular weight of solute. For a dilute solution, $W_1/M_1 + W_2/M_2$ approaches the value of W_1/M_1, and Eq. (5) becomes

$$M_2 = W_2 M_1 P^0/W_1(P^0 - P) \tag{6}$$

Due to the assumptions made, and the fact that Roault's law is most closely followed in dilute solutions, Eq. (6) only holds for low solute concentrations. In addition, the vapor pressure decrease is usually small and difficult to measure accurately, thereby limiting the use of this approach.

One interesting version of this technique is the "isopiestic" method. This is based on the fact that when two vessels containing two different solutes in the same solvent are confined in a closed space, the solvent distills from the solution of higher vapor pressure and condenses in the solution of lower vapor pressure. This continues until the vapor pressures of both solutions are equal. This distillation is accompanied by a decrease in the volume of one solution and a corresponding increase in the volume of the other(38). At equilibrium

$$M_2 = W_2 M_1 V_1 / W_1 V_2 \qquad (7)$$

where M_2 is the molecular weight of unknown, M_1 is the molecular weight of standard, W_2 is the weight of unknown, W_1 is the weight of standard, V_2 is the volume of unknown, and V_1 is the volume of standard.

c. Boiling Point Elevation

If the addition of a solute to a solvent lowers the vapor pressure of the solution, the temperature at which the vapor pressure of the solution equals the atmospheric pressure (boiling point) must be higher than that of the pure solvent(39). In this case also, the change in the boiling point (elevation) is proportional to the mole fraction of solute. The relationship between molecular weight and the boiling point elevation is

$$M_2 = K_e 1000 W_2 / \Delta T_e W_1 \qquad (8)$$

where M_2 is the molecular weight of unknown, W_2 is the weight of unknown, W_1 is the weight of solvent, ΔT_e is the boiling point elevation, and K_e is the molal elevation constant.

The constant K_e is defined as being equal to $RT_0^2 / 1000le$, where R is the gas constant, T_0 is the boiling point of pure solvent in °K, and le is the latent heat of vaporization per gram of solvent. The solute should have no effect on this constant, and, in general, the larger this constant the smaller W_2 / W_1 can be made and the closer the experiment will approach ideal conditions. Various solvents have widely differing values of K_e, and the selection depends on solubility and noninteraction considerations. Typical solvents are listed in Table 3.

TABLE 3

Molal   Elevation   Constants   of   Typical
Solvents

Solvent	Boiling point, °C	K_e
Methanol	64.96	0.83
Ethanol	78.5	1.19
Acetone	56.2	1.73
Benzene	80.1	2.60
Chloroform	61.2	3.85
Carbon tetrachloride	76.75	5.02

d. Freezing Point Depression

Another property of solutions related to the lowering of the vapor pressure is the well-known fact that dissolved substances lower the freezing point of solvents. This includes depression of the melting points for solutions that are solid at ambient temperatures. Equation (9) is almost identical to Eq. (8) for boiling point elevation, and is (39)

$$M_2 = K_f 1000 W_2 / \Delta T_f W_1 \tag{9}$$

where M_2 is the molecular weight of unknown, W_2 is the weight of unknown, W_1 is the weight of solvent, ΔT_f is the freezing point depression, and K_f is the molal depression constant.

The constant K_f is defined as being equal to $RT_0^2/1000 l_f$, where R is the gas constant, T_0 is the freezing point of pure solvent in °K, and l_f is the latent heat of fusion per gram of solvent. The solvents here are selected on the same basis as noted for boiling point elevation. However, the large magnitude of the K_f values exhibited by several solvents (Table 4) gives the freezing point depression method certain advantages.

TABLE 4

Molal Depression Constants of Typical Solvents

Solvent	Freezing point, °C	K_f
Acetic acid	16.7	3.9
Benzene	5.5	5.1
Naphthalene	80.2	7.0
Cyclohexane	6.5	20.2
Cyclopentadecanone	65.6	21.3
Camphene	49	31.0
Borneol	206	35.8
Cyclohexanol	24.6	37.7
Camphor	176–180	40.0
Pinene dibromide	170	80.9

2. HIGH MOLECULAR WEIGHT METHODS

Molecular weight techniques such as vapor density are obviously inapplicable to macromolecules. In addition, the usual colligative properties that are employed for low molecular weight determinations are also inapplicable. This is due to the fact that macromolecules are rarely free from contaminating ions, since their most common sources are man-made polymerization reactions and chemical isolations from natural products. Since these colligative effects are on an ion or molecular unit basis, a small amount of contaminant would cause a significant error in molecular weight determination. Osmotic pressure, however, due to its method of measurement, is an exception to this generalization.

The molecular weights of macromolecules, even naturally occurring ones such as proteins, can be quite large. Hemocyanin, for example, has been reported as having a molecular weight of 6.7×10^6 (40). It is to be expected, then, that "solutions" of such macromolecules would exhibit colloidal behavior and that many of the methods used to determine molecular weight would be based on the fact that they are essentially very small particles in suspension.

The molecular weight picture of macromolecules is complicated by the fact that the various techniques give results based on different averaging methods. With some techniques the value is a number average, with others, a weight average. As illustrated in Fig. 1, the two methods may yield differing results for a system containing a distribution of molecular weights.

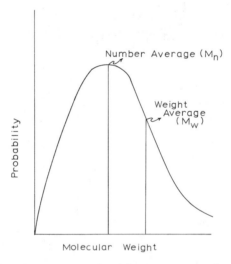

Fig. 1. Relation of number average and weight average molecular weights in a skewed distribution.

If equal statistical weight is given to all the molecules of a sample, whether large or small, the average molecular weight calculated will be the number average molecular weight, M_n, which may be defined as follows (41):

$$M_n = \sum_i n_i M_i \Big/ \sum_i n_i \tag{10}$$

where n_i is the number of molecules of molecular weight M_i, and the sum is taken over all values of i.

If a molecular weight is calculated by weighting each molecular class by the weight fraction it occupies in the sample, then the weight average molecular weight, M_w, is obtained (41). This function may be defined as follows:

$$M_w = \sum_i W_i M_i \Big/ \sum_i W_i = \sum (n_i M_i) M_i \Big/ \sum n_i M_i$$

$$= \sum n_i M_i^2 \Big/ \sum n_i M_i \tag{11}$$

where W_i is the weight fraction of component i. Number average molecular weights result from osmotic pressure measurement or other colligative property measurements. Weight average molecular weights result from the use of such techniques as light scattering.

The weight average molecular weight must always be greater than the number average molecular weight, except for a sample in which all the molecules have the same weight, so that $M_n = M_w$. Johnson (41) shows that for a sample containing 10% by weight of a polymer of $M = 10,000$ and 90% by weight of $M = 100,000$, the weight average molecular weight would be 91,000 while the number average would be 52,500.

All things considered, there is remarkable agreement among the molecular weights of proteins as determined by different methods. This is illustrated by the results shown in Table 5.

a. Osmotic Pressure

Osmotic pressure can be defined as "the excess pressure which must be applied to a solution to prevent the passage into it of solvent when they are separated by a perfectly semipermeable membrane" (39). A semipermeable membrane is one which allows the free passage of only solvent molecules.

Osmotic pressure is the only colligative property which offers a practical method for the determination of molecular weights in the range above 10,000. This is due to the fact that the effect of low molecular weight substances can be eliminated through the use of modified semipermeable

TABLE 5
The Molecular Weight of Proteins[a]

		Method			
Protein	Osmometry	Sedimentation equilibrium	Sedimentation velocity	X-ray analysis	Light scattering
Pepsin	36,000	39,000	35,500	39,300	—
B-Lactoglobulin	—	38,000	41,500	40,000	—
Ovalbumin	44,000	40,000	44,000	—	38,000
Hemoglobin[b]	67,000	68,000	68,000	66,700	—
Serum albumin[b]	73,000	68,000	70,000	< 82,800	74,000
Serum globulin[b]	174,000	150,000	167,000	—	—
Excelsin	214,000	—	295,000	305,000	280,000
Amandin	206,000	330,000	330,000	—	330,000
Insulin	48,000	41,000	35,000	36,000	—

[a]From Refs. *40,41*.
[b]Equine source.

membranes. With these, low molecular weight materials can distribute themselves equally on each side of the membrane, and thus not contribute to the pressure that is measured in the determination.

The equation relating osmotic pressure and molecular weight is in the form of the ideal gas law.

$$M_n = WRT/\pi V \tag{12}$$

In Eq. (12), W is the weight of substance (g), V is the volume of solution (liters), π is the osmotic pressure (atm), T is the temperature (°K), and R is the gas constant (liter atm).

As might be expected, Eq. (12) only holds for very dilute solutions. The relationship for more concentrated solutions is (*40*)

$$M_n = RTc/(\pi - Kc^2) \tag{13}$$

where c is the concentration in grams per liter (replacing w/v), and K is a constant of the system under study. In using Eq. (13), a plot of π/c versus c is extrapolated to zero concentration. At that point, the value of π is assumed to be the osmotic pressure of an ideal solution. As previously described, molecular weight results obtained on the basis of osmotic pressure measurements are number averages.

b. Rate of Diffusion

When a solution of a macromolecular material is placed in contact with the pure solvent, the macromolecules diffuse into the solvent (40). According to Fick's law, the quantity of a substance dS diffusing through an area A in time dt is

$$dS = -DA(dc/dx)dt \qquad (14)$$

where (dc/dx) is the concentration gradient, and D is called the diffusion coefficient. The latter may in turn be defined as

$$D = RT/N_0 f \qquad (15)$$

where R is the gas constant, T is the temperature in °K, N_0 is Avogadro's number, and f is the frictional constant. For spherical molecules of radius r, the frictional constant is defined in Stoke's law as

$$f = 6\pi n r \qquad (16)$$

where n is the coefficient of viscosity of the medium.

Molecular weight is defined as

$$M = V\sigma \qquad (17)$$

where V is volume and σ is the density. On a molar basis (N_0 particles), the molecular weight of spherical particles of volume $\frac{4}{3}\pi r^3$ is

$$M = \frac{4}{3}\pi r^3 \sigma N_0 \qquad (18)$$

The value of r can be determined by combining Eqs. (15) and (16) and solving for r. Thus

$$r = RT/N_0 6\pi n D \qquad (19)$$

and the overall equation for molecular weight would be

$$M = \frac{4}{3}\pi \sigma N_0 (RT/N_0 6\pi n D)^3 \qquad (20)$$

Similar equations can be developed for regular-shaped particles such as ellipsoids, cylinders, or rods. If shape is unknown, the method cannot be employed. Due to the nature of the process, this technique is frequently used for macromolecules too small for molecular weight estimations by sedimentation techniques.

The use and accuracy of the method are determined by experimental observation of the diffusion coefficient, D. This may be done by forming a sharp horizontal boundary between the solution and solvent, and measuring the rate of spread of the boundary by absorption spectrophotometry or refractometry.

c. Sedimentation Equilibrium

When a solution containing macromolecules is centrifuged at sufficient speeds to develop forces up to $10,000\,g$, the macromolecules, due to their higher specific gravity, will tend to sediment. Thus, the concentration of the solution will vary from the center to the periphery of the centrifuge. The centrifugal force is counteracted by the tendency of macromolecules in a more concentrated region to diffuse toward regions of lesser concentration. If this process is carried out until equilibrium is established, the molecular weight, M_w, may be defined as (40,42)

$$M_w = 2RT \ln (c_2/c_1)/(1-\rho/\sigma)\omega^2(x_2^2-x_1^2) \tag{21}$$

where R is the gas constant, T is the temperature in °K, σ is the density of the solute, ρ is the density of the solvent, ω is the angular velocity, and c_2 and c_1 are, respectively, the concentrations of solute at distances x_2 and x_1 from the center of rotation.

This method is based on the use of very high speed contrifuges called "ultracentrifuges." The concentrations are measured while the centrifuge is in operation, usually through transparent chambers by means of refractive index changes using a Schlieren system.

For substances having molecular weights over 5000, the method was seldom used because of the long time necessary to achieve equilibrium. A simple change in location for determining equilibrium concentrations [c_1 and c_2 in Eq. (21)] in the centrifugal field resulted in a more rapid determination of molecular weight (42). This modification of the equilibrium method has greatly increased its useful scope.

The molecular weight determined is a Z average, which is defined as (41)

$$M_z = \sum_i n_i M_i^3 \Big/ \sum_i n_i M_i^2 \tag{22}$$

d. Sedimentation Velocity

If a solution of a macromolecule is subjected to centrifugation at higher speeds than in the equilibrium method (i.e., approximately $250,000\,g$), the molecules will sediment at a much higher rate, giving rise to another method for estimating molecular weight (40).

The molecular weight may be calculated from the relationship

$$M = RTS/D(1-\rho/\sigma) \tag{23}$$

where S is the sedimentation constant, D is the diffusion constant, and the rest of the terms are the same as in Eq. (21).

The sedimentation constant is defined as

$$S = (dx/dt)\omega^2 x \tag{24}$$

where x is the distance of the moving boundary from the center of rotation at time t, and ω is the angular velocity of the centrifuge in radians per second (that is, 2π times the number of revolutions per second). The value S is determined in the ultracentrifuge in a manner similar to that used for the calculation of equilibrium. For proteins, S falls in the range of 10^{-13} to 200×10^{-13} sec. The unit of 10^{-13} sec is called a "Svedberg."

The magnitude of the diffusion coefficient, D, is a function of the frictional drag, f. Therefore, the presence of D in Eq. (23) eliminates the need to calculate f, and this, in turn, obviates the need to know particle shape. The diffusion coefficient is determined by the same method briefly discussed at the end of Sec. IV.A.2.b.

e. Viscosity

Mark (43) suggested that viscosity and molecular weight are related by

$$[n] = KM^\alpha \tag{25}$$

where $[n]$ is the intrinsic viscosity and K and α are constants. Intrinsic viscosity is related to both solution viscosity (n) and solvent viscosity (n_0), since it is the limit of $[(n/n_0)-1]/c$ as c approaches zero (infinite dilution).

For a given solvent-solute system, the constants in this equation must be determined through measuring molecular weight by another method. Thus, this equation does not provide a distinct method for determining molecular weight. Its main use has been with homologous series of synthetic or natural polymers. The molecular weight average determined by its use is (41)

$$M_\alpha = \left(\sum_i n_i M_i^{\alpha+1} \Big/ \sum_i n_i M_i \right)^{1/\alpha}. \tag{26}$$

For α values between 0.5 and 1.0, M_α falls in between a number and a weight average.

f. X-Ray Diffraction

When X rays pass through a crystal, they are reflected by the various crystal planes and give rise to a diffraction pattern. Patterns can be obtained with large single crystals or with powders of crystalline materials.

Through application of small angle scattering to single crystals, it has been possible to estimate the molecular weight of certain protein molecules(40). The difficulty of this method lies in the preparation of large single crystals of macromolecules.

Where X-ray determinations can be made, they provide the size of the crystal unit cell (molecular volume). From this and a density determination, the molecular weight can be calculated from

$$M = V\rho N_0 \qquad (27)$$

where V is the molecular volume (cm³), ρ is the density, and N_0 is Avogadro's number.

g. Light Scattering

When light passes through a solution of high molecular weight substances, part of the incident light is scattered, a phenomenon known as the "Tyndall effect." The intensity of the scattered light (T) increases with both the number and size of the scattering units.

For macromolecules that are small compared to the wavelengths of light used, the molecular weight relationship is(40)

$$M = T/HC \qquad (28)$$

where C is the concentration (g/cm³) and H is a proportionality constant. In turn, H is defined as

$$H = 32\pi^3\mu_0^2/3N\lambda^4(\mu - \mu_0/c)^2 \qquad (29)$$

where N is Avogadro's number, λ is the wavelength of light employed, and μ_0 and μ are the refractive indices of the solvent and solution, respectively.

Sometimes the size of a macromolecule is so large that it is comparable with the wavelength of the incident light. At this point interference occurs and the intensity of scattered radiation is no longer equally distributed about the right angle position. In this case, I_{45} and I_{135}, the intensities of the radiation scattered at 45 and 135° (i.e., forward and backward from the 90° angle), are found to differ. On this basis, a function called the "dissymmetry coefficient," q, is defined as

$$q = (I_{45}/I_{135}) - 1 \qquad (30)$$

As the concentration of macromolecules in solution decreases, q approaches a limiting value called the "intrinsic dissymmetry coefficient." This function is related to the spherical extension of the macromolecule and can be used to determine molecular size. If molecular size is coupled with density determination, molecular weight can be calculated.

h. Optical Behavior

In addition to the light-scattering technique, the "Tyndall effect" forms the basis of another method for molecular weight determination. When a powerful beam of light is focused in a macromolecular solution, the light is reflected by the particles. If this system is observed through a microscope at a right angle to the incident beam, the particles will appear as small points of light *(44)*. In this manner, the particles can be counted and the number per liter (N_L) then calculated by

$$N_L = \text{no. of particles/field/field volume in liters} \qquad (31)$$

If it is possible to determine or know through preparation the total weight of particles per liter (W), the molecular weight [weight of Avogadro's number (N_0) of particles] can be calculated as follows:

$$M = WN_0/N_L \qquad (32)$$

Furthermore, if assumptions are made as to the geometry of the macromolecule in question (i.e., whether it is spherical, ellipsoid, cylindrical, etc.), its dimensions can also be calculated.

In general, the resolution attained in this technique is determined by the intensity and wavelength of the light employed. For particles of low molecular weight, it is frequently necessary to use ultraviolet light. In addition, the refractive indices of particles must be different from that of the solvent. If the difference is sufficiently large, the detection of particles down to 4 nm in diameter is possible.

For smaller particles, resolution can only be achieved through utilization of a beam of electrons under constant voltage as the radiant energy source. The electron microscope, which uses this principle, can give a magnification of more than 100,000 times. Results of this technique are usually in the form of a photographic image. Thus, the geometry of a macromolecule is apparent and its size can be determined through knowledge of the magnification employed. This technique, then, can be used to calculate the volume of a molecule, and from density measurement, the molecular weight can be determined as described previously in Eq. (27).

B. Molecular Formula

One of the necessary steps in determining the structure of an organic molecule is the determination of its molecular formula, that is, the number and types of atoms making up the molecule. To determine the molecular formula one must know the molecular weight and the per cent of each element in the compound. Within the limitations imposed by the accuracy of the determinations, the sum of the percents of each element present should equal 100, provided all elements have been detected and estimated.

These data are put in tabular form and the per cent of each element present is divided by the atomic weight of the element. The smallest number representing these ratios is then divided into each ratio, and the result, to the nearest whole number, is the simplest formula. This procedure can be outlined as follows:

Element	Per cent	%/at. wt.	(%at. wt.)/1.365
C	49.351	4.112	3.013
H	9.603	9.603	7.035
O	21.841	1.365	1.000
N	19.205	1.372	1.005

The simplest formula that represents these data is C_3H_7ON, and the simplest formula weight is 73. The molecular weight should be a simple whole number multiple of this figure, and the subscript of each element should be multiplied by the molecular weight/simplest formula weight. Thus, if in this case the molecular weight were 146 (corresponding to the amino acid lysine), the formula would be $C_6H_{14}O_2N_2$.

Steps subsequent to formula determination involve the assignment of structure to the basic unit. For example, equivalent weight determinations might reveal the fact that there is one carboxylic and one amino group in the molecule. The ascription of further structure requires use of the techniques described in subsequent sections of this chapter.

For the purpose noted above, a pure compound is required. However, these analyses can be used on simple mixtures and, as noted later, on complex mixtures of organic compounds in aqueous systems. Since analyses of pure and mixed systems serve different purposes, they are discussed in separate sections.

1. ELEMENTAL ANALYSIS FOR MOLECULAR FORMULAS

a. Carbon and Hydrogen

Organic carbon can be converted to carbon dioxide by heating in a stream of pure oxygen. At the same time, hydrogen is converted to water. The CO_2 formed can be absorbed by NaOH on asbestos after the water has been absorbed by anhydrous magnesium perchlorate(45). The increase of weights of these absorbents is equal to the amounts of CO_2 and H_2O absorbed. The reactions are

$$\text{organic carbon} \xrightarrow[O_2]{\Delta} CO_2 \tag{33}$$

$$\text{organic hydrogen} \xrightarrow[O_2]{\Delta} H_2O \tag{34}$$

and the respective calculations are

$$\%C = (A_1 - A_2)F_1 100/S \tag{35}$$

$$\%H = (A_3 - A_4)F_2 100/S \tag{36}$$

where A_1 is the initial weight of NaOH absorbent (g), A_2 is the final weight of NaOH absorbent (g), A_3 is the initial weight of perchlorate absorbent (g), A_4 is the final weight of perchlorate absorbent (g), F_1 is the weight fraction of carbon in CO_2 (12/44), F_2 is the weight fraction of hydrogen in H_2O (2/18), and S is the sample weight (g).

Since organic elements such as N and F would interfere in the gravimetric portion of this procedure, some provision must be made to remove them and other interferences during the combustion step. For this reason, the combustion tube always contains silver, platinum, copper oxide–lead chromate mixture, and lead peroxide. The platinum aids in the combustion of any methane formed by degradation during heating, the silver removes oxides of sulfur and the halogens, the copper oxide–lead chromate mixture is another oxidant, and the oxides of nitrogen are converted into nitrates by the lead peroxide.

b. *Nitrogen*

One of the classic methods of determining nitrogen in organic compounds is the "Dumas procedure." The method is based on the production of nitrogen gas when an organic compound containing nitrogen is heated with copper oxides in a gas stream of carbon dioxide. Part of the organic nitrogen is converted into oxides of nitrogen, and these oxides react with copper to form nitrogen gas (45). The reactions are

$$\text{organic nitrogen} \xrightarrow[\text{CuO}]{\Delta} N_2 + \text{oxides of } N + O_2 + Cu^0 + H_2O \qquad (37a)$$

$$\text{oxides of } N + Cu^0 \longrightarrow CuO + N_2 \qquad (37b)$$

If there is an excess of oxygen present in the compound, it will be removed by the following reaction:

$$2Cu^0 + O_2 \longrightarrow 2CuO \qquad (38)$$

Theoretically, this method should be good for both reduced (amine, etc.) and oxidized (nitro, nitroso, etc.) nitrogen. However, many types of compounds give erroneous results unless special precautions are taken.

The liberated nitrogen is collected over a KOH solution and its volume measured. The temperature and barometric pressure are recorded and the latter is corrected for the vapor pressure of water in the measuring apparatus. The gas weight is then calculated from the gas law, and the per cent nitrogen is estimated. The calculation is given by Eq. (39):

$$\%N = 28P_cV100/SRT \qquad (39)$$

Here, P_c is the corrected pressure (atm), V is the volume (liters), S is the sample weight (g), R is the gas constant (liter-atm/°K), and T is the temperature (°K).

In addition to this basic procedure, organic nitrogen is also readily determined by the Kjeldahl method, which is covered in a subsequent section.

c. *Oxygen*

Organic oxygen has been a rather difficult element to determine. In fact, until recently, it was frequently estimated by difference. Now, however, it can be determined through the thermal decomposition of the organic compound in an inert gas stream containing excess carbon (45). The water–gas reactions liberate carbon monoxide, which reacts with iodine pentoxide to yield iodine. The liberated iodine is augmented through reaction with bromine and hydrogen iodide, and the resulting iodine is determined by titration with sodium thiosulfate using starch indicator. Halogens and acidic oxides are removed from the gas stream by KOH pellets. The schematic reactions are

$$\text{organic oxygen} \xrightarrow[\text{C}]{1120°\text{C}} \text{CO} \tag{40a}$$

$$5CO + I_2O_5 \longrightarrow 5CO_2 + I_2 \tag{40b}$$

$$I_2 + Br_2 \longrightarrow 2IBr \tag{40c}$$

$$2IBr + 4Br_2 + 6H_2O \longrightarrow 2HIO_3 + 10HBr \tag{40d}$$

$$2HIO_3 + 10HI \longrightarrow 6I_2 + 6H_2O \tag{40e}$$

$$6I_2 + 12S_2O_3^{2-} \longrightarrow 12I^- + 6S_4O_6^{2-} \tag{40f}$$

The calculation is typical for a titration:

$$\%O = NVE/10S \tag{41}$$

where N is the normality of thiosulfate solution, V is the volume of thiosulfate titrant (ml), E is the equivalent weight of oxygen in this reaction (which is $5/12 \times 16 = 6.667$), and S is the sample weight (g).

d. Sulfur

Organic sulfur can be converted to sulfate by heating under pressure with fuming HNO_3 in the presence of salts of the alkali or alkaline earth metals (45). If $BaCl_2$ is present, a precipitate of $BaSO_4$ is formed that can be determined gravimetrically. The reaction is

$$\text{organic sulfur} \xrightarrow[\text{HNO}_3]{\text{BaCl}_2, \Delta} CO_2 + H_2O + \underline{BaSO_4} \tag{42}$$

and the calculation is

$$\%S = WF100/S \tag{43}$$

Here, W is the precipitate ($BaSO_4$) weight (g), S is the sample weight (g), and F is the weight fraction of sulfur in $BaSO_4$, which is 0.1374.

In addition to this (Carius) method, organic sulfur may be determined through alkaline peroxide oxidation to sulfate in the Parr bomb (46), or by reduction to H_2S followed by iodimetric titration (47).

e. Phosphorus

Organic phosphorus can be determined by conversion to phosphates through oxidation of the organic matter in a hot H_2SO_4–HNO_3 mixture (45). The reaction is similar to that involved in the determinations of sulfur:

$$\text{organic phosphorous} \xrightarrow[\text{H}_2\text{SO}_4-\text{HNO}_3]{\Delta} H_3PO_4 + H_2O + CO_2 \tag{44}$$

In classical organic microanalysis, the phosphate so formed is determined gravimetrically by precipitation with ammonium molybdate. The reaction is

$$H_3PO_4 \xrightarrow[\text{HNO}_3]{\text{(NH}_4\text{)}_2\text{MO}_4} (NH_4)_3PO_4 \cdot 14MoO_2 \tag{45}$$

The per cent phosphorus is calculated as follows:

$$\%P = WF100/S \tag{46}$$

where W is the precipitate weight (g), S is the sample weight (g), and F is the weight fraction of P in the precipitate, which is 0.0160.

This formula and factor are empirical rather than theoretical, and other versions of the factor exist(45). Phosphates might better be determined colorimetrically through the development of the phospho-molybdenum blue complex in the presence of reducing agents(30).

f. Halogens

Organic halogen compounds can be oxidized with HNO_3 to their respective halides, and these halides can be determined gravimetrically as the silver salts(45). This determination is similar to the Carius method for sulfur. The reaction is as follows:

$$\text{organic halide} \xrightarrow[\text{HNO}_3]{\text{AgNO}_3 \cdot \Delta} AgX + CO_2 + H_2O \tag{47}$$

The calculation is given by

$$\%X = WF100/S \tag{48}$$

where W is the precipitate weight (g), S is the sample weight (g), and F is the weight fraction of X in AgX. For AgCl, $F = 0.2474$; for AgBr, $F = 0.4256$; and for AgI, $F = 0.5405$.

Organic halogens can also be converted to halides by the alkaline peroxide Parr bomb technique(48). If more than one halide species is present, differential analyses must be run. This may be done through amperometric titration(49) or ion exchange chromatographic separation(50).

2. ELEMENTAL ANALYSIS OF MIXED ORGANIC SUBSTANCES IN AQUEOUS SYSTEMS

In addition to being used for determining formulas of pure compounds, elemental analyses can also be undertaken on the organic materials in aqueous systems. In this usage, they are run directly on the natural mixtures of compounds or on fractions isolated from the environment. The purpose of these analyses is to determine the concentrations of organic matter as represented by organic carbon, organic nitrogen, etc. In this sense, the purpose is analogous to that of estimating "organic matter" content through determination of biochemical oxygen demand, chemical oxygen demand, or volatile solids(30).

The organic forms of the elements are determined after conversion to their inorganic forms. For example,

$$
\begin{aligned}
\text{organic carbon} &\longrightarrow CO_2 \\
\text{organic nitrogen} &\longrightarrow NH_3 \\
\text{organic phosphorus} &\longrightarrow PO_4^{3-} \\
\text{organic sulfur} &\longrightarrow SO_4^{2-} \\
\text{organic halides} &\longrightarrow X^- \text{ (i.e., } F^-, Cl^-, Br^-, I^-)
\end{aligned}
\tag{49}
$$

Hydrogen is not included in this list, as it is oxidized to water, which obviously cannot be determined in an aqueous medium.

Both the inorganic and organic forms of the listed elements are found in nearly all aquatic systems. The general procedure, therefore, is to determine inorganic species, then convert the organic species to inorganic species, and finally, determine the total as the inorganic form. The concentration of the organic element is then

$$
\text{organic} = \text{total} - \text{inorganic}
\tag{50}
$$

The success or failure of this approach is a function of the relative levels of the inorganic and organic species present. In most aquatic systems, the concentrations of the organic species of sulfur, phosphorus, and the halogens are low, while the concentrations of the corresponding inorganic species are high, (e.g., phosphorus as phosphate in treatment plant effluents). This means that in the described procedure, the concentration of organics, a small number, is the result of the subtraction of a large number (inorganic) from a slightly larger number (total) using Eq. (50). The errors inherent in such a procedure, along with the analytical problems involved in such low level determinations, have precluded widespread determination of the organic forms of such elements.

The two elements whose organic forms are commonly determined are carbon and nitrogen. These two elements are exceptions to the noted generalization primarily because their concentrations in natural aquatic systems are larger and their inorganic conversion forms, CO_2 and NH_3, are volatile. The volatile nature of CO_2 and NH_3 permits their removal before conversion of the organic species. As carbon and nitrogen represent the only organic elements of present analytical significance, the following discussion is limited to them.

a. Organic Carbon

Organic carbon is determined in aqueous systems through oxidation followed by some method of estimating the liberated CO_2. The general reaction is

$$\text{organic carbon} \xrightarrow{\text{oxidation}} CO_2 + H_2O \tag{51}$$

Most methods involve classic wet oxidation through the use of chemical oxidants at high temperature. Typical oxidizing solutions include sulfuric–chromic acid(51–53), potassium persulfate(54), and the Van Slyke reagent (chromium trioxide, potassium iodate, phosphoric acid, and fuming sulfuric acid)(45,55,56). However, oxidation by ultraviolet irradiation catalyzed by ferrous ions has also been employed(57).

The liberated CO_2 is usually scrubbed from the oxidation mixture by a stream of inert gas. It may then be determined gravimetrically as $BaCO_3$ following absorption in NaOH(56), by indicator titration after similar absorption(51), by conductometric titration after absorption in $Ba(OH)_2$(52), or directly by gas chromatography(58) and infrared absorption(59). Liberated CO_2 can also be estimated by mass spectrophotometry following collection in a liquid nitrogen cold trap(57), or manometrically in the Van Slyke "blood gas" apparatus(45). Total oxidation can also take place in sealed ampoules, with determination of CO_2 by gas chromatography(60).

In general, the limitations of wet combustion techniques are in the quantitative oxidation of organic carbon to CO_2, rather than in the estimation of the CO_2 evolved. Wet combustion techniques seem incapable of quantitatively oxidizing all organic compounds(61). This fact has led to the development of dry combustion techniques, in which the sample is first evaporated to dryness and then burned in oxygen(58,62).

Most of the recent development, however, has been concerned with extremely rapid methods that involve "wet" oxidation in a nonclassical sense. These latter techniques involve the use of very small sample volumes (i.e., microliter quantities) which are injected into a stream of oxygen that enters a catalytic furnace. Here the organic carbon is oxidized to CO_2 in a manner somewhat similar to the dry systems, and the CO_2 formed is determined by absorption in the infrared region (61, 63–65). The sample can also be injected into a hot copper oxide bed in a stream of inert (helium) gas and the liberated CO_2 determined by gas chromatography. Liberated CO_2 can also be reduced to methane and determined by flame-ionization gas chromatography (66).

Inorganic carbon may be present in aquatic systems as carbonate, bicarbonate, and free carbon dioxide. It is possible to convert all inorganic carbon species to carbonates which can then be removed by barium precipitation. This procedure cannot be used, however, in the presence of particulate organic matter. Therefore, major emphasis has been placed on the fact that acidification will convert all inorganic species to CO_2, which is volatile and can be removed from the system. Removal is accomplished by boiling or by bubbling an inert gas through an acidified sample to scrub out the CO_2 (61,62). In such procedures, however, volatile constituents can be lost. One answer to this problem is to remove CO_2 after acidification in a closed vessel by allowing it to diffuse to a well filled with alkali. After all CO_2 is absorbed by the alkali, the organic carbon in the liquid and gas phases is determined by combustion and infrared absorption (67).

b. Organic Nitrogen

The Kjeldahl determination remains the most widely used method for the estimation of organic nitrogen. This method is based on the fact that organic nitrogen is converted into ammonium sulfate when organic matter is oxidized (digested) by hot H_2SO_4.

$$\text{organic nitrogen} + H_2SO_4 \xrightarrow[\text{catalyst}]{\Delta} NH_4HSO_4 + H_2O + CO_2 + SO_2 \qquad (52a)$$

To speed the reaction and to ensure the completeness of the oxidation and recovery, catalysts such as HgO, and inert salts (to raise the boiling point of the mixture) such as K_2SO_4 are added to the reaction vessel along with the H_2SO_4. The reaction requires about 4 hr at a vigorous boil (340°C) to reach completion, which is frequently, but not always, indicated by the reaction mixture becoming colorless (45,68).

The acid mixture is then made alkaline with a $NaOH–Na_2S_2O_3$ mixture, and the ammonia is removed from the system by distillation. The reaction is

$$(NH_4)HSO_4 + 2NaOH \longrightarrow NH_3 + Na_2SO_4 + 2H_2O \qquad (52b)$$

Sodium thiosulfate is required to decompose the complex formed between ammonia and the mercury catalyst(45). This reaction is

$$Hg(NH_3)_2SO_4 + Na_2S_2O_3 + H_2O \rightarrow \underline{HgS} \downarrow + Na_2SO_4 + (NH_4)_2SO_4 \qquad (52c)$$

As ammonia is volatile, it is necessary to distill it into an acidic solution. This may be a standard acid solution, in which case the ammonia can be determined by back titration of excess acid with NaOH(69). It may also be a boric acid solution, in which case the ammonia can be determined directly by titration with standard HCl(45), H_2SO_4(70), or potassium acid iodate(68), with indicators such as methyl red, methylene blue, or methyl red–brom cresol green(45). The ammonia can also be determined colorimetrically by the Nessler reaction(30) or by reaction with a triketohydrindine hydrate reagent(71).

The digestion reaction noted above will be quantitative only for nitrogen in its lowest valence state. Nitro and nitroso groups are usually lost as oxides of nitrogen during acid digestion. To include such groups as the nitro (—NO_2), nitroso (—N=O), azo (—N=N—), and hydrazo (—NHNH—), it is necessary to reduce them to the lowest valence state before digestion. This can be done using sucrose(72) or a mixture of zinc and iron in a hydrochloric-formic acid mixture(73).

If inorganic nitrogen in the form of ammonia is present, it can be removed by distillation prior to digestion. For such a distillation, the system is usually buffered at about pH 7.4 with phosphate(30) or borate(74). Nitrates and nitriles can be reduced to nitrous oxide by ferrous sulfate in an acid medium and then removed by boiling(74). It must be noted, however, that in the removal of ammonia by distillation, certain volatile amines could also be lost and therefore not reported as organic nitrogen.

C. Equivalent Weight

Unlike molecular weight, equivalent weight is a chemical rather than a physical property. Equivalent weight is based on the chemical reactivity of specific functional groups and is, therefore, essentially a quantitative estimation of an organic functional group. There are many such determinations available; those for the following groups are typical: carboxylic (—COOH), acetyl (CH_3CO—), ester (—COOR), hydroxy (—OH), methoxy and ethoxy ($-OCH_3$, $-OC_2H_5$), epoxy ($-CH-CH-$),

$$\overset{\diagdown}{\underset{O}{\diagup}}$$

carbonyl ($\overset{\diagdown}{\underset{\diagup}{C}}=O$), primary and secondary amino (RNH_2, R_2NH),

methyl and ethyl amino ($\overset{\diagdown}{\underset{\diagup}{N}}CH_3$, $\overset{\diagdown}{\underset{\diagup}{N}}C_2H_5$), α-amino acid nitrogen

($-CHNH_2COOH$), thiol ($-SH$), and carbon–carbon unsaturation ($-C=C-$, $-C\equiv C-$).

Quantitative data obtained for these groups can be used in two ways. First, the data can be used to calculate the per cent of the group in the organic material; this is a frequent use of these data and it is especially applicable to fractions of mixed organic materials isolated from the environment. Second, the equivalent weight of the organic material in reference to the reacting group can be calculated. In conjunction with information on the molecular weight, this indicates the number of such functional groups in the molecule (number of functional groups = molecular weight/equivalent weight). This second approach is theoretically sound only with regard to pure organic compounds. Nevertheless, it has been usefully applied in a somewhat empirical manner to humic materials isolated from soil(36) and water(21).

The same quantitative data are obtained for each of the two uses; only the calculations differ. For example, in the direct titration of a functional group, F, in compound C, the following equations would hold:

$$\%F = ENV/10S \tag{53a}$$

$$\text{equivalent weight of } C = 1000S/NV \tag{53b}$$

Here, E is the equivalent weight of functional group F, S is the sample weight (g); N is the normality of titrant, and V is the volume of titrant (ml).

1. CARBOXYLIC

As the organic carboxylic group ($-COOH$) has a replaceable hydrogen, an organic compound containing this group is an acid and can be titrated with a base such as NaOH. If the acid is poorly soluble in water, as are the higher aliphatic (fatty) acids, then ethanol or other organic liquids may be used as the solvent instead of water(45,75). The general reaction is

$$RCOOH + NaOH \longrightarrow RCOONa + H_2O \tag{54}$$

The end point of the titration is usually determined by phenolphthalein indicator, but conductometric titrations may also be employed (30).

Phenol (C_6H_4OH) and hydroxy groups (—OH) do not interfere with the determination (76). However, the presence of just one aliphatic amino group (—CH_2NH_2) or more than one aromatic amino group (—$C_6H_4NH_2$) does interfere. In such cases, the basic character of the amino group must be destroyed, as in the formal titration of amino acids (40).

The calculation for carboxylic equivalent weight is given by

$$\text{equivalent weight} = 1000S/NV \qquad (55)$$

Here, S is the sample weight (g), N is the normality of titrant (NaOH), and V is the volume of titrant (ml).

2. ACETYL

An acetyl group (CH_3CO) may be attached to a nitrogen or oxygen atom in an organic molecule, giving rise to compounds called amides or esters, respectively. These compounds can be hydrolyzed by either acids or bases, depending upon their solubility.

$$CH_3COOR + NaOH \longrightarrow CH_3COONa + ROH \qquad (56)$$

$$CH_3CONHR + HCl + H_2O \longrightarrow CH_3COOH + RNH_3Cl \qquad (57)$$

When hydrolysis is complete, the mixture is acidified and the acetic acid formed is removed by steam distillation (77). The acetic acid in the distillate is determined by the method described for carboxylic acids, and the calculation of equivalent weight is also the same.

3. ESTERS

In the presence of a base, such as KOH, esters will react to form an alcohol and the salt of the acid group (—COOK). This reaction, known as saponification, is quite slow at room temperature. To drive it to completion within a reasonable period of time, it is necessary to reflux the ester with excess alkali at elevated temperatures. The usual solvents are alcohols such as methanol or ethylene glycol (75) in which both the ester and KOH are soluble. After saponification is completed, the excess alkali is determined by titration with standard HCl or H_2SO_4, using phenolphthalein as the indicator. The saponification reaction is as follows:

$$RCOOR + KOH \longrightarrow RCOOK + ROH \qquad (58)$$

The titration is simply the reaction of a strong acid with a strong base.

$$KOH + HCl \longrightarrow KCl + H_2O \tag{59}$$

Since this method is based on a back titration, the calculation is more complex than that for the carboxylic group.

$$\text{equivalent weight} = 1000S/[(N_1V_1) - (N_2V_2)] \tag{60}$$

Here, S is the sample weight (g), N_1 is the normality of KOH solution, V_1 is the volume of KOH solution (ml), N_2 is the normality of acid titrant, and V_2 is the volume of acid titrant (ml).

4. HYDROXYL

The hydroxyl group (—OH) can be esterified with excess acetic anhydride in pyridine. After esterification is complete, the mixture is added to water and the excess acetic acid titrated with a standard alkali (*78*). The schematic reactions are

$$ROH + (CH_3CO)_2O \longrightarrow ROCOCH_3 + CH_3COOH \tag{61}$$

$$(CH_3CO)_2O + H_2O \longrightarrow 2CH_3COOH \tag{62}$$

The hydrolysis of the acetate ester is too slow under these conditions to interfere in the determination. The equivalent weight may be calculated as follows:

$$\text{equivalent weight} = S/[(A/51) - (NV/1000)] \tag{63}$$

Here, S is the weight of sample (g), A is the weight of acetic anhydride (g) and 51 its equivalent weight, N is the normality of titrant (NaOH), and V is the volume of titrant (ml).

If necessary, the equivalent weight of the anhydride can be checked by the same general titration procedures.

5. ALKOXYL

Methoxyl (OCH_3) and ethoxyl (OC_2H_5) groups can be split from esters or ethers by treatment with boiling HI. In this process the groups are converted into methyl or ethyl iodide. The alkyl iodides formed react with alcoholic silver nitrate to form a double salt, $AgI \cdot AgNO_3$. The double salt in turn is split by the addition of water and nitric acid, and the silver iodide thus formed is determined gravimetrically (45). The schematic reactions for the production of AgI from a methoxy group are as follows:

$$RCOOCH_3 + HI \longrightarrow RCOOH + CH_3I \qquad (64a)$$

or

$$ROCH_3 + HI \longrightarrow ROH + CH_3I$$

then

$$CH_3I + AgNO_3 \longrightarrow AgI \cdot AgNO_3 \qquad (64b)$$

$$AgI \cdot AgNO_3 + H_2O + HNO_3 \longrightarrow \underline{AgI} \downarrow + AgNO_3 \qquad (64c)$$

The reactions for the ethoxy group are similar.

This method gives good results for volatile compounds and for compounds containing more than one alkoxy group. Since one mole of AgI is formed for each alkoxyl group present, the calculation of equivalent weight is as follows:

$$\text{equivalent weight} = SE/WF \qquad (65)$$

Here, S is the sample weight (g), E is the equivalent weight of iodine, W is the precipitate (AgI) weight (g), and F is the weight fraction of I in AgI, which is 0.5405.

6. EPOXY

The epoxy group will react with HCl to form a chlorohydrin. The excess HCl can then be titrated with NaOH to the phenolphthalein end point (79). The reactions are

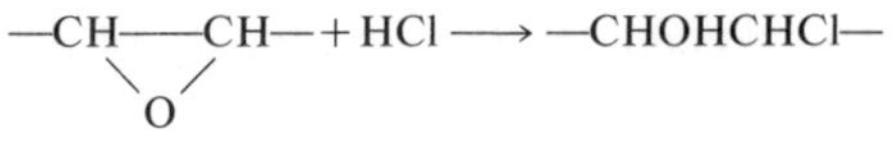

$$-CH\underset{\diagdown_{O}\diagup}{\underline{\qquad}}CH- + HCl \longrightarrow -CHOHCHCl- \qquad (66a)$$

and

$$HCl + NaOH \longrightarrow NaCl + H_2O \qquad (66b)$$

The determination is a back titration with the calculation similar to that for esters:

$$\text{equivalent weight} = 1000S/[(N_1V_1) - (N_2V_2)] \qquad (67)$$

where S is the sample weight (g), N_1 is the normality of HCl solution, V_1 is the volume of HCl solution added (ml), N_2 is the normality of NaOH titrant, and V_2 is the volume of NaOH titrant (ml).

7. CARBONYL

The carbonyl group (C=O) in aldehydes and ketones reacts with 2,4-dinitrophenylhydrazine to form insoluble hydrazones. The hydrazones can be removed by filtration, washed, dried, and then weighed(80). The reaction, using a ketone as an example, is as follows:

$$R_2CO + H_2NNHC_6H_3(NO_2)_2 \longrightarrow R_2CNNHC_6H_3(NO_2)_2 + H_2O \tag{68}$$

The calculation in this case is

$$\text{equivalent weight} = 180S/(W-S) \tag{69}$$

Here, S is the sample weight (g), W is the hydrazone precipitate weight (g), and 180 is a factor necessary to adjust changes in equivalent weights of sample and derivative as a result of their reaction [see Eq. (68)].

8. PRIMARY AND SECONDARY AMINO

Primary and secondary amines (RNH_2, R_2NH) can be determined by titration with perchloric acid in glacial acetic acid. The titration is carried out in nonaqueous solvents, such as benzene, with methyl violet as the indicator(81). The reaction is

$$RNH_2 + HClO_4 \longrightarrow RNH_3^+ + ClO_4^- \tag{70}$$

The calculation is as follows:

$$\text{equivalent weight} = 1000S/NV \tag{71}$$

Here, S is the sample weight (g), N is the normality of perchloric acid titrant, and V is the volume of perchloric acid titrant (ml).

Primary and secondary amines can also be determined through acetylation by a method similar to that described for hydroxy groups.

9. METHYL AND ETHYL AMINO

The N-methyl ($-NCH_3$) and N-ethyl ($-NC_2H_5$) groups in primary, secondary, and tertiary amines react with hydriodic acid to form the corresponding methyl and ethyl ammonium iodides. These quaternary ammonium compounds are then decomposed by heating to give methyl and ethyl iodides(45). In turn, the iodides are determined as noted under the determination of alkoxy groups. The schematic reactions involved in this determination are

$$NR + HI \longrightarrow NR \cdot HI \tag{72a}$$

and

$$NR \cdot HI \xrightarrow{\Delta} RI \tag{72b}$$

where R is an ethyl or methyl group. The calculations are also similar to those for the alkoxy groups.

$$\text{equivalent weight} = SE/WF \tag{73}$$

Here, S is the sample weight (g), E is the equivalent weight of iodine, W is the precipitate weight (g), and F is the weight fraction of I in AgI, which is 0.5405.

10. α-Amino Acid Nitrogen

Primary organic amines react with nitrous acid to form water, nitrogen gas, and an alcohol. With the amino acids, this reaction is so rapid that it can be used to determine amino acid nitrogen ($-CHNH_2COOH$) in the presence of ammonia, primary amines, purines, etc., all of which react too slowly to constitute an interference(45). The reaction is

$$RCHNH_2COOH + HONO \longrightarrow RCHOHCOOH + H_2O + N_2 \tag{74}$$

The nitrogen so liberated can be determined manometrically in the Van Slyke "blood gas" apparatus(82,83). In this procedure, the pressure increase due to the overall reaction is determined. This value is then corrected for the liberation of gases other than nitrogen through use of a blank. The calculation is as follows:

$$\text{equivalent weight} = 14S/Pf \tag{75}$$

Here, S is the sample weight (mg), P is the pressure increase (mm), and f is the factor required to convert pressure change to milligrams of amino acid N. This can be calculated or obtained from tables as a function of sample volume and temperature(45).

11. THIOL

Thiols (—SH) are oxidized to disulfides by cupric ions. The resultant cuprous ions in turn react with unoxidized thiol to form copper mercaptide. The excess cupric ions [Eq. (76c)] then react with iodide to form free iodine which is titrated with thiosulfate using starch indicator(84). The reactions for this back titration are as follows:

$$2Cu^{2+} + 2RSH \longrightarrow 2Cu^+ + RSSR + 2H^+ \tag{76a}$$

$$2Cu^+ + 2RSH \longrightarrow \underline{2CuSR \downarrow} + 2H^+ \tag{76b}$$

$$2Cu^{2+} + 4I^- \longrightarrow \underline{2CuI \downarrow} + I_2 \tag{76c}$$

$$I_2 + 2S_2O_3^{2-} \longrightarrow 2I^- + S_4O_6^{2-} \tag{76d}$$

One may calculate the equivalent weight as follows:

$$\text{equivalent weight} = 500S/[(N_1V_1) - (N_2V_2)] \tag{77}$$

Here, S is the sample weight (g), N_1 is the normality of cupric (butyl phthalate) solution, V_1 is the volume of cupric (butyl phthalate) solution added (ml), N_2 is the normality of thiosulfate titrant, V_2 is the volume of thiosulfate titrant (ml), and 500 is a stoichiometric factor.

12. GENERAL UNSATURATION

Unsaturated carbon bonds (—C=C—, —C≡C—) in an organic molecule can be determined by hydrogenation in the presence of a catalyst such as platinum. The reaction is as follows:

$$\ce{>C=C< + H2 ->[Pt] H-C-C-H} \tag{78}$$

The saturation of the double bond with hydrogen causes a decrease in the volume of gas above the solvent containing the organic compound. The volume difference is corrected for absorption of hydrogen by the catalyst, for temperature and barometric pressure differences, and for solvent vapor pressure. It is then used to calculate equivalent weight through the following equation(85):

$$\text{equivalent weight} = 22.4(S/V) \tag{79}$$

where S is the sample weight (mg) and V is the corrected volume of reacted hydrogen (ml).

It should be noted that an equivalent weight equal to one-half the molecular weight could mean the presence of two carbon–carbon double bonds, or a single carbon–carbon triple bond.

13. ACETYLENIC UNSATURATION

Acetylenes (RC≡CH) will form single salts when they react with ammoniacal silver nitrate. The salt thus formed is insoluble and is removed by centrifugation. Then, after acidification with HNO_3, the excess silver is titrated with thiocyanate using ferric indicator(86). The schematic reactions are

$$RC{\equiv}CH + Ag(NH_3)_2^+ \longrightarrow \underline{RC{\equiv}CAg \downarrow} + NH_3 + NH_4^+ \qquad (80a)$$

$$Ag^+ + CNS^- \longrightarrow \underline{AgCNS \downarrow} \qquad (80b)$$

This is essentially a back titration procedure, and the calculation is therefore

$$\text{equivalent weight} = 1000S/[(N_1V_1) - (N_2V_2)] \qquad (81)$$

Here, S is the sample weight (g), N_1 is the normality of silver nitrate solution, V_1 is the volume of silver nitrate solution (ml), N_2 is the normality of thiocyanate titrant, and V_2 is the volume of thiocyanate titrant (ml).

This procedure is quite valuable when used in conjunction with the method for general unsaturation described in the previous section.

V. Physical Identification Methods

There are many physical properties of organic compounds that are related to the nature of the organic molecules rather than or in addition to their number. These properties, therefore, may be of use in the identification of organic compounds and in ascribing molecular structure. In general, methods based upon physical properties are nonconsumptive, meaning that they do not consume or alter the organic compounds. This makes physical methods a logical choice, for example, for estimating the distribution of keto and enol forms in compounds exhibiting this type of tautomerism, since they should not affect the equilibrium. However, different methods may give different results. This is illustrated in Table 6 for the distribution of keto–enol forms of acetylacetone.

TABLE 6

Distribution of Keto–Enol Forms in Acetylacetone as Determined by Various Methods

Keto, %	Method	Type	Reference
24	Bromination	Chemical	87,88
27.6	Refrachor	Physical	87
18.6	NMR spectra	Physical	89
15.0	NMR spectra	Physical	90
24.0	NMR spectra	Physical	91

$$CH_3CCH_2CCH_3 \rightleftharpoons CH_3CCH=CCH_3$$

keto   enol

The difference among the four results obtained by physical methods indicates that caution must be used in interpretation, especially when such sophisticated procedures as nuclear magnetic resonance spectrophotometry (NMR) are employed.

At present, there are a large number of physical tests available to the organic analyst, and the trend is toward the use of these methods. Physical methods range from simple tests like boiling point determinations to various spectrophotometric procedures. The methods also range from those yielding little structural information to those that are invaluable in assigning structural configurations. In addition, physical methods can be used to determine isomer distributions (previously noted), to assist in formulating reaction mechanisms, and to explain many of the physical and chemical properties of organic compounds.

A. Simple Properties

All pure organic compounds have certain "simple" physical properties which are characteristic of them. Such properties include melting point, boiling point, density, refractive index, viscosity, and surface tension. These properties, mainly applicable to liquids rather than solids, can be and have been used to confirm the suspected identity of organic compounds. By themselves the properties are usually of little value in obtaining structural information, although many of them can be used to calculate certain additive properties which are related to structure. They are widely used as equipment for their determination is readily available and the procedures are simple (45,75).

B. Additive Properties

An additive property may be defined as a property that is the sum of the individual constituent properties. The only strictly additive property is molecular weight, which is equal to the sum of the atomic weights of each atom in the molecule. Most physical properties have both additive and constitutive aspects. A constitutive property is one that depends mainly on the arrangements of the atoms within the molecule. Refraction is mostly an additive property, while boiling point is mainly a constitutive property.

Both of these types of properties can be used to confirm suspected compound identities, or to obtain structural information about the molecules in question. This is especially true for additive properties with strong constitutive aspects. In almost all instances, the molecular weight and formula must be known or at least suspected.

1. MOLAR VOLUME

The "molar volume" (Vm) of a compound may be defined by

$$Vm = M/d \tag{82}$$

where M is the molecular weight and d is the density (g/liter).

Kopp(*39*) measured the molar volumes of organic compounds at their atmospheric pressure boiling points. He noted that the values were, to a considerable extent, additive functions of the volume equivalents of constituent atoms.

This property, however, is not strictly an additive one, but has certain constitutive aspects. Oxygen has a different volume equivalent depending upon whether it is single- or double-bonded. In a similar manner, the volume contribution of a benzene ring is less than would be expected from the constitutive atoms. Table 7 lists the values of the volume equivalents for a number of common atoms.

Thus, if the formula is known or suspected, the molar volume can be calculated as the sum of the atomic or group volumes, and compared to the actual value for the compound determined at, or extrapolated to, the boiling point. It is notable that associated organic solvents do not show abnormalities in molar volumes, and therefore association is not a source of error. Molar volume values can be influenced by structural factors and must be used with care. In addition, use of the property is limited by the fact that isomers like methyl propionate, ethyl acetate, and propyl formate all give the same value when the atomic volume equivalents are summed.

TABLE 7
Atomic and Functional Group Volume
Equivalents[a]

Atom or group	Volume equivalents	
	Kopp	LeBas
Hydrogen	5.5	3.7
Carbon	11.0	14.8
Chlorine	22.8	22.1
Bromine	27.8	27.0
Iodine	37.5	37.0
Oxygen (—O—)	7.8	7.4
Oxygen (C=O)	12.2	12.0
Sulfur	22.6	—
Benzene ring	—	−15.0

[a]Reprinted from Ref. *39* by courtesy of D. Van
Nostrand, New York.

2. THE PARACHOR

Sugden(*92*) developed an additive function that is related to the
molecular weight, density, and surface tension. This function, called the
"parachor," $[P]$, was defined as

$$[P] = [M/(D-d)]\lambda^{1/4} \tag{83}$$

where M is the molecular weight, D is the density of liquid (grams/liter),
d is the density of vapor (usually ignored), and λ is the surface tension
(dynes/cm). All of these properties are measured at the same temperature.
If the vapor density is ignored, then

$$[P] = (M/D)\lambda^{1/4} = Vm\lambda^{1/4} \tag{84}$$

If λ is set equal to unity, then $[P] = Vm$, or the parachor is the molar
volume at a surface tension of unity. Thus, a comparison of the parachor
of different compounds is equivalent to the comparison of molar volumes
at identical surface tensions (identical intermolecular attractions).

Although it is an additive property, the parachor has strong constitutive aspects. This is evident in Table 8.

Since certain bond types and ring configurations have assigned values,
the constitutive aspects of the parachor can lead to certain structural
information. Thus a value can be calculated from the formula and the

TABLE 8
Atomic and Functional Group Parachors[a]

Atom or functional groups	Parachor equivalents	Atom or functional groups	Parachor equivalents
Carbon	4.8	C=C	23.2
Hydrogen	17.1	C≡C	46.6
Nitrogen	12.5	3-membered ring	16.7
Phosphorus	37.7	4-membered ring	11.6
Oxygen	20.0	5-membered ring	8.5
Sulfur	48.2	6-membered ring	6.1
Fluorine	25.7	Esters	−3.2
Chlorine	54.3	Coordinate linkage	−1.6
Bromine	68.0		
Iodine	91.0		

[a]Reprinted from Ref. 93 by courtesy of Academic Press, New York.

assumed bonding or geometrical configuration, and then compared to the experimentally determined parachor. In this manner, certain assumptions about molecular structure can be verified.

Parachor values of strongly associated organic solvents tend to increase with temperature. In addition, this property is known to give erroneous information as to the structure of organic azides and organometallic compounds. Therefore, the parachor, like any other of the additive properties, must be used with caution.

3. MOLAR REFRACTION

The "molar refraction" $[R]$ is defined by the Lorentz–Lorentz equation as follows (39):

$$[R] = [(n^2 - 1)/(n^2 + 2)][M/d] \qquad (85)$$

where n is the refractive index, M is the molecular weight, and d is the density (g/liter).

The property is related to the molar volume, M/d. For radiation of infinite wavelength, it is considered to be the true volume of the molecules in 1 mole of compound, as distinct from the apparent volume.

For convenience, the wavelengths employed in measuring refractive index are the α, β, or λ hydrogen lines, or the sodium D line. This property is both additive and constitutive. Thus, as in the case of molar volume, tables of atom and group values have been worked out, from which molecular values can be calculated. Molar refractions for the sodium D line $[R_D]$ are listed in Table 9.

TABLE 9
Atomic and Functional Group Molar Refractions[a]

Atom or functional group	$[R_D]$	Atom or functional group	$[R_D]$
Carbon		Nitrogen	
Alkanes	2.42	Amines:	
Cycloalkanes:		primary	2.32
3-membered ring[b]	0.71	secondary	2.50
4-membered ring[b]	0.55	tertiary	2.84
8- to 15-membered ring[b]	0.55	Nitriles	3.12
Double bond[b]	1.73	Imides	3.78
Triple bond[b]	2.40	Oximes	3.90
		Hydrazones	3.46
Oxygen		Cyano group (—C≡N)	5.42
Hydroxy	1.53	Isocyano group (—N=C)	6.14
Ether	1.64	Secondary amides	2.75
Carbonyl	2.21	Tertiary amides	2.51
Peroxide	2.19	Urethanes	2.32
		Alkylidene amines	4.10
Halogens		Alkyl nitro group	6.72+
Chlorine in RCl	5.97	Aromatic nitro group	7.3+
Chlorine in RCOCl	6.34		
Chlorine in RCHClCOOR	6.70	Phosphorus	
Bromine in RBr	8.75	Esters (phosphoric)	3.75
Bromine in RCOBr	9.60	Esters (phosphorus)	7.04
Iodine in RI	13.90	Trialkylphosphine	9.14
		Dialkylarylphosphine	10.0
Hydrogen	1.10	Alkylphosphonic esters	4.27
Sulfur		Miscellaneous	
Alkylsulfides	8.00	Alkyl borons	3.67
Alkylthiocyanate	8.13	Alkylgermanium	9.00
Alkylaryl sulfides	9.20	Alkyltin	14.0
Thiophene	7.26	Trialkyl arsines	11.55
Disulfides	7.92	Trialkyl arsenites	9.52
Thioketones	9.7	Trialkyl arsenates	6.97
SO group:			
alkylsulfoxides	9.07		
SO₂ group:			
sulfanes	8.87		
SO₃ group:			
alkanesulfonates	10.35		
sulfites	11.13		
SO₄ group:			
sulfates	11.18		
Alkane thiols	7.81		

[a]Reprinted from Ref. 94 by courtesy of John Wiley (Interscience), New York.
[b]For compounds with these structures these values must also be added.

The measurements of density and refractive index are simple and easily performed. As usual, the molecular formulas must be known. Uses of $[R_D]$ in identification include the differentiation between alternate terpene structures and between the keto–enol forms of keto esters(94). Both differentiations are made through the observation that conjugated double bonds give a higher measured $[R_D]$ than would be calculated using values from Table 9.

4. REFRACHOR

The "refrachor" $[F]$ is a property related to the parachor and the refractive index. It may be defined as(87)

$$[F] = [P] \log (N_D - 1) \qquad (86)$$

where $[P]$ is the parachor and N_D is the refractive index using the sodium D line. Using techniques similar to those used to obtain the parachor and the molar refraction equivalent, atomic and functional group refrachor equivalents can be determined. Typical results are presented in Table 10. These values have been used to determine the per cent enol form in β keto esters and ketones, and also in distinguishing between aliphatic and aromatic halides.

TABLE 10
Atomic and Functional Group Refrachors[a]

Atom or group	$[F]$	Atom or group	$[F]$
Carbon		Halogens	
Carbon in CH_2	−25.27	Chlorine, aliphatic	2.05
CH_2 group	12.89	Chlorine, aromatic	25.53
		Bromine, aliphatic	16.51
Hydrogen in alkanes	19.08	Bromine, aromatic	22.80
		Iodine, aliphatic	9.67
Oxygen		Iodine, aromatic	15.94
In aliphatic esters	6.62		
OH group, aliphatic	11.09	Structure	
Carbonyl group (ketones)	7.88	Double bond	28.70
COO group, esters	17.64	Triple bond	60.18
COOH group	25.6	3-carbon ring	22.55
Nitro group (NO_2)	25.84	4-carbon ring	21.48
Nitrite group (ONO)	35.64	5-carbon ring	17.15
		6-carbon ring	13.05
		Methyl group	31.83
		Phenyl group	41.94

[a]From Ref. 87.

5. MOLAR MAGNETIC ROTATION

Faraday (39) discovered that a plane of polarized light is rotated when a transparent substance is placed in a magnetic field. Subsequent investigators have found that the angle of rotation (α) depends on the nature of the substance, the length, l, of the column of substance transversed by the light, and the strength of the magnetic field, H. The relationship is expressed by (39)

$$(\alpha) = WlH \qquad (87)$$

where W is the Verdet constant for the substance under study, and is the angle of rotation in a field of 1 G for a 1-cm pathlength. For water, the value is 0.01308 min of arc for the sodium D line. Using water as a standard, Perkin (39) defined the molar magnetic rotation $[M]$ of an organic compound as

$$[M] = M(\alpha)P'/M'(\alpha')P \qquad (88)$$

where M is the molecular weight, (α) is the angle of rotation (min), P is the density (g/liter), and prime values refer to water.

The molar magnetic rotation is both an additive and consitutive property. Typical values are given in Table 11.

Conjugation of double bonds results in a decrease in $[M]$ values from those expected on the basis of calculation. For example, the magnetic rotation of benzene should be the sum of 6×0.515 for 6C, 6×0.254 for 6H, and 3×1.11 for three double bonds; a sum of 7.94. This leaves a discrepancy of 3.34 units, when compared to the observed value of 11.28. Conjugated double bonds also give lower values and in approximately similar proportions. Thus, this property is useful in detecting conjugation and aromaticity, as well as keto–enol tautomerism.

TABLE 11
Magnetic Rotation Equivalents[a]

Hydrogen	0.254	Oxygen (—OH)	0.191
Carbon	0.515	Oxygen (—CHO)	0.261
Chlorine	1.734	Oxygen (⟍C=O)	0.375
Bromine	3.562	Nitrogen (—NH$_2$)	0.483
Iodine	7.757	Double bond	1.11

[a]Reprinted from Ref. 39 by courtesy of D. Van Nostrand, New York.

6. MAGNETIC SUSCEPTIBILITY

According to their behavior in a magnetic field, substances may be classified as ferromagnetic, paramagnetic, or diamagnetic(39). For two magnetic poles, m_1 and m_2, separated by a distance r, the force F between the poles is defined by

$$F = m_1 m_2 / \mu r^2 \tag{89}$$

Here μ is the magnetic permeability of the medium, and in a vacuum or air is assumed to be unity. In the case of a paramagnetic substance, μ is slightly greater than unity; for a ferromagnetic substance, μ is on the order of 1000 or more; and for a diamagnetic substance, μ is less than 1.

If the strength of the magnetic field is H, then

$$\mu = (4\pi I/H) + 1 \tag{90}$$

where I is the intensity (magnetic moment) of the magnetic field for a unit volume in the direction of the field. The "specific susceptibility", X, is defined as

$$X = (\mu - 1)/4\pi\rho \tag{91}$$

where ρ is the density.

For diamagnetic substances, X is usually negative, as μ is less than 1. Usually the value of X is multiplied by the atomic or molecular weights to give the atomic, X_a or molar, X_m, susceptibility. In an ideal solution, the susceptibility of a mixture of compounds a and b would be

$$X = f_a X_a + f_b X_b \tag{92}$$

where f_a and f_b are the weight fractions of a and b, respectively.

For organic compounds, the molar susceptibility is both an additive and constitutive property. Using techniques similar to those employed for the other constitutive properties, a table of atomic and structural susceptibility equivalents can be derived. Typical values are given in Table 12.

7. THE BOILING POINT NUMBER

The boiling point of an organic compound is usually considered to be a simple physical property. To obtain structural information from this property, an arbitrary self-consistent system of numerical atomic and

TABLE 12
Atomic and Functional Group Magnetic
Susceptibilities[a]

Atom or functional group	Susceptibility equivalent
Hydrogen	-2.85×10^{-6}
Carbon	-6.0×10^{-6}
Oxygen:	
In ethers, alcohols	-4.6×10^{-6}
In ketones	$+1.73 \times 10^{-6}$
In acids, esters	-3.4×10^{-6}
Double bonds:	
C=C	$+5.47 \times 10^{-6}$
C=N	$+8.2 \times 10^{-6}$
N=N	$+1.8 \times 10^{-6}$
Benzene ring	-1.5×10^{-6}
Triple bonds:	
C≡C	$+0.77 \times 10^{-6}$
C≡N	-0.8×10^{-6}
Halogens:	
Fluorine	-11.5×10^{-6}
Chlorine	-20.1×10^{-6}
Bromine	-30.6×10^{-6}
Iodine	-44.0×10^{-6}
Sulfur	-15.0×10^{-6}
Amino nitrogen	-5.6×10^{-6}

[a]Reprinted from Ref. 39 by courtesy of D. Van Nostrand, New York.

group contributions had to be devised. Kinney(95–97) assigned such numerical values for atoms and groups to a parameter called the boiling point number (b.p.n.). The relationship between this parameter and the boiling point is

$$b.p. = 230.14(b.p.n.)^{1/3} - 543 \tag{93}$$

$$\log b.p.n. = 3[\log (b.p. + 543) - 2.24] \tag{94}$$

As with the other properties discussed, the b.p.n. of compounds is the algebraic sum of the atomic and group numbers, each of which has been multiplied by the number of times it occurs in the molecule. The b.p.n. is a very strongly constitutive property, perhaps even excessively so. However, this does mean that considerable structural information can be obtained from it, even though its use may be complicated. The basic measurement, the boiling point, is a very simple one, and it can be used to check the assumed formula and structural detail. The complexities of the system can be observed from Tables 13 and 14.

C. Absorption Spectra

The total energy of an organic molecule may be described as follows:

$$E(\text{total}) = E(\text{electronic}) + E(\text{vibrational}) + E(\text{rotational}) + E(\text{translational})$$

With the practical exception of the energy of translation, these forms of molecular energy are quantized. Thus, a molecule has associated with it a series of discrete energy levels, and to rise to a higher energy state it must absorb an amount of electromagnetic energy equal to the difference between the initial (base) and final (excited) states. For light to be absorbed by an organic molecule, the radiation must be of that wavelength which corresponds to the difference of energy between the base and excited states in the molecule.

The relationship between energy (E) and wavelength (λ) is

$$E = hc/\lambda \tag{95}$$

where h is Planck's constant $(6.63 \times 10^{-27}$ erg-sec), and c is the velocity of light $(3.00 \times 10^{10}$ cm/sec).

Other fundamental equations of spectroscopy relate energy to frequency, ν, by

$$E = h\nu \tag{96}$$

and define the wave number $(\bar{\nu})$ as the reciprocal of wavelength:

$$\bar{\nu} = 1/\lambda \tag{97}$$

The relationships between the various spectroscopic functions are illustrated in Table 15. The boundaries noted for each spectral region are those recommended by the Joint Committee on Nomenclature in Applied Spectroscopy (98).

As can be seen from the noted energies, the rotational energy levels are "close together," the vibrational levels "further apart," and the electronic levels "much further apart." For example, light absorption in the visible region causes a "jump" from a given vibrational–rotational

TABLE 13
Carbon and Hydrogen Atomic and Group Boiling Point Numbers[a]

Atom or functional group	b.p.n.	Atom or functional group	b.p.n.
Main chain carbon	0.8	Types of olefinic linkage	
Main chain hydrogen	1.0	CH_2=CH	1.2
Main chain groups:		RCH=CH_2	1.5
Methyl	3.05	RCH=CHR	1.9
Ethyl	5.5	R_2C=CHR	2.3
Propyl	7.0	R_2C=CR_2	2.8
Butyl	9.7	Groups, unsaturated,	
2,2-dimethyl	−0.4	attached to main chain	
Two or three alkyls		Methylene	4.4
attached to adjacent		Ethylene	7.0
carbons of main chain		Vinyl	5.4
(sat.) of six carbons		Propylidene	9.0
or less	0.5	Butylidene	10.4
Four or more alkyls		Types of acetylenic linkage	
attached to adjacent		HC≡CH	4.0
carbons of main chain		RC≡CH	4.4
(sat.) of six carbons		RC≡CCH_3	5.4
or less	1.0	RC≡CR	4.8
Cyclic groups:		Diolefin types:	
Add 0.8 for each carbon,		Alkenes	4.8
1.0 for each hydrogen,		Conjugated, the	
the indicated values for specific		normal value plus	0.8
types of unsaturation, and the		Triolefin types:	
following for rings:		All bonds conjugated,	
Cyclopropyl	2.1	the normal value plus	2.4
Cyclobutyl	2.3	Two bonds conjugated,	
Cyclopentyl	2.5	the normal value plus	0.8
Cyclohexyl	2.7	Diacetylene types:	
Cycloheptyl	3.4	1, 3, normal only	—
Cyclooctyl	3.9	Other conjugated, the	
Cyclononyl	4.4	normal value plus	3.0
Cyclodecyl	4.9	Enyne types:	
Cyclohendecyl	5.4	Conjugated, the	
Cyclododecyl	5.9	normal value plus	0.8
Cyclotridecyl	6.4	Dienyne types:	
Cyclotetradecyl	6.9	Conjugated, the	
Cyclopentadecyl	7.4	normal value plus	2.4
Cyclohexadecyl	7.9		
Cycloheptadecyl	8.4		

[a]Reprinted from Ref. 93 by courtesy of Academic Press, New York.

TABLE 14
Functional Group Boiling Point Numbers[a]

Group	b.p.n.	Group	b.p.n.
Acids (—COOH)		Alcohols (—OH)	
HCOOH	20.0	CH_3OH	12.8
CH_3COOH	20.0	RCH_2OH	10.8
RCH_2COOH	19.3	R_2CHOH	8.8
$R_2CHCOOH$	18.6	R_3COH	6.8
R_3CCOOH	17.9	Ethers (—O—)	
		$(CH_3)_2O$	3.8
Esters (R'—OC—R) with O double bond		$(RCH_2)(CH_3)O$	2.9
HCOOCH$_3$	9.4	$(R_2CH)(CH_3)O$	2.9
CH_3COOCH_3	9.4	$(R_3C)(CH_3)O$	2.9
RCH_2COOCH_3	8.5	$(RCH_2)_2O$	2.0
CH_3COOCH_2R	8.5	$(R_2CH)(RCH_2)O$	2.0
$HCOOCH_2R$	8.5	$(R_3C)(RCH_2)O$	2.0
$HCOOCHR_2$	7.6	$(R_2CH)_2O$	1.1
$R_2CHCOOCH_3$	7.6	$(R_3C)(R_2CH)O$	1.1
RCH_2COOCH_2R	7.6	$(R_3C)_2O$	0.2
$CH_3COOCHR_2$	7.6		
$HCOOCR_3$	6.7	Aldehydes —C=O with H	
$R_3CCOOCH_3$	6.7	HCHO	8.8
$R_2CHCOOCH_2R$	6.7	CH_3CHO	8.8
$RCH_2COOCHR_2$	6.7	RCH_2CHO	8.2
CH_3COOCR_3	6.7	R_2CHCHO	7.6
RCH_2COOCR_3	5.8	R_3CCHO	7.0
$R_2CHCOOCHR_2$	5.8	Tertiary amines (≡N)	
$R_3CCOOCH_2R$	5.8	$(CH_3)_3N$	2.5
$R_2CHCOOCR_3$	4.9	$RCH_2(CH_3)_2N$	2.0
$R_3CCOOCHR_2$	4.9	$(R_2CH)(CH_3)_2N$	1.5
$R_3CCOOCR_3$	4.0	$(RCH_2)_2(CH_3)N$	1.5
		$(R_3C)(CH_3)_2N$	1.25
Ketones C=O		$(R_2CH)_2(CH_3)N$	1.25
		$(RCH_2)_3N$	1.25
$(CH_3)_2CO$	8.5	Cyanides (—CN)	
$(RCH_2)(CH_3)CO$	8.0	CH_3CN	15.2
$(R_2CH)(CH_3)CO$	7.5	RCH_2CN	14.0
$(RCH_2)_2CO$	7.5	R_2CHCN	12.8
$(R_3C)(CH_3)CO$	7.0	R_3CCN	11.6
$(R_2CH)(RCH_2)CO$	7.0	Isocyanides (—NC)	
$(R_3C)(RCH_2)CO$	6.5	CH_3NC	13.3
$(R_2CH)_2CO$	6.5	RCH_2NC	12.2
$(R_3C)(R_2CH)CO$	6.0	R_2CHNC	11.1
$(R_3C)_2CO$	5.5	R_3CNC	10.0
Primary amines (—NH$_2$)		Chlorides (—Cl)	
CH_3NH_2	8.4	RCH_2Cl	7.5

TABLE 14—*continued*

Group	b.p.n.	Group	b.p.n.
RCH$_2$NH$_2$	7.3	R$_2$CHCl	6.5
R$_2$CHNH$_2$	6.2	R$_3$CCl	6.0
R$_3$CNH$_2$	5.1		
Secondary amines (—NH)			
(CH$_3$)$_2$NH	6.0		
(RCH$_2$)(CH$_3$)NH	5.0		
(R$_2$CH)(CH$_3$)NH	4.0		
(RCH$_2$)$_2$NH	4.0		
(R$_3$C)(CH$_3$)NH	3.5		
(R$_2$CH)(RCH$_2$)NH	3.5		
(R$_2$CH)$_2$NH	3.0		
(R$_3$C)(RCH$_2$)NH	3.0		

^{a}Reprinted from Ref. *93* by courtesy of Academic Press, New York.

substate of the base electronic state to a given vibrational–rotational substate of an excited electronic state. In contrast, the near infrared excited molecules causes a jump from a given rotational substate of the base vibrational state to a given rotational substate of an excited vibrational state.

Due to the discrete nature of energy levels in molecules, as a system of molecules is subjected to light energy of changing wavelength, some wavelengths will be absorbed and others will be transmitted. A plot of the fraction of light energy either absorbed or transmitted can be made as a function of the wavelength. Such a plot is called an absorption (or transmission) spectrum and consists of a series of connected maxima and minima.

Spectra are employed in obtaining qualitative information about organic molecules and provide one of the most valuable tools available for this purpose. For theoretical and practical reasons, absorption spectrophotometry may be divided into visible-ultraviolet and infrared spectrophotometry. These two techniques are treated separately in the following discussion.

1. VISIBLE AND ULTRAVIOLET SPECTRA

Visible and ultraviolet spectra are conveniently treated together for two reasons: (1) Both involve transitions to a higher electronic energy level, and (2) the optical system and cells which are transparent to ultraviolet radiation are also transparent to visible radiation. Hence, deter-

TABLE 15
Relationship of Spectroscopic Functions

Wavelength, nm	Spectral region	Spectra origin (type of molecular excitation)	Wavenumber, cm⁻¹	Energy kcal/mole	eV
300,000			33.3	9.66×10^{-2}	4.2×10^{-2}
	Far infrared	Rotation			
30,000			333	9.66×10^{-1}	4.2×10^{-1}
	Middle infrared	Rotation-vibration			
3,000			$3,333$	9.66	4.2
	Near infrared				
780			$12,800$	37.16	1.61
	Visible	Rotation-vibration			
380			$26,300$	75.87	3.29
	Near ultraviolet	electronic (valence			
		electrons)			
200			$50,000$	143.43	6.22
	Far ultraviolet				
100			10^6	2868.6	124.4

minations involving the two spectra can be performed by the same instrument.

The "usable" visible-ultraviolet range extends from 780 nm down to boundary between the near and far ultraviolet regions. The lower limit is set by the fact that oxygen absorbs strongly at wavelengths shorter than 200 nm. The following discussion therefore relates only to the visible-near ultraviolet spectrum.

Most visible-ultraviolet spectrophotometry is carried on with solutions of the compound under study. Fortunately, there are many solvents of widely varying polarity which are transparent to radiation over most of the region. These solvents include hexane, methanol, ethanol, water, and chloroform.

In a given solvent, the visible-ultraviolet spectrum of an organic compound is characteristic of the compound and can be used to aid in its identification. However, due to solvent and other effects, visible-ultraviolet spectra tend to be relatively structureless with very few maxima. Thus, unlike infrared spectra, they do not provide characteristic "fingerprints" of individual compounds. A visible-ultraviolet spectrum is illustrated in Fig. 2. Furthermore, it is rarely possible to attribute strong absorption in the visible-ultraviolet region to any specific functional group or bonding. There are, however, certain generalizations that can be made, and these will be discussed.

To compare absorption spectra without actually presenting them, two pieces of information are required. One is the wavelength at which maximum absorption occurs (λ_{max}); the second is the molar absorptivity, ϵ, at λ_{max}. The latter is defined by

$$\epsilon = A_s/bc \qquad (98)$$

where A_s is the absorbance [$\log_{10}$ (1/transmittance)], b is the pathlength (1 cm), and c is the concentration (moles/liter).

For ϵ to be determined, the molecular weight must be known. If the molecular weight is unknown, then the absorbance of a 1-g/liter solution is determined for $b = 1$ cm, and this is designated as a_s [see Eq. (229)].

Absorption in the visible-ultraviolet region is associated with compounds containing π electrons, or in different terms, with compounds containing unsaturated bonds. These bonds can be in simple groups called "chromophores" which include C=C, C=O, C=N, N=N, C≡C, and N=N. The λ_{max} for the strong absorption region of such unconjugated bonds is near or below 200 nm (99). The carbonyl group (C=O) does exhibit a weak absorption maximum around 290 nm for aldelydes and ketones, but not for acids, amides, or esters (99). The azo group (—N=N—) has a weak maximum around 240 nm (100).

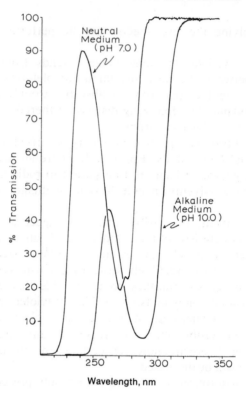

Fig. 2. The ultraviolet spectrum of *m*-cresol at two different pH values [courtesy of Dr. O. Aly, Rutgers University].

Conjugation of double bonds shifts absorption to longer wavelengths ("bathochromic shift") and frequently results in an increase in ϵ at λ_{max}. As illustrated by the examples in Table 16, progressive extension of a conjugated system produces regular shifts to longer wavelengths.

Bathochromic shifts also result when electron donor groups are conjugated to a double bond. Such groups contain unshared (nonbonding) electrons such as —O—, —S—, or —N=, and are called auxochromes since their effects are auxiliary to those of chromophores.

Another example of bathochromic behavior is found in aromatic compounds in the benzene condensed ring family. The data in Table 17 show a progressive bathochromic shift from benzene through naphthacene, along with a corresponding increase in ϵ at λ_{max}. The substitution of nitrogen for carbon atoms in aromatic rings has little effect on λ_{max}, but the effect on ϵ is variable.

Electron donor substituents on the benzene ring act as auxochromes and cause a shift in the benzene ring absorption at 255 nm. Substituents

TABLE 16

Bathochromic Shift of Conjugated Ethylenic Chromophores[a]

System	Name	Color	λ_{max}, nm	ϵ
$(C{=}C)_2$	Butadiene	None	217	21,000
$(C{=}C)_3$	Hexatriene	None	258	35,000
$(C{=}C)_4$	Dimethyloctate-traene	Pale yellow	296	52,000
$(C{=}C)_5$	Decapentene	Yellow	335	118,000
$(C{=}C)_8$	Dihydro-β-carotene	Orange	415	210,000
$(C{=}C)_{11}$	Lycopene	Red	470	185,000
$(C{=}C)_{15}$	Dehydrolycopene	Violet	504	150,000

[a]From Ref. 99.

TABLE 17

Bathochromic Shift in the Benzene Condensed Ring Family; Band 2[a]

Compound	Number of rings	Color	λ_{max}, nm	ϵ
Benzene	1	None	202	6,920
Naphthalene	2	None	275	5,620
Anthracene	3	None	375	7,940
Naphthacene	4	Yel-orange	473	11,200
Pentacene	5	Viol-blue	580	15,800

[a]From Ref. 88.

such as —OH, —OCH$_3$, —NH$_2$, and —N(CH$_3$)$_2$ greatly intensify the band and shift it toward the visible region (for the un-ionized forms). A similar effect is produced by groups having double bond (e.g., —CH=CH$_2$, —CHO, —NO$_2$) that can conjugate with those of the benzene ring. Alkyl and halogen substituent groups have comparatively little effect on benzene ring absorption (88).

Solvent effects on absorption in the visible-ultraviolet region may be significant. When a polar solvent interacts with a polar chromophore, absorption may be shifted to a lower wavelength. This phenomenon is called a "hypsochromic" shift. In addition, many organic compounds undergo shifts in λ_{max} due to structural changes resulting from redox or acid–base transitions. Nitrophenols, for example, shift from yellow to colorless as the pH decreases.

Another solvent effect is the elimination of fine vibrational spectral structure through molecular interactions. This, along with the basic problem of resolving fine structure in such a narrow spectral region, renders most visible-ultraviolet spectra relatively featureless. The lack of fine structures severely limits the use of such spectra for definite molecular identifications. Furthermore, conjugation of chromophores with other chromophores and auxochromes allows many different functional group systems to exhibit the same λ_{max}. This severely limits the use of spectra for functional group analysis.

In cases where different materials have the same λ_{max}, the ϵ value may sometimes provide an aid to identification. This is shown by the compounds in Table 18, which have similar λ_{max} values, but which could be easily distinguished on the basis of their molar extinction coefficients.

TABLE 18
Absorption Characteristics of
Several Compounds of
Similar λ_{max}[a]

Compound	λ_{max}, nm	ϵ
Benzene	255	230
Pyridine	250	2,000
Phenathrene	252	63,000
Anthracene	252	200,000

[a]From Ref. 99.

Despite the fact that exact structural knowledge usually cannot be determined, visible-ultraviolet spectra do give certain types of information. Lack of absorption between 200 and 800 nm narrows the probable identity of a compound to a group consisting of aliphatic or alicycle hydrocarbons, carboxylic acids, esters, alcohols, ethers, amines, nitriles, chlorides, and fluorides. Absorption between 250 and 300 nm with some fine structure suggests the presence of a benzene ring. Weak absorption with lack of fine structure may be due to unconjugated carbonyl, bromide, or iodide groups. The λ_{max} values for ethylenic conjugation (Table 16) have considerable interpretive value. Visible-ultraviolet spectra can also be used to differentiate between cis–trans isomers (99). Finally, absorption in the visible region is a general indication of four or more conjugated chromophores and auxochromes, especially in quinoid structures.

In some cases, visible-ultraviolet spectrophotometry can provide qualitative information through molecular weight determinations. Where a compound forms a derivative with a reagent which has a λ_{max} far removed from its own, the ϵ value of the derivative frequently is almost the same as that of the reagent. The amine picrates, for example, have a λ_{max} of 380 nm, with an ϵ of 13,400. The latter value is almost identical to that of the picric acid reagent. To determine the amine molecular weight, the a_s value [see Eq. (229)] of the derivative is first obtained. Then, since $a_s \times M = 13{,}400$, the derivative molecular weight M is $13.4 \times 10^3/a_s$. The molecular weight of the amine can then be determined by assuming a 1 : 1 addition and subtracting the molecular weight of picric acid from that of the derivative(101).

2. INFRARED SPECTRA

Infrared spectra provide a powerful means for obtaining information about the types of organic substances isolated from the environment. There are two principal ways in which the spectra are used for characterization studies. First, spectra are so characteristic for individual organic compounds that they provide a "fingerprint" for identification. An example is presented in Fig. 3. Second, the spectra permit functional group characterization, since most groups have associated with them a rather narrow range of absorption frequencies. A detailed list of groups with their associated absorption bands (group frequencies) is presented in Fig. 4.

The absorption frequency of a particular group varies slightly with small differences in molecular structure. This association has given rise

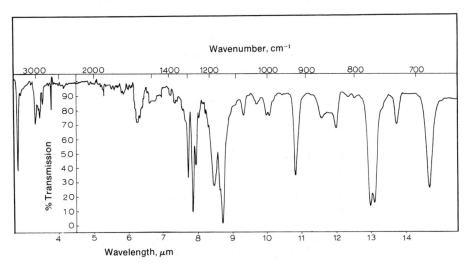

Fig. 3. Infrared spectrum of m-cresol [courtesy of Dr. O. Aly, Rutgers University].

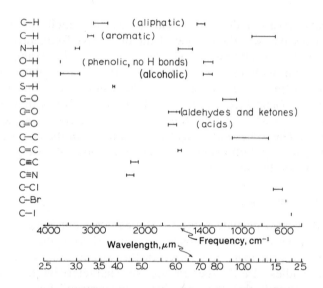

Fig. 4. Correlation of structure and infrared spectral bands.

to the so-called "neighboring group effect." Structural effect on carbonyl group absorption has been studied in detail and is summarized in Table 19(*35*). Due to the broadness of the infrared region the different types of carbonyl groups usually can be observed. Thus, the effect can be considered more an aid to identification than a problem in interpretation.

Much infrared spectrophotometric analysis is performed with solutions.

TABLE 19
Carbonyl Absorption
Frequencies[a]

Group	Frequency, cm^{-1}
Acids	1620–1754
Esters	1735
Acid andydrides	1754–1815
Acyl halides	1792
Amides	1662
Aldehydes	1720–1780
Ketones	1710

[a]Reprinted from Ref. *35* by courtesy of John Wiley (Interscience), New York.

Due to the fact that all solvents absorb in some part of the most useful region (2500–25,000 nm), certain group absorptions may be swamped by solvent effects. In addition, this also necessitates the use of short pathlengths. As the molar extinction coefficients for infrared absorption maxima are of the order of 10^3, as opposed to visible-ultraviolet values of 10^4–10^5, fairly concentrated solutions must be employed.

The tendency is to use solvents that do not have major absorption bands in the particular region of interest. Typical solvents are carbon disulfide, carbon tetrachloride, Nujol, cyclohexane, and chloroform. Water both absorbs heavily and attacks the salt cells (e.g., KCl, CsBr) usually employed in infrared spectrophotometry. If water is used as a solvent, absorption must be studied in a cell made of such materials as AgCl, BaF_2, CaF_2, some organic polymers, etc. These materials add their own absorption bands, and this causes some problems in work with highly polar compounds. Infrared spectra are so valuable a tool, however, that considerable modifications of the basic procedure have been made so that qualitative information can be obtained. As a result, infrared spectra can be determined on gases, liquids, thin films, solids in Nujol mulls, substances absorbed on filters and in salt pellets, and even on particulates by reflection techniques.

Attempts have been made to relate the molar extinction coefficients to groups in the infrared region. This is usually unsuccessful because, for narrow infrared bands, the peak height is also a function of instrumental resolution. This means that general values of ϵ would be of little or no use.

It has been noted, however, that the area under an absorption curve maximum is much less dependent on instrumental resolution. This area is obtained by integrating $\ln(I_0/I)$ over the frequency spread of the absorption band. A function $\bar{A}$, the integrated area, can then be defined as

$$\bar{A} = (1/C_m l) \int (\ln I_0/I)\, d\nu \qquad (99)$$

where I_0 is the intensity of the incident light, I is the intensity of the transmitted light, C_m is the molar concentration, ν is the frequency, and l is the pathlength(88). The area can be determined by many of the means used to determine peak area in gas chromatography.

This type of analysis reveals that there is some relationship between group identity and $\bar{A}$, as demonstrated in Table 20(88). The data refer specifically to one solvent system, but they can be used to check the suspected identity of an absorption band obtained in another solvent (see Chap. 30).

TABLE 20
Values of $\bar{A}$ in Carbon Tetrachloride Solution[a]

Functional group	Class of compounds	Approximate wave number	$\bar{A} \times 10^{-3}$ [b]
—OH	Primary alcohol	3630	2.1–2.4
	Secondary alcohol	3620	ca. 2.0
	Phenols	3600	5.0–6.0
	Oximes	3600	5.5–6.5
	Cyanohydrins	3600	2.0–3.0
—NH	Dialkylamines	3400	ca. 0.04
	Anilines	3400	0.8–4.5
	Anilines	3480	1.0–2.5
	Methyl anilines	3450	1.5–4.3
	Diarylamines	3450	ca. 2.0
	Pyrroles, indoles	3450	4.5–5.5
—CH₂—	Hydrocarbon	2900	ca. 3.3
	Hydrocarbon	1460	ca. 0.2
	—COCH₂—	2900	ca. 0.6
	—COCH₂—	1460	ca. 1.0
—CH₃—	Hydrocarbon	2900	ca. 3.8
	Hydrocarbon	1375	ca. 0.35
	—COCH₃—	2900	ca. 0.6
	—COCH₃—	1375	ca. 1.7
—C≡N	Aliphatic nitrile	2250	0.22–0.35
	Aromatic nitrile	2225	0.22–4.00
—N=C=N—	Carbodiimide	2130	45–50
—N=C=O	Phenylisocyanate	2270	62–74
—S—C≡N	Phenylthiocyanate	2170	1.2–2.3
—N=C=S	Alkylisothiocyanate	2100	17–52
—N=C=S	Aromatic isothiocyanate	2050	65–90
H | —C=O	Aliphatic aldehyde	1730	ca. 8.0
O H H ‖ | | —C—C=C	Unsaturated ketones and acetophenones	1690	7.5–9.5
R / —O—C=O	Aliphatic esters	1735	10–14
φ / —O—C=O	Benzoates	1720	12–16
R / —O—C=O	Esters	1200	ca. 13
\ C=O /	Alkyl ketones	1200	ca. 2.0

[a]Reprinted from Ref. 88 by courtesy of Elsevier, New York.
[b]See Eq. (99).

D. Fluorescence Spectra

As noted in the section on visible-ultraviolet spectra, the absorption of radiation in this region raises a molecule to a higher electronic energy state. Once absorption has occurred, the excited molecule may lose energy through a number of different mechanisms.

Usually the energy is given up as infrared radiation (gas phase) or degraded into heat by a stepwise loss of vibrational energy (solution phase). This can occur spontaneously or assisted by thermal collisions. As radiation in the visible-ultraviolet region has sufficient energy to break chemical bonds (i.e., in the range of 100 kcal/mole for about 280 nm radiation), the molecule may break down into free radicals. If sufficient energy, due to far ultraviolet radiation, is present (i.e., in the range of > 200 kcal/mole), the molecule may even break down into ions (99).

Another way the energy may be lost is through a spontaneous decrease in electronic energy accompanied by an emission of visible or ultraviolet light. The wavelength of the emitted radiation is virtually always longer than that of the absorbed radiation. This is due to the loss of some vibrational energy before emission occurs, and also to the fact that the molecule returns to a higher vibrational ground state than the one occupied prior to excitation. Two kinds of light emission, fluorescence and phosphorescence, are shown by organic molecules. The two are distinguished by great differences in their rates of emission (lifetimes). Fluorescence lifetimes range between 10^{-6} and 10^{-9} sec, whereas phosphorescence lifetimes are 10^{-4} sec or longer. The following discussion concerns only fluorescence spectra.

As illustrated in Fig. 5, two curves are usually presented for fluorescence spectra. Figure 5 represents the intensity of emitted (fluorescent) light as a function of wavelength. Here, the exciting radiation is that wavelength that yields maximum fluorescence, and the result is called the fluorescence spectrum per se. Also shown is the intensity of emitted light, at the wavelength of peak fluorescent emission, as a function of wavelength of the exciting radiation. Theoretically, this curve, called the excitation spectrum, should be the same as the visible-ultraviolet absorption spectrum. In practice, however, it may be slightly different.

Since fluorescence of molecules is dependent upon the absorption of visible or ultraviolet light, it is associated with the presence of π electrons. Substituent groups that increase the freedom of π electrons enhance fluorescence; those that localize π electrons diminish fluorescence. Benzene, which itself is weakly fluorescent, is a good example. Orthopara directing groups tend to activate the benzene ring, increase the freedom of the π electrons, and therefore, enhance fluorescence. Meta directing groups tend to inactivate the benzene ring, decrease freedom of the

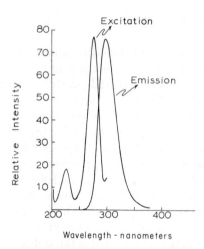

Fig. 5. Fluorescence excitation and emission spectra of *m*-cresol [courtesy of American Instrument Company].

π electrons, and inhibit fluorescence(*102*). There are, however, exceptions to this generalization, as is shown in Table 21.

In combination, the behavior of substituent groups is rather complex. Weakly ortho–para or meta directing groups may actually increase fluorescence when associated with strongly ortho para directing groups. Fluorescence is also strongly affected by such environmental factors as solvent nature, presence of buffers, pH, and temperature. For example, the fluorescence of sulfanilamide in hexane is only 42.8% and in dioxane only 24% of its value in water, while in ethanol it is 107% of this value (*102*). Buffers as such may inhibit (quench) fluorescence, but the general effect of pH is due to the relative fluorescence of compounds in their ionized and un-ionized forms. For example, both phenol and aniline fluoresce, but the corresponding phenylate and anilate ions do not(*102*). Increases in temperature, in general, tend to decrease fluorescence.

Many naturally occurring organics fluoresce, including certain vitamins, alkaloids, hormones, antibiotics, plant pigments, and metabolic by-products(*103,104*). Fluorescence can be used in approximately the same way as visible-ultraviolet spectra, that is, as a characteristic of the pure compound. Unfortunately, at the present time, there seems to be even less correlation of fluorescence spectra with structural makeup than occurs with visible-ultraviolet spectra. Of course, certain correlations do exist, as is shown in Tables 22 and 23 for phenols and indoles(*104*). Even here, however, the hydroxy benzoic acids fluoresce and are excited at wavelengths similar to the indoles. Furthermore, fluorescence at higher

TABLE 21
The Effect of Substituent Groups on
Benzene Fluorescence[a]

Substituent group	Effect	Orientation
Amino ($-NH_2$)	Strongly increasing	ortho-para
Alkylamino ($-NHR$)	Strongly increasing	ortho-para
Dialkylamino ($-NR_2$)	Strongly increasing	ortho-para
Hydroxy ($-OH$)	Strongly increasing	ortho-para
Alkoxy ($-OR$)	Strongly increasing	ortho-para
Nitrile ($-CN$)	Strongly increasing	meta
Alkyl ($-R$)	Weakly increasing	ortho-para
Fluoro ($-F$)	Weakly increasing	ortho-para
Bromo (Br)	Decreasing	ortho-para
Chloro ($-Cl$)	Decreasing	ortho-para
Iodo ($-I$)	Decreasing	ortho-para
Amido ($-NHCOR$)	Decreasing	ortho-para
Nitro ($-NO_2$)	Decreasing	meta
Carboxylic ($-COOH$)	Decreasing	meta
Sulfonic (SO_3H)	Decreasing	meta
Aldehyde ($-CHO$)	Decreasing	meta

[a]From Ref. 27.

TABLE 22
Fluorescent Characteristics of Phenols[a]

Compound	Activation maximum,[b] nm	Fluorescent maximum, nm	pH for maximum fluorescence
Phenol	270 (215)	310	1
o-Cresol	275 (220)	305	1–7
m-Cresol	280 (220)	305	1–7
p-Cresol	285 (230)	315	1–7
2:4-Xylenol	285	310	7
2:5-Xylenol	275 (230)	305	7
3:4-Xylenol	280 (230)	310	7
Catechol	270 (220)	325	7
Resorcinol	265 (215)	315	7
Quinol	285 (220)	340	1

[a]From Ref. 104.
[b]Figures in parentheses represent minor activation peaks.

TABLE 23
Fluorescent Characteristics of Indoles[a]

Compound	Activation maximum, nm	Fluorescent maximum, nm	pH for maximum fluorescence
Indole	280	350	7
Skatole	290	370	–
Oxindole	300	345	–
Indoxyl	310	395	–
2-Methylindole	280	355	–
2,3-Dihydoxyindole	315	400	–
Tryptophane	285	365	11
Tryptamine	290	360	7
5-Hydoxyindoleacetic acid	300	355	7
5-Hydroxyindole	290	355	1

[a]From Ref. 104.

wavelengths presents an even more confused picture. Thus, in its present state, fluorescence does not provide significant information on the nature of functional groups, and it is, therefore, of very limited use in qualitative studies of the aquatic environment. However, as a trace detection technique it is very sensitive (see Chap. 31).

E. Raman Spectra

When monochromatic light is scattered by a transparent material, the light scattered at right angles contains frequencies not present in the incident radiation. These shifts in frequency constitute the Raman spectrum of the transmitting medium. Each spectral peak represents the difference $\Delta\nu$ between the frequency of incident radiation ν and that of scattered radiation ν_s. A Raman spectrum is thus a series of lines in terms of their displacement from the frequency of incident radiation.

The incident radiation (visible or ultraviolet) may lose energy in exciting a molecule from its ground state to a higher vibrational-rotational state, giving rise to what are called "Stokes" lines. Conversely, if the molecule already is in an excited state, the radiation may gain in energy as the molecule returns to its ground state, giving rise to what are called "anti-Stokes" lines. Raman spectra are thus related to infrared spectra, as both involve vibrational–rotational transformations (105a).

In a manner similar to infrared spectra, Raman spectra are so characteristic for individual organic compounds that they provide a "fingerprint" for identification. Furthermore, the spectra can be employed to detect specific functional groups in organic molecules(*39,105a*) and can therefore be used to obtain certain information about mixtures. This is especially true for CH, CD, C=O, C=C, C≡C, and C≡CH groups, in which the lines fall in a region where other Raman lines seldom appear. Table 24 presents examples of Raman spectra for these groups.

TABLE 24
Raman Data for Selected Functional
Groups[a]

Functional group	Raman displacement, cm^{-1}
CH	3050–3080
CD	2260
C=O	1780
C=C	1640–1680
C≡C	2230
C≡CH	2110–2120

[a]Reprinted from Ref. *105a* by courtesy of Academic Press, New York.

In many cases, Raman data complement data from infrared spectra. For example, absorption by sulfur functional groups is frequently weak in the infrared, but strong Raman emissions are noted for the SH (2575 cm^{-1}), CS (640–695 cm^{-1}), and SS (510 cm^{-1}) groups(*35*).

Unlike the infrared, Raman spectra are mainly applicable to liquid systems. The component concentration is reasonably high ($> 1\%$)(*35*). However, water is an excellent solvent for Raman spectra, while it absorbs heavily in the infrared region. Despite several minor advantages, Raman spectrophotometry remains a less popular (and more expensive) technique than the infrared.

F. Mass Spectra

Mass spectrometry involves the passing of a beam of electrons through a system of organic molecules in the vapor phase, with the pressure adjusted so low that intermolecular collisions are infrequent. If the energy of the electrons in the beam is 1–15 eV, the molecule will lose an electron and become a positively charged molecular ion. The process is as follows:

$$m + e^- \rightarrow m^+ + 2e^- \tag{100}$$

If the energy of the electrons in the beam is increased to 50–70 eV, the positively charged molecular ion will break down into fragments:

$$m^+ \rightarrow m_1^+ + m_2 \tag{101}$$

One molecular ion will break into many fragments of different masses. The function of the mass spectrophotometer is to separate and identify the masses and relative abundance of these fragments.

This is done by accelerating the ions at a potential E of several thousand volts, and then allowing the accelerated ions to enter a magnetic field H. Here they are deflected at a right angle to their path, and describe a path of radius r. The radius is defined for a particle of mass m_1 by

$$r_1^2 = (m_1/e)(2E/H^2) \tag{102}$$

where e is the mass of an electron. At given values of E and H, each particle will describe a radius depending on its mass-to-charge ratio (which is usually 1). If the potential, E, of the accelerating field is continuously increased (or decreased), each charged fragment in turn will describe a circle of radius r. If these ions are suitably detected, a spectrum will be obtained of the relative abundance of fragments of different masses as a function of m/e. An example of a mass spectrum is presented in Fig. 6. In such a spectrum, peak location defines the mass of an ion (i.e., m/e), and peak height gives the corresponding abundance.

In certain instances, especially at higher voltages, a double ionization may occur:

$$m_1 + e^- \rightarrow m_1^{2+} + 3e^- \tag{103}$$

Here the apparent mass is $m/2e$, or one-half of the actual mass. Double ionization is easy to recognize if m is an odd number, as the result will be a fractional mass. A fragment where m is an even number is more difficult to recognize.

Information can be obtained from mass spectra in a number of ways. If the ionizing voltage is held at 9–15 eV, almost all the ions produced will be the singly ionized form of parent molecule P^+. The spectrum will show only one major peak, and this will represent Mp/e for the parent. Therefore, it is possible to determine the whole number molecule weight of the parent molecule (100).

Another method of obtaining information is through the isotope effect. For example, with a given compound, a certain portion of the molecules might contain a deuterium instead of a hydrogen atom, or a ^{13}C instead

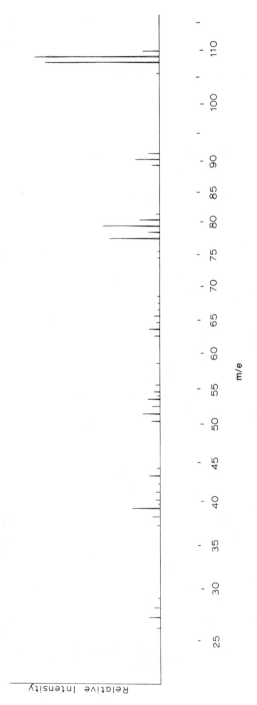

Fig. 6. Mass spectrum of *m*-cresol [courtesy of Varian Associates].

of a ^{12}C atom. In either case, there will be a minor peak at $Mp+1$, and if a few molecules happen to contain both heavier isotopes, a very small peak will occur at $Mp+2$. This effect is even more prominent for chlorine- and bromine-substituted organic compounds, as can be seen from the list of isotope ratios in Table 25.

TABLE 25
Relative Natural Abundance of
Certain Isotopes[a]

Isotopes	Relative abundance	Isotopes	Relative abundance
$^2D/^1H$	1.6×10^{-4}	$^{33}S/^{32}S$	7.8×10^{-3}
$^{13}C/^{12}C$	1.12×10^{-2}	$^{34}S/^{32}S$	4.4×10^{-2}
$^{15}N/^{14}N$	3.8×10^{-3}	$^{37}Cl/^{35}Cl$	3.2×10^{-1}
$^{17}O/^{16}O$	4.0×10^{-4}	$^{81}Br/^{79}Br$	9.8×10^{-1}
$^{18}O/^{16}O$	2.0×10^{-3}		

[a]From Ref. *100*.

Due to the abundance of the ^{37}Cl and ^{81}Br isotopes, compounds containing either element will have a strong peak at $Mp+2$ (one heavy atom/molecule) and a weak peak at $Mp+4$ (two heavy atoms/molecule). Such peaks are so characteristic that they represent one of the major methods for identification of chlorine- and bromine-substituted organic compounds.

As previously described, at a potential of 50–70 eV, an ionized molecule will split into numerous fragments. Various recombinations and rearrangements then occur, so the resultant spectrum is quite complex. In many cases, the parent peak at Mp will be quite small. Therefore, for high voltage spectra, the largest (base) peak is used as a reference, and the remaining peaks are expressed as percentages of this height. This technique provides a pattern of the percentage of each ion with reference to the commonest ion. The pattern, known as the cracking pattern, is complex and very characteristic of the parent. It thus permits definitive identification of organic compounds in a manner similar to that of infrared spectra.

In addition to use in establishing the identity of given compounds, mass spectra can also provide certain structural information because different types of organic compounds yield different fragment ions. The information is obtained by comparing the spectral peaks of such ions with data tabulated for known structures. A brief outline of such data is represented in Table 26.

Interpretation of mass spectra remains, however, a complex problem.

TABLE 26
Mass Spectra Data for Various Fragment Ions[a]

Compound type	M/e peak[b]	Peak origin[c]
Hydrocarbons:		
Aliphatic	$n \times 14 + 1$	$[(CH_2)_n H]^+$
	$27 + n \times 14$	$[CH_2{=}CH(CH_2)_n]^+$
Alicyclic	M_p	$[P]^+$
Olefinic	41	$[H_2C{=}CHCH_2]^+$
Aromatic	M_{pa}	$[P_a]^+$
	91	$[C_6H_5CH_2]^+$
	78	$[C_6H_6]^+$
	77	$[C_6H_5]^+$
	79	$[C_6H_7]^+$
	65	$[(C_6H_5CH_2){-}(C_2H_2)]^+$
	39	$[C_3H_3]^+$
Carboxylic acids	M_p	$[P]^+$
	$M_p - 17$	$[P-(OH)]^+$
	$M_p - 18$	$[P{-}(H_2O)]^+$
	$M_p - 45$	$[P-(COOH)]^+$
	60	$[CH_3COOH]^+$
Esters	M_{pa}	$[P_a]^+$
	$M_p - (31 + n \times 14)$	$[P{-}O(CH_2)_n CH_3]^+$
	$15 + n \times 14$	$[CH_3(CH_2)_n]^+$
	$59 + n \times 14$	$[COO(CH_2)_n CH_3]^+$
	$73 + n \times 14$	$[CH_2COO(CH_2)_n CH_3]^+$
	$31 + n \times 14$	$[O(CH_2)_n CH_3]^+$
	$43 + n \times 14$	$[CH_3(CH_2)_n CO]^+$
Alcohols	$M_p - 18$	$[P{-}H_2O]^+$
Primary	31	$[CH_2OH]^+$
Secondary	45	$[CH(CH_3)OH]^+$
Tertiary	59	$[C(CH_3)_2OH]^+$
Ethers	$15 + n \times 14$	$[CH_3(CH_2)_n]^+$
	$31 + n \times 14$	$[CH_3O(CH_2)_n]^+$
Aldehydes	29	$[CHO]^+$
	31	$[CH_2OH]^+$
	43	$[CH_2CHO]^+$
	$44 + n \times 14$	$[CH_3(CH_2)_n CHO]^+$
	45	$[CH_2CH_2OH]^+$
	$M_p - 1$	$[P{-}H]^+$
	$M_p - 18$	$[P{-}H_2O]^+$
	$M_p - 28$	$[P-(CO)]^+$
	$M_p - 29$	$[P-(CHO)]^+$
	$M_p - 43$	$[P-(CH_2CHO)]^+$
	$M_p - 44$	$[P-(CH_3CHO)]^+$
Ketones	M_p	$[P]^+$
	$15 + n \times 14$	$[CH_3(CH_2)_n]^+$
	$43 + n \times 14$	$[CH_3(CH_2)_n CO]^+$
	$17 + n \times 14$	$[(CH_2)_n OH]^+$
Methyl	58	$[CH_3COCH_3]^+$

TABLE 26 – *continued*

Compound type	M/e peak	Peak origin
Ethyl	$M_p - 29$	$[P-(CH_3CH_2]^+$
Amines	M_{pa}	$[P_a]^+$
	18	$[NH_4]^+$
	$30 + n \times 14$	$[CH_2(CH_2)_n NH_2]^+$
	$31 + n \times 14$	$[CH_3(CH_2)_n NH_2]^+$
Amides	M_p	$[P]^+$
	$M_p - 16$	$[P-(NH_2)]^+$
	$M_p - 17$	$[P-(NH_3)]^+$
	$M_p - 44$	$[P-(CONH_2)]^+$
	59	$[CH_3CONH_2]^+$
Nitriles	41	$[CH_3CN]^+$
	$26 + n \times 14$	$[(CH_2)_n CN]^+$
Nitro	30	$[NO]^+$
	46	$[NO_2]^+$
	$M_{pa} - 30$	$[P_a-(NO)]^+$
	$M_p - 58$	$[P-(NO+CO)]^+$
	$M_p - 16$	$[P-(O)]^+$
Mercaptans	M_p	$[P]^+$
	47	$[CH_2SH]^+$
	61	$[CH_2CH_2SH]^+$
	$M_p - 33$	$[P-(SH)]^+$
	$M_p - 34$	$[P-(H_2S)]^+$

[a]From Refs. *100, 101.*

[b]M_p and M_{pa} are the masses of the parent and aromatic parent molecule ions, respectively, while other values represent masses of fragments and certain groups.

[c]Peaks arise from parent (P^+) and aromatic parent (P_a^+) ions, parent ions less certain groups, or various fragments.

The major limitations of the method are the necessity of being able to vaporize the compound, and interferences arising from trace impurities (see Chap. 29).

G. *Nuclear Magnetic Resonance Spectra*

Atomic nuclei with spins have magnetic moments associated with this activity. These are usually oriented randomly, giving a system with no net moment. When the system is placed in a powerful magnetic field, however, the magnetic moments become aligned with the field applied. An introduction of energy tends to change the orientation of the molecules so that they are out of alignment with the field.

This orientation of magnetic nuclei in an external magnetic field is quantized. That is, when energy is applied and a new orientation follows, there is a discrete energy level representing the orientation. If the intensity of the magnetic field is H_o, then the energy difference ΔE would be described as follows:

$$\Delta E = \mu H_0 / I \qquad (104)$$

where μ is the magnetic moment and I is the spin number, which would be 1/2 for a proton. The significance of this equation lies largely in the fact that the energy difference is proportional to the magnetic moment. For a proton the magnetic moment, μ, is 1.410×10^{-23} erg G^{-1} in a magnetic field of 14,000 G (100). Since $\Delta E = h\nu$, the frequency corresponding to this difference can be calculated to be 60×10^6 cycles/sec (100), which is in the radio frequency region. Thus, if electromagnetic energy of this frequency is supplied, a proton at its lower energy state could absorb this energy and "jump" to the higher energy state. This process of energy absorption is called "magnetic resonance," and as the nucleus is involved it is more commonly called nuclear magnetic resonance (NMR). As $\Delta E = h\nu$, Eq. (104) can be changed to the following (100):

$$2\pi\nu = 2\pi\mu H_0 / hI \qquad (105)$$

where the value 2π is introduced to convert linear to angular momentum. The term, $1\pi\mu/hI$ which is the ratio of the magnetic moment to the angular momentum, can be defined as γ, the magnetogyric ratio, which is a characteristic of the nucleus.

If the field H_0 is fixed and the frequency is varied, strong absorption of the applied energy will occur at a frequency characteristic of the nucleus. Thus, an NMR spectrum may consist of a plot of absorption intensity versus the frequency of the energy applied, in which each nucleus will have its characteristic absorbance peak. The area under an NMR peak is proportional to the number of protons giving rise to it. Thus, the proton will "resonate" (undergo resonance absorption) at 60 Mc, ^{19}F at 56.4 Mc, ^{31}P at 24.3 Mc, ^{11}B at 19.25 Mc, and ^{13}C at 13.1 Mc. Nuclei such as ^{12}C, ^{16}O, and ^{32}S, which have even numbers of protons and neutrons, are inert magnetically and do not yield NMR peaks, as mentioned earlier.

If NMR absorbance peaks were restricted to ^{1}H, ^{19}F, ^{31}P, ^{11}B, ^{13}C, etc., they would not be of great value in organic molecule identification. Of much greater interest is the behavior of protons in organic molecules, where the electrons in the molecule interact with the applied field. Their response is diamagnetic, and this small field set up in opposition to the external field shields the nuclei. The nucleus therefore is found in an actual field which is smaller than the applied field.

Methanol, for example, exhibits two proton types, one associated with the CH_3 group and one with the OH group. Of the two nonproton atoms, oxygen is the more electronegative. Therefore, there is less electron cloud around the oxygen proton than around the methyl protons, and thus the CH_3 group exhibits a greater shielding effect than the OH group.

There are two ways of observing this effect. If the radio frequency (100) applied is kept constant, H_0 may be varied to observe the two absorbance peaks. Here, the H_0 for the CH_3 group will be larger than that of the OH group. If the field is kept constant, then the applied frequency may be varied to observe the two peaks. In this case, the OH group finds itself in a greater effective field than the CH_3 group, and therefore a higher frequency is required to cause the OH proton to resonate.

The absolute magnitude of the shift is small. For example, a methyl proton in a field of 14,000 G will absorb at 60,000,054 cycles rather than at 60,000,000 cycles(100). In addition, the magnitude of the shift is also proportional to the magnetic field H_0. Both of these problems can be successfully solved by comparative rather than absolute measurements.

This problem, therefore, is overcome by using a reference compound and employing a ratio between the shifts of the compound of interest and the reference. A commonly employed relationship(100) would be,

$$\delta = [H_0 \text{ (reference)} - H_0 \text{ (sample)}] \times 10^6 / H_0 \text{ (reference)} \qquad (106)$$

in which δ is a dimensionless ratio describing the chemical shift in parts per million. As the spectra are usually in frequency units, this ratio takes the form of

$$\delta = [\nu \text{ (sample)} - \nu \text{ (reference)}] \times 10^6 / \nu \text{ (reference or oscillator frequency)} \qquad (107)$$

Typical reference compounds are presented in Table 27. The usual reference is tetramethylsilane, in which the proton peak yields a resonance line or absorption peak at a very high applied field, well beyond those of most other protons. It has a frequency shift much lower than almost all other protons. This allows for positive values of δ in most cases. In addition, it is soluble in many organic solvents and so can readily be used as an internal standard.

TABLE 27
Chemical Shifts of Common Standards Relative to
Cyclohexane[a]

Standard	Relative shift, ppm	Standard	Relative shift, ppm
Chloroform	6.1	Dioxane	2.2
Benzene	5.3	Acetone	0.8
Methylene chloride	4.2	Cyclohexane	0
Water	3.6	Tetramethyl silane	−1.5

[a]Reprinted from 88 by courtesy of Elsevier, New York.

A typical NMR spectrum is presented in Fig. 7. In NMR work, dilute (2–10%) concentrations are employed. The ideal solvent, of course, should contain no protons. Carbon tetrachloride and deuterated chloroform or benzene are employed, as well as deuterium oxide for compounds showing appreciable solubility only in water. However, it must be noted that the solvent may also influence the shift, especially if it interacts with the sample.

The structural implications of NMR spectra lie in the fact that the chemical shift of the proton is greatly influenced by the structure of the functional group to which it is attached, as well as adjacent groups.

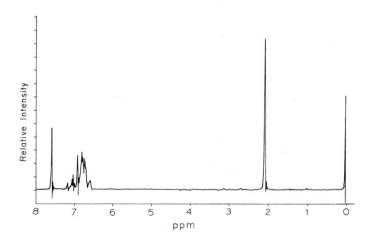

Fig. 7. NMR spectrum of m-cresol [courtesy of Varian Associates].

Thus, we can have chemical shifts the magnitude of which is associated with structural considerations. Such shifts are presented in Tables 28 and 29 (*88*).

The nature of the process itself indicates that NMR would not be as powerful a tool for general functional group analysis as IR spectrophotometry. However, the large separation of peaks makes this a valuable tool for the identification of paraffinic, olefinic, and aromatic OH group protons (*88*). This is also true for OH and NH_2 proton shifts from aromatic and aliphatic compounds. It is an excellent method for selecting among alternate structures arrived at by many of the other methods of structural identification.

TABLE 28
Chemical Shifts of Proton in CH Groups[a]

Group	Chemical shift,[b] ppm
CH_3 in CH_3CH_2—	−3.4 to −4.3
CH_2 in CH_3CH_2—CH_2	−3.5 to −4.5
CH_2 in a ring	−3.2 to −3.9
CH_3 in CH_3-aryl	−2.9 to −3.7
CH_3 in CH_3—C≡	−2.7 to −3.5
CH in H—$C(C)_3$	−3.3 to −3.7
CH_2 in C—CH_2CO—	−3.0 to −3.5
CH_3 in CH_3CO—	−2.8 to −3.4
CH_2 in C—CH_2C=C	−2.7 to −3.3
CH_3 in CH_3—N—	−2.5 to −3.2
CH_2 in —CH_2NH_2	−2.3 to −2.8
CH in CH≡C—	−2.0 to −2.7
CH_2 in —CH_2Cl	−1.8 to −2.4
CH_3 in CH_3O—	−1.2 to −1.8
CH_2 in —CH_2OH	−1.5 to −2.0
CH_2 in —CH_2NO_2	−1.0 to −1.5
CH in *cis*-dialkyl olefins	+0.1 to −0.7
CH in monoalkyl olefins	+0.7 to −1.0
CH in trialkyl olefins	+0.3 to −0.4
CH in *trans*-dialkyl olefins	+0.3 to −0.2
CH in $HC(C)_3$	+0.4 to −0.4
CH in benzene ring	+2.7 to +1.2
CH in HCOO—	+3.2 to +2.3
CH in HCOC—	+5.0 to +4.2

[a]From Ref. *88*.
[b]Water is used as a reference.

TABLE 29
Chemical Shifts of Proton in Electronegative Groups[a]

Group	Chemical shift,[b] ppm
—NH in alkylamines	−4.4 to −4.9
—NH$_2$ in alkylamines	−3.2 to −4.4
—SH in mercaptans	−3.0 to −3.6
—NH in arylamines	−1.7 to −2.1
—SH in thiophenols	−1.6 to −2.2
—NH$_2$ in arylamines	−1.0 to −1.8
—NH$_2$ in amides	+0.8 to −2.0
—OH in polyhydric alcohols	−0.3 to −0.9
—OH in alicyclic alcohols	−0.3 to −0.9
—OH in monohydric alcohols	+0.1 to −0.5
—OH in *ortho*-substituted phenols	+1.1 to −0.2
—OH in other phenols	+2.8 to +1.6
—OH in aliphatic acids (liquid)	+4.0 to +2.3
—OH in sulfonic acids (liquid)	+1.1 to +6.2
—OH in aromatic acids (depends on solvent)	+8.6 to 0.0

[a]From Ref. *88*.
[b]Water is used as a reference.

This is particularly true for the fine line structure or peak splitting exhibited by certain proton NMR peaks. This splitting occurs because the neighboring protons are also spinning magnets, and so the resonance position of a given proton should be affected by this spin of the nearby protons. As the number of the fine structure lines is a function of the number and orientation of the influencing protons, this is of value in the identification of the protons adjacent to the influenced peak. It must also be noted that under certain conditions group interaction can occur, causing the appearance of a larger number of spectral bands than anticipated by the types of protons present, and yielding a spectrum difficult to interpret.

H. X-Ray Diffraction

X-rays consist of short wavelength electromagnetic radiation, the spectrum of which extends from the lower limit of the vacuum ultraviolet region (previously defined as 100 nm) down to 0.005 nm. Wavelengths toward the middle of this range correspond to the spacings between planes in crystalline materials and thus provide the basis for diffraction analysis.

When X rays pass through a single crystal rotated about one of its main axes, the diffraction pattern produced consists of a large number of individual diffractions of different positions and intensities(105b,c). Each crystal plane is defined by its Miller indices (hkl), and the position of any given reflection from a crystal plane is determined by the angle of reflection, θ. In turn, θ is governed by the spacing of the plane d and the wavelength of the radiation λ. According to the Bragg equation:

$$\lambda = 2d_{hkl} \sin \theta \qquad (108)$$

Reflection will occur from all possible planes (hkl) out of a limit of $d(hkl) = \frac{1}{2}\lambda$, which happens at an angle of 90°. In general, the values of λ are selected to be much less than the maximum spacing, so that it is possible to observe reflections from a large number of planes. The intensity of each reflection varies from plane to plane and is a function of θ and the distribution of atoms in the plane.

Complex mathematical analysis of single-crystal data can reveal considerable structural information. Such analysis has been used to select among alternate structures proposed for steroids and penicillin, and to detect the occurrence of hydrogen bonding(106). In addition, as previously noted, it can be used to determine the size of unit cells and thus molecular weights.

A simpler method, for purposes of identification and characterization, is the "powder method" of Debye and Scherrer(105b). In this technique, the compound is powdered, placed into a thin-walled capillary tube and scanned in an X-ray beam. Random orientation of crystals will provide for every possible reflection. Thus, from each plane a cone of rays will be produced instead of a single reflected beam. These rays have a semi-vertical angle of 2θ, giving the locus of the reflected beam for all possible crystal positions. These curves are revealed as circles on a flat photographic plate set perpendicular to the incident beam. From the photograph the positions and the intensity of the circles can be determined. The pattern obtained is so completely characteristic of the compound that it provides a "fingerprint" for its identification. The ASTM provides standard powder patterns convenient for this use.

The powder method is applicable to simple mixtures as well as to pure compounds. However, diagnostic lines may overlap in mixtures making it impossible to identify all the compounds that are present. Mixtures of organic substances in natural waters are far too heterogenous to show diffraction patterns. When methods of organic separation are sufficiently perfected, however, it will be possible to obtain diffraction patterns of any relatively pure crystalline organic material isolated from water.

I. Specific Dispersion

The index of refraction of a substance is a function of the wavelength used in its measurement. It is generally determined with the sodium D line, but certain other wavelengths are also used. The dispersion, D, is the difference between refractive indices measured at two wavelengths:

$$D = n_{\lambda 1} - n_{\lambda 2} \qquad (109)$$

Using the refractive index for the sodium D line (N_D) and the refractive indices for the hydrogen F and C lines, the specific dispersion, $[D_s]$, can be defined as (107)

$$[D_s] = (n_D - 1)/(n_F - n_c) \qquad (110)$$

This value, sometimes referred to as the Abbé number, can be used in the identification of organic compounds.

Specific dispersion is not an additive property in the same sense as molar refraction $[R_D]$ (where numerical values are assigned to atoms and groups). Rather, in a manner similar to spectral determinations, various types of organic molecules have characteristic specific dispersions. An example of this is presented in Table 30 (107).

TABLE 30
Abbé Number as a Function of Structure[a]

Compound type	Abbé number
Polycyclic aromatics	18.7 to 20.8
Styrenes	23.6 to 30.8
Benzene homologs	29.7 to 34.9
Cyclic dienes, conj.	29.7 to 31.8
Cyclic dienes, unconj.	34.7 to 37.8
Aliphatic dienes, conj.	28.0 to 31.8
Aliphatic dienes, unconj.	39.6 to 42.0
Cyclic dienes	40.8 to 47.0
Olefins	44.0 to 52.0
Cycloolefins	45.6 to 48.8
Acelylenes	39.7 to 50.0
Aliphatic amines	47.9 to 54.0
Aliphatic ketones	51.0 to 56.9
Aliphatic aldehydes	53.5 to 55.4
Aliphatic hydrocarbons	56.0 to 59.5

[a]From Ref. 107.

In general, the refractive index of a substance increases as the wavelength of incident light decreases. This is due to the fact that as photon energy increases and approaches that required for electronic excitation, molecules become more effective at reducing the velocity of incident light.

For this reason, the value of $n_F - n_c$ in Eq. (110) is greater for substances which absorb in the near ultraviolet (e.g., polycyclic aromatics) than for those which absorb in the far ultraviolet (e.g., aliphatic hydrocarbons). Hence, lower $[D_s]$ values are expected for polycyclic aromatics than for aliphatic hydrocarbons, a fact substantiated by the data in Table 30.

J. Solubility

In addition to solubility classification by reactive solvents, which is described in a subsequent section, the solubility of a compound in an inert solvent has certain identification aspects. Due to the nature of the solubility process, this information is of even less importance than solubility in reactive solvents insofar as the structural information imparted. It does, however, supply some structural insight.

One approach to the solubility of organic compounds is through the use of Hildebrand's solubility parameter, δ (108). The definition is

$$\delta = (-E/\bar{v})^{1/2} \tag{111}$$

where $-E$ is the energy required for vaporizing the sample to the gas phase at zero pressure, and $\bar{v}$ is the molal volume of the liquid. It is thus a measure of internal pressure. The solubility of a compound in a solvent will be a function of how close the solubility parameters of compound and solvent are to each other. The major importance of the solubility parameter is in selection of solvent systems for organic materials. However, there is some structural significance to the solubility parameter.

The solubility parameter may be predicted from the structure by the formula of Small (109):

$$\delta = d\Sigma G/M \tag{112}$$

where d is the density, ΣG is the sum of the molar attraction constants in the molecule, and M is the molecular weight. Values of G are given in Table 31. From solubility studies in solvents of known δ, such as those presented in Table 32, a value of δ for the compound may be experimentally determined, and it can be compared with the calculated value for the suspected molecular structure using Eq. (112). This approach is

TABLE 31

Molar Attraction Constants, G, of the More Common Groups[a]

Group	G	Group	G
—CH$_3$	214	CO ketones	275
—CH$_2$— single-bonded	133	COO esters	310
—CH$\diagup\diagdown$	28	CN	410
$\diagdown\diagup$ C $\diagup\diagdown$	−93	Cl (mean)	260
CH$_2$=	190	Cl single	270
—CH= double-bonded	111	Cl twinned as in $\diagdown$CCl$_2\diagup$	260
$\diagdown$ C= $\diagup$	19	Cl triple as in —CCl$_3$	250
CH≡C—	285	Br single	340
—C≡C—	222	I single	425
Phenyl	735	CF$_2$ n-fluoro	150
Phenylene (o, m, p)	658	CF$_3$ carbons only	274
Naphthyl	1146	S-sulfides	225
Ring, 5-membered	105–115	SH thiols	315
Ring, 6-membered	95–105	ONO$_2$ nitrates	440
Conjugation	20–30	NO$_2$ (aliphatic)	440
H (variable)	80–100	PO$_4$ (phosphates)	500
O ethers	70	Si (1N silicones)	−38

[a]Reprinted from Ref. 86 by courtesy of Academic Press, New York.

very limited, and it cannot be used for compound-solvent systems in which hydrogen bonding occurs (i.e., alcohols, amines, carboxylic acids, phenols, etc.).

K. Dielectric Constants

Organic liquids may be considered insulating or nonconducting materials and therefore be classified as dielectrics. When such a material is placed between plates of a capacitor, the capacitance is increased by a factor K, which is called the dielectric constant. Thus, if the capacitance with a vacuum is C_0, and with the liquid it is C_1, then (49):

$$K = C_1/C_0 \qquad (113)$$

TABLE 32

Solubility Parameters of Various Solvents[a]

Compound	δ
Hexane	7.3
Benzene	9.2
Carbon tetrachloride	8.6
Chloroform	9.3
Chlorobenzene	9.5
o-Dichlorobenzene	10.0
Carbon disulfide	10.0
Acetone	10.0
Methyl ethyl ketone	9.3
Diethyl ether	7.4
Dioxane	9.9
Ethyl acetate	9.1
Ethyl lactate	10.0
Ethylene carbonate	14.7
Pyridine	10.7
Nitromethane	12.6
Ethanol	12.7
Ethylene glycol	14.2
Glycerol	16.5

[a]Reprinted from Ref. 86 by courtesy of Academic Press, New York.

The dielectric constant of organic liquids may be considered another physical characteristic of the compound, and in this respect it is similar to boiling point, refractive index, density, etc. In addition, there is a considerable spread for these values; for example, 1,4-dioxane has a dielectric constant of 2.2 while formamide has one of 109.5.

The structural implications here arise from one of the instruments used to determine the dielectric constant—the high frequency oscillator. It has been observed that the frequency of the oscillator changes with changes in the nature of the core material of the tank circuit coil. If this coil is hollow, various organic liquids may be placed in a cell within the coil. When this is done, there is a shift in the beat frequency, and it has been noted that these shifts are somewhat related to organic structure (110). Examples of this relationship are presented in Table 33.

TABLE 33

Beat Frequency Changes Induced by Various Organic
Compounds[a]

Compound	Beat frequency change, cycles/sec
n-Alkanes, $C_5 \rightarrow C_7$	$660 \rightarrow 697$
Alkyl alcohols, $C_1 \rightarrow C_8$	$5560 \rightarrow 2550$
Alkyl ketones, $C_3 \rightarrow C_7$	$4397 \rightarrow 3062$
Alkyl acids, $C_2 \rightarrow C_6$	$2190 \rightarrow 1005$
Alkyl esters (methyl) $C_3 \rightarrow C_5$	$2135 \rightarrow 1687$
Benzene, methyl benzene	$810 \rightarrow 830$
Benzene, halogen and nitrogen substituted	$1675 \rightarrow 5720$

[a]From Ref. *110*.

VI. Chemical Identification Methods

In conjunction with the few simple physical tests previously listed, chemical techniques represent the classic approach to qualitative organic analysis. As equipment required for the advanced physical methods became available, certain of these chemical techniques fell into disuse. Due to their simplicity and didactic impact, however, interest in their application has continued. Unlike most of the physical tests, the chemical methods are usually consumptive, the organic matter under study being altered or destroyed during their use.

A. *Classification Schemes*

One method of obtaining preliminary information on the nature of organic compounds is through the use of classification schemes.

1. SOLUBILITY SYSTEMS

Solubility classification entails the division of organic compounds into classes based on their solubility in various chemical reagents and solvents (*9*). A representative scheme is that of Shriner and Fuson(*75*). This system is based on the solubility of organic compounds in water, ethyl ether, 5% sodium hydroxide, 5% sodium bicarbonate, 5% hydrochloric acid. The general scheme is presented in Fig. 8. In a majority of cases, chemical reactions are involved in the noted solubilities, the organic compounds usually reacting through a functional group.

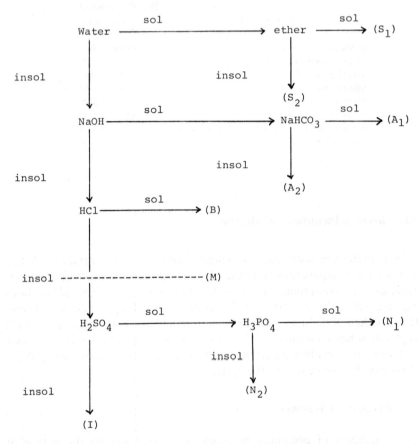

Fig. 8. Solubility classification scheme. See the text for the meaning of the designations S_1, S_2, A_1, A_2, B, M, N_1, N_2, and I.

The types of organic compounds found in these classes can be generally outlined as follows:

a. Water-Soluble Compounds

Class S_1: Low molecular weight (less than 4–5 carbon atoms) acids, acid chlorides, alcohols, aldehydes, ketones, amines, esters, nitriles, and ethers

Class S_2: Salts of acids or bases, polyhydroxyl compounds, polybasic acids, amino acids, hydroxy acids, certain amines and amides including some sulfur-containing compounds, and other types of highly polar organics

b. Acidic Compounds

Class A_1: Acids of lesser polarity than found in S_1 and S_2, and negatively substituted phenols

Class A_2: Weakly acidic compounds such as phenols, oximes, imides, higher amino acids, enols, lower mercaptans, certain primary and secondary nitro compounds, and sulfonamides of primary amines

c. Basic Compounds

Class B: All amines except those which are very poorly basic, such as diaryl and triaryl amines. Amphoteric compounds may be classified as either A_1 or A_2 in Class B

d. Neutral Compounds

Class M: Neutral compounds that contain atoms in addition to carbon, hydrogen, oxygen, and halogens. Examples are nitriles, nitrites, nitro compounds, azo and hydrazo compounds, negatively substituted amines, sulfones, sulfonamides, mercaptans, and thio ethers

Class N_1: Alcohols, aldehydes, ketones, and esters with less than nine carbon atoms but of a higher molecular weight would allow classification in S_1

Class N_2: Alcohols, aldehydes, ketones, and esters containing more than nine carbon atoms. Also quinones, ethers, and unsaturated hydrocarbons

Class I: This group contains unreactive compounds such as aliphatic hydrocarbons, some aromatic hydrocarbons, and halogen and other derivatives of both

These groupings are, of necessity, only approximate. They are of value in obtaining confirmative as well as preliminary information.

2. SOLVOCHROMIC AND THERMOCHROMIC SYSTEMS

Organic compounds can be classified according to their ability to either promote or interfere with chelate formation in an indicator system. One such system is the reaction between an iron salt and a gallate, in the presence of *o*-chloroaniline, to form a blue complex ion(*111*). The reaction is:

$$\underset{\text{(yellow)}}{\text{ferric chloride}} + \underset{\text{(colorless)}}{\text{propyl gallate}} + \underset{\text{(colorless)}}{o\text{-chloroaniline}} = \underset{\text{(blue)}}{\text{chelate}} + \underset{\text{(colorless)}}{\text{chloroanilinium ion}} \quad (114)$$

This reaction is reversible and can be driven to the left or right by temperature changes, yielding shifts in color. The effect of organic compounds on the temperature at which the shift occurs forms the basis of the classification system.

If the color of the compound-indicator mixture is yellow, the mixture is cooled until it turns blue, and the temperature is recorded. If the color of the compound-indicator mixture is blue, the mixture is heated until it turns yellow, and the temperature is recorded. The lower and upper temperature ranges are at -10 and $125°C$, respectively. The classes obtained by this system can be summarized as follows:

Class	Temperature range in which shift occurs, °C
I^S	< -10
I^W	-10–20
N	20–40
P^W	40–100
P^S	> 125

The superscripts S and W stand for strong and weak, respectively, and the class symbols represent inhibitory (I), neutral (N), and promoter (P) action of the compound being tested on the reaction given in Eq. (114) for the formation of the blue chelate.

A compound may be tested by using it as the solvent, or by preparing a dilute solution of it in bromobenzene. To differentiate between the two, the subscripts M and L are used, respectively. The types of organic compounds that are found in these classes are as follows:

a. Compound as the Solvent (M)

P_M^S: Aliphatic, aromatic, and heterocyclic amines, epoxides, phenols, and aliphatic amides

P_M^W: Alcohols

N_M: Hydrocarbons, diaryl ethers, mercaptans, and carboxylic acids

I_M^W: Aromatic nitriles and nitro compounds, alkylaryl ethers

I_M^S: Aldehydes, esters, trialkyl phosphates, sulfonic acids, phosphonic acids, carboxylic acids, chlorides, saturated heterocyclic ethers, aliphatic ketones, acid anhydrides and polyethers, arylalkyl ketones, alkylaryl sulfonate esters, aliphatic nitro compounds, nitriles, and monoethers

b. *Bromobenzene as the Solvent (L)*

P_L^S: Aliphatic amines and epoxides

P_L^W: Aromatic and heterocyclic amines

N_L: Phenols, carboxylic acids, diaryl ethers, hydrocarbons, mercaptans, aromatic nitro compounds, alkylaryl ethers, aliphatic nitro compounds, alkylaryl sulfonate esters

I_L^W: Alcohols, aromatic nitriles, aliphatic monoethers and nitriles, esters, and alkylaryl ketones

I_L^S: Trialkyl phosphates, sulfonic acids, saturated heterocyclic ethers, phosphonic acids, carboxylic acid chlorides, aliphatic acid anhydrides, ketones and polyethers, esters, and aliphatic amides

As in any such system, there are exceptions to the examples presented as being typical of a given class. In general, the exceptions are caused by reduced basicity of the compounds. This may result from halogenation near an oxygenated group, or through dilution of the effect of a group by a long hydrocarbon chain.

B. *Derivatives*

One of the classical methods for the identification of unknown organic compounds is the preparation of "derivatives." A "derivative" may be defined as a solid product formed by the reaction of an organic compound with some reagent. The melting points and boiling points of such products formed through the reaction of numerous organic compounds with suitable reagents have been classified and tabulated (75, 86, 112). Thus, an unknown compound of suspected identity can react with a specific reagent and the melting or boiling point of the product can be obtained. This value can then be compared with data tabulated from derivatives of known compounds with the same reagent. If several compounds form derivatives with similar melting or boiling points, various reagents must be employed.

Of course, for this technique to be applicable, an unknown must have listed derivatives. In addition, a considerable amount of information must be known about the chemical characteristics of the unknown, so that suitable reagents can be employed.

The unknown organic compound should be as pure as possible, and the reaction to form the derivative should be rapid and easily carried out. The derivative should be produced in good yield and in a pure or readily purified state. The melting point of the derivative should differ from that of the parent compound by at least 50°C and, in general, lie between 50 and 250°C (75).

The following sections describe the derivatives that are generally formed for different classes of compounds (75).

1. ACIDS

Acids usually react with thionyl chloride to form the acid chloride, and the acid chloride with an amine to form an amide. The reactions are:

$$RCHOOH + SOCl_2 \longrightarrow RCOCl + SO_2 + HCl \qquad (115)$$

$$RCOCl + R'NH_2 \longrightarrow RCONHR' + HCl \qquad (116)$$

where the R' is usually an alkyl- or halogen-substituted aromatic group.

In another reaction, the acids are condensed with o-phenylene diamine to form 2-alkylbenzimidazoles:

$$(117)$$

Also, the salt of the acid may be prepared by neutralization and, in turn, allowed to react with a substituted benzyl chloride to form an ester:

$$RCOOH + NaOH \longrightarrow RCOONa + H_2O \qquad (118)$$

$$(119)$$

The substituents, R, on the benzyl chloride ring may be phenyl or halogen groups.

2. ALDEHYDES AND KETONES

These carbonyl compounds may be treated with phenylhydrazine to form hydrazones:

$$(120)$$

$$R_2CO \; + \; H_2NH\langle\bigcirc\rangle \; \longrightarrow \; R_2CNNH\langle\bigcirc\rangle \tag{121}$$

For the lower molecular weight aldehydes and ketones, mono- and dinitro-substituted phenylhydrazones, halogenated phenylhydrazones, and naphthylhydrazones are prepared, rather than the unsubstituted phenylhydrazones.

Aldehydes and ketones can form semicarbazones by reaction with semicarbazide:

$$RCHO + H_2NNHCONH_2 \longrightarrow RCHNNHCONH_2 + H_2O \tag{122}$$

$$R_2CO + H_2NNHCONH_2 \longrightarrow R_2CNNHCONH_2 + H_2O \tag{123}$$

Many of the semicarbazones are prepared with semicarbazide in which one of the hydrogens is replaced by a substituted benzene or naphthylene group.

Oximes are prepared by reaction with hydroxylamine:

$$RCHO + H_2NOH \longrightarrow RCHNOH + H_2O \tag{124}$$

$$R_2CO + H_2NOH \longrightarrow R_2CNOH + H_2O \tag{125}$$

3. ALCOHOLS

The most common derivatives of alcohols are urethans and esters. The urethans are prepared by treating the alcohol with substituted aryl isocyanates:

$$ROH \; + \; \overset{R'}{\underset{}{\langle\bigcirc\rangle}}NCO \; \longrightarrow \; ROCONH\langle\bigcirc\rangle^{R'} \tag{126}$$

The substituent, R', on the benzene ring may be a nitro, chloro, or methyl group; or the benzene itself may be replaced by a naphthyl group.

Tertiary alcohols are not suitable for urethan formation, but esters can be prepared in the following manner:

$$R_3COH + C_2H_5MgBr \longrightarrow R_3COMgBr + C_2H_6 \tag{127}$$

$$R_3COMgBr + (R'CO)_2O \longrightarrow R'COOMgBr + R_3COOR' \tag{128}$$

The RCOOMgBr may be hydrolyzed to the free acid:

$$RCOOMgBr + H_2O \longrightarrow RCOOH + MgOHBr \tag{129}$$

Following the formation of the acid, the ester is separated from the mixture by extraction. If the acid anhydride $(RCO)_2O$ is a substituted phthalic anhydride, the reaction will result in only one organic product, thereby simplifying purification.

Esters of monohydroxy alcohols are usually prepared by treating the alcohol with a substituted benzoyl chloride:

$$ROH + ClCOCH_2 \langle\bigcirc\rangle^R \longrightarrow ROCOCH_2 \langle\bigcirc\rangle^R + HCl \quad (130)$$

The substituents, R, on the benzene ring are usually nitro groups.

4. PHENOLS

Since phenols are hydroxy compounds, some of their common derivatives are similar to those of alcohols. For example, phenols were added to naphthyl isocyanate or diphenyl carbamyl chloride to form naphthyl- or diphenylurethans, respectively.

$$\langle\bigcirc\rangle OH + \langle\bigcirc\bigcirc\rangle^{NCO} \longrightarrow \langle\bigcirc\bigcirc\rangle^{NHCOO}\langle\bigcirc\rangle \quad (131)$$

$$\langle\bigcirc\rangle OH + (\langle\bigcirc\rangle)_2NCOCl \longrightarrow (\langle\bigcirc\rangle)_2NCOO\langle\bigcirc\rangle + HCl \quad (132)$$

As in the case of the alcohols, nitro-, chloro-, and methyl-substituted phenylurethans can be prepared.

Phenols can also be identified by their bromination products:

$$\langle\bigcirc\rangle OH + 3Br_2 \longrightarrow Br\langle\bigcirc\rangle^{Br}_{Br} OH + 3HBr \quad (133)$$

Finally, the sodium salts of phenols can react with chloroacetic acid to give phenoxyacetic acids:

$$\langle\bigcirc\rangle ONa + ClCH_2COOH \longrightarrow \langle\bigcirc\rangle OCH_2COOH + NaCl \quad (134)$$

5. CARBOHYDRATES

Since carbohydrates are polyhydroxy compounds, they will react like alcohols to form esters. Usually, the acetate or benzoate ester derivatives are formed.

6. ETHERS

In general, ethers are poorly reactive and must be cleaved before or during derivative formation.

$$O_2N\text{-}C_6H_3(O_2N)\text{-}COCl + ROR \longrightarrow O_2N\text{-}C_6H_3(O_2N)\text{-}COOR + RCl \qquad (135)$$

If both R groups are the same and of low molecular weight, separation is easy. If not, then a mixture of four materials is formed which is usually difficult to separate.

7. AMINES

Primary and secondary amines react with organic isothiocyanates and isocyanates to form thioureas and ureas. As the reactions are the same, only the primary amine reactions are shown:

$$RNH_2 + C_6H_5NCS \longrightarrow RNHCSNHC_6H_5 \qquad (136)$$

$$RNH_2 + C_6H_5NCO \longrightarrow RNHCONHC_6H_5 \qquad (137)$$

As noted in Sec. VI.B.1 on derivatives for organic acids, primary and secondary amines will form amides with organic acid halides. The reactions may be of the following types:

$$RNH_2 + C_6H_5COCl \longrightarrow RNHCOC_6H_5 + HCl \qquad (138)$$

$$RNH_2 + C_6H_5SO_2Cl \longrightarrow RNHSO_2C_6H_5 + HCl \qquad (139)$$

In all four reactions [Eqs. (136)–(139)] there may be alkyl, nitro, or halogen substitution on the phenyl groups.

Tertiary amines, because of their structure, do not enter into substitution reactions. Therefore, since they are bases, derivative reactions generally involve salt formation.

Quaternary ammonium salts may be formed as follows:

$$NR_3 + CH_3I \longrightarrow [CH_3NR_3]^+ + I^- \tag{140}$$

$$NR_3 + (CH_3)_2SO_4 \longrightarrow [CH_3NR_3]^+ + CH_3SO_4^- \tag{141}$$

Salts also are commonly prepared from picric acid, sulfonic acids, and chloroplatinic acids in an analogous manner.

$$NR_3 + HOC_6H_2(NO_2)_3 \longrightarrow HNR_3^+ + [OC_6H_2(NO_2)_3]^- \tag{142}$$

8. Amino Acids

Amino acid derivatives are generally formed through reaction of the amino rather than the carboxyl group. The derivatives are usually similar to those of the primary and secondary amines.

9. Hydrocarbons

Alkanes and cycloalkanes are poorly reactive and are not usually identified through derivative formation. Unsaturated hydrocarbons may be identified as their addition products, which are formed as follows:

$$RHC{=}CRH + Br_2 \longrightarrow RHBrCCRHBr \tag{143}$$

$$RHC{=}CRH + HOCl \longrightarrow RHClCCRHOH \tag{144}$$

Unsaturated hydrocarbons may also be broken down by ozone to produce aldehydes or ketones and can be identified accordingly.

Alkynes can be converted to acids by ozonization, or if the acetylene group is terminal, they can form mercury derivatives:

$$2RC{\equiv}CH + K_2HgI_4 + 2KOH \longrightarrow (RC{\equiv}C)_2Hg + 4KI + 2H_2O \tag{145}$$

Conjugated dialkenes can form Diels–Alder condensation products with maleic anhydride or α-naphthoquinone:

$$(146)$$

$$(147)$$

Aromatic hydrocarbons can be identified directly through formation of their nitration products:

$$(148)$$

or indirectly as an amine derivative, following conversion to the amine:

$$(149)$$

$$(150)$$

Aromatic hydrocarbons can also react in the Friedl–Crafts condensation with phthalic anhydride:

$$(151)$$

Some aromatic hydrocarbons form molecular addition compounds with picric acid, but many of these are unstable and cannot be purified.

10. HALIDES

Aliphatic halides can be converted into organic amides through the reaction of the halides with magnesium to form Grignard reagents, followed by reaction with organic isocyanates:

$$RX + Mg \longrightarrow RMgX \qquad (152)$$

$$RMgX \; + \; \text{C}_6\text{H}_5\text{NCO} \longrightarrow \text{C}_6\text{H}_5\text{NCROMgX} \qquad (153)$$

$$\text{C}_6\text{H}_5\text{NCROMgX} \; + \; H_2O \longrightarrow \text{C}_6\text{H}_5\text{NHCOR} \; + \; MgOHX \qquad (154)$$

As usual, the phenol group of the isocyanate may be substituted.

Polyhalogenated compounds or tertiary halides are not usually identified through derivative formation. Halogenated aromatics can be identified through formation of their nitro derivatives (similarly to aromatic hydrocarbons) or, as follows, through formation of organic sulfonyl chlorides and reaction with amines to form sulfonamides:

$$X\text{C}_6\text{H}_5 \; + \; 2ClSO_3H \longrightarrow X\text{C}_6\text{H}_5SO_2Cl \; + \; HCl \; + \; H_2SO_4 \qquad (155)$$

$$X\text{C}_6\text{H}_5SO_2Cl \; + \; RNH_2 \longrightarrow X\text{C}_6\text{H}_5SO_2NHR \; + \; HCl \qquad (156)$$

The R is frequently a phenyl or substituted phenyl group.

11. MERCAPTANS

Both aryl (aromatic) and alkyl mercaptans may be converted into thioethers by the reaction of their salts with an aryl halide:

$$RSNa \; + \; Cl\text{C}_6\text{H}_5 \longrightarrow RS\text{C}_6\text{H}_5 \; + \; NaCl \qquad (157)$$

The corresponding thioesters can be prepared through reaction of the salts with aroyl halides:

$$RSNa \; + \; ClOC\text{C}_6\text{H}_5 \longrightarrow RSOC\text{C}_6\text{H}_5 \; + \; NaCl \qquad (158)$$

As is generally the case, the phenyl group in aroyl halides may be substituted.

12. MISCELLANEOUS

Organic compounds such as acetals, esters, nitriles, and amides can be identified through formation of derivatives with their hydrolysis products. The hydrolysis reactions, usually acid or base catalyzed, are as follows:

$$\begin{matrix} R \\ \diagdown \\ \\ (H)R \end{matrix} \begin{matrix} OR' \\ \diagup \\ C \\ \diagdown \\ OR' \end{matrix} + H_2O \longrightarrow 2R'OH + RCOR(H) \tag{159a}$$

$$ROOOR' + H_2O \longrightarrow RCOOH + R'OH \tag{159b}$$

$$RC\equiv N + 2H_2O \longrightarrow RCOONH_4 \tag{159c}$$

$$RCONHR' + H_2O \longrightarrow RCOOH + R'NH_2 \tag{159d}$$

In each case, except for the nitriles, hydrolysis results in a mixture of two compounds which must be separated before formation of derivatives. Separation is accomplished according to the nature of the hydrolysis products (acids, alcohols, and amines).

Organic nitro (NO_2), nitroso (NO), azo (—N=N—), hydrazo (—NH—NH—), and azoxy ($\overset{\overset{\textstyle O}{\uparrow}}{—N}$=N—) compounds are determined following their reduction to primary amines:

$$RNO_2 + 3H_2 \longrightarrow RNH_2 + 2H_2O \tag{160a}$$

$$RNO + 2H_2 \longrightarrow RNH_2 + H_2O \tag{160b}$$

$$RN\!\!=\!\!NR + 2H_2 \longrightarrow 2RNH_2 \tag{160c}$$

$$RNHNHR + H_2 \longrightarrow 2RNH_2 \tag{160d}$$

$$\overset{\overset{\textstyle O}{\uparrow}}{RN}\!\!=\!\!NR + 3H_2 \longrightarrow 2RNH_2 + H_2O \tag{160e}$$

If the two R groups in the azoxy, azo, or hydrazo compounds are different, mixtures will result which must be separated before synthesis of the amine derivatives.

C. Functional Group Tests

Before such physical methods of functional group identification as infrared and nuclear magnetic resonance spectrophotometry became commonplace, functional groups were detected through characteristic chemical reactions. These usually were the production of a precipitate, decolorization, color formation, etc. Most significant for reasonably reactive functional groups, and greatly influenced by steric and inductive effects, such reactions can give considerable structural information.

Although now largely replaced by more exact nonconsumptive physical methods, they have considerable historic importance, and many such reactions are still being employed for spot detection and tentative identification in paper and thin layer chromatography. There are a large number of such reactions, and the ones included in this section are only indicative of the use of this technique. As the actual reactions involved in many cases are not completely delineated, no reactions are detailed in this section. Instead, they are noted in tabular form along with certain significant details in Table 34.

D. Degradation

Another approach to the chemical elucidation of organic structure is through degradation procedures, that is, the breaking down of the original compound into smaller structural units. This is particularly useful when the degradative procedure is reasonably specific and when complex parent compounds may be broken down into simpler compounds that may be more readily identified. There are many organic reactions that fall into this category, and the purpose of this discussion is to demonstrate this method using some common examples.

1. HYDROLYSIS

a. Esters

Esters can be hydrolyzed to the free acid and alcohol using acids or bases. It is common to use alkaline hydrolysis with the hydroxide dissolved in water or organic solvents such as ethanol, depending on ester solubility (75,86). The reaction is:

$$RCOOR' + KOH \longrightarrow RCOOK + R'OH \tag{161}$$

TABLE 34
Chemical Tests for Functional Group Detection

Compound	Reagent	Observation	Remarks	Reference
Acids	pH indicators	Various colors	Organic acids and bases	112
	Ferric hydroxamate (after esterification)	Bluish-red color	Phenols and enols may react	112, 113
Alcohols	Hydrochloric acid	Immiscible liquid formation	Primary rapid, secondary slow, tertiary does not react	75, 86
	Carbon disulfide	Yellow precipitate	Primary and secondary only	112, 113
	Ferric hydroxamate (after esterification)	Blue-red color	Phenols and enols may react	112, 113
	N-Bromosuccinimide	Color reaction	Permanent orange = primary; temporary orange = secondary, no color = tertiary	112
	Periodic acid: Br₂	CO₂ evolution	Polyhydroxy alcohols including carbohydrates	113
Aldehydes	Sodium bisulfite	Precipitate, heat	Also methyl ketones	75, 86, 113
	2,4-Dinitrophenyl hydrazine	Yellow-orange precipitate	Also ketones	75, 113
	Ammoniacal AgNO₃	Silver mirror, or brown precipitate		75, 86, 112
	p-Rosaniline	Wine purple color		86
	o-Dianisidine	Various colors, mainly red, orange		113
Amines	Sodium nitroprusside-acetaldehyde	Blue-violet color	Secondary aliphatic	113
	Carbon disulfide-silver nitrate	Black silver sulfide	Primary and secondary aliphatic	113
	Chloroform-KOH	Carbylamine odor	Primary	86
	Nitrous acid	Nitrogen evolution	Primary	112

TABLE 34 (Contd)

Compound	Reagent	Observation	Remarks	Reference
	Tetrachloroquinone	Yellow-red color	Primary and secondary	112
	β-Naphthol-HNO_2	Color formation	Primary aryl	112
	Glutaconic aldehyde	Red color	Primary aryl	112,113
	Concentrated HNO_3	Yellow or blue color	Secondary aryl (free *para* position)	113
Amino acids	n-Bromosuccinimide	Orange precipitate	Tertiary amines	112
	Ninhydrin	Deep blue color	Primary and secondary amines also	113
Carbohydrates	Potassium sulfocyanate	Evolution of H_2S		113
	α-Naphthol	Violet-purple color		86,112
	Cupric ions	Red-brown precipitate	Also aliphatic aldehydes	75,112
	2,3,5-Triphenyltetrazolium chloride	Colored precipitate		86
	Anthrone	Blue-green color		86
	Phloroglucinol	Various colors	Pentoses, mainly red	86,112
	Aniline acetate	Pink-red color		86,112
	Resorcinol	Red color	Ketones	112
Hydrocarbons	Br_2 in CCl_4	Decolorization	Unsaturation; also phenols and amines	75,86,112
	$KMnO_4$ in H_2O	Decolorization	Unsaturation; also aldehydes, alcohols	75,86
	$AlCl_3$ and azoxybenzene	Color formation, red, green, blue	Aromatic	75
	Picric acid	Yellow precipitate	Aromatic, also amines and phenols	112
	Formaldehyde-H_2SO_4	Color formation	Aromatic	112
	Cuprous chloride-NH_3	Precipitate	Acetylenes with free hydrogen	86

	Reagent	Result	Notes	Refs.
Ketones	Sodium hypoiodite	Iodoform	Methyl ketones, ethanol, ethanal	75, 112
	Sodium nitroprusside	Various colors, mainly red, purple, blue	Methyl ketones	113
Mercaptans	Isatin	Green color		112
	Lead acetate	Yellow precipitate		112
Nitrates	Diphenylamine	Blue color	Also nitriles	112
Nitro	Nitrous acid	Red or blue color	Primary (red) and secondary (blue) only	112
Nitroso	Phenol-conc. H_2SO_4	Red color	Aliphatic or aromatic	112, 113
Phenols	Phthalic anhydride condensation	Various colored indicators		112
	Nitrous acid	Various colors	o and m substitution only	113
	Ferric chloride	Various colors including red, blue, purple, green	Enols also react	75, 86, 112
	4-Amino antipyrene	Red-orange color	Certain p substitution. Phenols do not react	112
	Nitrous acid and mercuric nitrate	Red color	o and m substitution. Phenols do not react	112, 113
Sulfides	Sodium plumbite and free sulfite	Yellow precipitate	Mercaptans give black precipitate	112
	Sodium nitroprusside	Red color	Mercaptans give blue-red color	112

b. Amides

As in the previous example, amides also can be hydrolyzed with acids or bases. An example would be (86):

$$RCONHR' + H_2O + HCl \longrightarrow RCOOH + R'NH_3^+Cl^- \tag{162}$$

c. Nitriles

Organic cyanides can also be hydrolyzed by strong acid (6NHCl) or bases (20% alcoholic KOH) to the acid and ammonia (86) as follows:

$$RC{\equiv}N + 2H_2O + HCl \longrightarrow RCOOH + NH_4Cl \tag{163}$$

$$RC{\equiv}N + H_2O + KOH \longrightarrow RCOOK + NH_3 \tag{164}$$

d. Sulfonic Acids

Sulfonic acids undergo desulfonation at elevated temperatures in the presence of dilute acids (86):

$$RSO_3H + H_2O \xrightarrow{\text{HCl}} RH + H_2SO_4 \tag{165}$$

Alkaline fusion of aromatic sulfonic acids yields phenols:

$$C_6H_5SO_3Na \xrightarrow[\Delta]{\text{NaOH}} C_6H_5ONa + NaHSO_4 \tag{166}$$

e. Ethers

These are not particularly reactive substrates, but ethers can be split by concentrated boiling HI or HBr (75). The reactions are as follows:

$$ROR + HI \longrightarrow ROH + RI \tag{167}$$

$$ROH + HI \longrightarrow RI + H_2O \tag{168}$$

f. Natural Polymers

Proteins (polyamides) and polysaccharides can be reduced to amino acids and monosaccharides, respectively, using acid hydrolysis (86). Alkaline hydrolysis is usually not employed due to extensive degradation and racemization of the hydrolytic products.

Proteins also can be hydrolyzed in a stepwise fashion using carboxypeptidase (86). This enzyme splits off the terminal amino acid, permitting a study of the amino acid sequence in the protein molecule.

2. REDUCTION

a. Hydrocarbons

Alkanes are already in a highly reduced form, so this approach is not applicable. Unsaturated hydrocarbons can undergo catalytic hydrogenation (114). For example:

$$\text{$\Large$}C{=}C\text{$\Large$} + H_2 \xrightarrow{Pd} \text{$\Large$}CHCH\text{$\Large$} \tag{169}$$

$$-C{\equiv}C- + 2H_2 \xrightarrow{Pd} -CH_2CH_2- \tag{170}$$

b. Halides

Halides may be dehalogenated by hydrolysis of the Grignard reagent (114):

$$RX + Mg \longrightarrow RMgX \tag{171}$$

$$RMgX + H_2O \longrightarrow RH + MgXOH \tag{172}$$

In addition, catalytic dehalogenation through hydrogenation can be employed (115).

c. Aldehydes and Ketones

These can be reduced to the corresponding primary and secondary alcohols using catalytic hydrogenation or lithium aluminum hydride (86, 114).

$$RCHO + H_2 \xrightarrow{Pt} RCH_2OH \tag{173}$$

$$R_2CO \xrightarrow[H_3O^+]{LiAlH_4} R_2CHOH \tag{174}$$

The catalytic reduction can also be used to eliminate the oxygen from the compound completely, allowing easier identification of the parent compound (116). The Clemmenson reduction will reduce carbonyl groups to methylene groups (114):

$$R_2CO \xrightarrow[HCl]{Zn-H_2} RCH_2R \tag{175}$$

Aldoses and ketoses (carbohydrates) can be reduced to the alcohols by use of sodium amalgam (86).

d. Acids

Acids can be reduced directly or as esters to the corresponding alcohol by $LiAlH_4$ (86):

$$RCOOH \xrightarrow{\text{LiAlH}_4} RCHO \xrightarrow[\text{H}_3\text{O}^+]{\text{LiAlH}_4} RCH_2OH \qquad (176)$$

In addition, reductive decarboxylation can be achieved through the destructive distillation of calcium salts (86):

$$Ca(OOCCH_3)_2 \xrightarrow{\Delta} CH_3COCH_3 + CaCO_3 \qquad (177)$$

e. Hydroxy Compounds

Aromatic hydroxy compounds may be reduced using zinc dust distillation (86,114).

$$C_6H_5OH + Zn \xrightarrow{\Delta} C_6H_6 + ZnO \qquad (178)$$

f. Nitro and Nitriles

Organic cyanides can be reduced to the amines using lithium aluminum hydrides (114):

$$R-C\equiv N \xrightarrow[\text{H}_3\text{O}^+]{\text{LiAlH}_4} RCH_2NH_2 \qquad (179)$$

Nitro compounds can be reduced to amines using catalytic hydrogenation employing Pt, Pd, or Ni catalysts, or by acid reducing agents (86,114).

$$C_6H_5NO_2 \xrightarrow[\text{HCl}]{\text{Zn}} C_6H_5NH_2 \qquad (180)$$

Amines represent the most reduced form of organic nitrogen. Further reduction with catalytic hydrogenation results in dentrification to the parent compound (117).

g. Natural Products

Reductive techniques have been employed in studies of the structure of humic acid (36) and lignin (118). Sodium amalgam and catalytic hydrogenation have been widely employed, and the resultant materials are approximately as complex as the parents, with certain exceptions.

3. OXIDATION

a. Hydrocarbons

Alkanes are relatively unreactive materials. However, alkanes of six carbons or more do undergo dehydrocyclization to alkylaryl compounds (114):

$$CH_3(CH_2)_5CH_3 \xrightarrow[430°C]{Cr_2O_3-Al_2O_3} \text{⟨◯⟩}CH_3 \qquad (181)$$

Alicyclic compounds behave in a similar fashion when treated with catalysts similar to those used in hydrogenations (114):

(182)

The alkyl chain in alkylaryl compounds is more subject to oxidation than the aromatic ring (114). The reaction is:

$$\text{⟨◯⟩}R \xrightarrow[\text{alkali}]{KMnO_4} \text{⟨◯⟩}COOH \qquad (183)$$

Alkenes are fairly easily oxidized. Cool dilute $KMnO_4$ will cause glycol formation (75):

$$3 \underset{}{\succ}C\!=\!C\underset{}{\prec} + 2KMnO_4 + 4H_2O \longrightarrow 3 \underset{\underset{H}{\overset{O}{|}}}{\succ}C\!-\!C\underset{\underset{H}{\overset{O}{|}}}{\prec} + 2MnO_2 + 2KOH \qquad (184)$$

Heated stronger permanganate solutions will cause cleavage at the double bond (75,86).

$$R_2C\!=\!CHR' \xrightarrow[\text{heat}]{KMnO_4} R_2C\!=\!O + R'COOH \qquad (185)$$

Acetylenes under these conditions result in the formation of organic acids (75):

$$R\!-\!C\!\equiv\!C\!-\!R' \xrightarrow[\text{heat}]{KMnO_4} RCOOH + R'COOH \qquad (186)$$

Another method of cleaving double bonds in aromatics or alkenes, and triple bonds in alkynes, is ozonalysis. The addition of ozone is only the first step in the reaction, producing a cyclic peroxide called ozonide. The ozonides when hydrolyzed under oxidative conditions produce ketones or organic acids(*114*):

$$R_2C=CHR' \xrightarrow[\text{hydrolysis}]{O_3} R_2CO + RCOOH \tag{187}$$

Bromine in inert solvents such as carbon tetrachloride gives addition products (dibromides) with alkenes, or tetrabromides with alkynes(*75*); this reaction is of little significance, as products are no simpler than starting materials.

b. Hydroxy Compounds

Under strong oxidative conditions tertiary alcohols yield fragments of little use in compound identification(*86*). Under less vigorous conditions they are unreactive. With such oxidizing agents as $KMnO_4$ in acetic acid or chromic acid, primary alcohols are oxidized to acids and secondary alcohols to ketones(*86,114*).

$$RCH_2OH \xrightarrow[\text{HAc}]{KMnO_4} RCOOH \tag{188}$$

$$R_2CHOH \xrightarrow[\text{HAc}]{KMnO_4} R_2C=O \tag{189}$$

Both lead tetracetate and periodic acid will oxidize 1,2 diols (glycols) to carbonyl compounds cleaving the carbon—carbon bond(*75,114*):

$$R-CHOHCHOH-R + HIO_4 \longrightarrow 2RCHO + H_2O + HIO_3 \tag{190}$$

$$R_2COHCOHR_2 + HIO_4 \longrightarrow 2R_2CO + H_2O + HIO_3 \tag{191}$$

Periodic acid will also oxidize and cleave adjacent (1,2)hydroxycarbonyl compounds or even 1,2 carbonyl compounds. In such cases, the carbonyl group is oxidized to an acid (COOH), and the hydroxy to an aldehyde or ketone(*75*).

c. Aldehydes and Ketones

Aldehydes are so easily oxidized that they will be oxidized slowly to acids by atmospheric oxygen at room temperature. Thus, dilute oxidants such as permanganate are quite effective(*86*).

$$RCHO \xrightarrow{KMnO_4} RCOOH \tag{192}$$

Aldoses are also readily oxidized, but the products to a certain extent reflect the oxidizing conditions (86):

$$CH_2OH(CHOH)_4CHO \xrightarrow{Br_2, H_2O} CH_2OH(CHOH)_4COOH \tag{193}$$

$$CH_2OH(CHOH)_4CHO \xrightarrow[dil]{HNO_3} HOOC(CHOH)_4COOH \tag{194}$$

$$CH_2OH(CHOH)_4CHO \xrightarrow{HIO_4} CH_2OH(CHOH)_3CHO \tag{195}$$

Ketones are oxidized with considerably more difficulty. Hot nitric or chromic acid oxidizes ketones to three or four acid fragments. Sodium hypochlorite will also oxidize methyl ketones to a mixture of acids.

$$(196)$$

Compounds containing a methyl group adjacent to a carbonyl group or a CHOH group undergo the iodoform reaction, which results in the cleavage of the methyl group and formation of an acid (75).

$$RCHOHCH_3 + NaOI \longrightarrow RCOCH_3 + NaI + H_2O \tag{197}$$

$$RCOCH_3 + 3NaOI \longrightarrow RCOCI_3 + 3NaOH \tag{198}$$

$$RCOCI_3 + NaOH \longrightarrow RCOONa + CHI_3 \tag{199}$$

Methylene ketones may be oxidized at α position as follows (114):

$$(200)$$

d. Amines

The usual result of amine oxidation is unspecific discoloration. Hydrogen peroxide, however, does give amine oxides (86, 114) with tertiary amines:

$$R_3N + H_2O_2 \longrightarrow R_3NO + H_2O \tag{201}$$

This is significant since pyrolysis of the tertiary amine oxide leads to the formation of an unsaturated compound and a dialkyl hydroxyl amine.

$$R_3NO \xrightarrow{\Delta} R\!-\!CH\!=\!CH_2 + R_2NOH \tag{202}$$

Aromatic amines can be converted to the corresponding phenols by diazonium salt formation:

$$\text{⟨O⟩}\!-\!NH_2 \ + \ HONO \ \longrightarrow \ \text{⟨O⟩}\!-\!N\!=\!N\!-\!OH \ + \ H_2O \tag{203}$$

$$\text{⟨O⟩}\!-\!N\!=\!NOH \ + \ H^+ \ \longrightarrow \ \text{⟨O⟩}\!-\!\overset{+}{N}\!\equiv\!N \ + \ H_2O \tag{204}$$

$$\text{⟨O⟩}N_2^+ \ + \ H_2O \ \xrightarrow{\Delta} \ \text{⟨O⟩}OH \ + \ N_2 \ + \ H^+ \tag{205}$$

e. Thiols

Mercaptans are easily oxidized to disulfides. This can be done by atmospheric oxygen, iodine, Fe^{3+}, etc.(*114*). The reaction is

$$2RSH + I_2 \longrightarrow RSSR + 2HI \tag{206}$$

Concentrated nitric acid will convert mercaptans to sulfonic acids(*86*):

$$RSH \xrightarrow[\text{conc}]{HNO_3} RSO_3H \tag{207}$$

f. Thioethers

Sulfides may be oxidized to the sulfoxide by concentrated HNO_3 or 30% H_2O_2(*86*):

$$R\!-\!S\!-\!R \xrightarrow[H_2O]{H_2O_2} RSOR \tag{208}$$

Fuming nitric acid, $KMnO_4$ in acetic acid, or 30% H_2O_2 in acetic acid will oxidize the sulfide to the sulfone:

$$R\!-\!S\!-\!R \xrightarrow[HA_c]{H_2O_2} RSO_2R \tag{209}$$

Thioethers may be cleaved by chlorine gas in acetic acid(*86*) to sulfonyl and alkyl chlorides:

$$R\!-\!S\!-\!R \xrightarrow[HA_c]{Cl_2} RSO_2Cl + RCl \tag{210}$$

or with cyanogen bromides to give an alkyl isothiocyanate and an alkyl bromide:

$$R—S—R \xrightarrow{\text{BrCN}} RSCN + RBr \tag{211}$$

g. Natural Products

Oxidative procedures have been employed in studies on the structure of humic acids in soils(36) and colored materials in water(119,120). In the humic acid studies, such oxidants have been peroxide, permanganate (alkaline), nitric acid, alkaline nitrobenzene, ClO_2, $NaIO_4$, and HIO_4. Various aromatic acids, mono and dicarboxylic acids, nitrophenols, nitro aromatic acids, and even anthroquinone were identified in the oxidation products(36).

Strong oxidizing agents such as Cl_2, ClO_2, or $KMnO_4$ gave no recognizable degradation products from organic substances isolated from colored waters(119). Milder oxidants such as alkaline CuO or nitrobenzene gave such substituted aromatics as vanillin, vanillic acid, syringic acid, catechol, resorcinol, protocatechuric acid, and 3,5-dihydroxybenzoic acid(120). These degradation products give insight into the type of structures that occur in these very complex organic compounds.

4. EXHAUSTIVE METHYLATION

This is a method of identifying the substituent groups on primary, secondary, and tertiary amines, which in this reaction are liberated as the corresponding alkenes if the carbon chain contains two or more carbons(34). This system will work for straight chain or cyclic amines as long as there is a hydrogen on the substituent group(113).

The steps in this reaction are: (1) the conversion of the amine to a tertiary amine, (2) the formation of the ammonium iodide from the tertiary amine, (3) the conversion of the iodide to the hydroxide, and (4) the heating of the hydroxide to form an alkene, water, and a tertiary amine in which a methyl group is substituted for one of the original groups.

As an example, let us take the degradation of the secondary cyclic amine, piperidine:

$$\text{piperidine} + CH_3I \longrightarrow \text{N—CH}_3 + HI \tag{212a}$$

$$\text{N—CH}_3 \xrightarrow{CH_3I} \overset{+}{N}(CH_3)_2 \; I^- \tag{212b}$$

$$\text{(structure) } \underset{CH_3}{\overset{+\,CH_3}{N}} \quad I^- \;+\; AgOH \;\longrightarrow\; \text{(structure) } \underset{CH_3}{\overset{+\,CH_3}{N}} \quad OH^- \;+\; AgI \qquad (212c)$$

$$\text{(structure) } \underset{CH_3}{\overset{+\,CH_3}{N}} \quad OH^- \;\xrightarrow{130^\circ C}\; \text{(structure) } \underset{CH_3}{\overset{+\,CH_3}{N}} \;+\; H_2O \qquad (212d)$$

$$\text{(structure) } N(CH_3)_2 \;+\; CH_3I \;\longrightarrow\; \text{(structure) } \overset{+}{N}(CH_3)_3 \quad I^- \qquad (212e)$$

$$\text{(structure) } \overset{+}{N}(CH_3)_2 I^- \;+\; AgOH \;\longrightarrow\; \text{(structure) } \overset{+}{N}(CH_3)_3 OH^- \;+\; AgI \qquad (212f)$$

$$\text{(structure) } \overset{+}{N}(CH_3)_3 OH^- \;\xrightarrow{130^\circ C}\; \text{(structure) } \;+\; (CH_3)_3N \;+\; H_2O \qquad (212g)$$

Thus the final products will be unsaturated substituents and trimethyl-amine. The general rule is that the least substituted ethylene will be eliminated first. Thus ethylpropylbutylamine would liberate ethylene, propylene, and butylene in that order. As in most reactions, there are exceptions (113), but this reaction has been useful in delineating the structure of such naturally occurring amines as the alkaloids.

E. Synthesis

When various physical and chemical tests result in a proposed structure that does not represent an organic compound previously known and studied, one method of structure proof would be to synthesize a compound of the proposed structure and compare it to the original. Although the information supplied by the various identification procedures may leave little doubt as to the structure, nothing is quite as satisfactory to the organic chemist as a proof of such deductions by an unequivocal synthesis of the compound. For this and more practical reasons, complex organics such as vitamins, alkaloids, hormones, antibiotics, and even peptides have been synthesized, resulting in a more definitive proof of structure.

As this chapter concerns analytical rather than synthetic methods, no generalizations on organic synthesis are presented. As many of the materials mentioned have been synthesized by rather specialized techniques, there would be little general value in their specific elucidation. This approach would be valuable in delineating structures for naturally occurring and waste water organic compounds, etc., for structural complexity similar to or simpler than those groups previously mentioned.

The use of synthesis to delineate the structure of lignin, humic acid, etc., does not seem probable in the near future.

F. Polarography

For the purposes of this chapter, polarography may be considered a method of analysis based on current voltage curves that arise at a dropping mercury microelectrode. When a plot is made of the current, I, carried as a function of voltage applied, V, the resulting curve may be divided into three segments. These are an initial region of low current with only a slight slope ($\Delta I/\Delta V$), a region of rapidly increasing slope, and a larger current region of slight slope (Fig. 9). The inflection point

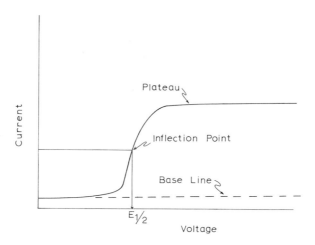

Fig. 9. Typical polarogram indicating half-wave potential.

of this current-voltage curve is defined as the half-wave, $E_{1/2}$, potential. This is essentially half the distance from the initial (residual current) line and the final (plateau current) line.

This half-wave potential is a function of the nature of the material undergoing reduction or oxidation, influenced strongly by the system in which this is occurring. Thus, it may be considered as a qualitative tool. Although, half-wave potential is usually considered a method for inorganic analysis, there are many types of organic compounds that will give characteristic half-wave potentials in proper systems, even though many of these oxidations or reductions are not thermodynamically reversible reactions.

As noted previously, the half-wave potentials of organic compounds are strongly system dependent. As hydrogen ions frequently enter into the electrode reactions, the use of buffer systems becomes necessary to obtain reproducible results[121]. The exact composition of the solvent system must also be known[121]. Fortunately, a wide variety of water-soluble solvents are available, such as alcohols, glycols, dioxane, and Cellosolve. For anhydrous media, acetic acid, ethanol, methanol, glycerol, ethylene glycol, and formamide may be employed. In addition, the presence of inorganic multivalent cations tends to shift the half-wave potential to a more positive potential, and this effect increases with increasing concentration[121].

Functional groups that can be reduced at a dropping mercury electrode include[121,122]:

(1) Carbon—carbon double bonds when conjugated with other carbon—carbon double bonds, aromatic rings, or other unsaturated groups
(2) Carbonyl groups in quinones, ketones, or aldehydes
(3) Nitrogen-containing groups such as azo, amine oxides, nitrite and nitrates, nitroso and nitro, nitrile, diazonium, phenylhydroxylammonium, and in quaternary ammonium compounds in general
(4) Organic halides
(5) Disulfides, polysulfides, and sulfones
(6) Epoxides and peroxides

Functional groups that can be oxidized at a dropping mercury electrode include[121,122] hydroquinones and ene-diols.

The half-wave potential in a carefully defined system is a function of

TABLE 35
Neighboring Group Effects on Half-Wave
Potential[a]

Compound[b]	Formula	Half-wave potential, volts
Styrene	$\phi CH{=}CH_2$	-2.35
1,1-Diphenylethylene	$\phi_2 C{=}CH_2$	-2.14
Triphenylethylene	$\phi_2 C{=}CH\phi$	-2.11
Tetraphenylethylene	$\phi_2 C{=}C\phi_2$	-2.05

[a]From Ref. 121.
[b]The compound is dissolved in the following solvent: 0.175 M tetrabutylammonium iodide in 3/1(w/w) dioxane/water solvent.

molecular structure. As more electron acceptor groups are substituted in an organic molecule, the half-wave potential is shifted toward lower values, indicating that the molecule is more easily reduced(121). The effect of neighboring groups on the half-wave potential is shown in Table 35. However, even though the structure of the compound influences the half-wave potential, given half-wave potentials or potential ranges cannot readily be attributed to specific functional groups. For example, a half-wave potential of -1.5 V could indicate a diiodide or a nitro group, -2.5 V naphthalene or a γ-lactone of a sugar acid, -0.4 V an azo or a thiol group, etc.(121). This, along with the fact that many functional groups do not respond in this procedure, has limited the qualitative aspects of organic polarography.

G. Chromatography

Chromatographic procedures involve the distribution of the materials to be separated between two phases, one mobile and one fixed. As the mobile phase containing these materials passes through the fixed phase, those materials which are most accommodated to the fixed phase spend proportionately more time in this phase and are thus retarded more than those materials which are less accommodated. Thus, depending on their degree of retardation, the materials present eventually draw away from each other and separation is achieved.

The mobile phase may be a gas or a liquid, the fixed phase a solid or a liquid supported by a solid. Using this division, the nature of the forces attracting the materials to the fixed phase and causing the retardation could be outlined as follows:

	Fixed phase	
Mobile phase	Liquid	Solid
Gas	Partition	Adsorption
Liquid	Partition	Adsorption
Liquid		Ion exchange
Liquid		Molecular sieve

Physically, the fixed phase is usually in the form of a column through which the mobile phase passes. However, in certain liquid mobile phase systems, the fixed phase may be a thin sheet of paper or silica gel or some other medium on an inert support.

Chromatography is an excellent separation procedure, enabling the separation of even closely related groups of compounds such as the hydrocarbons, phenols, organic acids, etc. Materials so separated may be estimated by a number of quantitative techniques. In addition, certain qualitative information can be obtained through chromatographic analysis. As chromatographic techniques fall into two general categories, they are covered separately. (Chapter 11 gives detailed use and application of a wide variety of chromatographic methods.)

1. PAPER AND THIN LAYER TECHNIQUES

These are liquid mobile phase systems in which the fixed phase is in the form of a thin sheet (saturated with a liquid phase) or layer of adsorbent. A typical operation would involve spotting a few microliters of a solution containing a mixture of solutes near one end of a paper or silica gel sheet. The solvent is then allowed to pass through the spot, traveling toward the other end of the sheet. Most frequently this is done vertically in an ascending (spot near bottom) or descending (spot near top) manner. It can also be done on circular sheets of filter paper by placing the solute drop at the center of the paper and allowing the solvent to move from the center outward. In either system, the procedure is operated so that the solvent front does not reach the end of the sheet. A ratio, the R_f value, can be defined as:

$$R_f = D_S/D_M \qquad (213)$$

where D_S is the distance the spot has moved from its original position, and D_M is the distance of the solvent front from the original position of the spot. The R_f value is a characteristic of a given compound on a specified fixed phase employing a defined mobile phase and at a stated temperature. In certain instances, small variations in the nature of any of these may cause the observed R_f value to differ from the reported R_f value. In this case, a standard compound may be run and the R_f values compared to this standard rather than employing an absolute value.

As it is a characteristic of a compound, the R_f value can be considered to be a tool in the identification of substances alone or in mixtures. It does not directly result in functional group identification, but phases can be selected so that even quite similar compounds can exhibit different R_f values. Examples of R_f values of closely related substances for thin layer and paper chromatography are shown in Tables 36 and 37. However, it should be noted that only the most tentative conclusions can be drawn from R_f values used alone.

Once the solvent front has approached the end of the sheet, it is removed from the mobile phase supply and dried, and the new location

TABLE 36
R_f Values of Phenols on Silica Gel Plates[a]

Compound	R_f values $\times$ 100	
	Solvent system 1[b]	Solvent system 2[c]
Phenol	55	93
Catechol	44	52
Resorcinol	35	25
Hydroquinone	35	26
Pyrogallol	32	12
Phloroglucinol	25	2
Gallic acid	23	—
Protocatechuic acid	35	10
Vitamin K$_3$	82	82

[a]From Ref. *127*.
[b]Benzene/methanol/acetic acid, 8/1/1 (v/v).
[c]Chloroform/acetic acid, 9/1 (v/v).

TABLE 37
R_f Values of Organic Acids on Paper[a]

Compound	R_f values $\times$ 100[b]			
	Solvent system 1[c]		Solvent system 2[d]	
	25°C	35°C	25°C	35°C
Adipic	85	82	82	80
Citric	32	27	35	35
Fumaric	61	60	78	76
Aconitic	45	42	71	65
Glutaric	78	78	79	77
Glycolic	56	60	54	54
Lactic	70	75	64	67
Maleic	50	58	38	37
Malic	42	41	46	46
Malonic	48	49	56	51
Oxalic	21	21	11	7
Succinic	72	68	72	72
Trihydroxyglutaric	25	24	16	24

[a]From Ref. *128*.
[b]On Whatman No. 1 paper.
[c]Phenol/water/90% formic acid (3 g/1 ml/1%).
[d]Isopropanol/t-butyl alcohol/water/benzyl alcohol/formic acid, 1/1/1/13/1% (v/v/v/v/%v).

of the spot is determined. Location techniques may be relatively general, such as heat charring, fluorescence, or quenching of fluorescence, or they may be quite specific for compounds with certain functional groups. Many reagents used for spot detection, therefore, are similar to those discussed in Sec. VI.A., "Classification Schemes." Typical identification reagents are described in Refs. 88 and 123. Thus, compounds with the same R_f value could be distinguished if specific spot identification methods are available for each. Also, the presumptive identity of a compound giving a spot can be further enhanced by the use of many such identification reagents. However, in separating and identifying the amino acids in a mixture from protein hydrolysis, there are so many compounds of similar nature that usual techniques result in spot overlap. Even if one employs two-dimensional chromatography, where the paper or plate is first developed in one direction and then at right angles to this direction, separation may not be complete (124).

Although there is a limit to how many compounds can be so separated and "tentatively" identified by R_f values, there are nevertheless many solvent systems and spot identification techniques available which have been most useful in qualitative analysis. Further identification can also be achieved by extracting the solute from the spot and subjecting it to many of the chemical and physical identification methods previously discussed.

2. Gas Chromatography

This is a gas mobile phase system in which the fixed phase (usually a solid, or a liquid adsorbed on a solid) is in the form of a column. In this technique, a sample is injected into a stream of carrier gas which passes through the column. The various components of the mixture are separated in the column and are detected as they exit from the column. The substances do not have to be gases but should exert a reasonable vapor pressure (approximately 1 mmHg) at column temperature (125).

Qualitative information may be obtained by measurement of the retention time (t_r) or the retention volume (v_r). The retention time is defined as the time required for the maximum of the solute band to pass through the column. The retention volume (v_r) is

$$v_r = t_r Q \tag{214}$$

where Q is the flow rate of the carrier gas. These values are dependent on the nature of the compound, the composition of the fixed phase, the carrier gas, temperature, flow rate, and pressure. The retention volume may be corrected for pressure as follows:

$$v_r(\text{corrected}) = v_r(\text{uncorrected}) [3(p_i/p_0)^2 - 1]/[2(p_i/p_0)^3 - 1] \quad (215)$$

Flow and temperature can be quite well controlled, but some variation in solid phases usually exists. Thus, retention volumes may be made relative to a standard compound run under experimental conditions.

The compounds issuing from the column are usually detected physically by a differential detector capable of responding to some property of the carrier gas stream affected by the solute. The most commonly used detectors are thermal conductivity, flame ionization, and electron capture. Of these, the first two respond to almost all organic compounds, but the last is virtually insensitive to carbonyls, hydrocarbons, and alcohols but particularly sensitive to alkyl halides, sulfides, nitriles, nitrates, phosphates, and organometals [126].

Compound identity can be substantiated, to some extent, through the use of different detector systems, derivative formation, or the independent

TABLE 38
Retention Volumes of Various Hydrocarbons[a]

Compound	Retention volumes[b]		
	Fixed phase 1[c]	Fixed phase 2[d]	Fixed phase 3[e]
Methane	0.8	0.7	0.75
Ethanol	1.2	1.0	0.75
Propane	2.7	1.5	0.75
n-Butane	7.5	3.0	0.75
Isobutane	5.2	2.2	0.75
Ethene	1.0	1.0	1.0
Propene	2.4	2.2	1.0
Butene-1	6.7	4.75	6.25
Isobutene	6.7	4.75	3.25
trans-Butene-2	8.0	5.89	1.75
cis-Butene-2	8.9	6.8	5.5
Acetylene	0.95	3.8	—
Methylacetylene	2.85	11.5	—
Butadiene	6.7	10.0	10.0

[a]From Ref. 125.
[b]Relative to that of ethene.
[c]Triisobutylene.
[d]Acetonyl alcohol.
[e]AgNO$_3$/glycol.

use of two or more liquid phases. None of these methods gives significant structural information, and for positive identification the compound must be collected after it leaves the column and subjected to certain of the physical and chemical tests previously described.

As would be expected of a chromatographic system, conditions can be adjusted so that even closely related compounds will have different v_r values. An example of this is presented in Table 38. Due to instrument sophistication, relative ease of quantitation (if the compound is known), and the qualitative aspects of the v_r values, this has been a most important technique. It should be stressed in closing, however, that chromatography is essentially a separation method, and that R_f, v_r, or t_r values can only result in approximate identification.

H. Dissociation Constants

In a field that is mainly noted for irreversible reactions, the dissociations of organic acids and bases are examples of simple reversible reactions.

$$RCOOH + H_2O = RCOO^- + H_3O^+ \tag{216}$$

$$RNH_2 + H_2O = RNH_3^+ + OH^- \tag{217}$$

These simple dissociations are equilibrium reactions and can therefore be described by equilibrium constants, as indicated in the following expressions:

$$K_a = [H_3O^+][RCOO^-]/[RCOOH] \tag{218}$$

and

$$K_b = [OH^-][RNH_3^+]/[RNH_2] \tag{219}$$

If we are dealing with thermodynamically correct values for the equilibrium constants, the quantities within the brackets represent the activities of the noted species. For present purposes, however, activity may be considered approximately equal to concentration.

For simplicity of presentation, it is customary to present dissociation constants in terms of their negative logarithms, that is, as pK_a and pK_b. For a given acid or base, these terms are related through the negative logarithim of the ionization constant of water (pK_w). Thus, for a given base, $pK_a = pK_w - pK_b$ for the acid and basic forms. A base, therefore, can also be described in terms of its pK_a value.

Dissociation constants can be determined experimentally by several different methods, but electrical conductance is most widely used due to its accuracy and simplicity. The degree of dissociation, α, at a given concentration is measured by the ratio of the equivalent conductance at that concentration (Λc) to the equivalent conductance at infinite dilution (Λo). Thus

$$\alpha = \Lambda c / \Lambda o \tag{220}$$

Once α is known, the ionization constant, K, is given by the expression

$$K = \alpha^2 / (1 - \alpha) V \tag{221}$$

where V is that volume of solution of concentration c which contains one equivalent of the compound. To obtain the true thermodynamic dissociation constant, however, the nonideal behavior of the ions must be taken into consideration.

For the dissociation noted in Eq. (218), the true dissociation constant is defined by (129)

$$pK_a = pH + \log (RCOOH/RCOO^-) + \log F_{RCOOH} - \log F_{RCOO^-} \tag{222}$$

where F_{RCOOH} is the activity coefficient of the un-ionized acid, and F_{RCOO^-} is the activity coefficient of the ionized acid. To solve this equation for pK_a, the pH is measured by a glass electrode, $(RCOO^-)$ is equal to (H^+), $(RCOOH)$ is equal to the initial acid concentration less the amount ionized, F_{RCOOH} is assumed to be unity, and F_{RCOO^-} can be calculated by using the Debye–Hückel equation (21). If this is done at the half-neutralization point, where $(RCOOH) = (RCOO^-)$, then Eq. (222) simplifies to

$$pK_a = pH - \log F_{RCOO^-}. \tag{223}$$

Sometimes the activity coefficients are ignored. In this case, the ionization constant is determined at a number of ionic strengths and then extrapolated to zero ionic strength.

In certain cases, pK_a values can also be determined spectrophotometrically. The requirements are that at least one of the two forms [see Eqs. (218) and (219)] must absorb in the visible-ultraviolet region, and, if both absorb, their absorption maxima must be sufficiently far apart to allow accurate measurement of each. The experiment is conducted by preparing a number of solutions of the compound in buffers of known pH.

The absorption of each is measured at λ_{max} of either the ionized or un-ionized form. A plot of absorption, A, versus pH, such as that shown in Fig. 10, is then prepared.

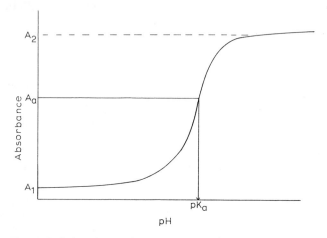

Fig. 10. Relation of pH and absorbance. A_a is equal to $\frac{1}{2}(A_1 + A_2)$.

TABLE 39
The Influence of Method of Determination on pK_a Values[a]

	Ultraviolet absorption	Electrometric titration
Phenol	9.87	4.8
p-Chlorophenol	9.02	9.05
p-Nitrophenol	7.00	6.85
p-Cresol	9.81	9.65
p-Aminophenol	—[b]	9.4
Aniline	4.67	4.5
p-Chloroaniline	4.05	4.05
p-Nitroaniline	1.13	—
p-Aminoaniline	6.2	6.2
Benzoic acid	4.16	4.15
p-Hydroxybenzoic acid	4.48	4.5
p-Aminobenzoic acid	4.82	4.85

[a]From Ref. 130.
[b]Sample unstable in alkali.

The value of pK_a is defined as the pH where absorbance, $A_a = \frac{1}{2}(A_1 + A_2)$. This determination can be extremely sensitive for organic acids and bases at concentrations as low as 10^{-4} M. A comparison of pK_a values determined by ultraviolet absorption and electrometric titration is presented in Table 39.

Every organic acid has a specific pK_a value. Although generalizations can be made, as shown in Table 40 different compounds may have the same pK_a value. Thus, although this method can give information on the types of compounds present, the considerable overlap of pK_a values has limited its use in chemical identification.

TABLE 40
pK_a Values as a Function of Compound Type[a]

Compound	pK_a range
Organic bases	
Aliphatic amines	9.7–11.0
β-Unsaturated (or polar subst.)	9.1–9.7
α-Unsaturated (or polar subst.)	6.4–9.3
α-Amino acids	9.2–10.2
Peptides	5.4–9.3
Aromatic nitrogen	3.5–7.2
Hetrocyclic nitrogen	
Non-aromatic	8.7–11.0
Aromatic	3.8–5.7
Guanidino	10.1–10.7
Monocarboxylic acids	
Formic	3.7–3.8
Aliphatic, saturated	4.8–5.1
Aliphatic, β-unsaturated	4.4–4.7
Aliphatic, γ-unsaturated	4.7–4.8
Aliphatic, β-phenyl	3.5–4.4
Aliphatic, α-polar subst.	
—NH$_2$	2.1–2.7
—NH—	3.1–3.9
—OH	2.8–4.2
—O—	2.5–3.3
—S—	2.5–4.0
—halogen	1.2–2.9
Aromatic carbon subst.	3.5–4.4
ortho, polar	2.2–3.8
meta, polar	3.8–4.6
para, polar	2.4–4.5

TABLE 40 (*Contd*)

Compound	pK_a range
Dicarboxylic acids	
Aliphatic	2.1–4.6, 5.2–7.5
Aromatic	3.0–3.6, 4.6–5.3
Polar, α-subst.	1.1–4.2, 3.5–5.3
Enols	4.2–6.8
Phenols	
Carbon-subst.	9.2–11.8
Polyphenols	9.7–11.4
Amino subst.	9.8–11.3
Halogen subst.	5.3–9.1
Nitro subst.	3.1–6.0
Hetero	7.3–10.5
Napthols	8.7–10.8
Indophenols	5.2–8.9
Other acidic compounds	
Oximes	9.9–11.7
Uric acids	3.9–5.9
Thiols	10.5–12.2
Organic arsenic acids	3.6–4.3, 8.3–9.2
Organic boric acids	9.1–10.7
Organic phosphoric acids	6.3–6.9
Carbohydrates	11.7–12.9

[a]From Ref. *131*.

VII. Quantitative Organic Analysis

In an aqueous system the quantity of an organic compound is estimated by establishing a numerical relationship between its concentration QA and the magnitude of some measurable property PA. This property may be of the system itself or one derived from it that is due to and proportional to the quantity of compound present. The general form of the functional relationship is

$$f(QA) = kf(PA) \qquad (224)$$

in which k is a constant and the functions may be linear, reciprocal, or logarithmic, among others.

Quantitative determinations should have certain characteristics. One is that a method should be sensitive, that is, $\Delta[f(PA)]/\Delta[f(QA)]$ should be as large as possible. Sensitivity is important because most natural aquatic systems contain numerous organic compounds that are present in only trace amounts. In fact, the limited sensitivity of available methods often necessitates preliminary concentration of samples.

An aspect even more important than sensitivity is that of selectivity. Selectivity implies that the system property measured should reflect the quantity of the specific organic compound of interest, and only this compound. Perfect selectivity is difficult, if not impossible, to achieve. Therefore, in an actual determination, the measured property frequently reflects the quantities of certain other compounds that are present.

The problem of selectivity is solved through the use of preliminary steps designed to overcome the effect of interfering compounds. These steps consist of many of the physical and chemical separation procedures previously noted. The nature and number of the steps required are determined by the inherent selectivity of the method, preliminary knowledge of what interfering chemical species are likely to be present, and the degree of certitude desired or required. Highly selective methods are not particularly common, and are typified by enzymatic determinations. When it is necessary to use a relatively nonselective property, highly refined separation procedures are required. Conversely, when highly refined separation procedures are employed, the quantitative property does not have to be particularly selective. A typical example of this is found in gas chromatography, in conjunction with its attendant "cleanup" procedure. Here quantitation can be achieved through a general nonselective property such as thermal conductivity, due to the presumed excellence of the separation procedures. There exists, therefore, an intimate relationship between the separation and quantitation steps in a quantitative organic analysis. In this section, only the quantitation steps are stressed, as they represent the more general aspects of quantitative analysis. Certain preliminary techniques may, however, be found in the examples given.

The properties used in quantitation include many of those presented in Sec. IV. In this case it is not a question of the amount of qualitative information supplied, but how well the property may be measured or quantitated, as well as the aspects of sensitivity and selectivity. As there are a large number of properties that can be used for this purpose, it is possible to determine an organic compound through many different techniques. Which technique to employ is decided on the basis of such practical considerations as knowledge of probable interferences, equipment availability, and the number of samples to be run, as well as those theoretical aspects previously noted.

For example, sodium *n*-dodecylbenzene (an alkylbenzene sulfonate, ABS) exhibits many properties that could be used to determine its concentration in aqueous solution. Methods based on 12 of these properties

$$C_{12}H_{25}\langle\bigcirc\rangle SO_3^- Na^+$$

and the attendant interferences of each are noted in Table 41. It is obvious that determinations 2, 5, and 6 are so unselective as to be worthless in mixtures. Methods 1, 3, 4, 7, and 8 are somewhat more selective, but not sufficiently so to permit their use on natural aquatic systems. Thus, the methods of choice would be 9, 10, 11, and 12.

TABLE 41

Comparative Methods for Sodium *n*-dodecylbenzene Quantitation

Method	Interference
1. Weight (evaporation and weighing)	Any nonvolatile organic or inorganic compound
2. Density	Any organic or inorganic compound
3. Conductivity	Any ionic organic or inorganic compound
4. Surface tension depression	Any surface-active agent
5. Organic carbon	Any organic compound
6. Disappearance of an oxidizing agent such as dichromate in a hot acid solution	Almost any organic compound plus oxidizable inorganic compounds
7. Oxidation followed by estimation of liberated sulfate	Any organic or inorganic compound containing sulfur
8. Direct ultraviolet spectrophotometry	Aromatic or other organic compounds absorbing in same region
9. Formation of a complex with a cationic dye; extractions of complex followed by visible spectrophotometry	Certain high molecular weight anionic organic substances
10. Extraction as amine salt and infrared spectrophotometry	Certain high molecular weight anionic organic compounds depending on infrared range employed
11. Two-phase titration with a cationic surfactant using an anionic dye as an indicator	Certain high molecular weight anionic organic compounds
12. Desulfonation followed by gas chromatography	Other volatile organic compounds having similar retention times

Many properties used for quantitative analysis are due to the presence of certain functional groups in molecules. Such properties, therefore, are not specific for one compound. This lack of specificity, however, is not necessarily a disadvantage, for it allows the determination of classes of compounds such as amino acids, fatty acids, hexoses, pentoses, etc. Group concentrations are usually reported as being due to one typical compound. For example, such compounds could be leucine for the amino acids, acetic acid for the lower fatty acids, glucose for the hexoses, etc. Results could also be reported as concentrations of the functional itself, in a manner analogous to the equivalent weight determination.

As noted in Table 41, the methods used for quantitative organic analysis vary widely in sensitivity and specificity. In this section, only the more specific methods are discussed. This does not imply, however, that such techniques as evaporation and weighing, density, surface tension, viscosity, conductivity, refractive index, optical rotation, etc., are not quite useful in simple systems, but only that their use in natural aquatic systems is fairly limited.

A. Titrimetry

In titrimetry, the volume of a reagent of known concentration required to stoichiometrically react with the unknown constituent is determined. The point at which the stoichiometric amount of reagent has been added is called the equivalence point. At this point, the number of equivalents of reagent added are equal to the number of equivalents of the constituent present, and it can be stated that:

$$W_c/E_q = N(V/1000) \qquad (225)$$

where W_c is the weight of the constituent present (g), E_q is the gram equivalent weight of the constituent present, and N and V are the normality and volume (ml), respectively, of the reagent employed in the determination.

Results from titrimetric procedures are usually reported in terms of their per cent (if the weight of the sample is known) or concentration in milligrams per liter (if the volume of the sample is known). The equations would be

$$\% = E_q NV/SW(10) \qquad (226)$$

$$\text{mg/liter} = E_q NV(1000)/SV \qquad (227)$$

where SW is the sample weight in grams and SV is the sample volume in milliliter.

Reactions that go to completion rapidly are usually employed for titrimetric procedures; common examples are neutralization, redox, complex ion, and precipitation reactions. Whatever the reaction, the success of the determination depends on the cessation of titrant addition once the equivalence point is reached, or some definite knowledge of what volume of the titrant is excess. Simple titrimetric procedures involve detection of the equivalence point by some obvious physical change in the titration system once the correct number of titrant equivalents have been added. Such signals include color changes, foam (as in the reaction between Ca^{2+} and stearate), etc. More complex procedures involve measuring some property of the system that changes as the titrant is added and that undergoes some obvious change at the equivalence point. Such properties as electrode potential, conductivity, current carried at an applied potential, temperature, turbidity, absorption of electromagnetic radiation, etc., have been employed in this fashion.

Titrimetric procedures are simple, low in cost, and moderately sensitive. Unfortunately, they are not particularly selective and thus frequently require preliminary separation steps. Typical of this is the determination of the lower organic acids by titration with an organic or inorganic base. Although crude estimates of these acids in high concentrations can be made by direct titration after purging the system of carbon dioxide(132), most procedures employ preliminary distillation(30,133), extractions (134,135), adsorption(136,137), or ion exchange(138,139). Naphthalene may be determined by titration with iodine after preliminary distillations, extraction, and picrate precipitation to remove interferences(140). Due to its anionic nature, ABS has been determined directly by two-phase titrations with cationic surfactants in the presence of cationic dyes(141, 142), in which the end point is observed by the transference of color to the nonaqueous phase, and by turbidimetric end point detection with similar titrants(143), with little or no interference control. Titrimetry has not been a widely employed method except after preliminary separations or when working with pure compounds in determining equivalent weights.

B. Polarography

Another analytically significant part of the voltage-current curve described in Fig. 9 is the upper region of slight slope, frequently described as the "plateau." The reason for the existence of this plateau is that the current carried in this region no longer directly follows the applied voltage according to Ohm's law. It is limited by the diffusion to the dropping mercury electrode of the species being oxidized or reduced. As this diffusion is concentration dependent, the height of the plateau over a line representing the residual current carried by the system without the reacting species is proportional to the concentration.

The Ilkovic equation relates this limiting diffusion current to the concentration (49):

$$I_d = [607nD^{1/2}m^{2/3}t^{1/6}]C \tag{228}$$

where I_d is the average diffusion current (μA), C is the concentration (mmoles/liter), n is the number of Faradays of electricity required per mole of the electrode reaction, D is the diffusion coefficient of the substance (cm^2/sec), m is the rate of flow of mercury from the capillary (mg/sec), and t is the drop time (sec). This equation is infrequently used in quantitation.

In general, external or internal standards are employed. In the first case, the plateau heights are determined for a number of standard solutions of the species undergoing oxidation or reduction, and a plot is made of plateau height (current) versus concentration. This is used to determine what concentration could give the plateau height of the substance in solution. If the current-concentration relationship is linear, a known concentration of the substance may be added to the system containing the unknown and a new plateau height determined. The unknown concentration can then be determined by a simple calculation.

This technique has not been too popular for quantitative analysis in this field. Polarographic reductions have been used to determine nitrogenous compounds such as nitrocyclohexane(144), nitriles(145), nitrobenzene derivatives(146), and benzene after conversion to nitrobenzene (147); aldehydes such as furfural and formaldehyde(148); and chlorobenzene derivatives(149). Polarographic oxidation has been used to determine cresols(150).

C. Chromatography

Although chromatography is, in essence, a separation procedure, the methods previously mentioned as qualitative techniques also have quantitative aspects that merit attention in this section. This is because the quantitation step is in some way an integral part of the procedure itself.

1. PAPER AND THIN LAYER CHROMATOGRAPHY

After separation of the various constituents of a mixture using paper or thin layer chromatography, the area containing the constituent can be removed and the constituent extracted and determined. However, this uses these techniques simply as separation schemes, rather than having the quantitation step as an integral part of the procedure. The latter methods include empirical relationships established between concentration and spot area, comparison to a standard or examination of the spot with a sequential detector (densitometer for colored spots, etc.), obtaining a curve of property versus distance, and establishing a relationship between concentration and peak area or height(151). Quantities as low as a few micrograms can be detected and estimated.

Much of the work in paper chromatography has involved the determination of the amino acids. Circular methods(152,153) and two-dimensional methods(154) have been used for sewage, sludges, and effluents. Circular methods have also been used for lipid determinations(155) in sewage sludges. Acidic substances have also been the subject of some research, mainly involving the lower fatty acids in sewages(12) and sludges(156), naphthenic acids(157), and phenols(158,159).

There has been considerable research on the detection and determination of phenols(160,161) and chlorophenols(162,163) by thin layer chromatography. Nonionic surfactants(164), amino acids(165), and oils (166) have also been determined by this technique.

2. GAS CHROMATOGRAPHY

The differential detectors that reveal, in the form of peaks, the substances separated from and leaving the chromatographic column can also be easily used to obtain quantitative results. Various concentrations of the substance in question may be passed through the column under the same conditions, and a relationship can be established between peak area or peak height and concentration. This technique is capable of detecting organic substances in the microgram (thermal conductivity) to picogram (flame ionization, electron capture) range. Thus, gas chromatography must be considered, along with activation analysis and isotope dilution, as representative of the most sensitive methods presently available for quantitative analysis.

Gas chromatography has been mainly used for volatile organic compounds, organic compounds that form volatile derivatives, or pryolysis products of nonvolatile organic substances. As with the other two chromatographic systems, organic acids have received considerable attention. The lower fatty acids (167–170), various carboxylic acids

(*171,172*), and higher fatty acids (10–20 carbon atoms)(*173*) have all been determined. Linear alkylbenzene sulfonates also have been determined by gas chromatography after desulfonation(*174*).

Phenols are another group that has received considerable attention. Mainly simple monohydric phenols(*175–180*) have been separated and determined, usually as ester or ether derivatives. In addition to organic acids and phenols, organic amines(*181*), aldehydes and ketones(*182*), sterols(*183*), methane(*184*), gasoline(*185*), and other petroleum products (*186,187*) have also been determined. It is evident, therefore, that this method has been the most useful and widely employed of the three chromatographic techniques discussed here.

D. Absorption Spectrophotometry

The relationship between concentration and the absorption of electromagnetic radiation is described by a variant of Eq. (98) as follows:

$$A_s = a_s bc \qquad (229)$$

where A_s is the absorbance (or $\log_{10} 1/\text{transmittance}$), b is the pathlength (cm), c is the concentration in any specified units (usually mg/liter), and a_s is the absorptivity. The wavelength at which A_s is measured is most frequently that at the absorption maximum for the compound in question or one of its derivatives. This wavelength is selected because it represents maximum sensitivity and also because at this wavelength A_s is least influenced by small changes (errors) in wavelength setting.

In general, the magnitude of a_s indicates the sensitivity of the method, and it is also obvious that sensitivity can also be increased by increasing the pathlength. Quantitative results are obtained by comparison between the absorbance of the system containing the compound in question and a previously determined plot of absorbance versus concentration, keeping conditions (including pathlength) constant. The absorbances are usually measured by comparison to that of distilled water or whatever solvent system is used, which is set equal to zero. To keep the photometric error at a minimum, it is best to arrange conditions so that the transmittance of the system being measured is between 20 and 65%.

1. NONDESTRUCTIVE TECHNIQUES

Absorption spectrophotometric techniques may be very roughly divided into nondestructive and destructive techniques, depending on whether the compound is essentially irreversibly altered in the determination or not. In general, direct visible-ultraviolet and infrared spectro-photometry do not so alter the compound, and therefore may be considered nondestructive techniques.

a. Visible-Ultraviolet

As both visible and ultraviolet spectra are reasonably simple, the selection of a wavelength at which to measure the absorbance is a relatively simple task. However, neither technique is particularly selective, and thus various types of preliminary separations are frequently employed. Solvents are relatively transparent in this region, and quantitation is a relatively simple process, with fairly good sensitivity.

One interesting aspect of absorbances is that they are additive. That is, at a given wavelength, the total absorbance measured is equal to the sum of the absorbances of the various constituents. Thus, it is possible to determine two or more constituents simultaneously, provided that the compounds do not absorb appreciably at each other's absorption maxima wavelengths.

The simple relationship would be

$$(A_1)_{\lambda_1} + (A_2)_{\lambda_1} = A_{\lambda_1} \tag{230}$$

$$(A_2)_{\lambda_2} + (A_2)_{\lambda_2} = A_{\lambda_2} \tag{231}$$

where A_{λ_1} and A_{λ_2} are the total absorbances at λ_1 and λ_2, the wavelengths of maximum absorbance for compounds 1 and 2, respectively. The absorbances within the parenthesis are due to that part of the total absorbance due to the individual compounds. Using Eq. (229), these equations become (for a unit pathlength)

$$(a_1)_{\lambda_1}(c_1) + (a_2)_{\lambda_2}(c_2) = A_{\lambda_1} \tag{232}$$

$$(a_1)_{\lambda_2}(c_1) + (a_2)_{\lambda_2}(c_2) = A_{\lambda_2} \tag{233}$$

where $(a_1)_{\lambda_1}$ is the absorptivity for compound 1 at a wavelength of λ_1, $(a_1)_{\lambda_2}$ is the same at a wavelength of λ_2, etc. Thus, if the four absorptivities are known, measurement of the total absorbance at λ_1 and λ_2 is all that is required to yield two equations with two unknowns. This technique has been used to simultaneously determine linoleic, linolenic, and arachidonic acids by ultraviolet absorption spectrophotometry after alkaline isomerization(188), and chlorophyll a, b, and c simultaneously by visible absorption spectrophotometry(189).

In the visible region, the use of this technique for quantitative determinations has been largely limited to chlorophylls and carotenoids (*189–191*). As might be expected, ultraviolet absorption spectrophotometry has been largely used in the determination of aromatic compounds. These have included alkylbenzene sulfonates(*192, 193*), phenols(*194, 195*), polynuclear hydrocarbons(*196, 197*), and acetophenone(*198*). In addition, there has been some use of the ultraviolet region in detection of general organic pollution(*199, 200*).

b. Infrared

Unlike the visible-ultraviolet region, infrared absorption spectra are quite complex and therefore result in selection problems for the wavelength at which the absorbance is to be measured, as well as the greater selectivity inherent in this choice. However, due to the short pathlengths employed and the relatively low absorptivities observed, the sensitivity of this region is poor compared to that of the visible-ultraviolet region. However, this is not quite the disadvantage it seems, as the various extractive procedures used to remove the constituent from water may also be employed in its concentration.

A more serious problem is accuracy. This is limited by the low energy of infrared radiation and the difficulty of establishing a true zero absorbance or 100% transmittance line. Although base line and ratio techniques are employed to more accurately measure the absorbance(*49*), the infrared region has been of only minor quantitative significance when compared to the visible-ultraviolet region. However, there has been considerable work on the determination of oils(*201–204*) and alkylbenzene sulfonates(*205*), including linear alkyl sulfonates(*206,207*), and some work on TNT(*208*), naphthalene(*209*), and phenols(*210*).

2. DESTRUCTIVE TECHNIQUES

The use of destructive techniques usually involves the reaction of the compound in question with certain added reagents to produce a material that absorbs in the visible-ultraviolet range. Also, the destruction of the compound in question is utilized to an extent sufficient to cause a loss of its ability to absorb radiation.

a. Colorimetric

The production of a characteristic color when a specific reagent is added to a solution containing the compound under investigation not only has significance in qualitative determinations but also in quantitative determinations as well. Indeed, many simple quantitation procedures use equipment other than spectrophotometers or apparatus for the measurement of absorbance, i.e., Nessler tubes, slide comparitors, Hehner cylinders, and Duboscq colorimeters(49). However, except for field work, etc., modern usage involves the measurement of the absorbance of the colored material formed at its wavelength of minimum per cent transmittance at maximum absorption.

Although the color may be due to a complex formation, as with the various cationic dye methods available for ABS determinations, most techniques involve condensation or other reactions that essentially destroy the original compound. Thus, colorimetric techniques are mainly destructive, unlike direct visible-ultraviolet or infrared methods.

Despite such minor limitations, the general applicability of the technique has resulted in its being the most widely used procedure in quantitative organic analysis. Individual determinations may vary from the highly selective estimation of a specific compound to the estimation of a group of related compounds in terms of one of the compounds of the group. As the color production is usually obtained through the reaction of a functional group, specificity for a compound is frequently obtained through preliminary separation steps. Due to the number of determinations involved in a review of only the more recent literature, these methods are presented in tabular form (Table 42).

TABLE 42

Colorimetric Methods of Quantitative Organic Analysis

Compound or class	Reaction	References
Alkylbenzene sulfonates	Methyl green	211,212
	Methylene blue	213–215
	Magenta	216
	Protein-cresol purple complex	217
	2,3,5-Tetrazolium chloride	218
Alcohol sulfates	Azure A	219
Polyethoxyalkylphenols	Cobalt thiocyanate	220
	Phosphomolybdic acid	221
	Phosphotungstic acid	222
Long chain fatty acid salts	Crystal violet	223
Cationic surfactants	Bromcresol blue	223
	Bromcresol purple	224
Phenols	4-Aminoantipyrene	225–229
	Pyramidone	230–232
	4-Aminophenezone	233,234
	Dimethylaminopyrene	235
	2,6-Dibromoquinone chlorimide	236

TABLE 42 (*Contd*)

Compound or class	Reaction	References
	p-Nitrosodimethylaniline	237,238
	α-Nitroso-β-naphthol	239
Amino acids	Ninhydrin	240–243
Proteins	Folin's reagent	244
	Biuret	245,246
Amines	Hypochlorite, starch-iodide	247
	Bromcresol purple	248
	Sulfonphthaleins	249
	Picric acid	250
Pyridine	Benzidine hydrochloride	251
	Barbituric acid	252
Creatine-creatinine	Picric acid	253,254
Adenine	Permanganate	255
	Zinc reduction, nitrite	256
DNA	Diphenylamine	257
Uric acid	Arsenophosphotungstic acid	258
Amino sugars	p-Dimethylaminobenzaldehyde	259
	Acetyl acetone	260
Carbohydrates	Anthrone	56,246,261
	p-Aminohippuric acid	262
	Resorcinol	263
Mercaptans	Bis(p-nitrophenyl) disulfide	264
Fatty acids (lower)	Hydroxylamine, $FeCl_3$	265
Aromatic hydrocarbons	Formaldehyde condensation	266,267
	Nitration, methyl ketone	268
Resin acids	Acetic anhydride, H_2SO_4	269
Lactic acid	Oxidation, p-hydroxyphenol	270
Lignin sulfonic acid	Nitrous acid	271
	Phosphotungstic, phosphomolybdic acid	272
Formaldehyde	Phenylhydrazine	273
Terpentine	Molybdophosphoric acid	274
Pine oil	Vanillin-HCl	275
Hydroquinone	Fe^{3+}, o-phenanthroline	276

b. Enzymes

Properly speaking, the use of enzyme-catalyzed reactions is not a part of absorption spectrophotometry. Many techniques, including titrations(277), polarimetry(277), and even light emission(278), can be used to detect product formation or disappearance due to the use of these specific catalysts. However, many of the methods so used do involve absorption spectrophotometry, and for this reason enzyme catalysis is mentioned here. As the usual techniques are employed, sensitivities are similar to those noted previously. The specificity of enzyme reactions is the main significance of this technique, but little use has been made of it in this field. However, uric acid has been determined by difference after conversion to allantoin by the enzyme uricase using ultraviolet absorption spectrophotometry(279,280), and glucose, fructose, and saccharose are converted to the phosphate derivatives by hexokinase. These derivatives are detected by visible absorption spectrophotometry(281) resulting in the determination of these sugars.

E. Fluorescence

As would be expected, the intensity of radiation emitted during fluorescence is proportional to the concentration. The general relationship is(49)

$$F = KI_0(1 - e^{-abc})$$ (234)

where F is the intensity of fluorescent radiant energy, I_0 is the incident radiant energy intensity, K is a proportionality constant, and abc are as in Eq. (229). For very low concentrations, e^{-abc} becomes approximately equal to $1 - abc$, and Eq. (234) then becomes

$$F = KI_0abc$$ (235)

Equation (235) indicates that the intensity of the exciting radiation must be kept constant or reference-ratio techniques must be employed. Also, fluorescence intensity tends to be somewhat temperature dependent. In addition, certain contaminants may quench fluorescence. Despite these minor disadvantages, this technique is sensitive in the parts per billion range and is one of the most sensitive methods for quantitative organic analysis.

This technique may be employed to measure intensity of emission of a fluorescent material or inhibition of such fluorescence (quenching). What use this technique has seen in this field has involved the determination of polynuclear hydrocarbons such as 3,4-benzopyrene(282,283) or 1,2-benzanthracene(284), and the antibiotic chlortetracycline(285) through emission measurement techniques. When it is considered that at liquid nitrogen temperatures concentrations of polynuclear hydrocarbons as low as 0.003 μg/liter can be determined(283), it is strange that so little has been accomplished with this most sensitive technique.

VIII. Summary and Conclusion

It has been the purpose of this chapter to briefly describe the problems inherent in studying aquatic organic substances and the theory behind organic analysis and its qualitative and quantitative aspects. In general, the major stress has been on qualitative rather than quantitative analysis. This does not reflect the quality or quantity of the quantitative organic analyses performed on aquatic organic substances. A considerable amount of excellent quantitative data exist in this area, but before any quantitative step can be taken, the identity of the specific compound or the general nature of a group of compounds must be known or suspected.

In addition to the prior knowledge necessary, the steps required to go from an initial observation (e.g., that when reagent A is added to compound B in solution, a specific color is formed) to an actual quantitative method are reasonably well established. They include a study of the effects of environmental factors (reagent dosage, time, temperature, pH, etc.) on the color intensity and stability, establishment of λ_{max}, and development of a quantitative relationship between absorbance and concentration. Such studies also frequently include a study of expected interferences and how these might be controlled in the environmental system to which the method is being applied.

Applications of quantitative approaches have been quite successful for waste water analysis(11,12) but quite unsuccessful for waste water treatment plant effluents(17,18). It is probable that considerable information must be obtained on the nature of those undetected organic substances before substantial quantitative progress can be achieved. Indeed, the complexity of such systems as aquatic color and soil humic acids renders simple quantitative approaches inappropriate(36,120).

To emphasize the future role of qualitative organic analysis, many physical and chemical techniques have been described. However, these should not all be considered to be of equal utility. In general, the physical qualitative techniques of greatest value have been infrared, nuclear magnetic resonance, and mass spectrophotometry. For certain structural detail, ultraviolet spectrophotometry and fluorescence, as well as some of the other techniques noted, have been of some value. The major application of chemical techniques for simple compounds has been in the development of characteristic functional group tests. For more complex organic compounds, chemical degradation procedures have been of great value.

The use of these techniques is best demonstrated by examples. A sterol (gorgosterol) was isolated from the gorgonian family. Mass spectrophotometry indicated a formula of $C_{30}H_{50}O$, and oxidation plus ultraviolet spectrophotometry, nuclear magnetic resonance spectroscopy and X-ray crystallography have revealed, or are expected to reveal, all the structural details (286). A material initially called geosmin, isolated from certain actinomycetes and a blue-green alga and characterized by a musty odor, was determined to be trans-1,10-dimethyl-trans-9-decalol. Physical identification techniques employed included infrared, Raman, nuclear magnetic resonance, and mass spectrometry. Also included were gas chromatographic and spectral analyses of breakdown products from treatment with acid, osmium tetroxide permanganate, and bromine. Even such classic chemical methods as the Lucas test for alcohols and 2,4-dinitrophenylhydrazine derivatives of carbonyl oxidation products (287,288) were employed.

Even more difficult problems are presented by such studies as those on the nature of the organic substances responsible for color in natural waters. Here, the structure is both large and complex, and destructive techniques have been required to obtain significant structural information (120). It can only be concluded, considering how little is actually known about the structure of many of the organic compounds present in effluents and receiving waters, that considerable attention in the future will be directed toward the partial solution of this problem.

REFERENCES

1. *Synthetic Detergents in Perspective*, Technical Advisory Council, The Soap and Detergent Association, New York, 1962.
2. S. D. Faust, *Clin. Pharmacol. Therap.*, **5**, 677 (1964).
3. Research Committee on Color Problems, *J. Am. Water Works Assoc.*, **59**, 1023 (1967).
4. R. Baker, *J. Am. Water Works Assoc.*, **55**, 913 (1963).
5. Advisory Committee on Review of USPHS 1946 Drinking Water Standards, *J. Am. Water Works Assoc.*, **53**, 935 (1962).
6. W. C. Hueper, *Proc. Conf. Physiol. Aspects of Water Quality, Washington, D.C.*, *1960*.
7. J. V. Hunter, Ph.D. Thesis, Rutgers Univ., New Brunswick, N.J., 1962.
8. J. R. Vallentyne, *J. Fisheries Res. Board Can.*, **14**, 33 (1957).
9. G. F. Lee, *Water Chemistry Seminars, 1965*, University of Wisconsin, Madison.
10. W. Rudolfs and J. L. Balmat, *Sewage Ind. Wastes*, **24**, 247 (1952).
11. J. V. Hunter and H. Heukelekian, *J. Water Pollution Control Federation*, **37**, 1142 (1965).
12. H. A. Painter and M. Viney, *J. Biochem. Microbiol. Tech. Eng.*, **1**, 143 (1959).
13. D. Rickert and J. Hunter, *J. Water Pollution Control Federation*, **39**, 1475 (1967).
14. J. Morris and W. Stumm, *Proc. 1st Rudolfs Conf., Rutgers Univ., New Brunswick, N.J., 1960*, p. 55.
15. S. Faust and M. Manger, *Intern J. Air Water Pollution*, **9**, 565 (1965).
16. J. Hunter and H. Heukelekian, *Proc. 15th Ind. Waste Conf., Purdue Univ., Lafayette, Ind., 1961*, p. 150.

17. R. Bunch, E. Barth, and M. Ettinger, *J. Water Pollution Control Federation*, **33**, 122 (1961).

18. H. Painter, M. Viney, and A. Bywaters, *Inst. Sewage Purif. J. Proc.*, 1961, p. 302.

19. E. Gjessing and G. Lee, *Environ. Sci. Technol.*, **1**, 631 (1967).

20. A. Burges, *Soil Sci.*, **2**, 128 (1960).

21. J. Shapiro, *Limnol. Oceanog.*, **2**, 161 (1957).

22. R. Baker, *J. Water Pollution Control Federation*, **37**, 1164 (1965).

23. S. Kobayashi and G. Lee, *Anal. Chem.*, **36**, 2197 (1964).

24. F. Middleton, A. Rosen, and R. Burttschell, *Manual for Recovery and Identification of Organic Chemicals in Water*, R. A. Taft Sanitary Engineering Center, Cincinnati, 1959.

25. P. Atkins, Jr., and N. Tomlinson, *Proc. 17th Ind. Waste Conf., Purdue Univ., Lafayette, Ind., 1962*, p. 764.

26. W. Weber, Jr., *J. Appl. Chem. (London)*, **14**, 565 (1964).

27. S. Faust and O. Aly, *J. Am. Water Works Assoc.*, **54**, 235 (1962).

28. S. Faust and I. Suffet, *Residue Rev.*, **15**, 44 (1966).

29. I. Murtaugh and R. Bunch, *J. Water Pollution Control Federation*, **37**, 410 (1965).

30. *Standard Methods for the Analysis of Water and Waste Water*, 12th ed., American Public Health Association, New York, 1965.

31. M. Schnitzer, J. Desjardins, and J. Wright, *Can. J. Soil Sci.*, **38**, 49 (1958).

32. R. Dubach and N. Melita, *Soils Fertilizers*, **26**, 293 (1963).

33. P. Busch and W. Stumm, *Environ. Sci. Technol.*, **2**, 49 (1968).

34. L. Fieser and M. Fieser, *Organic Chemistry*, 3rd ed., Heath, Boston, 1956.

35. J. Mitchell, Jr., in *Treatise on Analytical Chemistry* (I. M, Koltoff and P. J. Elving, eds.), Part II, Vol. 13, Wiley (Interscience), New York, 1966, p. 1.

36. G. T. Felbeck, Jr., *Advan. Agron.*, **17**, 327 (1965).

37. R. Packham, *Proc. Soc. Water Treat. Exam.*, **13**, 316 (1964).

38. F. Daniels, J. Mathews, J. Williams, P. Bender, G. Murphy, and R. Alberty, *Experimental Physical Chemistry*, McGraw-Hill, 6th ed., New York, 1962.

39. S. Glasstone, *Textbook of Physical Chemistry*, 2nd ed., Van Nostrand, New York, 1946.

40. F. Haurowitz, *Chemistry and Biology of Proteins*, Academic, New York, 1950.

41. P. Johnson, in *Determination of Organic Structures by Physical Methods* (E. Braude and F. Machod, eds.), Academic, New York, 1955, p. 25.

42. W. Archibald, *J. Phys. Chem.*, **51**, 1204 (1947).

43. H. Mark, in *Techniques in Organic Chemistry* (A. Weissberger, ed.), Vol. 1, 1st ed., Wiley (Interscience), New York, 1945, p. 135.

44. F. Daniels, *Outlines of Physical Chemistry*, Wiley, New York, 1948.

45. A. Steyermark, *Quantitative Organic Microanalysis*, 2nd ed., Academic, New York, 1961.

46. A. Elek and D. W. Hill, *J. Am. Chem. Soc.*, **55**, 3479 (1933).

47. *Standard Methods of Chemical Analysis* (N. J. Furman, ed.), 6th ed., Van Nostrand, New York, 1962.

48. J. Niederl and V. Niederl, *Micromethods of Quantitative Organic Elementary Analysis*, Wiley, New York, 1938.

49. H. Willard, L. Merritt, and J. Dean, *Instrumental Methods of Analysis*, 3rd ed., Van Nostrand, New York, 1958.

50. R. Degeiso, W. Rieman III, and S. Lindenbaum, *Anal. Chem.*, **26**, 1340 (1954).

51. W. Weber, Jr., and J. Morris, *J. Water Pollution Control Federation*, **36**, 573 (1964).

52. M. Effenberger, *Water Pollution Abstr.*, **38**, Abstr. No. 250 (1965).

53. J. Krey and K. H. Szekielda, *Z. Anal. Chem.*, **207**, 338 (1965); through *Chem. Abstr.*, **62**, 7502f (1965).
54. R. Wilson, *Limnol. Oceanog.*, **6**, 259 (1961).
55. T. Larson, F. Sollo, and B. J. Gruner, *Proc. 19th Ind. Waste Conf., Purdue Univ., Lafayette, Ind., 1964*, p. 761.
56. A. Neish, *Anal. Methods for Bacterial Ferment. Rept. No. 46-8-3* (2nd rev.), National Research Council of Canada, Saskatoon, 1952.
57. J. Beattie, C. Brocker, and D. Garvin, *Anal. Chem.*, **33**, 1890 (1961).
58. K. Szekielda and J. Krey, *Microchim. Ichnoanal. Acta*, **149** (1965); through *Chem. Abstr.*, **62**, 11529g (1965).
59. C. Oppenheimer, E. Concarau, and J. Van Arman, *Limnol. Oceanog.*, **8**, 487 (1963).
60. A. Fredericks and D. Hood, *Atomic Energy Comm. Publ. AD 618932*, Washington D.C., 1965.
61. C. Van Hall, J. Safranko, and V. Stenger, *Anal. Chem.*, **35**, 315 (1963).
62. H. Montgomery and N. Thom, *Analyst*. **87**, 689 (1962).
63. G. Kramig and R. Schaffer, *Ind. Water Eng.*, **2**, 16 (1965).
64. R. B. Schaffer, C. Van Hall, G. McDermott, D. Barth, V. Stenger, S. Sebesta, and S. Griggs, *J. Water Pollution Control Federation*, **37**, 1545 (1965).
65. C. Van Hall and V. Stenger, *Water Sewage Works*, **111**, 266 (1964).
66. E. Cooper, D. Heinekey, and A. Westwell, *Analyst*. **92**, 436 (1967).
67. C. Van Hall, D. Barth, and V. A. Stenger, *Anal. Chem.*, **37**, 769 (1965).
68. H. A. McKenzie and H. S. Wallace, *Australian J. Chem.*, **7**, 55 (1954).
69. I. Kolthoff and E. Sandell, *Textbook of Quantitative Inorganic Analysis*, 3rd ed., Macmillan, New York, 1952.
70. S. B. Morgan, J. Lackey, and F. Gilereas, *Anal. Chem.*, **29**, 833 (1957).
71. O. Holm-Hasen, *Limnol. Oceanog.*, **13**, 175 (1968).
72. R. Brodstreek, *Anal. Chem.*, **32**, 114 (1960).
73. A. Steyermark, B. McGee, E. Bass, and R. Kaup, *Anal. Chem.*, **30**, 1561 (1958).
74. M. Dudova, *Gidrokhim. Materialy*, **30**, 164 (1960); through *Chem. Abstr.*, **56**, 218e (1962).
75. R. Shriner and R. Fuson, *Systematic Identification of Organic Compounds*, 4th ed., Wiley, New York, 1956.
76. O. Kamm, *Quantitative Organic Analysis*, 2nd ed., Wiley, New York, 1964.
77. E. Clark, *Ind. Eng. Chem., Anal. Ed.*, **8**, 487 (1936).
78. J. Pelerson, K. Hedberg, and B. Christensen, *ibid.*, **15**, 25 (1943).
79. S. Siggia and R. Edsberg, *ibid.*, **20**, 762 (1948).
80. T. Ma, J. Logan, and P. Mazzella, *Microchem. J.*, **1**, 67 (1957).
81. R. Keen and J. Fritz, *Anal. Chem.*, **24**, 564 (1952).
82. D. Van Slyke, *J. Biol. Chem.*, **9**, 185 (1911).
83. *Ibid.*, **83**, 425 (1929).
84. H. Roth, *Mikrochim. Acta*, 766 (1958).
85. C. Ogg and F. Cooper,, *Anal. Chem.*, **21**, 1400 (1949).
86. F. Schneider, *Quantitative Organic Microanalysis*, Academic, New York, 1964.
87. S. Jashi and G. Tuli, *J. Chem. Soc.*, **60**, 3032 (1938).
88. M. C. St. Flett, *Physical Aids to the Organic Chemist*, Elsevier, New York, 1962.
89. L. Reeves, *Can. J. Chem.*, **35**, 1351 (1957).
90. C. Huggins, G. Pimental, and J. Schoalery, *J. Chem. Phys.*, **23**, 1244 (1955).
91. B. Bhar, *Arkiv Kemi*, **10**, 223 (1956).
92. S. Sugden, *J. Chem. Soc.*, **125**, 32 (1924).
93. H. Herbrandson and E. Nachod, in *Determination of Organic Structures by Physical Methods* (E. Braude and F. Nachod, eds.), Academic, New York, 1955, p. 3.

94. S. Lewin and N. Bauer, in *Treatise on Analytical Chemistry* (I. Kolthoff, P. Elving, and E. Sandell, eds.), Part I. Vol. 6, Wiley (Interscience), New York 1965, pp. 3922–3923.

95. R. Kinney, *J. Am. Chem. Soc.*, **60**, 3032 (1938).

96. R. Kinney, *Ind. Eng. Chem.*, **32**, 559 (1940).

97. R. Kinney and W. Spiethoff, *J. Org. Chem.*, **14**, 71 (1949).

98. Report No. 6, Joint Committee on Nomenclature in Applied Spectroscopy, *Anal. Chem.*, **24**, 1349 (1952).

99. E. Braude, in *Determination of Organic Structure by Physical Methods* (E. Braude and F. Nachod, eds.), Academic, New York, 1955, p. 131.

100. R. Pecsok and L. Shield, *Modern Methods of Chemical Analysis*, Wiley, New York, 1968.

101. K. Cunningham, W. Dawson, and F. Spring, *J. Chem. Soc.*, **1951**, 2305.

102. R. Williams and J. Bridges, *J. Clin. Pathol.*, **17**, 371 (1964).

103. J. Radley and J. Grant, *Fluorescence Analysis in Ultraviolet Light*, 4th ed., Chapman and Hall, London, 1954.

104. American Instrument Co., Luminescence Data Sheet No. 2392-11, 1966.

105a. F. Cleveland, in *Determination of Organic Structure by Physical Methods* (E. Braude and F. Nachod, eds.), Academic, New York, 1955, p. 235.

105b. B. D. Cullity, *Elements of X-Ray Diffraction*, Addison Wesley, Reading, Mass, 1956, Chap. 6.

105c. M. J. Buerger, *X-Ray Crystallography*, Wiley, New York, 1942.

106. J. Robertson, in *Determination of Organic Structure by Physical Methods* (E. Braude and F. Nachod, eds.), Academic, New York, 1955, p. 463.

107. S. Lewin and N. Bauer, in *Treatise on Analytical Chemistry*, (I. Kolthoff, P. Elving, and E. Sandell, eds.), Part I, Vol. 6, Wiley (Interscience), New York, 1965, p. 3927.

108. J. Hilderbrand and R. Scott, *The Solubility of Non-electrolytes*, 3rd ed., Reinhold, New York, 1949.

109. P. Small, *J. Appl. Chem.*, **3**, 71 (1953).

110. P. W. West and L. Brovassard, *Anal. Chem.*, **22**, 469 (1950).

111. S. Soloway and P. Rosen, *Anal. Chem.*, **29**, 1820 (1957).

112. N. Cheronis and J. Entrikin, *Semimicro Qualitative Organic Analysis*, 2nd ed., Wiley (Interscience), New York, 1957.

113. F. Feigl, *Spot Tests, Vol. II, Organic Applications*, Elsevier, New York, 1954.

114. D. Cram and G. Hammond, *Organic Chemistry*, 2nd ed., McGraw-Hill, New York, 1964.

115. C. Thompson, H. Coleman, C. Ward, and H. Rall, *Anal. Chem.*, **34**, 154 (1962).

116. C. Thompson, H. Coleman, R. Hopkins, C. Ward and H. Rall, *Anal. Chem.*, **32**, 1762 (1960).

117. C. Thompson, H. Coleman, C. Ward, and H. Rall, *Anal. Chem.*, **34**, 151 (1962).

118. F. Brauns, *The Chemistry of Lignins*, Academic, New York, 1952.

119. A. Black and R. Christman, *J. Am. Water Works Assoc.*, **55**, 897 (1963).

120. R. Christman and M. Ghassemi, *J. Am. Water Works Assoc.*, **58**, 723 (1966).

121. I. Kolthoff and J. Lingane, *Polarography*, Vol. II, Wiley (Interscience), New York, 1952.

122. P. Elving, in *Organic Analysis* (J. Mitchell, Jr., I. Kolthoff, E. Proskauer, and A. Weissberger, eds.), Vol. II, Wiley (Interscience), 1954, pp. 209–312.

123. Eastman Kodak Company, *TLC Visualization Reagents*, Rochester, N.Y., 1969.

124. H. Heukelekian and J. Balmat, *Sewage Ind. Wastes*, **31**, 412 (1959).

125. A. Keulemans, *Gas Chromatography*, 2nd ed., Reinhold, New York, 1959.

126. S. Dal Nogare and R. Juret, *Gas Chromatrography—Theory and Practice*, Wiley, New York, 1962.

127. O. Ali, *Z. Anal. Chem.*, **234**, 251 (1968).

128. J. Stark, A. Goodbau, and H. Owens, *Anal. Chem.*, **23**, 413 (1951).

129. H. Brown, D. McDaniel, and O. Häfliger, in *Determination of Organic Structure by Physical Methods* (E. Braude and F. Nachod, eds.), Academic, New York, 1955, p. 567.

130. J. Vanderbelt, C. Henrich, and S. Vanderberg, *Anal. Chem.*, **26**, 726 (1954).

131. T. Parke and W. Davis, *Anal. Chem.*, **26**, 642 (1954).

132. R. DiLallo and O. Albertson, *J. Water Pollution Control Federation*, **33**, 356 (1961).

133. E. L. Bykova, *Byul. Nauchn.-Tekhn. Inform. Gos. Geol. Kom. SSSR, Otd. Nauchn.-Tekhn. Inform., Vses. Nauchn.-Issled. Inst. Mineral'n. Syr'ya*, 94 (1964); through *Chem. Abstr.*, **64**, 439H (1966).

134. T. Kelley, *Anal. Chem.*, **37**, 1078 (1965).

135. J. Thomas, C. Wherry, and A. Pearson, *Proc. 10th Ind. Waste Conf., Purdue Univ., Lafayette, Ind., 1955*, p. 267.

136. F. Pohland and B. Dickson, *Water Works Wastes Eng.*, **1**, 54 (1964).

137. H. Mueller, T. Larson, and W. Lennarz, *Anal. Chem.*, **30**, 41 (1958).

138. H. Dostal and D. Dilley, *Bio-Science*, **14**, 35 (1964).

139. A. Semenov, V. Bryzgalo, and V. Datsko, *Gidrokhim. Materialy*, **38**, 137 (1964); through *Chem. Abstr.*, **62**, 159026 (1965).

140. G. Shirma, *Gigiena Sanit.*, **26**, 61 (1961); through *Chem. Abstr.*, **56**, 4544 (1962).

141. G. Edwards and M. Ginn, *Sewage and Ind. Wastes*, **26**, 945 (1954).

142. R. Cropton and A. Joy, *Analyst*, **88**, 516 (1963).

143. M. Frankhouser, University Microfilms, Ann Arbor, Mich., Order No. 63-3243; *Dissertation Abstr.*, **23**, 4512.

144. G. Vainshtein and P. Papsueva, *Anal. Abstr.*, **8**, 2178 (1961).

145. M. Bobrova and A. Kudasheva-Matueeva, *Zh. Obshch. Khim.*, **28**, 2929 (1958); through *Chem. Abstr.*, **53**, 5976 (1959).

146. P. Zaitsev, V. Dichenskii, and V. Krasnosel'skii, *Zavodsk. Lab.*, **32**, 800 (1966); through *Chem. Abstr.*, **65**, 13398 (1966).

147. M. Adamovsky, *Vodni. Hospodarstvi*, **16**, 102 (1966); through *Chem. Abstr.*, **65**, 6908 (1966).

148. N. Kuchamova, S. Verigo, and O. Mamontova, *Tr. Vses Nauchn. Issled. Inst. Pererobotke Ispol'z. Topliva*, **12**, 237 (1963); through *Chem. Abstr.*, **60**, 1428b (1964).

149. B. Fleszar, *Chem. Anal (Poland)*, **9**, 1075 (1964); through *Chem. Abstr.*, **63**, 6316h (1965).

150. O. Sheveleva and L. Buidakova, *Ochenye Zapiski Tseht., Nauchn.-Issled., Inst. Olovjan Prom.*, 1965; through *Chem. Abstr.*, **66**, 13905 (1967).

151. R. Block, E. Durrum, and G. Zweig, *A Manual of Paper Chromatography and Paper Electrophoresis*, Academic, New York, 1955.

152. C. Sastry, P. Subrahmanyam, and S. Pillai, *Sewage Ind. Wastes*, **30**, 1241 (1958).

153. P. Subrahmanyam, C. Sastry, P. Rao, and S. Pillai, *J. Water Pollution Control Federation*, **32**, 344 (1960).

154. J. Balmat, Ph.D. Thesis, Rutgers Univ., New Brunswick, N.J., 1955.

155. C. Viswanathan, M. Bai, and S. Pillai, *J. Water Pollution Control Federation*, **34**, 189 (1962).

156. R. Manganelli and F. Brofazi, *Anal. Chem.*, **29**, 1441 (1957).

157. E. Bykova, *Byul. Nauchn.-Tekhn. Inform., Gos. Geol. Kom. SSSR, Otd. Nauchn.-Tekhn. Inform., Vses. Nauchn.-Issled. Inst. Mineral'n. Syr'ya*, **87** (1964), through *Chem. Abstr.*, **62**, 12901d (1965).

158. Y. Lure and Z. Nikolaeva, *Zavodsk. Lab.*, **30**, 937 (1964); through *Chem. Abstr.*, **61**, 13035d (1964).

159. Y. Lure and Z. Nikolaeva, *Zavod. Lab.*, **30**, 8, (1964); through *Chem. Abstr.*, **60**, 13035 (1964).
160. H. Seeboth, *Water Pollution Abstr.*, **40**, Abstr. No. 52 (1967).
161. F. Dyatlovitskaya, L. Botvinova, and E. Maktaz, *Water Pollution Abstr.*, **41**, Abstr. No. 313 (1968).
162. M. Zigler and W. Phillips, *Environ. Sci. Technol.*, **1**, 65, (1967).
163. M. Gebott, *Solutions* (Ann Arbor, Mich.), **6**, 8 (1967).
164. S. Patterson, E. Hunt, and K. Tucker, *Inst. Sewage Purif. J. Proc.*, Part 2, 190 (1966).
165. Y. Chau and J. Riley, *Deep-Sea Res. Oceanog. Abstr.*, **13**, 1115 (1966).
166. G. Giebler, P. Koppe, and H. Kempf, *Gas Wasserfack*, **105**, 1093 (1964); through *Chem. Abstr.*, **61**, 1590e (1964).
167. R. A. Baker, *J. Gas Chromatog.*, **4**, 418 (1966).
168. L. Nemtseva, T. Kishkmova, and A. Semenov, *Gidrokhim. Materialy Akad. Nauk*, **41**, 121 (1966); through *Chem. Abstr.*, **67**, 93851z (1967).
169. J. Hrivwak, *Vodni. Hospodarstvi*, **14**, 394 (1964); through *Chem. Abstr.*, **62**, 8822 (1965).
170. A. Otsuki, *Nippon Kagaku Zasshi*, **84**, 798 (1963); through *Chem. Abstr.*, **60**, 2640e (1964).
171. H. A. Painter, M. Viney, and A. Bywaters, *Inst. Sewage Purif J. Proc.*, Part 4, 302 (1961).
172. L. Lamar and D. Goerlitz, *J. Am. Water Works Assoc.*, **55**, 797 (1963).
173. J. Slowey, L. Jeffrey, and D. Hodd, *Geochim. Cosmochim. Acta*, **26**, 607 (1962).
174. R. Swisher, *J. Am. Oil Chem. Soc.*, **43**, 137 (1966).
175. W. Navcke and F. Tarkmann, *Brennstoff Chem.*, **45**, 263 (1964); through *Chem. Abstr.*, **62**, 322 (1965).
176. S. Goren-Strul and A. Mostaert, *Anal. Chim. Acta*, **34**, 322 (1966).
177. L. Semenchenko and V. Kaplin, *Zavodsk. Lab.*, **33**, 801 (1967); through *Chem. Abstr.*, **67**, 102601 (1967).
178. R. Argoner, *Anal. Chem.*, **40**, 122 (1968).
179. R. Baker and B. Malo, *Environ. Sci. Technol.*, **1**, 997, (1967).
180. T. Hermann and A. Post, *Anal. Chem.*, **40**, 1573 (1968).
181. L. Namtseva, A. Pashanova, T. Kishkinova, and A. Semenov, *Gidrokhim. Materialy Akad. Nauk*, **41**, 121 (1966); through *Chem. Abstr.*, **67**, 93847 (1967).
182. M. Jaworski, T. Slareczek, and J. Bobinski, *Chimin. Analit.*, **11**, 685 (1967).
183. J. Murtaugh and R. Bunch, *J. Water Pollution Control Federation*, **39**, 404 (1967).
184. R. Navone and W. Genninger, *J. Am. Water Works Assoc.*, **59**, 757 (1967).
185. R. Jeltes and R. Veldink, *J. Chromatog.*, **27**, 242 (1967).
186. I. Adams, *Process Biochem.*, **2**, 33 (1967).
187. J. Sugar and R. Conway, *J. Water Pollution Control Federation*, **40**, 1622 (1968).
188. P. Mueller, Ph.D. Thesis, Rutgers Univ., New Brunswick, N.J., 1955.
189. A. Duxburg and C. Yentsch, *J. Marine Res.*, **15**, 92 (1957).
190. T. Parsons, *J. Marine Res.*, **21**, 164 (1963).
191. C. Lorenzen, *Oceanog. Limnol.*, **12**, 343 (1967).
192. E. Setzkorn and R. Huddleston, *J. Am. Oil Chem. Soc.*, **42**, 1081 (1966).
193. W. Weber, Jr., J. Morris, and W. Stumm, *Anal. Chem.*, **34**, 1844 (1963).
194. J. Martin, Jr., C. Orr, C. Kincannon, and J. Bishop, *J. Water Pollution Control Federation*, **39**, 21 (1967).
195. H. Eisenhauer, *J. Water Pollution Control Federation*, **36**, 1116 (1964).
196. P. Wegewood and R. Cooper, *Analyst*, **81**, 43 (1956).
197. R. Tye, M. Grof, and A. Horton, *Anal. Chem.*, **27**, 248 (1955).
198. Z. Boldina, *Gigiena Sanit.*, **32**, 65 (1967); through *Chem. Abstr.*, **67**, 120067g (1967).

199. P. Jones and G. Heinke, *Water Pollution Abstr.*, **41**, Abstr. No. 354 (1968).

200. M. Mrkva, *J. Water Pollution Control Federation*, **41**, 1923 (1969).

201. L. Lindgren, *J. Am. Water Works Assoc.*, **49**, 55 (1957).

202. F. Ludzack and C. Whitfield, *Anal. Chem.*, **28**, 157 (1956).

203. W. Fastabend, *Chem. Ingr. Tech.*, **37**, 728 (1965); through *Chem. Abstr.*, **63**, 9960e (1965).

204. M. Golubeva, *Lab. Deli*, **665** (1966); through *Chem. Abstr.*, **66**, 79455w (1967).

205. J. Vaughn, *J. Am. Water Works Assoc.*, **50**, 1343 (1958).

206. C. Hoehler, J. Cripps, and A. Greenboro. *J. Water Pollution Control Federation*, **39**, R92 (1967).

207. J. Fairing and F. Short, *Anal. Chem.*, **28**, 1827 (1963).

208. R. Morris and J. Daughertz, *Proc. 15th Ind. Waste Conf.*, *Purdue Univ., Lafayette, Ind., 1961*, p. 281.

209. S. Faingol'd and S. Senkovskaya, *Koks Khim.*, **8**, 32 (1957); through *Chem. Abstr.*, **55**, 2963 (1961).

210. R. Simard, I. Hasegawa, W. Bandaruk, and C. Headington, *Anal. Chem.*, **23**, 1384 (1951).

211. D. Abbott, *Analyst*, **88**, 240 (1963).

212. W. Moore and R. Kolbeson, *Anal. Chem.*, **28**, 161 (1956).

213. J. Longwell and W. Maniece, *Analyst*, **80**, 167 (1955).

214. P. Degens, H. Van du Zee, and J. Kimma, *Sewage Ind. Wastes*, **25**, 20 (1953).

215. Z. Grzbiela and E. Wysokinska, *Zesz. Nauk Politech. Slask. Chem.*, **34**, 91 (1966); through *Chem. Abstr.*, **68**, 1587u (1968).

216. R. Cropton and A. Jog, *Analyst*, **88**, 516 (1963).

217. F. Loomeijer, *Anal. Chim. Acta*, **10**, 147 (1954).

218. H. Renault and L. Bigot, *Water Pollution Abstr.*, **34**, Abstr. No. 1817 (1961).

219. J. Steveninck and J. Riemersma, *Anal. Chem.*, **38**, 1250 (1966).

220. N. Crabb and H. Persinger, *J. Am. Oil Chem. Soc.*, **41**, 752 (1964).

221. W. Krygielowa, H. Szmalowa, and J. Wasociez, *Przemyst Chem. (Poland)*, **45**, 516 (1966); through *Chem. Abstr.*, **66**, 98283 (1967).

222. P. Pitter, *Water Pollution Abstr.*, **41**, Abstr. No. 448 (1969).

223. V. Lovell and F. Sebba, *Anal. Chem.*, **38**, 1926 (1966).

244. J. Fogh, R. Rasmussen, and K. Skadhouge, *Anal. Chem.*, **26**, 392 (1954).

225. S. Faust and E. Mikulewicz, *Water Res.*, **1**, 405 (1967).

226. M. Ettinger, C. Ruchoft, and R. Lishka, *Anal. Chem.*, **23**, 1783 (1951).

227. S. Osaki, *Bunseki Kagaku*, **7**, 275 (1958); through *Chem. Abstr.*, **57**, 8370 (1962).

228. Z. Babeshkina, V. Kaplin, and N. Fesenko, *Gidrokhim. Materialy*, **35**, 207 (1963); through *Chem. Abstr.*, **58**, 11100f (1963).

229. M. Higashiura, *Kagaku To Kogyo*, **40**, 90 (1966); through *Chem. Abstr.*, **65**, 8533f (1966).

230. M. Csanady, *Hidrol. Kozl. (Hungary)*, **44**, 371 (1964); through *Chem. Abstr.*, **64**, 4790 (1966).

231. V. Kaplin, S. Panchenko, and N. Fesenko, *Koks Khim.*, **8**, 41 (1964); through *Chem. Abstr.*, **61**, 15826 (1964).

232. N. Sergeeva, *Novye Metody Analiza Khim. Sostava Podzemn. Vod.*, 123 (1967); through *Chem. Abstr.*, **68**, 71976m (1968).

233. W. Dozanska and C. Sikorowska, *Water Pollution Abstr.*, **41**, Abstr. No. 208 (1968).

234. D. Abbott, *Proc. Soc. Water Treat. Exam.*, **13**, 153 (1964).

235. V. Kaplin and N. Fesenko, *Gigiena Sanit.*, **25**, 41 (1960); through *Chem. Abstr.*, **54**, 25411d (1960).

236. G. Tallon and R. Hepner, *Anal. Chem.*, **30**, 1521 (1958).

237. R. Hill and L. Herndon, *Sewage Ind. Waste*, **24**, 1389 (1952).
238. I. Nusbaum, *Sewage Ind. Waste*, **25**, 311 (1953).
239. V. Kapisinska, *Vodni. Hospodarstvi*, **17**, 503 (1967); through *Chem. Abstr.*, **68**, 89791h (1968).
240. S. Moore and W. Stein, *J. Biol. Chem.*, **211**, 907 (1954).
241. E. Yemm and E. Cocking, *Analyst*, **80**, 209 (1955).
242. H. Rosen, *Arch. Biochem. Biophys.*, **67**, 10 (1957).
243. Y. Lee and T. Takahashi, *Anal. Biochem.*, **14**, 71 (1966).
244. W. Johnson and C. Woods, *Water Sewage Works*, **111**, 342 (1964).
245. C. Woods, *Proc. 20th Ind. Wastes Conf., Purdue Univ., Lafayette, Ind., 1965*, p. 501.
246. A. Gaudy, *Ind. Water Wastes*, **7**, 17 (1962).
247. G. Dahlgren, *Anal. Chem.*, **36**, 596 (1964).
248. L Nemtsova, A. Semenov, and V. Datska, *Gidrokhim. Materialy, Akad. Nauk*, **38**, 150 (1964); through *Chem. Abstr.*, **62**, 15902 (1965).
249. A. Pearce, *Chem. Ind.*, **24**, 825 (1961).
250. Y. Lur'e and V. Panova, *Zavodsk. Lab.*, **27**, 1333 (1961); through *Chem. Abstr.*, **56**, 8487 (1962).
251. R. Kroner, M. Ettinger, and W. Moore, *Anal. Chem.*, **24**, 1877 (1952).
252. G. Voigh and W. Stech, *Fortschr. Wasserchem. Ihrer Grenzg.*, **2**, 242 (1965); through *Chem. Abstr.*, **68**, 15919 (1968).
253. R. Bonsner and H. Toussky, *J. Biol. Chem.*, **145**, 581 (1945).
254. D. Abbott, *Analyst*, **87**, 494 (1962).
255. J. Davis and R. Morris, *Anal. Biochem.*, **5**, 64 (1963).
256. D. Woodhouse, *Arch. Biochem.*, **25**, 347 (1950).
257. W. Hattingh and M. Siebert, *Water Res.*, **1**, 197 (1967).
258. D. Norton, M. Plunkett, and F. Richards, *Anal. Chem.*, **26**, 454 (1954).
259. N. Boas, *J. Biol. Chem.*, **204**, 553 (1953).
260. N. Anita and C. Lee, *Limnol. Oceanog.*, **9**, 262 (1964).
261. D. Morris, *Science*, **107**, 254 (1948).
262. I. Iveleva, A. Semenov, and V. Datsko, *Gidrokhim. Materialy*, **38**, 144 (1964); through *Chem. Abstr.*, **62**, 15904c (1965).
263. J. Congdon and S. Zaiatz, *Tech. Papers Ann. Meeting, Sugar Ind. Technicians*, **20**, 30 (1961).
264. G. Ellman, *Arch. Biochem. Biophys.*, **74**, 443 (1958).
265. H. Montgomery, J. Dymock, and N. Thom, *Analyst*, **87**, 949 (1962).
266. L. Zanoni and A. Venturini, *Boll. Lab. Chem. Prov. (Italy)*, **18**, 869 (1967); through *Chem. Abstr.*, **69**, 21813g (1968).
267. Y. Lur'e and V. Panova, *Zavodsk. Lab.*, **29**, 293 (1963); through *Chem. Abstr.*, **59**, 1371h (1963).
268. F. Devlaminck, *Bull. Centre Belge Etude et Doc. Eaux (Leige)*, **101**, 135 (1959).
269. W. Carpenter, *Tappi*, **48**, 11 (1965).
270. H. Barnes and D. Finlaysen, *Limnol. Oceanog.*, **8**, 292 (1963).
271. V. Felicetta and J. McCarthy, *Tappi*, **46**, 337 (1963).
272. S. Komaki, *Bunseki Kagaku*, **11**, 1197 (1962); through *Chem. Abstr.*, **58**, 3195f (1963).
273. V. Stankovic, *Chem. Zvesti*, **16**, 683 (1962); through *Chem. Abstr.*, **58**, 7712a (1963).
274. Y. Lur'e and V. Panova, *Zavodsk. Lab.*, **29**, 33 (1963); through *Chem. Abstr.*, **59**, 308d (1963).
275. J. Sutton, *U.S. Bur. Mines, Rept. Invest. No. 4990*, 1953; through *Chem. Abstr.*, **47**, 89466 (1953).
276. Y. Lur'e and Z. Nikolaeva, *Zavodsk. Lab.*, **30**, 937 (1964); through *Chem. Abstr.*, **61**, 13035d (1964).

277. J. Neilands, in *Organic Analysis* (J. Mitchell, Jr., I. Kolthoff, E. Proskauer, and A. Weissberger, eds.), Vol. IV, Wiley (Interscience), New York 1960, p. 65.
278. O. Hansen and C. Booth, *Limnol. Oceanog.*, **11**, 510 (1966).
279. J. O'Shea and R. Bunch, *J. Water Pollution Control Federation*, **37**, 1444 (1965).
280. G. Kupchik and G. Edwards, *J. Water Pollution Control Federation*, **34**, 376 (1962).
281. J. Ruchti and D. Kunkler, *Schweiz. Z. Hydrol.*, **28**, 62 (1966); through *Chem. Abstr.*, **66**, 79456x (1967).
282. J. Jager and B. Kassowitzova, *Chem. Listy*, **62**, 216 (1968); through *Chem. Abstr.*, **68**, 72118w (1968).
283. L. Soholz, H. Altmen, and Z. Fresenius, *Anal. Chem.* (Ger.), **240**, 81 (1968); through *Chem. Abstr.*, **69**, 79994 (1968).
284. K. Ershova and I. Mints, *Gigiena Sanit.*, **33**, 52 (1968); through *Chem. Abstr.*, **69**, 109670 (1968).
285. I. Shabunin and V. Laurenchuk, *Gigiena Sanit.*, **30**, 63 (1965); through *Chem. Abstr.*, **62**, 14320f (1965).
286. Anon., *Chem. Eng. News*, April 6, 1970, p. 13.
287. L. Medsker, D. Jenkins, and J. Thomas, *Environ. Sci. Technol.*, **2**, 461 (1968).
288. N. Gerber, *Tetrahedron Letters*, No. 25 1968, p. 2971.

Chapter **21** **Inorganic Analytical Chemistry in Aqueous Systems**

Jack L. Lambert
DEPARTMENT OF CHEMISTRY
KANSAS STATE UNIVERSITY
MANHATTAN, KANSAS

I. Introduction

Natural waters, from the analytical standpoint, can be considered in three categories. Surface waters and most groundwaters usually are very dilute solutions containing cations, anions, and neutral species, and the analytical methods used are designed to eliminate interference from one or more substances that may also be present in low concentration. Seawater, on the other hand, contains several ions in high but approximately constant concentration. In many cases, the methods used for the dilute solutions encountered in surface waters and groundwaters can be adapted to the analysis of seawater with minor modifications. Estuarine waters require special consideration because of the variability in concentration of the ions that are essentially in constant concentration in seawater. Waste waters also present problems because of their variable composition and possible high concentrations of suspended solids.

The inorganic substances to be considered in this chapter are ionic species and neutral compounds of the more common "representative" elements of the periodic table. Other elements and their compounds in trace concentrations are covered in Chap. 24. Specifically, the elements included in this chapter can be classified in the following abbreviated periodic table:

I	II	III	IV	V	VI	VII
H						
		B	C	N	O	F
Na	Mg		Si	P	S	Cl
K	Ca			As	Se	Br
						I

The range of substances is further limited to those that are stable in the presence of dissolved oxygen, except for those few that occur chiefly in certain waste waters. The arrangement of the elements is alphabetical, with ions and neutral species considered under the central, or most logical, element present.

Brief reference is made to each of the pertinent methods of water analysis contained in Ref. *1*. These methods are recognized as reliable and are followed almost universally in cases involving litigation and enforcement. Reference is also made to the most commonly used methods for seawater and estuarine water analysis; details of the methods can be found in the sources quoted.

The considerable literature of water analysis is thoroughly reviewed every odd-numbered year in the April "Annual Reviews" issue of *Analytical Chemistry*, which was used as a guide to most of the recent literature quoted herein. A review is also given each year in the annual literature review of the *Journal of the Water Pollution Control Federation*. All articles pertaining to water analysis presumably are abstracted in *Chemical Abstracts* and in *Analytical Abstracts*.

It is the purpose of this chapter to review briefly the accepted methods of water analysis now in use and to discuss the trends in the recent literature for each species with a view to possible directions for future research. Only methods for radioactivity are omitted from consideration.

In view of the present-day research emphasis on instrumental methods of analysis based on physical properties, as contrasted to the more traditional methods based on selective or specific chemical reactions followed by instrumental measurements, it is interesting that the latter still predominate. A count of the various methods described in Ref. *1* reveals that 34 of the methods are colorimetric (spectrophotometric or visual comparison), 18 are titrimetric, and four are gravimetric. Only eight are purely instrumental—flame photometric, glass electrode potentiometric, voltammetric, or spectrophotometric without extensive chemical pretreatment. The traditional methods prevail not only because of their proven trustworthiness, but also because they are peculiarly suited to the sensitivities and precisions required and to the economics of water laboratory operation.

II. Arsenic

Like selenium, arsenic mimics its congeners, nitrogen and phosphorus, in metabolic processes, and thus in most of its compounds it is toxic to plants and animals. Also like selenium, it rarely occurs in concentrations above 10 ppb in surface waters or groundwaters or 20 ppb in seawater. Industrial wastes and insecticide or herbicide residues may occasionally produce transient higher concentrations. The permissible limit in drinking water is 0.01 ppm; concentrations exceeding 0.05 ppm constitute grounds for rejection. Angino and co-workers(*1*) have detected 10 to 70 ppm in several common presoaks and household detergents, and 2 to 8 ppb in the Kansas River.

Arsenic is most advantageously converted for analysis to the hydride, arsine. *Standard Methods(1a)* describes two methods that utilize arsine generated by zinc in acid solution in a Gutzeit generator. In one method, the arsine is passed through glass wool impregnated with lead acetate to remove hydrogen sulfide and then permitted to react with silver diethyl-dithiocarbamate dissolved in pyridine to form a red complex that absorbs at 535 nm. Concentrations of arsenic as low as 30 ppb can be determined by this method, and only antimony interferes. The other method is the classic Gutzeit method, wherein the arsine reacts with a paper strip impregnated with mercury(II) bromide after passing through a glass wool–lead acetate scrubber. The length of the ascending yellow-brown stain is proportional to the concentration of arsenic in the sample. Again, the only interference is antimony. As little as 1 μg of arsenic in an aliquot of 25 ml can be detected.

Few new methods have been proposed in recent years for the determination of arsenic in water. The colorimetric determination of microgram quantities of arsenic(V) in seawater by reduction of molybdoarsenic(III) acid to a heteropoly blue, following reduction of arsenic(V) to arsenic(III) with ascorbic acid and cocrystallization with thionalide in acid solution, was described by Portman and Riley(2). Differential determination of arsenate, arsenite, and phosphate by selective solvent extractions of molybdophosphate and molybdoarsenate was reported by Sugawara and Kanamori(3). The arsenate and phosphate ions present are converted to the complex acids and extracted, after which the arsenic(III) is oxidized to arsenic(V) and the complex acid is extracted. The molybdenum in each fraction is released by treatment with sodium hydroxide and is determined spectrophotometrically with thiocyanate, which constitutes a 12-fold chemical amplification. Nakaya(4) described a method in which arsenic(V) was reduced to arsenic(III), complexed with thionalide, and extracted with ether. It was then back-extracted with aqueous hypobromite, oxidized to arsenic(V), and determined by the molybdenum blue method. Davidyuk(4a) reported a polarographic method for arsenic, in which arsenate ion is reduced chemically to arsenite prior to the determination.

III. Boron

More than one species of borate ion may exist in solution, but all soluble borates react in the same manner for analytical purposes. Drinking waters usually contain less than 0.1 ppm boron, but waters contaminated by cleaning compounds or industrial wastes may contain as much as 1–2 ppm, and plants may be adversely affected if such waters are used for irrigation. Seawater normally contains about 5 ppm boron.

Standard Methods(5) describes two colorimetric methods and one potentiometric titration method. Curcumin (tumeric) forms a red rosocyanine with borate when evaporated in the presence of oxalic and hydrochloric acids. The absorbance of the rosocyanine is measured at 540 nm in 95% ethanol solution. The method is applicable in the range 0.1–1.0 ppm, and the only interference is nitrate ion in concentrations above 20 ppm. For the range 1.0–10.0 ppm boron, the carminic acid method is recommended. The red color of carminic acid changes to purple or blue with increase in absorption at 585 nm in the presence of borates in concentrated sulfuric acid. No commonly encountered ions interfere in this method. In the potentiometric titration method, the acidity of boric acid is enhanced by complexation with excess mannitol. The acid thus generated is titrated with standard sodium hydroxide solution with a glass electrode pH meter as the end point indicator. Phosphate ion interferes in concentrations above 10 ppm.

Greenhalgh and Riley(6) modified the curcumin method for boron in seawater and found no interference from the ions customarily encountered. A titrimetric method for boron in seawater was described by Murakami(7) in which the enhanced acidity produced by mannitol with boric acid was used to produce iodine from potassium iodide and potassium iodate. The iodine generated was titrated with standard thiosulfate solution. Parker and Barnes(8) used the fluorescence of the borate–benzoin complex in alcoholic sodium carbonate solution for boron in seawater after removal of interferences by ion exchange.

The rate of publication of new methods for the determination of borates in water has decreased in recent years. Most new colorimetric methods, when compared to the carminic acid and curcumin methods, are found to be without significant advantage. Lishka(9) examined four reagents, 1,1'-dianthrimide, carminic acid, Victoria violet, and curcumin, and recommended the curcumin procedure. Powell and Poindexter(10,11) recommended 1,1'-dianthrimide over carminic acid because of its greater sensitivity. Numerous other chelating dyes have been investigated as colorimetric reagents for borate ion, among which are quinalizarin and alizarin red S, reported by Scher(11a), and quercetin, described by Hiiro(11b).

Colorimetric extraction procedures also have been suggested. Vasilevskaya and Lenskaya(12) extracted the ion association compound formed between the crystal violet cation and the boron–salicylate complex anion into benzene, and determined boron down to 1 ppm by measuring the absorbance of the extracted compound at 585 nm. Semenov(13) reported an extraction method with methylene blue as the counter cation that is sensitive to 0.05 ppm.

IV. Bromine

Bromide ion is the only form in which bromine exists in natural waters, but it may be converted to hypobromite during the chlorination of water supplies. It is nontoxic in the concentrations normally encountered, which are usually below 1 ppm; this is of little direct concern except that its presence in water supplies may indicate contamination by industrial wastes or intrusion of seawater. Seawater normally contains about 380 ppm bromide ion.

Standard Methods(*14*) describes a method involving the bromination of phenol red to produce a red to violet color at pH 5.0–5.4. Bromide ion is first oxidized to hypobromite ion by chloramine-T (sodium *p*-toluenesulfonchloramide). None of the commonly encountered substances in potable waters interferes. The method is recommended for the range 0.1–1.0 ppm bromide ion. Visual comparison techniques may be used, or the absorbance may be measured spectrophotometrically at 590 nm.

The selective reactions of bromine most often used are its bromination reactions with aromatic organic compounds and the complexing ability of bromide ion for mercury(II). Bromination reactions were reported by Podgornyi and Bezler(*15*) in a method in which free bromine produced by dichromate oxidation is used to form an extractable rose-violet dye from fuschin–sulfuric acid solution, and by Elbeih and El-Sirafy(*16*) in a method which used generated bromine to bleach Chromotrope 2B. Iwasaki and co-workers(*17,18*) reported two methods: one in which bromine from acid permanganate oxidation is extracted and reacted with alcoholic iron(III) aluminum sulfate–mercury(II) thiocyanate solution to produce red iron(III) thiocyanate complex, and the other in which the oxidation of iodide ion or iodine to iodate by permanganate ion in acid solution is catalyzed by bromide ion. The absorbance of the red complex in the first method is measured at 460 nm for bromide ion in the range 0.1–50 ppm. In the latter method, the solvent-extracted unreacted iodine is inversely proportional to the concentration of bromide ion; bromide ion can be determined in the range 0.005–0.13 ppm under the conditions described. Fishman and Skougstad(*19*) proposed modifications for this method and established optimum conditions.

A solid-state electrode selective for bromide ion has recently been made commercially available by Orion Research, Inc.(*20*). Interference is encountered from iodide, sulfide, and cyanide ions. Bystritskii et al.(*21*) used a silver–silver bromide electrode for the potentiometric determination of bromide ion in the parts per billion range.

V. Calcium

Calcium ion is a common constituent of natural waters, resulting from the dissolution of limestone, dolomite, and gypsum. While not usually important from a physiological standpoint, calcium ion is unique in being desirable in moderate concentrations and undesirable either in very low or very high concentrations. Small concentrations are necessary to form a protective coating on the inside of steel pipes and boilers, but high concentrations result in harmful scale formation and waste of soap by precipitation of insoluble fatty acid salts. Surface waters and groundwaters may contain up to 100 ppm calcium ion, while seawater normally contains about 400 ppm.

Calcium ion is determined in *Standard Methods* (*22*) by two procedures involving the precipitation of insoluble calcium oxalate: gravimetrically as the oxalate salt and titrimetrically by permanganate titration of the redissolved oxalate salt. In both methods, strontium, silica, aluminum, iron, manganese, phosphate, and suspended matter interfere. Strontium, if present, causes high results and must be determined separately by flame photometry. Procedures are described for removal of the other interferences. A third method—the EDTA (ethylenediaminetetraacetate) complexometric titration method of Schwarzenbach(*23,24*)—is also described in *Standard Methods* (*22*). The titration is carried out at pH 12–13, where magnesium has been largely precipitated as the hydroxide, with murexide, Eriochrome blue-black R, or Solochrome dark blue as indicators. Strontium and barium interfere, as do iron(II) and iron(III) above 20 ppm. Manganese, zinc, lead, aluminum, and tin interfere above 5 ppm, and copper interferes above 2 ppm. Orthophosphate anion must be absent.

Barnes(*25*) describes a modification of the oxalate titrimetric method for use in seawater analysis in which standard cerium(IV) sulfate solution is used to oxidize the oxalate ion and the excess ceric ion is titrated with standard iron(II) ammonium sulfate solution. Modifications of the EDTA complexometric titration method have been investigated for calcium (and magnesium) in seawater by a number of investigators.

Katz and Navone(*26*) reported a modified EDTA complexometric differential titration method for calcium and magnesium. Colorimetric methods have not been as successful for calcium as the titrimetric and gravimetric methods. This is probably due to the relatively high concentrations of calcium customarily encountered and to the lack of specificity of chelating reagents for calcium in the presence of its congeners. Close and West(27) synthesized a "chelate cage" reagent, cyclotris-7-(1-azo-8-hydroxynaphthalene-3,6-disulfonic acid) (Calcichrome), which reacts selectively with calcium at pH 12 for spot tests or as a complexometric end point indicator. Rehwoldt et al.(28) describe 2-chloro-5-cyano-3,6-dihydroxybenzoquinone as a new colorimetric reagent for calcium ion. Glyoxal bis(2-hydroxanil) has been investigated as a reagent for the colorimetric determination of very small concentrations of calcium ion. A considerable amount of research has been done on dye chelate indicators for titrimetric methods.

Flame photometry has been proposed by several investigators for the determination of calcium ion. Ruf(29) removed aluminum, iron, and titanium ions by precipitation as hydroxides, and sulfate and phosphate ions by anion exchange. Rowe(30) eliminated the depressive effects of silica, sulfate, aluminum, and phosphate ions without chemical separation by adding magnesium ion, with perchloric acid added to eliminate bicarbonate ion interference. Ion exchange procedures have been proposed for the removal of interferences in flame photometric methods as well as in titrimetric and gravimetric methods.

Selective electrodes for calcium ion have been made commercially available by Corning Glass Works(31) and by Orion Research, Inc.(32). The Corning electrode uses a liquid ion exchanger with a porous glass membrane and is linear in response from 10^{-1} to $10^{-4} M$ calcium ion when log concentration of calcium is plotted against millivolts between pH 5 and 11. Selectivity is 100:1 for calcium ion over barium, strontium, nickel, and magnesium ions and 1000:1 for calcium ion over sodium and potassium. Orion research describes its electrode as $0.1 M$ calcium didecylphosphate in di-n-octylphenylphosphonate in contact with $0.1 M$ calcium chloride in 20% agar gel against a silver–silver chloride electrode. The response of log calcium ion vs. millivolts is linear down to $10^{-5} M$ in the pH range 5.5–11. The electrode is 50 times as sensitive to calcium ion as to magnesium and 1000 times as sensitive to calcium ion as to potassium or sodium ions.

VI. Carbon

The only inorganic compounds of carbon that are of interest in water analysis are carbon dioxide, carbonate and bicarbonate ions, and cyanide ion. With increasing amounts of carbon monoxide in air pollution, this species may be of more concern in water in the future than it is at present. The ratio of the carbonate–bicarbonate pair is determined by the pH of the solution; that of the bicarbonate–carbon dioxide pair is determined by pH, temperature, and the partial pressure of carbon dioxide in the air. Cyanide ion almost always is the result of man-made contamination.

A. Carbon Dioxide

Water in equilibrium with the atmosphere rarely exceeds 10 ppm carbon dioxide, but groundwaters that have been in contact with limestone or dolomite may have reacted to produce higher concentrations. *Standard Methods* (33) offers a titrimetric method in which a freshly obtained sample is titrated with standard sodium hydroxide solution to the phenolphthalein end point, and a nomographic determination for free carbon dioxide content can be made when pH, total alkalinity, temperature, and total mineral content are known.

Carbon dioxide evolved from solution can be absorbed in alkaline solution and determined by titrimetric, gravimetric, or conductivity methods. Gaunt and Shanks (34), using gas chromatography, measured carbon dioxide evolved from acidified boiler feed water up to 0.1 ppm. Waterman (35) used an infrared gas analyzer to determine carbon dioxide in surface seawater.

B. Bicarbonate–Carbonate–Hydroxide Alkalinity

As alkalinity in natural water is commonly due to one or a combination of these ions, it is convenient to discuss analytical methods for all three ions in one place. Surface waters and groundwaters vary widely in bicarbonate ion concentration; often it is the predominate ionic species present. The traditional titrimetric method is that described under Alkalinity in *Standard Methods* (36). Hydroxide ion concentration and half of the carbonate ion concentration are measured when a sample is titrated with standard acid solution to the disappearance of the pink phenolphthalein color at pH 8.3. Half of the carbonate ion concentration and all of the bicarbonate ion concentration are measured by titration with standard acid solution to the methyl orange or bromcresol green–methyl red mixed indicator color change at pH 5.1–4.5. Differing amounts of alkalinity, expressed as calcium carbonate, require slightly different end points, varying from pH 5.1 for 30 ppm to 4.5 for 500 ppm. Methyl orange is used between pH 4.6 and 4.0, while the mixed indicator is used between pH 5.2 and 4.5.

Older analytical methods used for seawater, which is alkaline, were based on methods used for natural waters of much lower salinity. Barnes (37) recommends two methods, one involving the addition of excess acid and determination of alkalinity by calculation of the pH, and the other involving back titration of the excess acid by standard carbonate-free base with methyl red or methyl orange indicator. In the pH calculation method, the effect of varying chlorinities on the activity coefficient of hydrogen ion must be taken into account. In the back titration method, in which the carbon dioxide formed is removed by refluxing, boric acid from dissolved borates is not included in the alkalinity value obtained. Smith and Hood(38) recommend tris(hydroxymethyl)aminomethane as a buffer to eliminate drift in pH measurements of seawater. Park et al.(39) report that the inductive conductivity meter used to determine the salinity of seawater can be used to determine its alkalinity, as titration with 1.000 N HCl produces little conductivity change until the total alkalinity has been neutralized. Their results differed no more than 0.9% from those obtained by the pH method.

Herce and Luppi(40) have reviewed the various methods for the determination of alkalinity. Roberson and co-workers(41) concluded that determinations of pH, alkalinity, and specific conductance made in the field were generally higher, but more representative of the actual conditions, than those made later in the laboratory on stored samples.

Carpenter(42) reported a gas chromatographic method, which measures the carbon dioxide evolved from solutions of carbonate in the range 0.2–10 ppm, that is rapid and 100 times more sensitive than gravimetric and titrimetric methods. Underwood(43) and Underwood and Howe(44) determined dissolved carbon dioxide, bicarbonate, and carbonate as carbonate ion by spectrophotometric measurement of absorbance at 235 nm, and also by conversion to carbon dioxide, which upon reaction with the enzyme carbonic anhydrase rapidly forms carbonic acid that is titrated by standard sodium hydroxide solution without fading false end points.

C. Cyanide Ion

Natural waters almost never contain detectable concentrations of cyanide ion, and so any amount found in water supplies is considered to be the result of man-made contamination. Cyanide ion is toxic to most forms of life, the threshold varying with the type and size of the organism. Fish are killed by long exposure to as little as 0.1 ppm, and 0.3 ppm is sufficient to kill the microorganisms responsible for self-purification. It is a strong complexing agent and therefore may be present in cationic, anionic, or neutral metal complexes, the stabilities of which relate inversely to their toxicities. The permissible limit in drinking water is 0.01 ppm; water containing cyanide in excess of 0.2 ppm should be rejected.

Standard Methods (*45*) describes two methods, one titrimetric and one colorimetric, that convert and determine as cyanide ion all forms except the very stable hexacyanocobaltate(III). Both methods require a preliminary distillation step with special treatments to remove interference from sulfide ion, fatty acids, and oxidizing agents. The volumetric method is a modified Liebig titration in which cyanide ion is titrated with silver ion to form the dicyanoargentate(I) anion, with dimethylaminobenzalrhodanine as the indicator. The method is sensitive to as little as 0.1 ppm cyanide ion. The colorimetric method is the Epstein pyridine–pyrazolone method. The alkaline distillate is adjusted to pH 6–7 with added sodium hydroxide and acetic acid, the cyanide ion is converted to cyanogen chloride with chloramine-T, and the blue dye formed by reaction of cyanogen chloride with the pyridine–pyrazolone reagent is measured at 620 nm. Greater sensitivity is possible with a butanol extraction, with absorbance measurement at 630 nm. The effective ranges are 0.04–0.2 ppm cyanide ion in the aqueous sample, or 0.02–0.2 ppm in the butanol extract.

Methods of analysis and the instrumentation for the determination of cyanide ion and cyanogen compounds produced by chlorination have been reviewed by Lancy and Zabban (*46*). Most of the recent publications on cyanide ion determination involve modifications of the standard colorimetric methods with emphasis on total (free plus complexed) cyanide ion concentration. Hikime et al. (*47*) developed an indirect colorimetric method for cyanide ion in the range 0.5–10.0 ppm which involves the decolorization by mercury(II) of the yellowish brown copper(II)-bis-2 (hydroxyethyl)dithiocarbamate complex. Guilbault and Kramer (*48, 49*) reported specific fluorescence methods for cyanide ion, in the range 0.2–50 μg/ml, which use quinone and quinone derivatives as reagents. The same workers (*50*) have also described a specific and extremely sensitive spectrophotometric method in which cyanide ion catalyzes the reduction of *o*-dinitrobenzene by *p*-nitrobenzaldehyde. An instrumental method sensitive to 27 ppb cyanide was developed by Schneider and Freund (*51*) in which air is brought to equilibrium with an acid solution containing cyanide ion and the hydrocyanic acid concentration is determined by gas chromatography. A cyanide-selective solid-state membrane electrode is supplied by Orion Research, Inc. (*51a*). The recommended concentration range for cyanide detection is 10^{-3} to 10^{-5} M for solution pH values of 0 to 14. Sulfide ion must be absent from the solutions and iodide concentrations may be no more than ten times that of cyanide. However, chloride and bromide concentrations of one million and five thousand times that of cyanide, respectively, can be tolerated. Complexation with metallic and hydrogen ions lower free cyanide ion concentrations. This condition must be considered when utilizing this sensor which is sensitive to free cyanide ion.

VII. Chlorine

Natural waters contain chlorine as chloride ion, and treated water supplies may, in addition, contain species of higher oxidation states such as hypochlorous acid and hypochlorites, chloramines, and chlorine dioxide. (The subject of analysis for chlorine residual in water is covered in Chap. 22.) Trace amounts of chlorate and perchlorate ions sometimes are present but are of little concern.

A. Chloride Ion

While physiologically of little concern (except in very high concentrations), it is interesting that the "saltiness" of water depends to some extent upon the counter-cation. A salty taste can be detected in 250 ppm sodium chloride but not in four times that concentration of chloride ion if calcium or magnesium are the counter-ions. In irrigation water, high concentrations of electrolytes, of which chloride ion is the most common constituent, may be harmful to crop plants and tend to accumulate in the soil; in boiler feed waters, high ionic concentrations contribute to the electrolytic corrosion of pipes and boilers. Surface waters and groundwaters vary widely in chloride ion concentrations and may contain up to several hundred parts per million; seawater is more nearly constant, with normally about 18,000–19,000 ppm. Drinking water should not contain more than 250 ppm chloride ion.

Chlorinity is the concentration of halide ions defined on a weight basis, and chlorosity is the concentration defined on a volume basis. Chlorinity can be related to total salt content, or salinity. For details on the methods and standards used, reference should be made to more extensive works on the analysis of seawater (see Chaps. 2 and 19).

For natural waters other than seawater, *Standard Methods*(52) recommends three titrimetric methods – an argentometric (Mohr) method, a patented mercurometric method, and a potentiometric method for colored or turbid solutions. In the argentometric titration, chloride, bromide, iodide, and cyanide ions are titrated with standard silver nitrate solution with potassium chromate as the end point indicator. Sulfide, sulfite, and thiosulfate interfere but can be destroyed by appropriate treatments with hydrogen peroxide. Orthophosphate ion interferes in concentration above 25 ppm, and soluble iron in excess of 10 ppm obscures the end point. The mercurometric method has the advantage that the non-ionized mercury (II) chloride formed during titration by standard mercury(II) nitrate solution is soluble. Excess mercury(II) ion immediately past the end point is indicated by formation of the purple mercury(II)-diphenylcarbazone. Iodide and bromide ions titrate as chloride, and sulfite, chromate, and iron(III) interfere in concentrations above 10 ppm. The potentiometric method uses a glass and silver–silver chloride electrode system with an electronic voltmeter. In addition to other halides, which titrate as chloride ion, ferricyanide, chromium(VI), and iron(III) interfere.

For the determination of chlorinity in seawater, Barnes (53) recommends the argentometric Mohr or Volhard methods, which differ in the end point indicator used. As both methods suffer interference from proteins, Barnes also recommends Somogyi's zinc hydroxide reagent to deproteinize the sample prior to the Volhard titration or, alternatively, Sendroy's iodo-metric method. Sendroy's method involves the heterogeneous reaction between solid silver iodate and halide ion, with subsequent reaction of the released iodate ion with iodide to form iodine which is titrated with stand-ard sodium thiosulfate solution to the starch end point. Strickland and Parsons (54) recommend the Mohr method and describe a high precision method for seawater and a low precision method for brackish or estuarine waters. Jenkins (55) recommends the mercury(II) nitrate method described in *Standard Methods* for determination of the chlorosity in estuarine waters, with the modification that $0.141\ N$ Hg(II) nitrate is used in place of $0.0141\ N$. The inductively coupled conductivity meter of Brown and Hamon (56) is also used for measurements of salinity of seawater.

Colorimetric methods for chloride ion include the decolorization of the blue-violet mercury(II)-diphenylcarbazone complex (also used as an end point indicator in the mercury(II) titration method), and the formation of the red iron(III) thiocyanate complex from thiocyanate ion released by chloride from mercury(II) thiocyanate complex. Both are subject to a number of interferences. For very low concentrations of chloride ion in very pure water, Woelk (57) used periodate ion to oxidize chloride ion to chlorine and, in an adaptation of the colorimetric method for cyanide ion, reacted the chlorine with cyanide ion, pyridine, and barbituric acid. Marczenko and Choluj-Lenarczyk (58) also used periodate oxidation of chloride ion to chlorine to decolorize methyl red. A detector tube for rapid estimation of chloride ion in the range 5–40 ppm was developed by Kobayashi and Takeno (59). The tube is filled with potassium chromate and silver nitrate on silica gel, and forms a discolored zone as chloride ion solution is drawn through. Insoluble mercury(II) chloranilate, reported by Bertolacini and Barney (60), is a convenient colorimetric reagent which exchanges chloranilate anion (absorption maximum 332 nm) for chloride ion.

Two chloride-selective electrodes, which respond to chloride ion, are available commercially from Orion Research, Inc. (61). One is a solid-state electrode, which is also responsive to bromide, iodide, and cyanide ions; sulfide ion must be absent. The other is a liquid ion exchange electrode, which also responds to bromide, iodide, nitrate, and perchlorate ions. Both have extremely wide concentration response and measure activity rather than concentration.

B. Hypochlorous Acid–Hypochlorite Ion

Drinking water supplies can be disinfected by treatment with ozone or chlorine. The choice is made on the basis of economics and convenience. Residual chlorine (hypochlorite) has no objectionable taste, but taste problems do arise due to reaction products of chlorination, such as chloramines and chlorinated phenolic compounds. Chlorine demand is measured as the rate of uptake of standard hypochlorite ion solution as determined by one of the methods of residual chlorine analysis. As chloramines also are oxidizing agents, methods are available for the differentiation of mono-, di-, and trichloramines along with hypochlorite chlorine.

Standard Methods (62) offers eight methods for the determination of residual chlorine. These include an iodometric titration, four ortho-tolidine colorimetric methods, two amperometric titration methods, and an iron(II) titrimetric method. The iodometric titration method is used as a standard and measures total available residual chlorine by the reaction of hypochlorites and chloramines with iodide ion in acetic acid solution at pH 3–5 and titration with standard sodium thiosulfate solution to the starch end point. This method will detect as low as 0.04 ppm chlorine but is most useful for high residual chlorine samples. Ortho-tolidine (3,3′-dimethylbenzidine) measures both free and combined available chlorine by reaction to give a yellow holoquinone which absorbs at 435 nm. Extremely detailed directions are given for the simple ortho-tolidine test up to 10 ppm, the flash test for "free" chlorine up to 0.3 ppm, the orthotolidine–arsenite (OTA) method for differentiation of free available and combined available chlorine, and the drop dilution method for field measurements greater than 10 ppm. The amperometric titrations utilize an easily polarizable noble metal-indicating electrode with a silver–silver chloride electrode and a microammeter. Phenylarsine oxide solution is the titrant and reacts to depolarize the indicator electrode, which is originally polarized by the residual chlorine. The end point is taken as the minimum meter reading. By titration at pH 7 with a phosphate buffer, free available chlorine is measured, and by further titration at pH 4 after potassium iodide and acetate buffer have been added, combined available chlorine is measured. If potassium iodide is added after the end point for free available chlorine is reached at pH 7, subsequent titration will measure monochloramine, and then titration at pH 4 will measure dichloramine. The serial titration method with standard iron(II) ammonium sulfate solution, with orthotolidine as the indicator, will first measure free available chlorine, trichloramine, and chlorine dioxide at pH 6.3–6.5. Continued titration after potassium iodide has been added will measure monochloramine, and further titration after acidification and reneutralization will measure dichloramine. Trichloramine and chlorine dioxide may be extracted with carbon tetrachloride and determined differentially by the decrease in the amount of free available chlorine as determined by the first titration. Alternatively, oxalic acid may be used to destroy hypochlorite ion, in which case the first titration step will measure only trichloramine.

In an evaluation of eight methods for the determination of free chlorine, which included 38 references, Nicolson(63) found the colorimetric procedure based on the reaction of N,N-diethyl-p-phenylenediamine to be the best in the presence of combined chlorine. In the absence of combined chlorine, a modification of the Zincke–Donig reaction for cyanide ion, in which chlorine reacts with cyanide to form cyanogen chloride and produces a color with barbituric acid in pyridine, was found to be the best. Spectrophotometric methods were found to be more reproducible than the titrimetric. Katz and Heukelekian(64) compared results obtained with the cyanogen chloride method with those from other methods and found that they varied widely from those obtained by the amperometric method. While the superiority of neither was demonstrated, both compared favorably with the starch–iodine and orthotolidine methods for tap water. Starch–iodide titrimetric methods were used for residual chlorine in water (64a) and wastes(64b,64c,64d). Bossy(65) described an automatic photometric method for chlorine in the range 0–0.5 ppm which is based on the starch–triiodide reaction. Residual chlorine was determined spectrophotometrically by Taras(65a), Sollo and Larson(65b), and Athavale et al.(66) by its bleaching effect on methyl orange, measured at 510 nm.

Berndt(67) suggested an iodometric method with three thiosulfate titrations for chlorine, chlorite, and chlorine dioxide. Karge(68) proposed an alternate method because of the limitations of Berndt's method, with two titrations by phenylarsine oxide solution to determine, first, chlorine and one-fifth of the chlorine dioxide and, second, chlorine, chlorine dioxide, and chlorite. Chlorine is then determined separately by the cyanide–pyridine–barbituric acid modifications of the Zincke–Donig method. For the differentiation of chlorine, chloramine, chlorine dioxide, and chlorite, Meier-Ewert and Bruenner(69) used four titrations with arsenious acid and thiosulfate.

C. Chlorine Dioxide

This yellow, volatile gas is used to destroy certain types of tastes and odors and is prepared for immediate use by the reaction of hypochlorite with chlorite in acid solution.

The methods described in *Standard Methods*(70) are modifications of those used for free and combined available chlorine. Temporary standards prepared with chlorine dioxide are standardized by the iodometric titration method for total available residual chlorine above. In the ortho-tolidine–oxalic acid (OTO) method, oxalic acid is added to the sample to destroy residual chlorine, and acidic orthotolidine solution is added to determine chlorine dioxide. The reaction is arrested after color development by sodium arsenite solution, which also minimizes interferences. Free available chlorine (hypochlorite) present can be determined by the OTA method. The amperometric titration method, by four separate titrations when phenylarsine oxide solution, can be used to differentiate hypochlorite chlorine, chloramines, chlorite, and chlorine dioxide. The methods of Palin(71), Berndt(67), and Karge(68) include determination of chlorine dioxide in mixtures. Hodgden and Ingols(71a) and Kerenyi and Kuba(72) discussed the use of tyrosine for the determination of chlorine dioxide in the presence of hypochlorite chlorine, which is eliminated by monoethylamine.

VIII. Fluorine

Fluorine exists in water only as the fluoride ion. It is a constituent of the apatite portion of the tooth structure and hence may be considered a necessary nutritional trace element. Because of the narrow limits between efficacy and the disfigurement of tooth enamel when fluoride ion is supplied via drinking water, more research probably has been done on the determination of fluoride ion than on any other single species other than hydrogen ion encountered in water. Permissible concentrations in drinking water vary between 0.7 and 1.2 ppm, depending on the temperature; concentrations exceeding 1.4–2.4 ppm, depending on the temperature, constitute cause for rejection.

Until the recent advent of selective membrane electrodes, almost all analytical methods for determining fluoride ion in water employed competitive complexometric reactions that are followed colorimetrically. *Standard Methods*(73) describes two methods: one with the zirconium complex of 1,2-dihydroxyanthraquinone-3-sulfonate (alizarin red S) with measurement by visual comparison up to 1.4 ppm or spectrophotometrically up to 2.5 ppm, and the other with the zirconium complex of 2-(sulfophenylazo)-1,8-dihydroxy-3,6-naphthalenedisulfonate (SPADNS) by spectrophotometric measurement. In both reagent complexes, zirconium is chelated by two adjacent oxygen atoms on the ligand, and the ligand is released with consequent change in hue by formation of a more stable zirconium–fluoride complex. Direct distillation and steam distillation steps are described for removal of interferences. Aluminum, iron(III), phosphate, hexametaphosphate, and sulfate ions interfere to varying degrees. The alizarin red S method requires about 1 hr for color development, while the SPADNS method is very rapid.

Numerous modifications of the above methods have been reported, and a number of other chelating dyes have been investigated as the chromophoric ligand. One of the more interesting reagents is the cerium (III) complex of 3-dicarboxymethylaminomethyl-1,2-dihydroxyanthraquinone (Alizarin Complexan), reported by Yamamura et al.(74), which adds fluoride ion and changes hue. Murakami and Uesugi(75) used the lanthanum complex of the same ligand to determine fluoride ion in seawater up to 1.0 ppm. Several investigators have recommended ion exchange as a method for removing interferences, in place of older distillation procedures. Aluminum, which is an interference in most of the colorimetric methods, was found by Bhakuni and Sharma(76) not to interfere in the zirconium–Eriochrome cyanine method if the solution is maintained at 35°C. In addition to complexes of zirconium and lanthanum, dye compounds and chelates of thorium also have been suggested as colorimetric reagents for fluoride ion. Insoluble thorium chloranilate was reported by Hensley and Barney(77) as an insoluble reagent in water–methyl Cellosolve solution at pH 4.5. The released colored chloranilate anion absorbs at 540 nm and more strongly at 330 nm. The methyl Cellosolve is used to suppress the blank which otherwise is appreciable. Phosphate and a number of cations interfere and must be removed.

A fluoride-sensitive electrode which has recently become commercially available from Orion Research, Inc.(78,79) measures activities of fluoride ion from 1 to $10^{-6} M$. As fluoridated drinking water supplies contain about $5 \times 10^{-5} M$ fluoride ion and at this level activity is approximately equal to concentration, the electrode should be of value in continuous monitoring of water supplies if it is capable of discriminating within the narrow range of permissible concentrations. The membrane is a single crystal of lanthanum fluoride doped with europium, and because of its low solubility ($K_{sp} = 10^{-29}$) it does not contribute significant amounts of fluoride ion to the solution. Its useful range is between pH 4 and 9. Below pH 4 fluoride ion is lost as un-ionized hydrogen fluoride, and above pH 9 hydroxide ion (which is similar to fluoride ion in size and charge) is present in sufficiently high concentration to produce high readings. No interference is observed from other commonly encountered ions.

IX. Hydrogen

Hydrogen undergoes no hydrolysis reactions and has a very low solubility in water, where the partial pressure of the gas is practically zero. The few methods reported in the literature of recent years indicate an interest in dissolved hydrogen in boiler waters and other closed systems where leakage or corrosion may introduce the gas.

A. Dissolved Hydrogen

Swinnerton et al. (80), Massart and Missa (81), and Hissel (82) reported gas chromatographic methods for hydrogen and several other gases in low concentration. Faber and Brand (83) described an indirect method for hydrogen involving the decrease in oxygen concentration after treatment with platinum catalyst; the Winkler method is used if the hydrogen content is between 3 and 15 ml/liter, and the o-toluidine method is used for hydrogen content down to 0.3 ml/liter.

B. Hydrogen Ion

The concentration of hydrogen ion (defined in such a manner as to include the various protonated water species) is measured as the instantaneous concentration, or activity, or as available hydrogen ion concentration by titration to a predetermined end point that allows for hydrolysis of the counter-anion. The instantaneous concentration has for some years been measured almost entirely by the glass electrode–calomel potential and is too well known to merit further consideration here. Hydroxide ion concentration is calculated from the hydrogen ion concentration, as the two are related through the ion product constant for water.

The ability of aqueous solution to donate protons (acidity) or to accept protons (alkalinity) measures both the actual concentration of hydrogen of hydroxide ions and that available from undissociated acids or bases. For this type of measurement, titrimetry is almost universally used. The equivalence point of the titration is determined by the color change of an indicator (usually a dye) that is a weak acid or weak base, by potentiometric change as sensed by a glass electrode, or by other instrumental means such as abrupt change in conductance of the solution. The classic titrimetric methods are described in *Standard Methods*(*84*).

The sources of acidity in most natural waters are carbon dioxide and pollution by mineral acids and salts with cations that hydrolyze to produce protons. The sources of alkalinity may be hydroxide, carbonate, or bicarbonate ions, the latter two producing hydroxide ions by hydrolysis. Borate ion also is an occasional source of alkalinity.

Few colorimetric methods exist for the determination of acidity or alkalinity. Reactions which quantitatively depend upon the available hydrogen ion concentration, such as the iodate-iodide reaction, have been used for this purpose but have no advantage over conventional methods.

X. Iodine

Iodine occurs in natural waters only in trace concentrations as iodide ion and, although physiologically important, is now of little interest nutritionally because of the availability of seafoods and iodized salt. Like bromide ion, its presence may indicate industrial pollution or intrusion of seawater into freshwater supplies. Seawater normally contains about 50 ppb iodine.

Standard Methods(*85*) describes only one method, which is tentative at the time of the 12th edition. Iodide ion catalyzes the oxidation of arsenious acid by cerium(IV); the reaction is permitted to proceed under controlled conditions for a given period of time, after which iron(II) sulfate is added to destroy unreacted cerium(IV). The iron(III) produced is determined colorimetrically as the red thiocyanate complex, which is related inversely to the original iodide ion concentration. The reaction mixture is swamped with chloride ion to reduce its effect and the inhibiting effects of metals that form stable iodo complexes. The method is applicable for iodide in the 0–14 ppb range.

Most direct colorimetric, titrimetric, and gravimetric methods for iodide ion are not of sufficient sensitivity for its determination in concentrations normally encountered in natural waters. In a review with 38 references, Pavlova and Shishkina(*86*) found that a thiosulfate titration following bromine oxidation to iodate was preferred. Proskuryakova et al.(87) compared titrimetric, colorimetric, and kinetic (catalytic) methods and found the thiocyanate–nitrite catalytic method and the brilliant green extraction method of Lapin and Reis(88) to be the most useful from the standpoints of precision and sensitivity. Catalytic reactions form the basis for several methods reported. Jungreis and Gedalia(89) determined iodide ion in the range 2.5–20 ppb by its effect on the reaction between chloramine-T and tetra base acetate to form a colored compound, while Ramanauskas(90) employed its catalytic effect on the reaction between chloramine-B (sodium benzenesulfonchloramine) and tetramethyl-diaminodiphenylmethane (tetra base). Yonehara(91) used a catalytic method for the determination of iodide and iodate in seawater based on the fact that the rate of fading of the red iron(III) thiocyanate complex in the presence of 2000 ppm chloride ion is effected equally by either ion. In a more direct and sensitive method, Novikov(92) reported a method based on the formation of the blue linear starch–triiodide complex.

Mariani and Maura(93) discussed the use of anion exchange resins to collect iodide ion from large volumes of water. Rastegina(94) used an alumina column with absorbed mercury(II) ion to collect iodide ion as insoluble mercury(II) iodide, which subsequently is dissolved in ether and measured by a spot test procedure.

Orion Research, Inc.(95) has made an iodide-selective membrane electrode commercially available. The membrane is solid state and the interferences are listed as sulfide, thiosulfate, and cyanide ions. The electrode responds to iodide ion from saturation levels to 10 ppb over the pH range 0–14; at the lower levels of response, activities are almost identical to concentration.

XI. Magnesium

Unlike calcium, which is physiologically without effect in moderate concentrations, magnesium ion in concentrations above 125 ppm can produce cathartic and diuretic effects. Like calcium, it causes scale formation in pipes and boilers. Magnesium ion concentration in groundwaters and surface waters may vary from zero to several hundred parts per million but rarely exceeds 30 ppm; seawater normally contains approximately 1200 ppm.

In *Standard Methods*(96), magnesium ion is determined gravimetrically following gravimetric or titrimetric determination of calcium ion by the oxalate methods, or colorimetrically by adsorption of brilliant yellow dye on magnesium hydroxide. In the gravimetric method, oxalate is destroyed by oxidative evaporation with nitric acid and the magnesium is precipitated as magnesium ammonium phosphate. By two consecutive precipitations as magnesium ammonium phosphate, the nitric acid destruction of oxalate may be avoided. Interferences are the same as for calcium and would have been removed in the prior calcium determination. In the colorimetric method, calcium does not have to be removed, but its effect and that of aluminum are minimized by swamping to make their influence constant and predictable. Tolerable concentrations for iron, chloride, orthophosphate, and fluoride ions are given; manganese(III) and zinc must be absent. Aliquots containing 0.1–0.8 mg of magnesium are determined after treatment in 100 ml of solution at 525 nm.

For magnesium ion in seawater, Barnes(97) recommends precipitation with 8-hydroxyquinoline followed by bromination of the 8-hydroxyquinoline with acidified bromate–bromide solution. The excess bromine is titrated iodometrically. Calcium ion does not interfere and the method is less tedious than precipitation as magnesium ammonium phosphate. Copper(II) is an almost negligible interference. Modifications of the EDTA complexometric titration method have been investigated for calcium and magnesium ions in seawater by Skopintsev and Kabanov(98), Pate and Robinson(99,100), and Rial and Molins(101).

Literature reports on methods for magnesium ion and differential methods for magnesium and calcium, or for both as hardness, have been numerous in recent years. The majority of the methods reported involve complexometric titration procedures with a number of end point indicator dyes suggested. Katz and Navone(102) described a method for the differential determination of calcium and magnesium in which calcium is titrated at pH 12–13 with standard EDTA solution with Eriochrome blue SE as the indicator, followed by titration at pH 10.1 with EDTA using Eriochrome black T as the indicator. Marti and Herrero(103) proposed an ion exchange procedure for the separation of magnesium and calcium prior to EDTA titration. Fabregas et al.(104) used barium EGTA [ethylene glycol bis-(β-aminoethyl)-N,N'-tetraacetate] to eliminate interference from calcium in the EDTA titration of magnesium with methyl red–Eriochrome black T indicator.

Rai et al.(105) developed a spectrophotometric method for low concentrations of magnesium in water with Azovan blue dye. Other colorimetric reagents for magnesium include 1-azo-2-hydroxy-3-(2,4-dimethylcarboxanilido)naphthalene-1'-(2-hydroxybenzene) suggested by Peaslee (106), Titan yellow suggested by Petrukhin(107) and van Schouwenburg (108), and xylidyl blue II suggested by Ogata and Hiroi(109).

Diehl et al.(*110*) found *o,o'*-dihydroxyazobenzene to be the best fluorometric reagent for the determination of magnesium in the presence of calcium. The reagent may also be used spectrophotometrically, but the fluorometric method is more sensitive. Disalicylideneethylenediamine was reported by Serebryakova et al.(*111*) for the fluorometric determination of magnesium.

Blaedel and Laessig(*112*) developed a continuous potentiometric EDTA titrator for determining concentrations of calcium and magnesium (as total hardness) as low as $10^{-5} M$. The flame photometer was used by Turkin and Svistov(*113*) for magnesium in concentrations as low as 0.5 ppm through the use of the 285.2-nm line. Atomic absorption was suggested by McPherson(*114*) for magnesium and calcium at concentrations as low as 0.01 and 0.1 ppm, respectively. No specific membrane electrodes have as yet been developed for magnesium, although the divalent cation electrode marketed by Orion Research, Inc.(*115*) is equally sensitive for calcium and magnesium in determining water hardness.

XII. Nitrogen

The gaseous element is present in all natural waters that have been exposed to air, but, because of its inertness and lack of toxicity, its concentration is rarely determined. In its compounds, it is found in a greater variety of oxidation states than any other element. As ammonia it is a product of microbiological action on organic matter, and its presence can be taken as evidence of sanitary pollution in raw surface water supplies. Biologic oxidation or oxidative treatment of water supplies converts proteins, amines, and ammonia to nitrite or nitrate ions. Hypochlorite treatment of water supplies converts amines and ammonia initially to chloramines. All forms of nitrogen eventually return to the very stable elemental form and hence to the atmosphere. Surface waters rarely contain more than trace concentrations of nitrite and nitrate ions, but groundwaters may contain potentially harmful concentrations of nitrate. The permissible limit for nitrate ion in drinking water is 45 ppm; higher concentrations cause infant methemoglobinemia. Seawater normally contains 10–700 ppb of chemically combined nitrogen. Cyanide ion is also a nitrogen-containing species found in waters polluted by man-made wastes; it is discussed under carbon compounds.

A. Ammonia

Standard Methods (*116*) recommends the Nessler method for ammonia. Nessler's reagent, tetraiodomercurate(II) anion in alkaline solution, is used either with distillation from solution buffered to pH 7.4 for low concentrations of ammonia, or without distillation for concentrations exceeding 0.2 ppm. Nessler's reagent reacts with ammonia to form the yellow to reddish brown iodide of Millon's base, which, although insoluble, may be determined by visual comparison techniques or, with proper regard for the change in hue with concentration, photometrically. A number of organic compounds and sulfide ion interfere by altering the color or causing a turbidity, and calcium ion above 250 ppm reacts with the phosphate buffer of the distillation step to lower the pH to interfere with the distillation procedure.

In seawater, calcium and magnesium ions cause turbidities with the Nessler method. Barnes(*117*) favors the Witting-Bach modification, which uses barium chloride and sodium hydroxide to precipitate these ions, and the Wattenberg–Cooper modification which complexes calcium and magnesium with tartrate (Rochelle salt). A third method involves distillation, reaction with alkaline hypobromite to destroy the ammonia, and titration of the excess hypobromite with standard acid naphthyl red solution. Strickland and Parsons(*118*) recommend the method of Kruse and Mellon(*119*) adapted to seawater by Strickland and Austin(*120*), in which chloramine-T, pyrazolone, and pyridine react with ammonia to form a color measurable in the range 0.2–10 μg-atom/liter. Jenkins(*121*) found the pyrazolone–pyridine method recommended by Strickland and Parsons(*118*) unsuitable for estuarine waters because of the wide variation in salt concentration. He recommends instead the distillation–Nesslerization method of *Standard Methods* (*116*) with sodium carbonate in place of the phosphate buffer. Richards and Kletsch(*122*) oxidized ammonia and labile amino compounds in seawater to nitrite ion with hypochlorite in strongly basic solution; sulfanilamide is then diazotized in acid solution and coupled with N-naphthylethylenediamine, and the azo dye measured spectrophotometrically at 540 nm.

Of the methods and modifications reported in the recent literature, very few involve the Nessler method, which probably indicates the maturity of the method. The indophenol method and its various modifications likewise have received less attention very recently, although a number of articles appeared in the years immediately preceding 1965. Roskam and de Langen(*123*) recommended thymol in acetone in place of phenol for ammonia in seawater. Kaplin and co-workers(*124,125*) used hypobromite as the oxidant in the indophenol method and combined this with a chloroform extraction and aqueous alkaline reextraction to remove interferences. Jenkins et al.(126) used hypobromite in alkaline solution to oxidize ammonia to elemental nitrogen, which is then determined by gas chromatography. Zitomer and Lambert(126a) used hypochlorite to convert ammonia to trichloramine, and determined the trichloramine by colorimetric iodometry.

B. Nitrite Ion

Although nitrite ion is a transient species in the nitrogen cycle, its value as an indicator of pollution by organic material makes nitrite analysis necessary. Standard Methods(127) lists only the diazotization-coupling method, for which trichloramine is the only positive interference, copper (II) the only negative interference, and iron(III) an interference because of precipitation under the conditions of the method. Nitrite ion can be determined in the range 0.001–2 ppm, with a modification for still higher concentrations. The method is so sensitive that precautions must be taken to use nitrite-free water and reagents. Sulfanilic acid is almost universally used as the aromatic amine to be diazotized; sulfanilamide and p-aminobenzoic acid are sometimes used, but the only new, readily diazotizable amine is 4-aminophenyltrimethylammonium ion, proposed by Lambert and Zitomer(128). Standard Methods(127) recommends 1-naphthylamine as the coupling agent, but N(1-naphthyl)-ethylenediamine and N,N-dimethyl-1-naphthylamine are also widely used. A variation on the usual diazotization practice is the use of Crosby(129) of Cleve's acid, 1-naphthylamine-7-sulfonic acid, which is effective for nitrite in the range 0–20 ppb.

Barnes(130) uses the reagents recommended by Standard Methods for seawater, and Jenkins(131) also recommends the same regents for nitrite ion in estuarine waters. Strickland and Parsons(132) employ sulfanilamide and N(1-naphthyl)-ethylenediamine for nitrite ion in seawater.

A number of differential methods for nitrite and nitrate nitrogen have been proposed which utilize the diazotization-coupling reaction to determine nitrite ion concentration directly and nitrite plus nitrate concentration following controlled reduction of nitrate ion to nitrite. The reduction procedure has been used by a number of investigators for the determination of nitrate ion only, and hence the reduction step is discussed under nitrate ion determination. Vaccaro et al.(133) proposed a modification for seawater analysis in which nitrite ion is first oxidized by ozone to nitrate and the nitrate ion reduced by hydrazine to nitrite. Nitrite ion concentration is determined directly by a diazotization-coupling method. The advantage of this sequence of reactions is that it avoids reduction of the nitrite ion originally present during the reduction of nitrate ion.

C. Nitrate Ion

In the absence of microbiological reduction, nitrate ion tends to accumulate in natural waters, and its determination is made difficult by its lack of selective reactions. Standard Methods(134) describes four methods: two colorimetric following reaction, one direct spectrophotometric, and one polarographic. One of the colorimetric methods involves the nitration of phenoldisulfonic acid to a yellow nitro derivative that absorbs at 410 nm. Nitrite ion above 0.2 ppm interferes in an unpredictable manner, and chloride ion is a serious interference above 10 ppm. West and Ramachandran(135) have reported a similar method, with chromotropic acid in place of phenoldisulfonic acid, which is tolerant of high chloride ion concentration. Brucine oxidation is the other colorimetric method described in Standard Methods, and the compound likewise absorbs at 410 nm. All strong oxidizing and reducing agents interfere. A tentative method in Standard Methods which is sensitive to 0.02 ppm takes advantage of the nitrate ion absorbance at 220 nm in the ultraviolet for direct measurement without treatment of sample. A second measurement is made at 275 nm to correct for the presence of dissolved organic matter that may absorb at the lower wavelength. Nitrite ion, chromium(VI), and surfactants may be corrected for by individual correction curves. The polarographic method, likewise tentative, for nitrate ion concentrations above 0.5 ppm nitrate nitrogen is based on the reducibility of nitrate ion at the dropping mercury electrode in the presence of uranium(VI) acetate. Nitrite ion, and phosphate ion in high concentrations of nitrate ion, interfere, as does iron(III) to a slight extent. Nitrite ion must be determined by another method, while phosphate ion interference often can be eliminated by dilution of the sample.

The phenoldisulfonic acid method cannot be used in seawater because of the high chloride ion concentration. Barnes(136) recommends two methods that depend upon the oxidizing ability of nitrate in sulfuric acid solution for reduced strychnine or diphenylbenzidine. He also recommends diazotization following reduction of nitrate to nitrite by hydrazine in alkaline solution with a copper(II) catalyst; the 24-hr reduction period is greatly shortened. A polarographic method restricted to land-based laboratory use is also described. Strickland and Parsons(137) also recommend reduction to nitrite followed by diazotization. Jenkins (131) describes a modification of the Standard Methods brucine method

(134), in which the temperature is controlled and the decreased response due to chloride ion is stabilized by addition of a large amount of sodium chloride. A sensitive spectrophotometric method involving the extraction into toluene of the reaction product of nitrate ion with 2,6-xylenol was reported by Andrews(138); nitrite and chloride interfere but are removed by chemical means.

Brewer and Riley(139), in an automated method, used reduction of nitrate to nitrite by a column of cadmium metal and formation of an azo dye by the method of Bendschneider and Robinson(140). Morris and Riley(141) found the reduction by amalgamated cadmium to be about 91% efficient. Chow and Johnstone(142) suggested the reduction with zinc dust in ammoniacal solution with a manganese(IV) catalyst. Henriksen(143) used a diazotization procedure originally published by Mullin and Riley(144), in which hydrazine at 70°C with a copper catalyst shortens the reduction time from 24 hr to 15 min, in an automated method for nitrate and nitrite ions in fresh and saline waters.

Although there have been numerous publications on the subject of nitrate ion analysis in recent years, little that is fundamentally new has been proposed. Most of the methods have been based either on nitration/oxidation of organic compounds or on reduction to nitrite ion for which the nearly specific diazotization-coupling reaction sequence is available. The latter has served as the basis for several differential methods for nitrate and nitrite. Complete reduction to ammonia and determination with the Nessler reagent have also been explored.

A basically new method was proposed by Bloomfield et al.(145,146) which involves the interference of nitrate ion with the formation of colored complexes of rhenium with α-furildioxime and syn-phenyl-2-pyridyl ketoxime in acid solution. Orion Research, Inc.(147) has recently made commercially available a nitrate-selective electrode which incorporates a liquid water-immiscible ion exchanger. The response is greater to perchlorate, chlorate, and iodide ions than to nitrate, and somewhat less to bromide and sulfide ions. The effective range is 0.6–6000 ppm in solutions of pH 2–12.

XIII. Oxygen

The inorganic forms of oxygen in water, not including hydrogen peroxide, hydroxide ion, and protonated water species, consist of dissolved oxygen and ozone. Water in contact with air will contain varying concentrations of oxygen, depending upon the oxygen demands made by dissolved reducing substances or aquatic life forms and upon the rate of recharge from the air. The amount of ozone in water from natural sources is negligible, but its use as a disinfectant in water supplies necessitates analytical methods for its control.

A. Dissolved Oxygen

The basic method for the determination of dissolved oxygen (DO) is the Winkler method, in which manganese(II) in alkaline solution is oxidized to manganese(III), which upon acidification will oxidize iodide ion to iodine. The iodine generated is measured titrimetrically with standard thiosulfate solution. *Standard Methods* (148) describes seven modifications of this method and one tentative polarographic method. One modification is designed to circumvent interference from nitrite, which is destroyed by azide ion. Another uses permanganate in acid solution to oxidize iron(II) to iron(III), which is complexed with fluoride ion. Excess permanganate is destroyed with oxalic acid. Still another modification involves alkaline hypochlorite oxidation to destroy sulfite, thiosulfate, and polythionate anions; the excess hypochlorite is destroyed by titration with iodide ion solution to the starch–triiodide end point, to which the subsequent Winkler reaction returns at the end of the thiosulfate titration. Two other modifications are described—a "short" method for convenience and another for high dissolved oxygen or high organic content. An alum flocculation modification is described for the removal of suspended solids that consume appreciable amounts of iodine in acid solution, and a copper(II) sulfate–sulfamic acid modification is described for removal of biologic flocs found in activated sludge mixtures. The polarographic method has few advantages over the titrimetric Winkler method except in some waste waters. The electrodes used are usually solid metal electrodes immersed in electrolyte solution inside a plastic membrane permeable to oxygen. Activity, or percentage of saturation value, is measured.

Barnes(149) describes the standard Winkler titrimetric method for dissolved oxygen in seawater, as well as the azide, alkali-hypochlorite, and iodine difference modifications. Strickland and Parsons(150) recommend the traditional Winkler titrimetric method for normal concentrations and an absorptimetric method for high-precision analysis of extremely low concentrations. Ivanoff(151) recommends photometric determination of iodine in place of the usual titration. Swinnerton et al. (80) recommend a gas chromatographic method that was modified by Park and Catalborno(152) for seawater analysis.

The numerous recent publications regarding the determination of dissolved oxygen in water reflect the current concern about increasing amounts of reducing pollutants in water supplies, and problems of corrosion. An extensive review by Mancy and Jaffe of analytical methods for dissolved oxygen in natural and waste waters is available as a U.S. Public Health Service publication(*153*). The majority of recent publications deal with modifications of the Winkler method. Much research has gone into electrometric methods, including polarographic, conductometric, and coulometric methods. An increased interest in automated and continuous methods is also noted (see Chap. 27).

Colorimetric methods for dissolved oxygen have received considerable attention. A review of such methods for boiler feed waters was made by Shtern(*154*). In addition to the oxidation of leuco bases of sodium indigodisulfonate, methylene blue, and Safranine red T, other chelating chromophoric ligands for manganese(III) have been proposed. Ostrowski et al.(*155*) reported the use of N,N'-bis(2-hydroxypropyl)-*o*-phenylenediamine for dissolved oxygen in the 0–5 ppm range for speed and good precision. According to Ungureanu(*155a*), dissolved oxygen oxidizes cerium(III) in alkaline solution to cerium(IV) and can be determined photometrically by the absorbance of the oxidation product between cerium(IV) and *o*-toluidine.

Wright and Lindsay(*156*) used a column of thallium metal and measured the change in specific conductance of the effluent due to thallium(I) produced by dissolved oxygen in the 10–200 ppb range.

B. Ozone

The only pair of allotropes of importance in water analysis are oxygen and ozone. Ozone in residual concentrations of 0.1–0.2 ppm finds use as a potent germicide that also hastens the oxidation of iron(II) and manganese(II) to the trivalent state. It has the advantages that its principal decomposition product is oxygen and that it does not form a series of persistent weaker oxidizing agents as chlorine (hypochlorite) does in forming chloramines with amine nitrogen. Ozone is not a constituent of natural waters.

One titrimetric and two related colorimetric methods are described in *Standard Methods*(*157*). In the titrimetric method, ozone is swept from a sample by a stream of air or nitrogen into potassium iodide absorber solution, and the liberated iodine is titrated by standard thiosulfate solution to the starch end point. The limit of detection is 30 ppb ozone. In the colorimetric methods, ozone oxidizes manganese(II) to manganese(III), which in turn oxidizes orthotolidine to a yellow compound. Ozone also oxidizes orthotolidine directly in the OTA method, with arsenite present to minimize interference from nitrite ion or iron(III). Both methods suffer interference from manganese dioxide and both have a lower limit of 20 ppb ozone. The yellow color produced absorbs at 535 nm.

In contrast to the numerous publications on methods for the determination of dissolved oxygen, few articles have dealt with ozone. A coulometric ozone sensor was investigated by Wartburg et al.(158) for laboratory and field use; peroxides, peroxy acids, and sulfites were found to interfere. In a colorimetric method, Koppe and Muhle(159) used N,N,N',N'-tetramethyl-*p*,*p*'-diaminodiphenylmethane and N,N,N',N'-tetramethyl-*p*,*p*'-diaminotriphenylmethane at pH 2, which are oxidized by ozone in the range 0.02–1.0 ppm in the presence of other oxidants to the corresponding fluorene. Following extraction into chloroform, the concentration of the fluorene is measured at 492 nm. Manganese dioxide, chlorine, and hydrogen peroxide interfere but are eliminated by permanganate. Galster(160) used N,N,N',N'-tetramethyl-*p*-phenylenediamine in citric acid solution in a similar manner. A somewhat different reaction, pyridine-4-aldehyde from ozone cleavage of 1,2-di(4-pyridyl) ethylene, was reported by Hauser and Bradley(161).

XIV. Phosphorus

In natural waters, phosphorus occurs as orthophosphate and polyphosphate anions and in trace amounts as organically bound phosphorus. Fertilizers and detergents add significant amounts of soluble phosphorus to drainage and waste waters and promote the growth of algae in reservoirs. As many phosphate salts and some organic phosphorus compounds have low solubility, phosphorus may also appear in the suspended matter or sludge in water samples. Seawater contains 1–100 ppb of chemically combined phosphorus.

Standard Methods(162) describes only the heteropoly blue method for the determination of soluble phosphates, together with an acid hydrolysis procedure for the determination of orthophosphate and condensed phosphates for total phosphate. Both colorimetric methods involve the formation of molybdophosphoric acid from an acid solution of ammonium molybdate with aminonaphtholsulfonic acid as the reductant for phosphate ion in the 0.1–30 ppm range, and with stannous chloride for the 0.05–3 ppm range. Under the conditions of the methods, silicates do not interfere in any ordinary concentrations. Arsenic and germanium must be absent, and sulfide ion must be destroyed. Soluble iron must be below 0.1 ppm in the sample. Beer's law is obeyed at 690 nm and at 650 nm with reduced sensitivity. Filter photometers with a red filter may also be used. The sensitivity of the stannous chloride method can be enhanced by extraction of the heteropoly blue compound into a 1:1 benzene–isobutanol solvent with spectrophotometric measurement at 625 nm. The effect of interferences also is minimized by the extraction process.

For orthophosphate ion in seawater, Barnes(163) recommends a modification of the stannous chloride reduction of molybdophosphoric acid in which the absorption of the heteropoly blue compound is measured at 705 nm. Even greater sensitivity can be achieved with metol at 100°C in place of stannous chloride. Strickland and Parsons(164) also recommend the stannous chloride reduction method. Jenkins(165) investigated the available methods for the determination of both soluble and insoluble orthophosphate, polyphosphate, and organic phosphate in estuarine waters. Soluble phosphate is defined as that which passes through a 0.45-μm membrane filter and is determined by extraction of the molybdophosphoric acid into a 1:1 isoamyl alcohol–benzene mixture followed by reduction with stannous chloride. Total soluble phosphorus is determined by the same method after persulfate digestion. Total insoluble phosphorus requires digestion of the filter membrane and a nitric acid–perchloric acid digestion, followed by the above method for orthophosphate anion. The results of a study for the preservation of the various phosphorus forms are included in Ref. 165. Jones and Spencer (166) recommend the molybdenum heteropoly blue method of Harvey (167) of five methods they compared for the determination of inorganic phosphate in seawater.

Most research in recent years on the determination of phosphates in water has involved variations of the reduced molybdophosphoric acid procedure, with a number of reducing agents and extraction solvent mixtures having been investigated. One of the most thoroughly studied methods is the single-solution reagent of Murphy and Riley(168), which contains ammonium molybdate, ascorbic acid, and a small amount of antimonyl potassium tartrate. Orthophosphoric acid reacts with the reagent to form a stable blue-violet compound that is measured spectrophotometrically at 882 nm. Soluble arsenic compounds, of the common constituents of natural waters, interfere. Jones(169) recommends this method for the estimation of phosphorus in seawater. Fishman and Skougstad(170) reported a simplified method based on the method of Murphy and Riley(168).

Breitling(171) reported a vanadate-molybdate reagent for orthophosphate in water with measurement of the absorbance at 365 nm without a reduction procedure. Variations of this procedure were investigated by Abbott and co-workers(172,173), Moeller(174), and Proft(175). Sugawara and Kanamori(3) described an exceedingly sensitive method in which the molybdenum in solvent-extracted molybdophosphoric acid was determined as the thiocyanate complex. The 12-fold chemical amplification permits the determination of microgram amounts of phosphate ion. This method is designed to differentiate and determine arsenite, arsenate, and phosphate.

XV. Potassium

Potassium ranks seventh, immediately below and approximately equal to sodium, in relative abundance in the lithosphere, but its concentrations in natural waters generally are much lower than sodium ion concentrations. Potable waters rarely exceed 20 ppm potassium; brines may reach 100 ppm, and seawater averages about 380 ppm. The presence of potassium in drinking water is physiologically unimportant.

Flame photometry is ideally suited to the determination of trace concentrations of potassium ion in water, but atomic absorption techniques have not been widely applied. *Standard Methods* (*176*) describes a flame photometric method applicable either to a direct reading or internal standard type of flame photometer at wavelength 768 nm. The lower limit of sensitivity is approximately 0.1 ppm with the better flame photometers and expert technique. With lithium as the internal standard, sodium-to-potassium ratios up to 5:1, calcium-to-potassium ratios up to 10:1, and magnesium-to-potassium ratios up to 100:1 are tolerated. Flame photometry is the predominate method among those reported in the recent literature.

Standard Methods also describes an indirect colorimetric method sensitive to 5 ppm in which potassium ion is first precipitated as dipotassium sodium hexanitritocobaltate(III). The nitrite ligand is destroyed by oxidation with standard potassium dichromate solution in the presence of sulfuric acid, and the excess dichromate is determined spectrophotometrically at 425 nm. A filter photometer equipped with a violet filter may also be used, or the solutions may be compared visually to prepared standards. The lower limit is 5 ppm with a 15-cm light path. Ammonium ion interferes and must be absent.

Barnes (*177*) recommends silver hexanitritocobaltate(III) for the gravimetric, titrimetric, or colorimetric determination of potassium ion in seawater. The titrimetric method is indirect; the potassium hexanitritocobaltate(III) precipitate is oxidized with excess standard cerium(IV) sulfate solution, and the excess is back-titrated with standard iron(II) ammonium sulfate solution.

The insolubility of potassium salts of hexanitritocobaltate(III), tetraphenylborate, and dicrylaminate anions has been proposed as a basis of methods for the determination of potassium ion. Kriventsov(*178*) described a turbidimetric method based on colloidal suspensions of potassium hexanitritocobaltate(III), while Costache(*179*) reported a turbidimetric micromethod based on potassium tetraphenylborate. Prochazkova (*180*) proposed a titrimetric method involving potassium tetraphenylborate in which the precipitated salt is dissolved in acetone and titrated with standard silver nitrate solution with potassium chromate as the indicator. An indirect titrimetric method was described by Murakami (*181*) for potassium ion in seawater. Potassium ion is precipitated with an excess of standard sodium tetraphenylborate solution and the excess titrated with standard benzalkonium chloride solution. Skinner and Docherty(*182*) reported an automated method in which tetraphenylborate anion in excess of that necessary to precipitate the potassium present is determined spectrophotometrically at 254 nm. A gravimetric method was suggested by Efendi and Sokolovich(*183*) for potassium ion in waters of low mineralization with dicrylaminate anion as the precipitant.

XVI. Selenium

Selenite and selenate ions are the forms most likely to be encountered in natural waters. There is some indication that selenium may be an essential trace element, but all types can be considered toxic to animals and are a matter of concern in concentrations exceeding 10 ppb. Drinking water having selenium concentrations above 10 ppb should be rejected. Selenium in seawater rarely exceeds several parts per billion. Selenium mimics its congener sulfur in metabolic processes in plants and animals.

Selenium in all its forms can be converted to selenium(IV) by oxidation to the selenate followed by reduction to the selenite, or evolved by distillation from a bromine-hydrobromic acid solution as selenium tetrabromide which hydrolyzes to selenious acid. *Standard Methods*(*184*) describes two colorimetric methods: one in which 3,3'-diaminobenzidine is used as the reagent following either the oxidation-reduction procedure or the bromine-hydrobromic acid distillation procedure, and the other in which hydroxylamine is used to reduce selenious acid to a selenium sol following the distillation procedure. Iron(III) is eliminated as an interference by use of EDTA, but iodide and bromide cause low results in the 3,3'-diaminobenzidine oxidation-reduction method. No common ions interfere in the 3,3'-diaminobenzidine/bromine-hydrobromic acid distillation method. In the selenium sol method, arsenic, antimony, and germanium may distill over in the bromine-hydrobromic acid distillation. The 3,3'-diaminobenzidine method yields a piazselenol which is extracted into toluene and measured at 420 nm to determine selenium in the range 0–50 ppb. The selenium sol method is not so sensitive, and samples with very low selenium concentrations must first be concentrated by evaporation.

Selenium in seawater was determined by Chau and Riley(185) after coprecipitation with iron(III) hydroxide with the 3,3'-diaminobenzidine reagent by using isotope dilution to correct the results for small losses due to the analytical procedure. Biswas and Dey(186) used a ring-oven technique with a magnesia mixture to remove arsenate, and thiourea to reduce selenium(IV) to the element. Selenium concentrations as low as 0.23 ppm can be determined.

Almost all recent publications concerned with the analysis of selenium in water report modifications of the 3,3'-diaminobenzidine method. Ariyoshi et al.(187) reported that o-phenylenediamine is superior to 3,3'-diaminobenzidine for the determination of selenium, although the absorbance in toluene extract must be measured in the near ultraviolet at 335 nm. Another variation on aromatic diamine reagents is the use by Parker and Harvey(188) of 2,3-diaminonaphthalene and the fluorimetric determination of the resulting Dekalin-extracted piazselenol.

XVII. Silicon

Silicon occurs in natural waters as soluble or colloidal silicates in concentrations normally in the range 0–10 ppm and in rare cases up to 60 ppm. Silicates must be removed from boiler feed waters because of hard scale formation. Diatoms, sponges, and radiolaria secrete silica in the form of opal. Silicon concentrations in seawater vary from 20 to 4000 ppb.

Standard Methods(189) describes a gravimetric method and two colorimetric methods that employ molybdosilicic acid; in one colorimetric method the yellow heteropoly acid is measured photometrically or by visual comparison, and in the other the molybdosilicic acid is reduced to a heteropoly blue compound. In the latter method, the molybdenum is reduced to the quinquivalent state by 1-amino-2-naphthol-4-sulfonic acid. The absorption of the blue compound is more intense than that of the parent yellow heteropoly acid and hence the method is more sensitive. In the gravimetric method, silicates and dissolved silica are precipitated as partially dehydrated silicic acid during evaporation and baking and are ignited to form silicon dioxide. Treatment with hydrofluoric acid converts the silicon dioxide to volatile silicon tetrafluoride, and the loss of weight is proportional to the silicon dioxide in the ignited residue. Silicate from glassware is the only interference and can be avoided by use of plastic containers and platinum laboratory ware.

In the colorimetric methods, phosphate is prevented from forming molybdophosphoric acid by use of oxalic acid. Iron compounds, sulfide ion, and preexisting color or turbidity also interfere. The limit of detection of the molybdosilicic acid method is approximately 1 ppm and that of the heteropoly blue method is about 0.02 ppm, both from visual measurements made with 50-ml Nessler tubes.

Jenkins(190) recommends the heteropoly blue method for the determination of silicates in estuarine waters. Barnes(191) and Strickland and Parsons(192) also favor the heteropoly blue method for seawater analysis.

Almost all recent research on methods for the determination of silicates in water has been on improvements and modifications of the molybdosilicic acid or the heteropoly blue colorimetric methods. Morrison and Wilson(193–195) made detailed studies of the heteropoly blue method and reported procedures for the determination of "reactive" silicates (monomeric and dimeric silicic acids) and total silica. Schink(196) suggested the extraction of the molybdosilicic acid by ethyl acetate and measurement of the absorbance at 335 nm for the determination of silica in seawater. Sono et al.(197) extracted the heteropoly blue compound with 5% polyoxyethylenated laurylamine and chloroform and measured the absorbance at 750 nm in an ultrasensitive method for silica in boiler water.

XVIII. Sodium

Sodium ion is found in most natural waters and varies in concentration from 1 ppm or less to several hundred parts per million in surface waters and groundwaters. Seawater has a high and approximately constant concentration of about 10,000 ppm sodium ion as one of the "conservative" elements. Sodium is one of the few elements for which no colorimetric method is recommended.

Standard Methods(198) describes a flame photometric method applicable either to a direct reading or internal standard type of flame photometer at 598 nm. The lower limit of sensitivity is approximately 0.1 ppm, which can be extended to 0.01 ppm with modifications of technique. With lithium as the internal standard, potassium-to-sodium ratios up to 5:1, calcium-to-sodium ratios up to 10:1, and magnesium-to-sodium ratios up to 100:1 are tolerated. Sodium chloride stock solutions are used to prepare calibration curves covering several concentration ranges. *Standard Methods* also describes a gravimetric method in which lithium ion and potassium ion (in large concentrations and particularly in the presence of sulfate ion) interfere. Sodium ion is precipitated as sodium zinc uranyl acetate hexahydrate with a large excess of zinc uranyl acetate solution saturated with the sodium triple salt. The gravimetric factor is very favorable. Barnes (*199*) recommends gravimetric determination with zinc uranyl acetate for sodium in seawater. Phosphate interferes and is removed with calcium hydroxide.

The insoluble sodium zinc uranyl acetate hexahydrate salt is used as the basis of several proposed indirect titrimetric methods. Mamedov and Emirdzhanova(*200*) converted the triple salt into an insoluble arsenate salt which is subsequently determined by an iodometric titration method. Dolezal et al.(*201*) determined sodium ion by complexometric titration of the uranium after precipitation with cobalt uranyl acetate and reduction with ascorbic acid. Nogina and Kobyak(*202*) recommended an indirect complexometric method with Complexon III in ammoniacal solution with Eriochrome black T indicator.

Atomic absorption has not been utilized to its full extent in sodium analysis because its extreme sensitivity is not needed. Glass membrane electrodes selective for sodium ion are available commercially from Corning Glass Works(*202a*) and Beckman Instruments, Inc.(*202b*). These electrodes generally have favorable selectivity for sodium in the presence of lithium, potassium, and ammonium ions. The response to hydrogen ion is roughly equivalent to that of sodium, but the electrode is more sensitive to silver ion than to sodium.

XIX. Sulfur

The species of interest in natural waters are sulfide, sulfite, and sulfate ions. As is true of other elements of major biological importance, sulfur exhibits a cycle. Sulfate sulfur accumulates because of its stability toward reduction. Its concentration varies widely in surface waters and groundwaters. Seawater normally contains about 800–900 ppm sulfur. Sulfide ion usually is the result of anaerobic bacteriological processes, although man-made wastes may also be a source. Sulfite ion is a transitory intermediate species in water.

A. Sulfide Ion

The species present below pH 13 are either HS⁻ or H_2S, which impart taste and odor to water when present in concentrations of hundredths of a part per million. *Standard Methods(203)* describes two methods for sulfide ion determination, a titrimetric and a somewhat more sensitive colorimetric procedure. Both can be used for the determination of dissolved and total sulfide ion (dissolved plus acid-soluble sulfide salts). In the titrimetric method, total sulfide is converted to hydrogen sulfide in the acidified sample and swept out with carbon dioxide into a zinc acetate solution. An excess of standard triiodide ion solution is added, the solution is acidified, and the remaining triiodide is titrated with standard thiosulfate solution to the starch end point. Dissolved sulfide is determined separately on a sample from which suspended solids have been removed by flocculation. The colorimetric method involves the formation of methylene blue dye cation from *p*-aminodimethylaniline, iron(III) chloride, and sulfide in acid solution. The color produced is compared visually to standards prepared by dilution of methylene blue stock solutions. The method is sensitive to 0.05 ppm. Dissolved sulfide is determined separately after removal of suspended solids.

Few new methods for sulfide sulfur have appeared in the recent literature. Kurzawa(204) described a sulfide-induced reaction of azide ion with iodine for trace concentrations. Orion Research, Inc.(205) has made commercially available a sulfide-selective solid-state electrode usable over the pH range 1–12. The electrode responds to sulfide activity from 1 to $10^{-17} M$ or, conversely, to silver ion in the range $1-10^{-7} M$.

B. Sulfite Ion

Sulfur dioxide is stable in air, but sulfite ion is easily oxidized to sulfate ion in aqueous solution. Both sulfur dioxide and sulfite ion usually are the result of man-made pollution and are of interest for that reason, except for the use of sulfite salts as deoxidants and anticorrosion agents in boiler waters. *Standard Methods(206)* describes only a titrimetric method, in which sulfite is oxidized by iodine produced from an acidified standard potassium iodide–potassium iodate solution and the excess titrated to the starch end point. The method is sensitive to 2 ppm sulfite ion and is not selective, as oxidizable organic matter and sulfide ion produce high results. Nitrite ion, which would destroy sulfite ion in acid solution, is removed by sulfamic acid. There appears to be little interest in the determination of sulfite ion in the recent literature.

C. Sulfate Ion

Concentrations of sulfate ion in natural waters vary from several to several thousand parts per million as the result of groundwater contact with gypsum deposits or the oxidation of pyrites. Because of its cathartic effect with sodium or magnesium as the counter-ion, potable waters should not exceed 250 ppm sulfate. *Standard Methods* (*207*) describes a gravimetric method with two modifications and a turbidimetric method for the determination of sulfate ion. In the gravimetric method, sulfate ion is precipitated by barium ion in hydrochloric acid solution and the precipitate is either ignited or dried. The precipitation of barium sulfate is subject to a number of offsetting variables. Interference leading to high results is encountered when barium salts are occluded or when suspended matter such as silica is precipitated with barium sulfate; interference leading to low results occurs because of coprecipitation of cations of lower atomic weight than barium. As precise results usually are not required, the methods described suffice for routine analytical work. The turbidimetric method is based on controlled particle size precipitation of barium sulfate in hydrochloric acid solution, and is applicable directly for the range 0–40 ppm. Detailed instructions are provided and must be followed for reproducible results.

Barnes (*208*) recommends ion exchange between the relatively insoluble barium iodate and sulfate ion in solution to release iodate ion. The iodate ion is reduced by iodide ion in acid solution to form iodine, which is titrated with standard thiosulfate solution to the starch end point. Insoluble barium chloranilate, which exchanges chloranilate anion for sulfate ion, has been reported by Bertolacini and Barney (*60*) as a convenient reagent for sulfate ion.

Much work has been done in recent years on complexometric titration methods, but none has been widely accepted. Kleber and Franke (*209*) recommended the conventional gravimetric barium sulfate method over EDTA chelometric titration methods. Koval'tsov and Konovalov (*210*) reviewed volumetric and physical–chemical methods to find one suitable for the determination of sulfate with automatic analyzers, but recommended turbidimetric and nephelometric methods. A number of other chelating reagents and end point indicators have been proposed. Ion exchange has been used by a number of investigators to remove interfering cations prior to complexometric titration methods. Standard lead nitrate solution has been used by Nechiporenko (*211–214*) as a titrant with end point indicators such as dithizone or diphenylcarbazone. Despite earlier reports by Pungor et al. (*215*) of sulfate-selective electrodes of the barium sulfate-impregnated rubber type, no satisfactory electrode has been made commercially available.

XX. General Reference Works

A number of books, manuals, and reviews of interest in water analysis have appeared in recent years in addition to those quoted previously in this chapter. Some cover a wide range of elements or compounds and analytical procedures, while others deal with specific analytical techniques.

For discussions of current methods and techniques for trace analysis with a view to future developments, the *National Bureau of Standards Monograph 100(216)* is highly recommended. *Trace Analysis: Physical Methods* by Morrison(*217*) is also of interest to water chemists. A work concerned exclusively with water analysis is *L'Analyse Chimique et Physico-Chimique de l'Eau*, by Rodier(*218*). A reliable source of tested methods of relevance to water analysis is the *Official Methods of Analysis of the Association of Official Agricultural Chemists*, 10th edition(*219*). A comprehensive review of methods for quantitative inorganic analysis through 1957 has been compiled by Kodama(*219a*).

Multivolume works of which new volumes appear regularly and which treat all phases of analytical chemistry are the *Treatise on Analytical Chemistry*, edited by Kolthoff, Elving, and Sandell(*220*), *Advances in Analytical Chemistry and Instrumentation*, edited by Reilley and McLafferty(*221*), *Standard Methods of Chemical Analysis* by Welcher(*222*), and the *Encyclopedia of Industrial Chemical Analysis*, edited by Snell and Hilton(*223*).

Recent specialized works relating to water analysis include *Atomic Absorption Spectroscopy* by Robinson(*224*) and *Amperometric Titrations* by Stock(*225*). Complexes and complexometric titrations are discussed in *Die Komplexometrische Titrations* by Schwarzenbach and Flaschka(*226*), and in *Chelates in Analytical Chemistry*, edited by Flaschka and Barnard(*227*). Mavrodineanu and Boiteux have authored a three-volume work, *Flame Spectroscopy(228)*; Mavrodineanu also compiled the *Bibliography on Flame Spectroscopy, Analytical Applications, 1800–1966(229)*.

Fluorescence: Theory, Instrumentation, and Practice by Guilbault (*230*) and *Fluorescence and Phosphorescence Analysis: Principles and Applications*, edited by Hercules(*231*), treat analytical aspects of phenomena that have not been thoroughly exploited in water analysis. Gas chromatography, which has had some applications in water analysis in recent years, is covered in *The Practice of Gas Chromatography*, edited by Ettre and Zlatkis(*232*). *Laboratory Handbood of Chromatographic Methods*, edited by Mikes(*233*), includes other aspects of chromatography. The present volume includes chapters on the application of radionuclide determination, gas chromatography, mass spectroscopy, luminescence techniques, and electroanalytical methods to water analysis (see Chaps. 25, 28, 29, 31, and 32, respectively). Research on ion-selective electrodes was reviewed in an article by Rechnitz(*233a*).

Ion exchange, a frequently used procedure for sample concentration and removal of interferences, is treated in *Analytical Applications of Ion Exchangers* by Inczedy(*234*), and in *Ion Exchange,* Vol. 1, edited by Marinsky(*235*) (see Chap. 11). Procedures for the determination of a number of elements in water by neutron activation are described by Landstrom and Wenner(*236*) and Leddicotte and Moeller(*237*). Recent advances in volumetric analysis are covered in *Newer Redox Titrants* by Berka, Vulterin, and Zyka(*238*). *Reagents and Solutions in Analytical Chemistry* by Perlman(*239*) is a basic work for analytical chemists.

Manuals for the analysis of natural waters include one by Bush and Higgins(*240*), which also includes sections on radiochemical analysis and analysis of silicate rocks and ores. Methods used by the U.S. Geological Survey are presented in a manual by Rainwater and Thatcher(*241*). Methods developed at the U.S. Bureau of Mines for a number of common constituents in water are reported by Collins et al.(*242*). A manual for the analysis of industrial water and wastewater was published by the American Society for Testing and Materials(*243*). Methods used in oceanographic studies were described by Goya et al. in a U.S. Department of Commerce report(*244*). Photometric methods for determining many of the common constituents in water are reported in a manual edited by Zimmerman(*245*). Mackereth(*246*) prepared a manual of water analysis especially for limnologists.

REFERENCES

1. E. E. Angino, L. M. Magnuson, T. C. Waugh, O, K. Galle, and J. Bredfeldt, *Science,* **168**, 389 (1970).

1a. *Standard Methods for the Examination of Water and Wastewater,* 12th ed., American Public Health Association, New York, 1965, p. 56.

2. J. E. Portman and J. P. Riley, *Anal. Chim. Acta,* **31**, 509 (1964).

3. K. Sugawara and S. Kanamori, *Bull. Chem. Soc. Japan,* **37**, 1358 (1964); through *Chem. Abstr.,* **61**, 14358g (1964).

4. S. Nakaya, *Bunseki Kagaku,* **12**, 241 (1963); through *Chem. Abstr.,* **59**, 7238d (1963).

4a. L. A. Davidyuk, *Dopovidi Akad. Nauk Ukr. RSR,* **1966**, 90; through *Chem. Abstr.,* **64**, 19192a (1966).

5. *Standard Methods for the Examination of Water and Wastewater,* 12th ed., American Public Health Association, New York, 1965, p. 60.

6. R. Greenhalgh and J. P. Riley, *Analyst,* **87**, 970 (1962).

7. T. Murakami, *Himeji Kogyo Daigaku Kenkyu Hokoku,* **14**, 108 (1961); through *Chem. Abstr.,* **58**, 924h (1963).

8. C. A. Parker and W. J. Barnes, *Analyst,* **85**, 828 (1960).

9. R. J. Lishka, *J. Am. Water Works Assoc.,* **53**, 1517 (1961).

10. W. A. Powell and E. H. Poindexter, *U.S. Atomic Energy Comm. CCC–1024–Tr–226,* 1957.

11. *Ibid., TR–1024–TR–229*, 1957.

11a. A. Scher, *Hidrol. Közl.*, **37**, 168 (1957); through *Chem. Abstr.*, **55**, 1974c (1961).

11b. K. Hiiro, *Bull. Chem. Soc. Japan*, **34**, 1748 (1961); through *Chem. Abstr.*, **56**, 12299f (1962).

12. A. E. Vasilevskaya and L. K. Lenskaya, *Zh. Analit. Khim.*, **20**, 747 (1965) (in Russian); through *Chem. Abstr.*, **63**, 15538h (1965).

13. V. A. Semenov, *Gigiena i Sanit.*, **31**, 62 (1966) (in Russian); through *Chem. Abstr.*, **65**, 1949b (1966).

14. *Standard Methods for the Examination of Water and Wastewater*, 12th ed., American Public Health Association, New York, 1965, p. 66.

15. L. N. Podgornyi and F. I. Bezler, *Byul. Inst. Biol. Vodokhranilishch, Akad. Nauk SSSR*, **1958**, 56; *Ref. Zh. Geofiz.*, **1959**, Abstr. No. 10250; through *Chem. Abstr.*, **56**, 15299d (1962).

16. I. I. M. Elbeih and A. A. El-Sirafy, *Chemist-Analyst*, **54**, 8 (1965).

17. I. Iwasaki, S. Utsumi, A. Tomonari, I. Morita, and M. Shiota, *Nippon Kagaku Zasshi*, **80**, 744 (1959); through *Chem. Abstr.*, **54**, 19278a (1960).

18. M. Shiota, S. Utsumi, and I. Iwasaki, *Ibid.*, 753 (1959); through *Chem. Abstr.*, **54**, 19278a (1960).

19. M. J. Fishman and M. W. Skougstad, *Anal. Chem.*, **35**, 146 (1963).

20. Orion Research, Inc., Cambridge, Mass., *Tech. Lit. Model 94–35*.

21. A. L. Bystritskii, V. B. Aleskovskii, and V. V. Bardin, *Izv. Vysshikh Uchebn. Zavedenii Khim. Khim. Tekhnol.*, **6**, 31 (1963); through *Chem. Abstr.*, **59**, 9323e (1963).

22. *Standard Methods for the Examination of Water and Wastewater*, 12th ed., American Public Health Association, New York, 1965, p. 70.

23. G. Schwarzenbach, U.S. Pat. 2,583,890 (1952).

24. G. Schwarzenbach, U.S. Pat. 2,583,891 (1952).

25. H. Barnes, *Apparatus and Methods of Oceanography, Part One: Chemical*, Wiley (Interscience), New York, 1959, p. 213.

26. H. Katz and R. Navone, *J. Am. Water Works Assoc.*, **56**, 121 (1964).

27. R. W. Close and T. S. West, *Talanta*, **5**, 221 (1960).

28. R. E. Rehwoldt, B. L. Chasen, and J. B. Li, *Anal. Chem.*, **38**, 1018 (1966).

29. F. Ruf, *Compt. Rend. Semaine Geol. Comite Natl. Malgache Geol.*, **1964**, 105 (in French); through *Chem. Abstr.*, **64**, 7364f (1966).

30. J. J. Rowe, *Geochim. Cosmochim. Acta*, **27**, 915 (1963); through *Chem. Abstr.*, **59**, 15027h (1963).

31. Corning Glass Works, Medfield, Mass., *Tech. Lit. Catalog No. 476230*.

32. Orion Research, Inc., Cambridge, Mass., *Tech. Lit. Model 92–90*.

33. *Standard Methods for the Examination of Water and Wastewater*, 12th ed., American Public Health Association, New York, 1965, p. 77.

34. H. Gaunt and C. Shanks, *Chem. Ind. (London)*, **1964**, 651.

35. L. S. Waterman, *Nature*, **205**, 1099 (1964).

36. *Standard Methods for the Examination of Water and Wastewater*, 12th ed., American Public Health Association, New York, 1965, pp. 48, 369, 438, 530.

37. H. Barnes, *Apparatus and Methods of Oceanography, Part One: Chemical*, Wiley (Interscience), New York, 1959, p. 200.

38. W. H. Smith, Jr., and D. W. Hood, *Recent Res. Fields Hydrosphere, Atmosphere Nucl. Geochem.*, **1964**, 185.

39. K. Park, M. Oliphant, and H. Freund, *Anal. Chem.*, **35**, 1549 (1963).

40. C. E. Herce and E. C. Luppi, *Rev. Asoc. Arg. Quim. Tec. Ind. Cuero*, **5**, 91 (1964) (in Spanish); through *Chem. Abstr.*, **65**, 5215c (1966).

41. C. E. Roberson, J. H. Feth, P. R. Seaber, and P. Anderson, *U.S. Geol. Surv. Profess. Paper 475-C, C212-5*, 1963.

42. F. G. Carpenter, *Anal. Chem.*, **34**, 66 (1962).

43. A. L. Underwood, *Anal. Chem.*, **33**, 955 (1961).

44. A. L. Underwood and L. H. Howe, III, *Anal. Chem.* **34**, 692 (1962).

45. *Standard Methods for the Examination of Water and Wastewater*, 12th ed., American Public Health Association, New York, 1965, p. 448.

46. L. E. Lancy and W. Zabban, *Am. Soc. Testing Mater. Spec. Tech. Publ. 337*, 1963, p. 32.

47. S. Hikime, H. Yoshida, and M. Yamamoto, *Bunseki Kagaku*, **10**, 508 (1961); through *Chem. Abstr.*, **58**, 6181h (1963).

48. G. G. Guilbault and D. N. Kramer, *Anal. Chem.*, **37**, 918 (1965).

49. G. G. Guilbault and D. N. Kramer, *Anal. Chem.*, **37**, 1395 (1965).

50. G. G. Guilbault and D. N. Kramer, *Anal. Chem.*, **38**, 834 (1966).

51. C. R. Schneider and H. Freund, *Anal. Chem.*, **34**, 69 (1962).

51a. Orion Research, Inc., Cambridge, Mass., *Tech. Lit. Model 94-06.*

52. *Standard Methods for the Examination of Water and Wastewater*, 12th ed., American Public Health Association, New York, 1965, pp. 85, 370.

53. H. Barnes, *Apparatus and Methods of Oceanography, Part One: Chemical*, Wiley (Interscience), New York, 1959, pp. 85, 217.

54. J. D. H. Strickland and T. R. Parsons, *A Manual of Sea Water Analysis, Bull. No. 125*, Fisheries Research Board of Canada, Ottawa, 1960, pp. 11, 19.

55. D. Jenkins, *J. Water Pollution Control Federation*, **39**, 161 (1967).

56. N. L. Brown and B. V. Hamon, *Deep-Sea Research*, **8**, 65 (1961); through *Chem. Abstr.*, **56**, 9891e (1962).

57. H. U. Woelk, *U.S. Atomic Energy Comm. HMI–B7*, 1959.

58. Z. Marczenko and L. Choluj-Lenarczyk, *Chem. Anal. (Warsaw)*, **11**, 1221 (1966) (in Polish); through *Chem. Abstr.*, **67**, 29073v (1967).

59. Y. Kobayashi and H. Takeno, *Bunseki Kagaku*, **13**, 1208 (1964) (in Japanese); through *Chem. Abstr.*, **62**, 7257h (1965).

60. R. J. Bertolacini and J. E. Barney, II, *Anal. Chem.*, **30**, 202 (1958).

61. Orion Research, Inc., Cambridge, Mass., *Tech. Lit. Models 92–17, 94–17.*

62. *Standard Methods for the Examination of Water and Wastewater*, 12th ed., American Public Health Association, New York, 1965, pp. 90, 375, 440.

63. N. J. Nicolson, *Analyst*, **90**, 187 (1965).

64. S. Katz and H. Heukelekian, *Sewage and Ind. Wastes*, **31**, 1022 (1959); through *Chem. Abstr.*, **53**, 22640b (1959).

64a. F. J. Hallinan and W. R. Thompson, *J. Am. Chem. Soc.*, **61**, 265 (1939).

64b. W. V. D. Tiedeman, *J. Am. Water Works Assoc.*, **15**, 391 (1926).

64c. D. Tarvin, *Sewage Ind. Wastes*, **24**, 1130 (1952).

64d. M. L. Koshkin, *J. Am. Water Works Assoc.*, **29**, 1761 (1937).

65. M. Bossy, *Tech. Sci. Munic.*, **57**, 201 (1962); through *Chem. Abstr.*, **61**, 4062b (1964).

65a. M. Taras, *Anal. Chem.*, **19**, 342 (1947).

65b. F. W. Sollo, Jr., and T. E. Larson, *J. Am. Water Works Assoc.*, **57**, 1575 (1965).

66. V. T. Athavale, C. V. Krishnan, and A. R. Subramanian, *Analyst*, **87**, 707 (1962).

67. H. Berndt, *Staedtehygiene*, **1960**, 224; through *Chem. Abstr.*, **61**, 8057g (1964).

68. H. Karge, *Z. Anal. Chem.*, **200**, 57 (1964); through *Chem. Abstr.*, **60**, 10386a (1964).

69. H. Meier-Ewert and H. Bruenner, *Staedtehygiene*, **1960**, 223; through *Chem. Abstr.*, **61**, 8057f (1964).

70. *Standard Methods for the Examination of Water and Wastewater*, 12th ed., American Public Health Association, New York, 1965, p. 116.

71. A. T. Palin, *Proc. Soc. Water Treat. Exam.*, **9**, 81 (1960); through *Chem. Abstr.*, **55**, 7714e (1961).
71a. A. W. Hodgden and R. S. Ingols, *Anal. Chem.*, **26**, 1224 (1954).
72. P. Kerenyi and P. Kuba, *Chem. Zvesti*, **17**, 146 (1963); through *Chem. Abstr.*, **59**, 12507a (1963).
73. *Standard Methods for the Examination of Water and Wastewater*, 12th ed., American Public Health Association, New York, 1965, p. 135.
74. S. S. Yamamura, M. A. Wade, and J. H. Sikes, *Anal. Chem.*, **34**, 1308 (1962).
75. T. Murakami and K. Uesugi, *Bunseki Kagaku*, **14**, 235 (1965) (in Japanese); through *Chem. Abstr.*, **62**, 15902h (1965).
76. T. S. Bhakuni and N. N. Sharma, *Environ. Health (India)*, **4**, 6 (1962).
77. A. L. Hensley and J. E. Barney, II, *Anal. Chem.*, **32**, 828 (1960).
78. Orion Research, Inc., Cambridge, Mass., *Tech. Lit. Model 94–09*.
79. M. S. Frant and J. W. Ross, Jr., *Science*, **154**, 1553 (1966).
80. J. W. Swinnerton, V. J. Linnenbom, and C. H. Cheek, *Anal. Chem.*, **34**, 483 (1962).
81. R. Massart and L. Missa, *Centre Belge Edtude et Document Eaux, Bull. Trimestr. CEBEDEAU*, **47**, 43 (1960); through *Chem. Abstr.*, **55**, 9734f (1961).
82. I. J. Hissel, *Tribune CEBEDEAU*, **16**, 60 (1963); through *Chem. Abstr.*, **59**, 2507e (1963).
83. P. Faber and W. Brand, *Kerntechnik*, **6**, 79 (1964); through *Chem. Abstr.*, **61**, 1610h (1964).
84. *Standard Methods for the Examination of Water and Wastewater*, 12th ed., American Public Health Association, New York, 1965, pp. 46, 367, 438, 529.
85. *Standard Methods for the Examination of Water and Wastewater*, 12th ed., American Public Health Association, Inc., New York, 1965, p. 152.
86. G. A. Pavlova and O. V. Shishkina, *Tr. Inst. Okeanol. Akad. Nauk SSSR*, **67**, 165 (1964); through *Chem. Abstr.*, **61**, 15834a (1964).
87. G. F. Proskuryakova, R. V. Shveikina, and M. C. Chernavina, *Izv. Vysshikh Uchebn. Zavedenii Khim. Khim. Tekhnol.*, **6**, 729 (1963); through *Chem. Abstr.*, **60**, 9017d (1964).
88. L. N. Lapin and N. V. Reis, *Lab. Delo*, **7**, 21 (1961); through *Chem. Abstr.*, **56**, 2289d (1962).
89. E. Jungreis and I. Gedalia, *Mikrochim. Acta*, **1960**, 145 (in English).
90. E. Ramanauskas, *Leituvos TSR Aukstuju Mokyklu Mokslo Darbai Chem. Chem. Technol.*, **5**, 9 (1964) (in Russian); through *Chem. Abstr.*, **61**, 10450b (1964).
91. N. Yonehara, *Bull. Chem. Soc. Japan*, **37**, 1101 (1964); through *Chem. Abstr.*, **61**, 13043b (1964).
92. G. V. Novikov, *Gigiena i Sanit.*, **25**, 62 (1960); through *Chem. Abstr.*, **55**, 856g (1961).
93. E. Mariani and G. Maura, *Ann. Chim. (Rome)*, **51**, 158 (1961).
94. P. V. Rastegina, *Zobnaya Bolezn Sb.*, **2**, 275 (1959); from *Ref. Zh. Khim.*, **1960**, Abstr. No. 73138; through *Chem. Abstr.*, **55** 7715a (1961).
95. Orion Research, Inc., Cambridge, Mass., *Tech. Lit. Model 94–53*.
96. *Standard Methods for the Examination of Water and Wastewater*, 12th ed., American Public Health Association, New York, 1965, p. 167.
97. H. Barnes, *Apparatus and Methods of Oceanography, Part One: Chemical*, Wiley (Interscience), New York, 1959, p. 215.
98. B. A. Skopintsev and V. V. Kabanov, *Tr. Morskogo Gidrofiz. Inst.*, **13**, 130 (1958); through *Chem. Abstr.*, **54**, 25401a (1960).
99. J. B. Pate and R. J. Robinson, *J. Marine Res.*, **17**, 390 (1958).
100. J. B. Pate and R. J. Robinson, *J. Marine Res.*, **19**, 12 (1961).

101. J. R. B. Rial and L. R. Molins, *Bol. Inst. Espan. Oceanog.*, No. 111 (1962); through *Chem. Abstr.*, **58**, 6572d (1963).

102. H. Katz and R. Navone, *J. Am. Water Works Assoc.*, **56**, 121 (1964).

103. F. B. Marti and C. A. Herrero, *Inform. Quim. Anal. (Madrid)*, **19**, 105 (1965) (in Spanish); through *Chem. Abstr.*, **64**, 10920g (1966).

104. R. Fabregas, A. Badrinas, and A. Prieto, *Talanta*, **8**, 804 (1961) (in German).

105. R. M. Rai, P. C. Pande, and B. N. Tripathi, *Indian J. Chem.*, **4**, 144 (1966) (in English); through *Chem. Abstr.*, **65**, 2996b (1966).

106. D. E. Peaslee, *Soil Sci. Soc. Am. Proc.*, **30**, 443 (1966); through *Chem. Abstr.*, **66**, 8066c (1967).

107. I. V. Petrukhin, *Tr. Smolensk. Nauchin.-Issled. Vet. Stantsii*, **1960**, 188; from *Ref. Zh. Khim. Biol. Khim.* **1961**, Abstr. No. 17S67; through *Chem. Abstr.*, **56**, 926e (1962).

108. J. C. van Schouwenburg, *Neth. J. Agr. Sci.*, **13**, 53 (1965) (in English); through *Chem. Abstr.*, **66**, 16319j (1967).

109. H. Ogata and K. Hiroi, *Bunseki Kagaku*, **8**, 21 (1959) (in Japanese); through *Chem. Abstr.*, **55**, 3296a (1961).

110. H. Diehl, R. Olsen, G. I. Spielholtz, and R. Jensen, *Anal. Chem.*, **35**, 1144 (1963).

111. G. V. Serebryakova, E. A. Bozhevol'nov, and G. S. Godlina, *Metody Analiza Khim. Reaktivov i Preparatov, Gos. Kom. Sov. Min. SSSR po Khim.*, **4**, 92 (1962); through *Chem. Abstr.*, **61** 3681a (1964).

112. W. J. Blaedel and R. H. Laessig, *Anal. Chem.*, **38**, 186 (1966).

113. Yu. I. Turkin and P. F. Svistov, *Tr. Glavnoi Geofiz. Observ. im A. I. Voeikova*, **1960**, 89; *Ref. Zh. Khim.*, **1961**, Abstr. No. 13D82; through *Chem. Abstr.*, **56**, 926f (1962).

114. G. L. McPherson, *At. Absorption Newsletter*, **4**, 186 (1965); through *Chem. Abstr.*, **66**, 8038g (1967).

115. Orion Research, Inc., Cambridge, Mass., *Tech. Lit. Model 92–32*.

116. *Standard Methods for the Examination of Water and Wastewater*, 12th ed., American Public Health Association, New York, 1965, pp. 186, 389, 498.

117. H. Barnes, *Apparatus and Methods of Oceanography, Part One: Chemical*, Wiley (Interscience), New York, 1959, p. 129.

118. J. D. H. Strickland and T. R. Parsons, *A Manual of Sea Water Analysis, Bull. No. 125*, Fisheries Research Board of Canada, Ottawa, 1960, p. 75.

119. J. M. Kruse and M. G. Mellon, *Anal. Chem.*, **25**, 1188 (1953).

120. J. D. H. Strickland and H. K. Austin, *J. Conseil Conseil Perm. Intern. Exploration Mer.*, **24**, 446 (1959).

121. D. Jenkins, *J. Water Pollution Control Federation*, **39**, 159 (1967).

122. F. A. Richards and R. A. Kletsch, *Recent Res. Fields Hydrosphere, Atmosphere Nucl. Geochem.*, **1964**, 65.

123. R. T. Roskam and D. de Langen, *Anal. Chim. Acta*, **30**, 56 (1964).

124. V. G. Datsko and V. T. Kaplin, *Sovrem. Metody Analiza Prirodn. Vod. Akad. Nauk SSSR*, **1962**, 118; through *Chem. Abstr.*, **58**, 9957d (1963).

125. V. T. Kaplin and N. G. Fesenko, *ibid.*, **1962**, 123; through *Chem. Abstr.*, **58**, 9957b (1963).

126. R. W. Jenkins, Jr., C. H. Cheek, and V. J. Linnenbom, *Anal. Chem.*, **38**, 1257 (1966).

126a. F. Zitomer and J. L. Lambert, *Anal. Chem.*, **34**, 1738 (1962).

127. *Standard Methods for the Examination of Water and Wastewater*, 12th ed., American Public Health Association, New York, 1965, pp. 205, 400.

128. J. L. Lambert and F. Zitomer, *Anal. Chem.*, **32**, 1684 (1960).

129. N. T. Crosby, *Proc. Soc. Water Treat. Exam.*, **16**, 51 (1967); through *Chem. Abstr.*, **67**, 25318f (1967).

130. H. Barnes, *Apparatus and Methods of Oceanography, Part One: Chemical*, Wiley (Interscience), New York, 1959, p. 126.

131. D. Jenkins, *J. Water Pollution Control Federation*, **39**, 164 (1967).

132. J. D. H. Strickland and T. R. Parsons, *A Manual of Sea Water Analysis, Bull. No. 125*, Fisheries Research Board of Canada, Ottawa, 1960, p. 71.

133. R. F. Vaccaro, J. C. Thunberg, and B. H. Ketchum, *Limnol. Oceanog.*, 7, 322 (1963); through *Chem. Abstr.*, **60**, 6622g (1963).

134. *Standard Methods for the Examination of Water and Wastewater*, 12th ed., American Public Health Association, New York, 1965, pp. 195, 392.

135. P. W. West and T. P. Ramachandran, *Anal. Chim. Acta*, **35**, 317 (1966).

136. H. Barnes, *Apparatus and Methods of Oceanography, Part One: Chemical*, Wiley (Interscience), New York, 1959, p. 113.

137. J. D. H. Strickland and T. R. Parsons, *A Manual of Sea Water Analysis, Bull. No. 125*, Fisheries Research Board of Canada, Ottawa, 1960, p. 61.

138. D. W. W. Andrews, *Analyst*, **89**, 730 (1964).

139. P. G. Brewer and J. P. Riley, *Deep-Sea Res. Oceanog. Abstr.*, **12**, 765 (1965); through *Chem. Abstr.*, **64**, 17250e (1966).

140. K. Bendschneider and R. J. Robinson, *J. Marine Res.*, **11**, 87 (1952); through *Chem. Abstr.*, **47**, 2641g (1953).

141. A. W. Morris and J. P. Riley, *Anal. Chim. Acta*, **29**, 272 (1963).

142. T. J. Chow and M. S. Johnstone, *ibid.*, **27**, 441 (1962).

143. A. Henriksen, *Analyst*, **90**, 83 (1965).

144. J. B. Mullin and J. P. Riley, *Anal. Chim. Acta*, **12**, 464 (1955).

145. R. A. Bloomfield, J. C. Guyon, and R. K. Murmann, *Anal. Chem.*, **37**, 248 (1965).

146. R. A. Bloomfield, J. C. Guyon, and R. K. Murmann, *J. Am. Water Works Assoc.*, **57**, 935 (1965).

147. Orion Research, Inc., Cambridge, Mass., *Tech. Lit. Model 92-07*.

148. *Standard Methods for the Examination of Water and Wastewater*, 12th ed., American Public Health Association, New York, 1965, pp. 218, 405, 505.

149. H. Barnes, *Apparatus and Methods of Oceanography, Part One: Chemical*, Wiley (Interscience), New York, 1959, p. 178.

150. J. D. H. Strickland and T. R. Parsons, *A Manual of Sea Water Analysis, Bull. No. 125*, Fisheries Research Board of Canada, Ottawa, 1960, p. 23.

151. A. Ivanoff, *Compt. Rend.*, **254**, 4493 (1962); through *Chem. Abstr.*, **57**, 8361h (1962).

152. K. Park and M. Catalborno, *Deep-Sea Res. Oceanog. Abstr.*, **11**, 917 (1964); through *Chem. Abstr.*, **62**, 15903f (1965).

153. K. H. Mancy and T. Jaffe, *Public Health Serv. Publ. No. 999–WP–37*, U.S. Dept. Health, Education, and Welfare, 1966.

154. O. M. Shtern, *Energetik*, **12**, 18 (1964) (in Russian); through *Chem. Abstr.*, **62**, 10222f (1965).

155. S. Ostrowski, Z. Jasinska, and Z. Zolendziowska, *Farm. Polska*, **21**, 743 (1965) (in Polish); through *Chem. Abstr.*, **65**, 1959a (1966).

155a. C. Ungureanu, *Energetica (Bucharest)*, **12**, 182 (1964); through *Chem. Abstr.*, **61**, 11755g (1964).

156. J. M. Wright and W. T. Lindsay, Jr., *Proc. Am. Power Conf.*, **21**, 706 (1959).

157. *Standard Methods for the Examination of Water and Wastewater*, 12th ed., American Public Health Association, New York, 1965, p. 219.

158. A. F. Wartburg, A. W. Brewer, and J. P. Lodge, Jr., *Air Water Pollution*, **8**, 21 (1964); through *Chem. Abstr.*, **60**, 13017b (1964).

159. P. Koppe and A. Muhle, *Z. Anal. Chem.*, **210**, 241 (1965); through *Chem. Abstr.*, **63**, 1587e (1965).

160. H. Galster, *Z. Anal. Chem.*, **186**, 359 (1962); through *Chem. Abstr.*, **57**, 9597d (1962).
161. T. R. Hauser and D. W. Bradley, *Anal. Chem.*, **38**, 1529 (1966).
162. *Standard Methods for the Examination of Water and Wastewater*, 12th ed., American Public Health Association, New York, 1965, p. 230.
163. H. Barnes, *Apparatus and Methods of Oceanography, Part One: Chemical*, Wiley (Interscience), New York, 1959, p. 151.
164. J. D. H. Strickland and T. R. Parsons, *A Manual of Sea Water Analysis, Bull. No. 125*, Fisheries Research Board of Canada, Ottawa, 1960, pp. 41, 47.
165. D. Jenkins, *J. Water Pollution Control Federation*, **39**, 165 (1967).
166. P. G. W. Jones and C. P. Spencer, *J. Marine Biol. Assoc. U.K.*, **43**, 251 (1963); through *Chem. Abstr.*, **59**, 6127g (1963).
167. H. W. Harvey, *ibid.*, **27**, 337 (1948); through *Chem. Abstr.*, **42**, 7464e (1948).
168. J. Murphy and J. P. Riley, *Anal. Chim. Acta*, **27**, 31 (1962).
169. P. G. W. Jones, *J. Marine Biol. Assoc. U.K.*, **46**, 19 (1966); through *Chem. Abstr.*, **65**, 1959b (1966).
170. M. J. Fishman and M. W. Skougstad, *U.S. Geol. Surv. Profess. Paper No. 525–B*, 1965, p. 167; through *Chem. Abstr.*, **63**, 8026g (1965).
171. V. Breitling, *Mitt. Ver. Grosskesselbesitzer*, **89**, 109 (1964); through *Chem. Abstr.*, **61**, 6777c (1964).
172. D. C. Abbott and G. E. Emsden, *Proc. Soc. Water Treat. Exam.*, **12**, 230 (1963); through *Chem. Abstr.*, **60**, 10387a (1964).
173. D. C. Abbott, G. E. Emsden, and J. R. Harris, *Analyst*, **88**, 814 (1963).
174. G. Moeller, *Z. Pflanzenernaehr. Dueng. Bodenk.*, **106**, 20 (1964) (in German); through *Chem. Abstr.*, **62**, 1446g (1965).
175. G. Proft, *Limnologica*, **2**, 407 (1964) (in German); through *Chem. Abstr.*, **63**, 9654e (1965).
176. *Standard Methods for the Examination of Water and Wastewater*, 12th ed., American Public Health Association, New York, 1965, p. 238.
177. H. Barnes, *Apparatus and Methods of Oceanography, Part One: Chemical*, Wiley (Interscience), New York, 1959, p. 211.
178. M. I. Kriventsov, *Sovren. Metody Analiza Prirodn. Vod. Akad. Nauk SSSR*, **1962**, 20 (in Russian); through *Chem. Abstr.*, **59**, 11101a (1963).
179. C. Costache, *Igiena*, **15**, 233 (1966) (in Romanian); through *Chem. Abstr.*, **66**, 68775z (1967).
180. V. Prochazkova, *Prace Vyzkum. Ustavu Cs. Naftovych Dolv. Publ. 16*, 1962, p. 137; through *Chem. Abstr.*, **55**, 16861i (1961).
181. T. Murakami, *Bunseki Kagaku*, **14**, 880 (1965) (in Japanese); through *Chem. Abstr.*, **64**, 3204a (1966).
182. J. M. Skinner and A. C. Docherty, *Talanta*, **14**, 1393 (1967).
183. M. E. Efendi and V. B. Sokolovich, *Izv. Tomsk. Politekhn. Inst.*, **102**, 138 (1959); through *Chem. Abstr.*, **58**, 4301d (1963).
184. *Standard Methods for the Examination of Water and Wastewater*, 12th ed., American Public Health Association, New York, 1965, p. 250.
185. Y. K. Chau and J. P. Riley, *Anal. Chim. Acta*, **33**, 36 (1965).
186. S. D. Biswas and A. K. Dey, *Analyst*, **90**, 56 (1965).
187. H. Ariyoshi, M. Kiniwa, and K. Toei, *Talanta*, **5**, 112 (1960).
188. C. A. Parker and L. G. Harvey, *Analyst*, **87**, 558 (1962).
189. *Standard Methods for the Examination of Water and Wastewater*, 12th ed., American Public Health Association, New York, 1965, p. 258.
190. D. Jenkins, *J. Water Pollution Control Federation*, **39**, 161 (1967).

191. H. Barnes, *Apparatus and Methods of Oceanography, Part One: Chemical,* Wiley (Interscience), New York, 1959, p. 163.

192. J. D. H. Strickland and T. R. Parsons, *A Manual of Sea Water Analysis, Bull. No. 125,* Fisheries Research Board of Canada, Ottawa, 1960, p. 55.

193. I. R. Morrison and A. L. Wilson, *Analyst,* **88,** 88 (1963).

194. I. R. Morrison and A. L. Wilson, *Analyst,* **88,** 100 (1963).

195. I. R. Morrison and A. L. Wilson, *Analyst,* **88,** 446 (1963).

196. D. R. Schink, *Anal. Chem.,* **37,** 764 (1965).

197. K. Sono, H. Watanabe, and Y. Mitsukami, *Bunseki Kagaku,* **12,** 352 (1963); through *Chem. Abstr.,* **59,** 6128b (1963).

198. *Standard Methods for the Examination of Water and Wastewater,* 12th ed., American Public Health Association, New York, 1965, p. 273.

199. H. Barnes, *Apparatus and Methods of Oceanography, Part One: Chemical,* Wiley (Interscience), New York, 1959, p. 210.

200. I. A. Mamedov and A. A. Emirdzhanova, *Izv. Vysshikh Uchebn. Zavedenii Neft i Gas.,* **6,** 84 (1963); through *Chem. Abstr.,* **60,** 10387e (1964).

201. J. Dolezal, I. Novozamsky, and J. Zyka, *Collection Czech. Chem. Commun.,* **27,** 1830 (1962) (in German); through *Chem. Abstr.,* **57,** 13179e (1962).

202. A. A. Nogina and G. G. Kobyak, *Uch. Zap. Permsk. Gos. Univ.,* **111,** 122 (1964) (in Russian); through *Chem. Abstr.,* **64,** 9432f (1966).

202a. Corning Glass Works, Medford, Mass., *Tech. Lit. Catalog No. 476210.*

202b. Beckman Instruments, Inc., Fullerton, Calif., *Tech. Lit. Catalog Nos. 39278 and 39046.*

203. *Standard Methods for the Examination of Water and Wastewater,* 12th ed., American Public Health Association, New York, 1965, pp. 293, 426.

204. Z. Kurzawa, *Chem. Anal. (Warsaw),* **5,** 555 (1960) (English summary); *Ref. Zh. Khim.,* **1965,** Abstr. No. 14G149; through *Chem. Abstr.,* **55,** 6739g (1961).

205. Orion Research, Inc., Cambridge, Mass., *Tech. Lit. Model 94-16.*

206. *Standard Methods for the Examination of Water and Wastewater,* 12th ed., American Public Health Association, New York, 1965, p. 294.

207. *Standard Methods for the Examination of Water and Wastewater,* 12th ed., American Public Health Association, New York, 1965, p. 287.

208. H. Barnes, *Apparatus and Methods of Oceanography, Part One: Chemical,* (Interscience), New York, 1959, p. 219.

209. W. Kleber and G. Franke, *Brauwelt,* **106,** 79 (1966) (in German); through *Chem. Abstr.,* **64** 13911f (1966).

210. V. A. Koval'tsov and G. S. Konovalov, *Gidrokhim. Materialy,* **37,** 118 (1964) (in Russian); through *Chem. Abstr.,* **62,** 11529b (1965).

211. G. N. Nechiporenko, *Gidrokhim. Materialy,* **33,** 143 (1961); through *Chem. Abstr.,* **57,** 10946e (1962).

212. G. N. Nechiporenko, *Gidrokhim. Materialy,* **33,** 151 (1961); through *Chem. Abstr.,* **57,** 10946f (1962).

213. A. A. Matveev and G. N. Nechiporenko, *Gidrokhim. Materialy,* **33,** 134 (1961); through *Chem. Abstr.,* **57,** 10946g (1962).

214. G. N. Nechiporenko, *Sovrem. Metody Analiza Prirodn. Vod. Akad. Nauk SSSR,* **1962,** 32; through *Chem. Abstr.,* **59,** 9675f (1963).

215. E. Pungor, J. Havas, and K. Toth, *Instr. Control Systems,* **38,** 105 (1965); through *Chem. Abstr.,* **63,** 18624a (1965).

216. *Trace Characterization, Chemical and Physical, Natl. Bur. Std. Monograph 100* (W. W. Meinke and B. F. Scribner, eds.), U.S. Govt. Printing Office, 1967.

217. G. H. Morrison, *Trace Analysis: Physical Methods*, Wiley (Interscience), New York, 1965, 582 pp.

218. J. Rodier, *L'Analyse Chimique et Physico-Chimique de l'Eau*, 3rd ed., Dunod, Paris, 1966.

219. *Official Methods of Analysis of the Association of Official Agricultural Chemists*, 10th ed. (W. Horwitz, Chmn., P. Chichilo, P. A. Clifford, and H. Reynolds, eds.), Association of Official Agricultural Chemists, Washington, D.C., 1965.

219a. K. Kodama, *Methods of Quantitative Inorganic Analysis*, Wiley (Interscience), New York, 1963.

220. *Treatise on Analytical Chemistry* (I. M. Kolthoff, P. J. Elving, and E. B. Sandell, eds.), Wiley (Interscience) New York, sequential volumes.

221. *Advances in Analytical Chemistry and Instrumentation* (C. N. Reilley and F. W. McLafferty, eds.), Wiley (Interscience), New York, sequential volumes.

222. F. J. Welcher, *Standard Methods of Chemical Analysis*, 6th ed., Van Nostrand, Princeton, N. J., sequential volumes.

223. *Encyclopedia of Industrial Chemical Analysis* (F. D. Snell and C. L. Hilton, eds.), Wiley (Interscience), New York, sequential volumes.

224. J. W. Robinson, *Atomic Absorption Spectroscopy*, Dekker, New York, 1966.

225. J. T. Stock, *Amperometric Titrations*, Wiley (Interscience), New York, 1965.

226. G. Schwarzenbach and H. Flaschka, *Die Komplexometrische Titrations*, Ferdinand Enke Verlag, Stuttgart, 1965.

227. *Chelates in Analytical Chemistry* (H. A. Flaschka and A. J. Bernard, Jr., eds.), Vols. 1–4, Dekker, New York.

228. R. Mavrodineanu and H. Boiteux, *Flame Spectroscopy*, Wiley, New York, 1965.

229. R. Mavrodineanu, *Bibliography on Flame Spectroscopy, Analytical Applications, 1800–1966, Natl. Bur. Std. Misc. Publ. 281*, Washington, D.C., 1967.

230. G. G. Guilbault, *Flourescence: Theory, Instrumentation, and Practice*, Dekker, New York, 1967.

231. *Flourescence and Phosphorescence Analysis: Principles and Applications* (D. M. Hercules, ed.), Wiley (Interscience), New York, 1966.

232. *The Practice of Gas Chromatography* (L. S. Ettre and A. Zlatkis, eds.), Wiley, New York, 1967.

233. *Laboratory Handbook of Chromatographic Methods* (O. Mikes, ed.,), Van Nostrand, Princeton, N. J., 1967.

233a. G. A. Rechnitz, *Chem. Eng. News*, **45** (25), 146 (1967).

234. J. Inczedy, *Analytical Applications of Ion Exchangers*, Pergamon, Long Island City, N. Y., 1966.

235. *Ion Exchange* (J. A. Marinsky, ed.), Vol. 1, Dekker, New York, 1966.

236. O. Landstrom and C. G. Wenner, *Aktiebolaget Atomenergi, Stockholm AE-204*, 1965 (in English); through *Chem. Abstr.* **66**, 12361g (1966).

237. G. W. Leddicotte and D. W. Moeller, *U.S.Atomic Energy Comm. CF-61-5-118*, 1961.

238. A. Berka, J. Vulterin, and J. J. Zyka, *Newer Redox Titrants*, Pergamon, Long Island City, N.Y., 1965.

239. P. Perlman, *Reagents and Solutions in Analytical Chemistry*, Franklin, Englewood, N.J., 1966.

240. W. E. Bush and L. J. Higgins, *U.S. Atomic Energy Comm. RMO-3001 (Rev. 1) (Suppl.)*, 1965.

241. F. H. Rainwater and L. L. Thatcher, *U.S. Geol. Surv. Water Supply Paper 1454*, 1960.

242. A. G. Collins, C. Pearson, D. H. Attaway, and J. W. Watkins, *U.S. Bur. Mines, Rept. Invest. 5819*, 1961.

243. *Manual on Industrial Water and Industrial Waste Water*, Tech. Publ. No. 148-I, 2nd ed., American Society for Testing and Materials, Philadelphia, 1966.

244. H. Goya, P. Zigman, M. Lai, and J. Mackin, *U.S. Dept. Commerce, Office Tech. Serv. AD 404, 884*, 1963.

245. M. Zimmerman, *Photometrische Metall- und Wasseranalysen*, 2 Auflage, Wissenschaftliche Verlagsgesellschaft, Stuttgart, 1961.

246. F. J. H. Mackereth, *Freshwater Biol. Assoc. Sci. Publ. 21*, 1963.

Chapter 22 Residual Chlorine Analysis in Water and Waste Water

Henry C. Marks
PENNWALT CORPORATION
KING OF PRUSSIA, PENNSYLVANIA

I. Introduction

The term residual chlorine is commonly understood to mean the sum of the concentrations of compounds containing active chlorine which remain at any given time after the addition of gaseous chlorine, chlorine solution, or a hypochlorite to water or waste water. Molecular chlorine will not be present since hydrolysis to hypochlorous acid is rapid and complete under usual conditions in water or waste water(1, 2). Hypochlorous acid, monochloramine, dichloramine, and nitrogen trichloride may all be found in chlorinated water, while monochloramine, dichloramine, and organic chloramines are the most common species in waste waters. Residual chlorine in the form of hypochlorous acid is termed free residual chlorine, while that existing as chloramines is known as combined residual chlorine.

Even in relatively unpolluted water some ammonia is usually present. If the residual chlorine is not in excess with respect to the ammonia, hypochlorous acid will exist only briefly, since the equilibria shown in Eqs. (1) and (2) are rapidly displaced far to the right in the pH range (6.0–9.0) usually encountered in chlorinated water(3).

$$HOCl + NH_3 \rightleftharpoons NH_2Cl + H_2O \qquad (1)$$

$$NH_2Cl + HOCl \rightleftharpoons NHCl_2 + H_2O \qquad (2)$$

If the ratio of chlorine to ammonia is sufficiently high, hypochlorous acid and one or more of the chloramines may be present for some time after chlorination. Table 1 illustrates this with data obtained after adding chlorine to buffered distilled water containing ammonia(4). This system will exhibit the so-called break point reaction (see Fig. 10 of Chap. 10) in which the chlorine gradually oxidizes nitrogenous compounds to a mixture of gaseous nitrogen and various nitrogen oxides so that ultimately either hypochlorous acid or chloramine remains(3). It is seen that some oxidation has already taken place in 5 min. The relative concentrations of the three chloramines will in any case be determined mainly by pH(3,5).

TABLE 1

Effect of Chlorine Dose on Concentrations of Residual Chlorine Compounds[a]

Chlorine dose, ppm	Residual chlorine, ppm, as	
	HOCl	Chloramines
2	0.0	1.7
3	0.2	2.0
4	0.6	1.6
5	1.2	1.2
6	1.8	0.8

[a]0.5 ppm NH_3, pH 7.9, 32°F, measurement 5 min after chlorine addition. Based on data in Ref. 4, p. 1235.

In the pH range of 7–8 monochloramine is the predominant species. Nitrogen trichloride is found only at low pH or when a large excess of hypochlorous acid is present, and then it represents only a small fraction of the total.

Most waste waters contain sufficient concentrations of nitrogen compounds to combine all of the residual chlorine as chloramines unless the pH is so high or so low that the equilibria shown in Eqs. (1) and (2) are displaced to the left by ionization of one of the reactants. It is safe to assume that the principle species are monochloramine, dichloramine, and organic chloramine. Since the identities and characteristics of the organic amines are practically never known, no a priori conclusions can be drawn about the distribution of the residual chlorine among the various forms. As discussed later, the analytical methods currently available do not permit this distinction to be made.

One of the main problems in residual chlorine determination arises from the number of different active chlorine compounds that may be present and the widely different degrees of chemical and biological activity that they exhibit. Hypochlorous acid is appreciably ionized at pH values of 7 and above(6), and hypochlorite ion contributes little disinfecting activity (3). Yet even under these circumstances it is far more active than the ammonia chloramines in the same pH range(7,8). While the many known organic chloramines present a very wide range of disinfecting activity, most of those usually encountered are substantially less active than the ammonia chloramines(9). It is evident that a determination of total residual chlorine may have limited utility and that some knowledge of the form of the residual chlorine is essential in most cases.

Another problem stems from the fact that the concentration or the nature of the residual chlorine may be altered by the very analytical procedure used to characterize it. Some procedures require acidification of the sample, which favors reactions of the type shown in Eq. (3):

$$NH_2Cl + H_3O^+ \longrightarrow NH_4^+ + HOCl \tag{3}$$

This places an inherent limit on the precision of such a method when it is used to distinguish free and combined residual chlorine, as is well illustrated by the orthotolidine–arsenite method(10).

Another consequence of acidification is actual loss of residual chlorine through consumption of hypochlorous acid by substances that would not react with the less active chloramines(11).

In iodometric procedures the liberated iodine is consumed by substances in wastes not previously oxidized by chloramines, so that low results are obtained unless the pH is low enough(12). Consequently, iodometric procedures that are useful for distinguishing the various species of chlorine compounds in relatively unpolluted water are not applicable to waste water.

All of the methods that have been proposed for residual chlorine determination are based on its properties as an oxidizing agent, with just one exception. This is based on the reaction with cyanide ion to form cyanogen chloride which in turn produces a colored complex with a suitable amine (13). This method, then, is the most nearly specific of all that have been proposed (see Sec. VIII), although it has not been sufficiently developed to be useful in practice.

The oxidizing action of residual chlorine has been used in analytical procedures in the following ways: (1) to convert a redox indicator to the colored oxidized form, (2) to bleach a colored compound as a result of oxidation, (3) to react in an oxidation-reduction titration, (4) to depolarize a cathode for endpoint detection in mechanism 3, and (5) to liberate iodine, which in turn acts in accordance with mechanism 1, 3, or 4.

The important examples(10) currently in use are: (a) the orthotolidine method based on mechanism 1 (see Sec. II), (b) iodometric titration using sodium thiosulfate based on mechanisms 3 and 5 (see Sec. XI), (c) amperometric titrations based on mechanisms 3, 4, and 5 (see Sec. XII), and (d) titration with ferrous ammonium sulfate based on mechanisms 1, 3, and 5 (see Secs. II and V).

The several forms of residual chlorine vary in the facility with which they oxidize a given substrate, and this is the basis of all the procedures for distinguishing one from the other. In the standard orthotolidine–arsenite method(10) the free residual chlorine reacts with the indicator in a matter of seconds, while the combined residual chlorine is much slower. Thus the latter can be removed with arsenite after the free chlorine has all reacted. The differences in reactivities among the various forms of combined chlorine are not sufficiently sharp in this case to permit distinguishing among them.

In the amperometric and the ferrous ammonium sulfate titrations(10) the titrating reagents react only with free chlorine at pH 7. If iodide is added the combined chlorine will also react; the monochloramine reacts at pH 7 and both monochloramine and dichloramine react at lower pH values. These methods, then, permit a more complete separation, provided organic chloramines are not present. (See Sec. XII for a discussion of this complication.)

As would be expected, other oxidizing agents consitute the most common cause of interference in determination of residual chlorine. Ferric ion and manganese in its higher oxidation states are the important ones, although occasionally dichromate is present. Nitrite ion may act either as an oxidizing agent in some iodimetric procedures or as a diazotizing agent in colorimetric methods. It is not often a problem. There are two general methods of dealing with the interference problem. In the one, exemplified by the orthotolidine and the ferrous ammonium sulfate titration methods, the concentrations of interfering substances are determined separately on a sample from which the residual chlorine has been removed with a reducing agent such as arsenite(10). In the second, exemplified by the amperometric titration, the procedure can be carried out in a sufficiently high pH range that the interfering substances are unreactive(10).

While relatively few methods for determining residual chlorine have come into general use, many have been proposed. The following sections are not restricted to those in use but also include those that have not been fully developed because of newness or plain lack of attention as well as those that have shown little promise.

Chlorine dioxide, which is sometimes used in water treatment, is not properly included in the term residual chlorine. However, it is either partially or wholly recovered by many of the residual chlorine methods and is sometimes expressed numerically in terms of residual chlorine(13a). Procedures that have been devised to determine it or to distinguish it from the various forms of residual chlorine are included in the appropriate sections of this chapter.

II. Methods Based on Orthotolidine

A. Background

The orthotolidine method, first developed by Ellms and Hauser[14], has been and continues to be the most extensively used method of residual chlorine determination. The primary method is based on the oxidation of the colorless benzidine structure (1) to the highly yellow-colored holoquinone structure (2) at low pH[15,16]. More recent modifications[17]

(1)   (2)

have made use of the oxidation at slightly higher pH to produce the blue meriquinone, the equimolecular complex of orthotolidine, and its holoquinone[16]. If insufficient orthotolidine is added, an unstable red chloroholoquinone is formed[18].

Certain precautions are recommended to anyone handling quantities of orthotolidine[18a].

B. Acid Orthotolidine Methods

1. DETERMINATION OF TOTAL RESIDUAL CHLORINE

a. Procedure

This commonly used procedure involves addition of the sample to a solution of the indicator in dilute hydrochloric acid and visual comparison of the color with an artificial standard[10]. It is based on spectrophotometric studies[19] that defined the conditions for quantitative production of the yellow holoquinone. Unless indicator concentration and acidity are properly controlled, mixtures of oxidation products are formed, resulting in unstable colors and lack of reproducibility[20–22]. Under such circumstances Beer's law will not apply[18,23] and artificial standards will not be a spectral match for the color being measured[18, 20]. The color resulting from oxidation to pure holoquinone conforms quite closely to Beer's law[19], facilitating comparison with permanent standards or use of a photometric method based on a calibration curve [10,24]. When combined chlorine is present, sufficient time must be allowed for full color development. The optimum time period depends on temperature, but in any case the color should be read within 5 min(*10*).

b. Interference

Manganese in its higher oxidation states produces quantitatively the same color intensity as an equivalent concentration of residual chlorine (*20,25*). The reaction with ferric ion is not quantitative, but more than 0.3 ppm may interfere under certain circumstances(*10,26*). Interference by nitrite ion is believed to be due to the production of a diazo compound rather than to oxidation(*16*). Since this reaction with orthotolidine is slow, and light is necessary, this interference is usually not important. It is of no concern when free residual chlorine is present(*27*).

Attempts to avoid manganese interference by removing it without affecting residual chlorine have been unsuccessful(*20,28*). Gilcreas and Hallinan(*29,30*) showed that the residual chlorine could be reduced with sodium arsenite without affecting the interfering compounds. The latter can then be determined by the orthotolidine method and used to correct the reading obtained by the normal procedure(*10*).

c. Application to Waste Water

In waste water the residual chlorine is usually all combined and very often includes both ammonia chloramines and organic chloramines. Many of the latter are relatively unreactive with orthotolidine. In an effort to make the recovery as complete as possible an increased amount of indicator solution is used for wastes(*30a,31*). However, recovery is usually very poor, not only because much of the residual chlorine is unreactive, but also because the hypochlorous acid produced by acidification (see Sec. I) is preferentially consumed by organic acceptors in the waste(*11*).

This was demonstrated by determining residual chlorine on chlorinated secondary effluent and also on chlorinated raw sewage by the orthotolidine method, by an iodometric method that had been shown to minimize both types of loss(*11*), and by the same iodometric method on samples which had been held at pH 1.5 and 30 sec before analyzing. Table 2 shows some of the residual chlorine values obtained at various time periods after treatment with chlorine. Comparison of columns 3 and 4 shows that in raw sewage most of the residual chlorine is lost through acidification, while in secondary effluent, which contains less oxidizable material, only part of the loss is attributable to this cause. By comparison of columns 2 and 4 it can be concluded that in this secondary effluent the combined chlorine includes compounds unreactive to orthotolidine.

Poor recoveries such as these mean that often correlation between

TABLE 2
Sources of Error in Residual Chlorine
Determination[a]

Contact time, min	Residual chlorine, ppm		
	o-Tolidine	Iodometric	After acidification
	Raw sewage		
15	0.1	4.8	0.6
30	0.0+	3.9	0.3
60	0.0	2.6	0.0
120	0.0	0.8	0.0
	Secondary effluent		
15	0.6	4.1	2.1
30	0.4	2.6	0.7
60	0.2	1.9	0.4
120	0.05	0.9	0.0

[a]Based on data in Ref. 11, p. 1199.

orthotolidine reading and disinfection is poor(12,32–34) and some disinfection can take place even when the orthotolidine method fails to show any residual chlorine(31,35,36). But in most cases where the dosage of chlorine is sufficient to bring about the desired degree of disinfection, some residual chlorine is recoverable with orthotolidine(31,33,36). Consequently, the method continues to be of great practical value in controlling chlorination of sewage as long as an understanding of the chemistry is brought to bear on the interpretation of the results. (For further discussion of this problem see Secs. XI.B and XII.C.)

2. ORTHOTOLIDINE-ARSENITE MODIFICATION FOR FREE AND COMBINED CHLORINE

As explained in Sec. I, the lower reactivity of combined chlorine provides the basis for distinguishing it from the free residual chlorine. An attempt has been made to utilize the difference in rates of formation of the red chlorinated holoquinone(37). However, a modification of the standard orthotolidine procedure gives the most reliable results. The color develops in 5 sec after mixing the sample and the indicator can be taken as the free chlorine to a good approximation(10). Gilcreas and Hallinan(29,30) showed that addition of sodium arsenite reduces any chlorine that has not reacted with the orthotolidine without affecting the color already produced. This provides sufficient time for careful reading of the free chlorine. Total residual chlorine is determined on a separate sample, as indicated in Sec. II.B.1. The combined residual chlorine is the difference between the total and the free residual chlorine.

A simpler modification that yields semiquantitative results consists simply of estimating the reading at the end of the 5 sec without addition of arsenite (*10,38*).

In either case the precision depends on the ratio of chloramine to free chlorine, since the method is based on a difference in reaction velocities. Chilling the sample to 1°C (*10*) reduces but does not eliminate the error.

The substances that interfere with this method are those that are operative in the determination of total residual chlorine, and a suitable correction is made by reducing the chlorine with arsenite and then determining the amount of interference. The interfering color must be read both after 5 sec and after 5 min and the corrections applied to the free chlorine and total chlorine readings, respectively (*10*).

C. Neutral Orthotolidine Methods

1. COLORIMETRIC PROCEDURE

As indicated in Sec. II.A, orthotolidine is oxidized to the blue meriquinone when the pH is only slightly below the neutral point. Under these conditions combined chlorine is unreactive (*39*) so that the basis for selectively determining free residual chlorine is presented.

However, the color thus produced is so unstable that it was of no practical value until Palin (*17*) discovered that sodium hexametaphosphate stabilizes the system. At the same time he found that in the presence of iodide both free chlorine and monochloramine are recovered. He therefore suggested a method that uses the neutral orthotolidine reagent without potassium iodide to give free chlorine, the same reagent with potassium iodide to give the sum of the free chlorine and the monochloramine, and the usual orthotolidine procedure to give the total residual chlorine. Standards are made by adding the necessary amounts of standard iodine solution to water containing the neutral orthotolidine reagent.

This method has not been found generally useful. Probably the main disadvantages have been lack of sensitivity, necessity of preparing fresh standards for every determination, and a rather involved procedure.

2. Ferrous Ammonium Sulfate Titration

The selective nature of the oxidation to the meriquinone has been used more advantageously in a titration procedure(10). Palin(17) found that in the presence of sodium hexametaphosphate the stabilized meriquinone could be quantitatively titrated with ferrous ammonium sulfate in the pH range of 6–7. After titration to disappearance of the initial blue color to give the free chlorine, iodide is added and the additional blue color is titrated as a measure of the monochloramine. It was found that acidification of the sample at this point followed by neutralization to the original pH gives a further increment which can be titrated to measure the dichloramine.

Nitrogen trichloride appears partly in the free chlorine fraction and at times partly in the dichloramine fraction. An extraction procedure can be used to determine nitrogen trichloride and also to prevent its interference with determination of the other components(17).

Oxidized manganese interferes and has to be determined separately after removal of the residual chlorine with arsenite in the presence of iodide. Fortunately, the hexametaphosphate prevents interference by the ferric ion produced in the titration.

To obtain reproducible results very strict attention must be paid to pH, temperature, amount of indicator, order of reagent addition, and composition of reagent solutions(40,41). While it is now a Tentative Standard Method(10) for distinguishing free chlorine, monochloramine, dichloramine, and nitrogen trichloride in water, it is not clear how generally useful it will be, and efforts are being made to replace the orthotolidine with other indicators (see Secs. IV and V). Because it depends on liberation of iodine, it is not applicable to wastes which contain iodine acceptors.

D. Chlorine Dioxide Determination

Chlorine dioxide reacts with the usual acid orthotolidine indicator to give the same colored compound as that given by the various forms of residual chlorine. However, only one of the five oxidizing equivalents per mole reacts immediately. Thus, chlorine dioxide will interfere with the determination of free chlorine residual, each part by weight of chlorine dioxide producing a color corresponding to approximately a half part by weight of free residual chlorine(13a,42). Chlorite ion, which is produced in this reaction as well as in other reactions in the water, reacts slowly and incompletely with the orthotolidine(13a). Whenever chlorine dioxide is used, chlorine is also applied, and the need for the proper analytical methods is apparent.

An attempt has been made(43) to treat the sample with excess of an ammonium salt to convert free chlorine to chloramine and then to take any free chlorine reading as chlorine dioxide. This does not give consistent results(42) because the high proportion of combined chlorine produces too great an error (see Sec. II.B.2).

Aston(42) has proposed a further modification of the orthotolidine–arsenite test based on the use of oxalic acid to selectively remove free chlorine residual. After allowing 10 min for this reaction, the regular orthotolidine-arsenite test as described in Sec. II.B.2 is run. The "free chlorine fraction" measures the sum of the chlorine dioxide and the interfering substances such as manganese. Since the interfering substances can be determined separately by addition of arsenite before orthotolidine, chlorine dioxide concentration is the difference between the two readings. By applying the standard orthotolidine-arsenite method without previous addition of oxalic acid, the sum of the free chlorine, the chlorine dioxide, and the interfering substance is found in the "free chlorine fraction." The standard orthotolidine test, itself, yields a value that includes everything. Individual species are then calculated by appropriate subtractions.

Although this method has been accepted as standard(13a) and is the best practical method available, it leaves much to be desired. Lack of precision is inherent when concentrations of individual species are represented by differences between readings, each of which is subject to considerable error. Indeed, extreme care must be exercised to achieve reasonable reproducibility, and it is highly recommended that the orthotolidine colors be measured photometrically rather than visually(13a). There seems to be some doubt that chlorite is completely without effect on the color produced(13a), and this may be the reason for reports of unreliability(44). A further complication may arise through precipitation of calcium oxalate when the method is applied to hard water(13a,44). It has been suggested that this could be avoided by substituting malonic acid for oxalic(44), and in the Wallace and Tiernan laboratories this has shown some promise(45). Much more work is needed to determine its usefulness.

III. Benzidine Derivatives Other Than Orthotolidine

The considerable number of positions that can be substituted in the benzidine molecule provides ample opportunity to try to improve on orthotolidine. It is not unreasonable to suppose that the proper substitution would change reactivity so that chloramines could not produce colored quinoid structures, thus providing sharper distinction between free and combined residual chlorine. Another objective might be to shift into the neutral range the pH at which highly colored, stable quinoid compounds are formed in order to avoid interference and to minimize alterations in the nature of the residual chlorine by the indicator itself.

In fact, a considerable number of benzidine derivatives substituted in the positions indicated in structure (3) have been studied, and one or two

(3)

have been proposed as indicators for residual chlorine. Unfortunately, none has presented advantages in any given application to provide sufficient incentive for further development.

The parent compound was shown at an early date to give a blue color at pH 4–5 and a green to yellow color at lower pH (46–48). These colors, which are probably a meriquinone and a holoquinone, respectively, are both relatively unstable. This together with poor sensitivity (49) were reasons why little further interest was shown in this compound.

Orthodianisidine [structure (3), in which E and F are methoxy and all others are hydrogen] had been shown to function as an indicator in oxidation-reduction titrations and also had been suggested for determining residual chlorine (50). McEwen and Grant (51), looking for a chlorine indicator that would not be oxidized by ferric ion, investigated this compound. The variations due to age of the indicator solution and the pH of the sample, as well as uncertainty as to which forms of residual chlorine are indicated by the method, tended to discourage further development.

Although the compound in which A, B, C, and D are methyl and the others hydrogen was shown in 1954 (49) to be among the most sensitive of the benzidine compounds, and although the claim was made that it was more accurate than the orthotolidine method (52), Nicolson (53) concluded that reproducibility was not satisfactory and in other respects this compound presented no advantages over orthotolidine.

Cursory testing on residual chlorine has been carried out on additional alkyl and aryl derivatives in which [refer to structure (3)] E is methyl; A is methyl; A and C are methyl; E and F are ethyl; A and C are ethyl; and A is phenyl. In each case the other positions are unsubstituted (16,49). None showed any advantage over indicators already in use. They were found to be oxidized to colored compounds by residual chlorine, and on the basis of the superficial examination given them, the possibility of a useful method cannot be ruled out.

On the other hand, the compounds in structure (3) where E and F are chlorine or where either E and F or G were sulfonic acid groups or nitro groups showed such little promise that further effort in this direction is probably not justified (*16*).

The compound 3,3'-dimethyl-4,4'-diamino-1,1'-dinaphthyl or 3,3'-dimethylnaphthidine (4) was suggested by Belcher et al. (*49*).

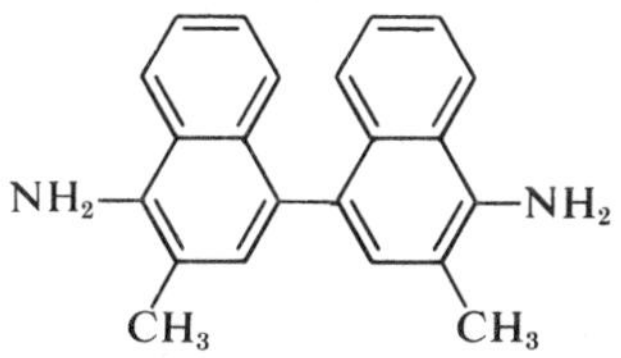

(4)

because it yields an especially stable purple-red meriquinone, is more sensitive than other benzidines, and obeys Beer's law. Nicolson (*53*) confirmed this and judged it one of the most applicable indicators for field work, although oxidized manganese interferes. Apparently there is also a problem with the quantitative recovery of combined chlorine. This is not surprising since a glacial acetic acid solution of the indicator was used with no attempt to control the pH at which the color was formed. From the data available it would seem that this indicator has sufficient promise to make it worthwhile to study the effects of pH, iodide, and other factors on the recoveries of the various forms of residual chlorine.

IV. Methods Based on *p*-Aminodimethylaniline

A. Reactions

This indicator, shown in structure (5) and first proposed by Kolthoff (*54*) and by Alfthan (*47*), is oxidized by residual chlorine to a red or purple meriquinone (*55*). This highly colored and fairly stable oxidation product is produced over a wide pH range (*55*) by both free and combined chlorine. The latter reacts much more slowly than free chlorine, and the higher

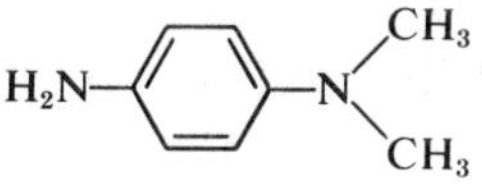

(5)

the pH the greater the difference. At pH 6.8 combined chlorine is practically without effect(56).

Below pH 6, and to an increasingly great extent as the pH is decreased, iron and manganese interfere(16,47,57,58). As the pH is increased above 6, interference from dissolved oxygen increases(58).

The indicator solution, itself, exhibits instability in all otherwise suitable solvents(53). This and the problems of interference have prevented general use of the indicator even though the following procedures have been developed for its use.

B. Procedures

1. Total residual chlorine is determined using a phosphoric acid-phosphate buffer to give a pH of about 3 and allowing 5 min for color development(55,57). No provision is made for correcting for interferences.

2. Free chlorine is determined by adding the indicator at pH 6 and reading the color as quickly as possible. The combined chlorine is then recovered by lowering the pH to 4 and allowing time for reaction(58,59). The method is only semiquantitative because some combined chlorine reacts at pH 6.

3. Free chlorine is determined by adding the indicator at pH 6.8 and reading the color. Then iodide is added at the same pH and the additional color is read as combined chlorine(56). The only provision for minimizing dissolved oxygen interference is to carry out the test as quickly as possible.

4. Palin(17) showed that this indicator could be used instead of orthotolidine in the ferrous ammonium sulfate titration for free chlorine, monochloramine, and dichloramine. The homologous compound, p-aminodiethylaniline, is preferred, however (see Sec. V).

V. Methods Based on p-Aminodiethylaniline

A. Reactions and Properties

This compound, the diethyl homolog of structure (5), is oxidized in the same way by residual chlorine. It is subject to the same interferences and is oxidized by free and combined residual chlorine at different rates. However, it is less reactive than the methyl derivative to combined chlorine and permits a sharper differentiation(60).

The same problem of instability of the indicator solution is encountered (53,60). Some improvement has been achieved by using the oxalate salt of the indicator(60), but this use is limited to procedures not requiring low pH where the oxalate reacts with the residual chlorine. It has therefore been recommended that it be used in the form of a powder or tablet(53, 60), in which forms it is said to be commercially available(61). Difficulty in dissolving the tablets has been reported(53).

B. Procedures

1. DETERMINATION OF FREE CHLORINE—COLORIMETRIC

In a simplified procedure for free residual chlorine alone, a solution of the oxalate salt of the indicator is used at pH 6.6 and the color is measured after 2 min(62). Interference from dissolved oxygen or combined chlorine is negligible. Permanganate interferes but is not ordinarily encountered. Manganese in intermediate oxidation states commonly found in chlorinated water has not been tested but probably would not interfere because of the pH. Other common interfering substances are without effect.

2. DETERMINATION OF FREE CHLORINE AND CHLORAMINES—COLORIMETRIC

Palin(60) showed that free and combined chlorine can both be determined by the procedure developed for the dimethyl homolog (see Sec. IV.B.3). The indicator was used in the form of the oxalate in a solution which also contained ethylenediamine tetraacetic acid to increase stability. This sequestering agent was also added to the phosphate buffer to minimize interference from metals and dissolved oxygen. Powdered mixtures or tablets containing the reagents can also be used(53,60).

It was also shown(60) that, following the free chlorine reading, addition of only 2–5 mg of potassium iodide per 100 ml of sample permits determination of monochloramine. After addition of 1 g of potassium iodide, the remainder of the combined chlorine including dichloramine is recovered. Very careful control of conditions is required to prevent combined chlorine from appearing with the free chlorine and dichloramine from appearing with the monochloramine(53,60). Nitrogen trichloride is recovered in the dichloramine fraction, and a method has been devised for estimating it separately(60).

A procedure was also worked out to correct for manganese in its higher oxidation states if this is found necessary(60). In the normal pH range of this procedure, manganese as it usually occurs in water should not interfere.

3. FERROUS AMMONIUM SULFATE TITRATION

The reagent solutions and the sequence of iodide additions given in Sec. V.B.2 can be used in a titration procedure(60). Instead of measuring each increment of color, a standard solution of ferrous ammonium sulfate is used to titrate the solution to a colorless end point, as in Sec. II.C.2. Again the three fractions represent free chlorine, monochloramine, and dichloramine, with the nitrogen trichloride appearing in the dichloramine fraction. When monochloramine is absent, the nitrogen trichloride can be estimated by taking the reading obtained after adding the small amount of iodide as the sum of the free chlorine and half the nitrogen trichloride(63).

The problem of monochloramine appearing in the free chlorine fraction and dichloramine appearing in the monochloramine fraction is even more acute than in the colorimetric procedure, apparently because of the time consumed in titration(53,60). For the same reason, dissolved oxygen can interfere slightly in this procedure, and some fading of the color can also occur. The manganese problem is the same in the two cases.

C. Applications

It is evident that this indicator should be useful as a very simple method for determining free residual chlorine only, or as a complete method for distinguishing between various forms of combined chlorine in water. It seems to have aroused most interest in England(53), apparently because of reproducibility and perhaps because of some concern about the toxicity of orthotolidine(18a). There is little evidence of any great use in the United States. The number of reagents and the critical nature of some of the conditions may be deterrents. More experience and more thorough study of some of the steps are needed.

A theoretical advantage over the acid orthotolidine methods lies in the fact that drastic pH change of the sample is not necessary, thus minimizing perturbations by the analytical method. However, this advantage is most significant in waste waters where procedures involving liberation of iodine at pH near neutrality are not applicable because of side reactions.

D. Chlorine Dioxide Determination

Palin(64) proposed a modification to be used if chlorine dioxide is present along with free and combined residual chlorine. One sample is acidified with sulfuric acid and iodide is added. Then, after addition of indicator and buffer to raise the pH, the color is titrated with ferrous ammonium sulfate. The method given in B.3 of this section is used on another sample and the concentrations of the several species are calculated by appropriate subtraction. Results are therefore based on differences between observed values subject to considerable error.

VI. The Crystal Violet Method

A. Basis and Characteristics

This method depends on the oxidation of 4,4',4"-methylidyne tris(N,N-dimethylaniline), the leuko form of crystal violet as shown in structure (6) (65,66). With iodine, for which the method was originally developed (65),

(6)

the oxidation reaction is stoichiometric. Residual chlorine produces a slightly different color that corresponds in intensity to less than stoichiometric oxidation (66). This has been interpreted on the basis of some chlorine substitution accompanying the oxidation. The result is that the ratio of indicator to chlorine is critical and that the color does not completely follow Beer's law.

Combined chlorine reacts more slowly than either free chlorine or iodine if mercuric chloride is present. This provides a basis for differentiation (66,67).

The indicator solution is stable indefinitely if protected from light.

B. Procedure (66,67)

For the determination of free residual chlorine the sample is buffered to pH 3.8–4.1 and the indicator solution is added in a prescribed manner to minimize side reactions. The indicator solution contains the dye precursor dissolved in dilute perchloric acid containing mercuric chloride. The color develops immediately and absorbance is measured at 592 mμ.

In the determination of total residual chlorine the buffered sample is first treated with potassium iodide and, after allowing time for reaction, the same procedure is followed as for free chlorine. For the reasons given in Sec. VI.A, it is necessary to construct separate absorbance curves.

C. Interferences and Reliability (66,67)

Combined chlorine will slowly produce color even in the absence of iodide. But if the free chlorine is read within 3 min, the error is no greater than 0.02 ppm, even in the presence of high concentrations of monochloramine. From measurements on samples that had been chlorinated beyond the break point (see Sec. I) it was concluded that the concentrations of dichloramine and nitrogen trichloride thus produced do not interfere with the free chlorine determination. By comparison with the orthotolidine and amperometric titration methods, the reliability of the method was demonstrated for monochloramine as well as for free chlorine. Adequate data for dichloramine and nitrogen trichloride recoveries are not given, but from Palin's data (60) it would be expected that the former would be quantitatively recovered and the latter partially recovered.

Ferric ion, nitrate, alkalinity, total hardness, or up to 600 ppm chlorides do not interfere, and nitrate interference is negligible. Oxidized manganese does interfere, and no means of avoiding it has been presented. It should be possible to devise a procedure using arsenite to determine the manganese correction to be applied.

The applicability of the method has been demonstrated only for water, and its suitability for wastes is doubtful because of possible reaction between the liberated iodine and the organics in the waste. It should have considerable utility as applied to water, because it affords differentiation between free and combined residual chlorine with less pH change than orthotolidine requires, and the stability of the indicator solution is better than that of p-aminodiethylaniline. There is need for greater study and experience with this method.

VII. Methyl Orange Method

The determination of chlorine by means of its bleaching action on methyl orange and methyl red was practiced in Germany many years ago (68). Apparently, methyl orange afforded greater sensitivity and was given more attention (69). Because chloramine does not bleach the dye, methyl orange was studied as an indicator to detect free residual chlorine specifically(28,70). Taras(71) worked out the details of a colorimetric method to determine free residual chlorine quantitatively, hydrochloric acid being used to adjust the sample to pH 3 where oxidized manganese interferes. He showed that two equivalents of residual chlorine react with each mole of dye. Immediately after addition of the indicator, the remaining color is compared to permanent standards made with methyl orange.

Solo and Larson(72) developed a colorimetric procedure for determining both free and combined residual chlorine. After the reading for free chlorine, bromide is added, which causes the combined chlorine to react(73). Using chloroacetic acid to bring the pH to 2, they showed that Beer's law is followed as long as sufficient excess of methyl orange is present, and that absorbance measurements at 510 mμ can be used provided the calibration curve is constructed at approximately the same temperature as the samples being measured. Manganese interferes and must be determined after removing residual chlorine with arsenite.

The method has not come into general use, mainly because dependence on bleaching has obvious disadvantages. Great care must be exercised with respect to the precise quantity of reagent added, and in order to avoid too much sacrifice of sensitivity the amount of reagent used in any one determination must be limited. This necessitates repeated tests if the residual chlorine happens to fall outside the limits of the initial quantity of reagent used. Furthermore, the highest sensitivity lies at the upper end of the concentration scale instead of at the lower end where it is most needed.

Attempts have been made(73,74) to overcome these disadvantages by using methyl orange as the reagent to titrate residual chlorine, but sensitivity and reliability appear to leave something to be desired. It is reported, on the one hand(74), that the last 0.05 ppm residual chlorine reacts very slowly with the reagent, and on the other(73), that the titration should be carried to completion as quickly as possible. In this connection Solo and Larson(72) report that side reactions occur unless an excess of methyl orange is present. This indicator has the important advantages of complete stability and good specificity for free chlorine in the presence of combined chlorine. Apparently, however, these advantages have not been sufficient to compensate for its basic defects.

VIII. Cyanogen Chloride Method

In this method the residual chlorine reacts with cyanide ion to form cyanogen chloride, which in turn reacts with pyridine to open the ring and give a dialdehyde. This then reacts with an amine, for example, to give a colored dianil(75). The sequence is shown schematically in Eqs. (4), (5), and (6).

$$KCN + HOCl \longrightarrow CNCl + KOH \qquad (4)$$

$$CNCl + pyridine \longrightarrow dialdehyde \qquad (5)$$

$$dialdehyde + amine \longrightarrow dianil \qquad (6)$$

In the method as first proposed(75), the amine was benzidine. Problems of stability of indicator solutions(53) and reliability led to the substitution of barbituric acid(76).

Since the fundamental reaction is not an oxidation, this method is specific for the halogens and is not subject to the interferences encountered with other methods. There is no inherent characteristic of the reaction which provides a basis for distinguishing among the various forms of residual chlorine. For this, additional steps, such as selective removal of the free chlorine, are necessary.

Three modifications of the method using different amines have been evaluated and are as follows:

A. Benzidine

a. The original method(75) consists of adding potassium or sodium cyanide to the sample, waiting for a specified time, adding the benzidine-pyridine reagent, again allowing time for reaction, and finally measuring the color photometrically. In attempts to improve reliability, variations have been made in the composition of the indicator solution, in the timing of the various steps, and in the solvent for the indicator(77,78). These attempts have not been completely successful(53,76).

b. A modification suitable for wastes has been made by Nusbaum and Skupeko(79), who extract the cyanogen chloride with butyl or amyl alcohol and then add the benzidine-pyridine reagent before separating the layers. After color development is completed the alcohol layer is separated and the color measured. They also suggested addition of arsenite to remove free chlorine before beginning the test, to permit a determination of combined chlorine specifically. They did not demonstrate the validity of this step.

    Katz and Heukelekian(80) evaluated this procedure on wastes and did not get good agreement with the amperometric titration method (see Sec. XII). Further investigation would be desirable.

B. Sulfanilic Acid

Nicolson(53) evaluated the procedure involving substitution of sulfanilic acid for benzidine and found it unreliable.

C. Barbituric Acid

A significant improvement results when the amine is barbituric acid (13,76) and due attention is paid to the timing of reagent addition and color measurement. This gives much better sensitivity and reliability, although the mixed reagent is not completely stable(53).

Since this is the only method available that is specific for residual halogen, further attempts to place it on a firmer footing for both water and wastes are justified. It is potentially capable of the necessary sensitivity and reliability.

IX. Miscellaneous Amino and Hydroxy Aromatic Compounds

A. Residual Chlorine

Tarvin et al.(16) screened a total of 40 compounds in addition to some of the benzidine structures referred to in Sec. III. As might be expected, many amino derivatives of benzene and especially of naphthalene are oxidized to highly colored compounds by residual chlorine. There were no new compounds that the authors considered promising, although the evaluations were not extensive and no effort was made to remedy defects such as manganese interference. In many cases the indications were so obviously unpromising that there is no need to consider them further.

Nicolson(53) made a more thorough study of one aromatic compound, namely, tetrakis (p-dimethylaminophenyl) ethylene (7), which had been proposed by Kul'berg et al.(81) because of sensitivity and wide pH range of color development. He showed that the color was stable with time, and

(7)

the sensitivity and reproducibility were among the best. The promise of this compound as a specific indicator for free residual chlorine is greatly decreased, however, by the apparent instability of the reagent solution (the solvent being acetone). Apparently no attempt has been made to overcome this disadvantage in spite of the fact that a stable solution of this compound might show interesting utility.

B. Chlorine Dioxide

In the search for a specific colorimetric method for chlorine dioxide, Hodgden and Ingols (82) studied its reaction with tyrosine, in which a colored oxidation product is formed. They recommended a procedure, carried out at pH 4.6, wherein the resulting color is measured photometrically or by comparison with cobalt nitrate standards. While free residual chlorine does not produce a color, it does react with tyrosine to prevent full color development by chlorine dioxide. In addition, both amount of indicator and temperature are critical. At lower temperatures color production is impractically slow, so that warming to 20°C and reading after 10 min is recommended. However, after 15 min the color begins to fade. Apparently these difficulties have been sufficient to prevent any substantial use of the method.

Post and Moore (44) developed a method based on the oxidation of 1-amino-8-naphthol-3,6-disulfonic acid (8) to a colored compound, the residual chlorine being first removed with malonic acid. While sensitivity

(8)

is satisfactory and the color is stable when ferric chloride is used, the indicator solution itself is not very stable under ordinary circumstances. Furthermore, chlorite is measured as chlorine dioxide. This interference is serious since some chlorite may be present at any time as a result of incomplete reaction in the generation of chlorine dioxide and as a result of reduction of chlorine dioxide by organic matter.

More recently (83) an attempt was made to utilize the bleaching action of chloride dioxide on acid chrome violet K or 1,5-bis-(4-methylphenyl-amino-2-sodium sulfonate)-9,10-anthraquinone, as shown in structure (9).

(9)

By using an ammonium chloride-ammonia buffer at pH 8.1–8.4, interference from residual chlorine, chlorates, chlorites, and other ions is avoided. Chlorine would interfere if it were not for the use of ammonia to convert it to chloramines. The reaction is fast and the color stable, but because the measurement is based on bleaching a color, sensitivity is limited, which means that precision is adequate only down to 0.2 ppm.

X. The α-Naphthoflavone-Iodide Method

Gilcreas and Hallinan(84) investigated this indicator system because they recognized that the requirements of pH range and iodide concentration made the conventional starch-iodide method vulnerable to interference by other oxidizing agents (see Sec. XI). While α-naphthoflavone had been used previously(85) as a titration indicator in iodometric methods, Gilcreas and Hallinan directed their work toward a colorimetric method operating at a pH high enough to prevent interference by manganese, iron, and nitrite.

With free residual chlorine in water their results were in good agreement with those obtained by the orthotolidine method. When some organic matter was present, the method gave higher values than the orthotolidine method; in sewage it was not considered applicable. Nicolson(53) rejected the method after preliminary screening because he experienced difficulty with precipitation and because there was a change in hue as well as color density with change in chlorine concentration.

Hallinan(86) modified the method for application to sewage and wastes on the assumption that the errors were due to color and turbidity in the sample. The iodine complex was filtered off and redissolved in alcohol and the color was then measured.

Although the method was used locally, it never came into general use. For one thing, the concentrations of indicator and iodide must be carefully controlled and regulated for each range of chlorine concentration, otherwise the desired blue color is not obtained. Of even greater importance are the limitations inherent in any iodide method carried out at a relatively high, fixed pH (10).

XI. Starch-Iodide Methods

A. Application to Water

Attempts to utilize the blue starch-iodine color resulting from

$$HOCl + 2I^- + H^+ \longrightarrow I_2 + Cl^- + H_2O \tag{7}$$

in a colorimetric method have been unsuccessful with respect to sensitivity and reliability $(28,87)$.

Low sensitivity limits the titration of the liberated iodine with thiosulfate using the starch end point to standardizing chlorine solutions and to determining higher values of residual chlorine in applications where there is no need to distinguish between free and combined chlorine (10). Maximum recovery is obtained in the pH range of 1–2 utilizing adequate iodide concentration, but manganese interference can be completely avoided only at pH 5 and above (88). As a compromise which gives good recovery and avoids some interference, the pH range of 3–4 is now recommended (10).

B. Application to Wastes

When this titration procedure is used on wastes, where usually all of the chlorine is combined, recovery depends even more critically on pH $(89–91)$. Low pH increases recovery both by converting the residual chlorine to hypochlorous acid (Sec. I) and by decreasing loss of iodine through side reactions (12). When the pH is sufficiently low the recovery may be severalfold higher than given by orthotolidine and it may be substantial even when orthotolidine shows zero (89). While antibacterial activity may be observed in the latter case $(89,90)$, there is evidence that not all of the residual chlorine recovered in the acid titration is of practical value in disinfection (91). Indeed, results on some wastes lead to the conclusion that adequate disinfection results only when some residual chlorine is shown by the titration in the neutral region (89).

In the procedure now recommended (*30a*) the titration is carried out at pH 4 to obtain reasonably complete recovery of those chlorine compounds that are of practical value and to decrease interference. In order to minimize loss of iodine in side reactions the sample is added to an excess of thiosulfate in the presence of iodide, and the unreacted thiosulfate is determined by titration with standard iodine (*30a,92*). It has been shown (*92*) that, in the absence of highly colored waste materials, the results are in reasonable agreement with those obtained using an amperometric end point (*11*), although the sensitivity of the latter end point is greater. It has also been shown that standard phenylarsine oxide (see Sec. XII) can be used instead of thiosulfate with the advantage of greater stability (*30a*). Problems of interfering substances and of correlating residual chlorine concentration with disinfection are, therefore, identical to those encountered in the amperometric titration method; they are discussed in Sec. XII.

C. Determination of Chlorine Dioxide

Iodometric titration at pH 8 of a solution containing no oxidizing agent other than chlorine dioxide will give a value corresponding to 1 eq/mole. After acidification with acetic acid this titration yields 4 additional equivalents per mole (*93*). If the two fractions are in a ratio of higher than 1 to 4, it can be assumed that residual chlorine is also present. Various attempts have been made to utilize this procedure to determine all species of residual chlorine, but the method with starch end point is not sufficiently sensitive. It is now regarded as useful only to standardize chlorine dioxide solutions for calibration purposes (*13a*) (also see Secs. II. D and XII. D).

XII. Amperometric Titrations

A. Background

This method is based on early work by Rideal and Evans (*94*) and others on the use of amperometry to record residual chlorine and on the work of Foulk and Bawden (*95*) in applying the "dead stop" end point to iodometric titrations. The latter made use of two platinum electrodes immersed in the sample with a voltage of 10–15 mV impressed, permitting current to flow when iodide is oxidized at the anode and iodine reduced at the cathode. Thus, changes in iodine concentration produce changes in current, which provides a basis for amperometric end point detection in any iodometric titration.

Kolthoff and Pan(96) elucidated the theory for the case where reversible equilibrium obtains at both electrodes and originated the term "biamperometric titration" for the case where both electrodes contact the sample. They pointed out that in amperometric titrations the anode could also be in the form of a standard half-cell connected by a salt bridge which they designated as an external anode.

The method was adapted to the determination of the low concentrations of residual chlorine normally found in water and wastes by providing a large cathode area and good agitation to afford the necessary sensitivity(4). A flat gold cathode and external anode were used. Platinum cathodes of various shapes have also been used, most frequently with an external anode(10, 92, 97–99).

The arrangement with two platinum electrodes immersed in the sample has been used to detect the end point in the iodometric titration of total residual chlorine in water(100), and in several examples it has been shown (101) to give satisfactory results in the amperometric titrations described in the following sections. Cell construction is simpler, and, with low impressed voltage(101), circuitry is less complicated. However, the nature of the reaction at the anode is not clear when no iodide is present. Further experience is necessary to determine whether circumstances can arise in which the anode reaction becomes a factor in controlling current flow, making end point determination more difficult.

B. Application to Water

This method as applied to water is useful mainly as a means of distinguishing between free and combined residual chlorine and between mono-chloramine and dichloramine. In the neutral range of pH and in the absence of iodide, only free residual chlorine is recovered(4, 10). The use of a controlled amount of iodide at the same pH affords recovery of monochloramine as well(10, 102). By decreasing the pH to 4 and adding additional iodide, dichloramine is included(10, 40, 102). If nitrogen trichloride is present, approximately two-thirds will appear in the free chlorine fraction and an additional increment in the dichloramine fraction.

Because the procedure includes titration with no iodide present, the work on which its validity is based was done mainly with an external anode. Likewise, the titrant must react quantitatively with hypochlorous acid. Sodium arsenite is satisfactory for the neutral pH range(4), but for the complete procedure it is preferable to use phenylarsine oxide(10, 12,

102), which reacts quantitatively with hypochlorous acid and with iodine at neutral or low pH(*103*). In the iodometric procedure for total residual chlorine, sodium thiosulfate can be used(*100*).

In each of the titrations each addition of titrant decreases the concentration of oxidizing agent, which results in a decrease in current flowing. Once the end point has been reached, further increments of titrant produce no further change in current when the titrant and its oxidation product behave irreversibly at the electrode. This is illustrated in Fig. 1, which shows a typical curve for the titration of free residual chlorine with arsenite using the apparatus of Marks and Glass(*4*). A curve of the same form is obtained in the other titrations described above.

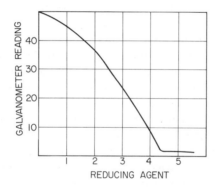

Fig. 1. Typical amperometric titration curve. Residual chlorine concentration is 0.46 ppm. [Reprinted from Ref. *4*, p. 1229, by courtesy of the *Journal of the American Water Works Association*.]

Interference from other oxidizing agents, notably manganese, can arise if the pH is permitted to fall below 3.5(*10*). Organic chloramines can appear in either the monochloramine or the dichloramine fraction, but this has little significance in water analysis.

There have been reports of low recoveries due to volatilization of chlorine compounds caused by the violent agitation used(*53, 100*). In this author's experience this is significant only during the titration of free chlorine in the presence of combined chlorine. Before the addition of iodide some of the chloramines can be aerated out of solution. Most of this loss can be avoided by determining free chlorine on one sample and then adding titrant equivalent to the free chlorine together with the required iodide to a second sample before beginning the agitation. The determination is then completed as usual.

C. Application to Waste Waters

In waste waters residual chlorine is practically always combined, so that iodide must be used. The liberated iodine, being more active chemically than combined chlorine, will be consumed by an iodine demand that exists over and above the chlorine demand. The lower the pH during the determination, the smaller the loss of iodine; but even at low pH it is necessary to add excess titrant before adding iodide and to back-titrate with standard iodine(12). At pH 4 and below, the iodine loss is negligible (30a). At this low pH arsenite cannot be used, so that the choice of titrant is between phenylarsine oxide(12) and sodium thiosulfate(92). For stability reasons the former is preferred(30a).

The accepted procedure (30a) is to mix the measured volume of standard titrant, potassium iodide, and the buffer in the titration vessel and then add the sample. Using one or the other form of amperometric titration apparatus, standard iodine is added. If the iodine solution is not added too rapidly, the current remains unchanged until the end point is approached. Near the end point a definite and permanent increase is produced by a small increment of iodine.

By carrying out the procedure above pH 3.5 interference from oxidized manganese and similar substances is avoided(103a, 34). Chromate ion interferes by slowly reacting with the excess phenylarsine oxide even at pH 4, but in a chromate waste relatively free of organics the method for the various forms of residual chlorine in water as described in Sec. XII. B is applicable(104). Trivalent chromium, cadmium, zinc, and divalent nickel do not interfere, but the procedure cannot be carried out in the presence of silver or cuprous ions(104). High concentrations of cupric ion cause difficulty. Color and turbidity are, of course, without effect, so that the amperometric end point can be used in many instances where the starch end point cannot (see Sec. XI. B).

Some chlorinated waste waters may contain relatively inert organic chloramines which react so slowly at pH 3.5 and above that quantitative recovery is not obtained and there may even be some uncertainty in the end point. When interference is not a problem, pH 2.5 or even pH 1 is recommended(12, 30a).

Regardless of which pH range is used, the recoveries are often rather large multiples of those given by the orthotolidine method as explained in Sec. II. B. l.c. On the basis of reproducibility and the relationship between dose and residual chlorine, there is reason to believe that the procedure carried out at a sufficiently low pH and based on either the amperometric or starch end points gives essentially complete recovery (12, 34, 105).

It has been shown that this procedure provides a reasonably satis-factory correlation between degree of disinfection and residual chlorine even when the chlorine dose is extended to the range where nothing is recovered by orthotolidine(12, 32–34). The relatively inactive chlorine compounds found in such cases can be of importance in minimizing the development of sulfide odors or other manifestations of septicity.

When it comes to destruction of organisms of sanitary significance, some of the less active chlorine compounds may be of little practical consequence. An attempt was made by Day et al.(34) to correlate disinfecting activity with ability to oxidize ethyl thioglycolic acid, but this did not prove useful. Further effort is justified to find methods of separating the combined chlorine compounds of different degrees of activity.

D. Chlorine Dioxide

Chlorine dioxide interferes with the amperometric titration of both free residual chlorine and combined chlorine. In the presence of iodide it liberates 1 eq of iodine/mole at pH 7 and 5 eq/mole in acid solution(98). Wherever chlorine dioxide is being used it is desirable to have a method to determine its concentration as well as to avoid the error in the deter-mination of residual chlorine.

Haller and Listek(98) devised an amperometric procedure for deter-mining free chlorine, combined chlorine, chlorine dioxide, and chlorite based on the disproportionation of chlorine dioxide at high pH.

Free chlorine is determined by amperometric titration with phenyl-arsine oxide at pH 7 on a sample which has been held at high pH for 10 min. The sum of the free and combined chlorine is determined on a second sample treated in the same way except that a relatively large amount of iodide is used. By titration of an untreated sample at pH 7 using the same amount of iodide the sum of the free and combined chlorine and one-fifth of the chlorine dioxide is obtained. The total of all species, including chlorite, is found by acidification in the presence of the iodide followed by titration at pH 7(13a).

The method has not been found to have the desired reliability(106, 107). The disproportionation reaction is not as fast and as reproducible as originally thought and dichloramine is at least partially destroyed along with the chlorine dioxide. Both effects tend to introduce positive errors which can become very large when the result is multiplied by five to convert to residual chlorine equivalents.

A brief account has been given recently of the attempt to remove free chlorine with malonic acid and then to titrate chlorine dioxide with manganous sulfate without interference from chloramine(107). It is stated that the method has some promise, with problems yet to be overcome, but no supporting data are given. It must be concluded that a completely satisfactory amperometric titration procedure for chlorine dioxide is not yet at hand.

XIII. Amperometry

Beginning with the work of Rideal and Evans(94) much use has been made of the current arising from cathode depolarization as a direct measure of residual chlorine. A variety of electrode arrangements and conditions of measurement have been used with varying degrees of success for the continuous recording of residual chlorine(97, 108–111). In no case were the conditions such that a limiting diffusion current could be reached at any reasonable value of impressed voltage. Usually the device was operated at some point on the current voltage curve chosen empirically on the basis of field experience. Consequently, none of this work has given rise to a true analytical method since all of the devices depend on very frequent calibration using a reliable method.

XIV. Voltammetry

Heller and Jenkins(112) studied the reduction of hypochlorous acid, hypochlorite ion, and the three ammonia chloramines at the dropping mercury electrode. At concentrations ranging from 0.001 to 0.020 N all except dichloramine were irreversibly reduced. Dichloramine was not reduced at all under the experimental conditions used. Diffusion currents were obtained which were proportional to concentrations but independent of pH. Hypochlorite and nitrogen trichloride showed half-wave potentials about 0.73 V more positive than those of monochloramine.

Marks and Bannister(97) found the dropping mercury electrode unsatisfactory and investigated rotated platinum electrodes of different surface areas. Even down to concentrations of 1 ppm and less, they obtained well-defined limiting currents linearly proportional to concentration of either hypochlorous acid or monochloramine but independent of pH. The chloramine was reduced in a distinctly more negative potential range than hypochlorous acid. The shapes of the current voltage curves did not correspond to a reversible process in which the current is controlled purely by concentration polarization. Likewise, considerable time periods were required for equilibrium after each voltage change.

Such a lengthy procedure is not suitable for regular analytical work, even if the necessary improvement could be made in sensitivity and precision. On the other hand, the method offers an alternative approach to an understanding of the reactivities of hypochlorous acid and chloramines, and there would seem to be justification for further attempts to apply it in the study of the various forms of residual chlorine.

XV. Use of Electrode Potentials

Because of variations in activity with pH, temperature, and the nature of the chlorine compounds present, it is apparent that measurement of residual chlorine concentration alone cannot predict the degree of disinfection to be expected. Theoretically, it is possible to predetermine the effects of temperature and pH and to take them into account. Practically, this is done in a qualitative manner when specifying values for residual chlorine to accomplish a given purpose, but a quantitative treatment is not possible. It is this situation that has led to attempts at various times to find an electrode potential measurement that would give a direct indication of disinfecting activity(113–116). The results of these efforts show the improbability of being able to establish an empirical relationship between potential and activity merely by making a very large number of measurements.

Perhaps this is the reason for attempts to provide a theoretical basis by relating the potential measurement to a very useful thermodynamic quantity, the standard oxidation-reduction potential. From the early work of Remington and Trimble(117) it could be seen that the hypochlorous acid, hypochlorite, chloride system does not establish reversible equilibrium with an inert metallic electrode, and this was well documented by the work of Chang(114), who included the various types of chloramines. The difficulties of interpretation and application of electrode potential measurements under these circumstances have been clearly stated by Chang(114) and also by Rohlich(118). It can be concluded on the basis of present knowledge that this method holds no promise for the quantitative determination of residual chlorine in its various forms or of effectiveness in disinfection.

On the other hand, there have been some special applications of electrode potential measurements on an empirical qualitative basis. Hood (119) showed that such measurements could be used to indicate changes in sewage at various stages in a plant, including those changes brought about by chlorination. Hoot(*120*) showed that potential measurements could be useful in odor control in sewage treatment plants. Rosenthal (*121*) mentioned their use in plating plants to control treatment of cyanides by chlorine or of chromates by reducing agents. In any instance where use can be made of an indication of a qualitative change from a reducing medium to an oxidizing medium, or vice versa, the method may find application.

REFERENCES

1. J. C. Morris, *J. Am. Chem. Soc.*, **68**, 1692 (1946).
2. R. E. Connick and Y. T. Chia, *J. Am. Chem. Soc.*, **81**, 1280 (1959).
3. G. M. Fair, J. C. Morris, S. L. Chang, I. Weil, and R. P. Burden, *J. Am. Water Works Assoc.*, **40**, 1051 (1948).
4. H. C. Marks and J. R. Glass, *J. Am. Water Works Assoc.*, **34**, 1227 (1942).
5. R. M. Chapin, *J. Am. Chem. Soc.*, **51**, 2112 (1929); **53**, 912 (1931).
6. J. C. Morris, *J. Phys. Chem.*, **70**, 3798 (1966).
7. C. T. Butterfield, E. Wattie, S. Megregian, and C. W. Chambers, *Public Health Rept.*, **58**, 1837 (1943).
8. C. T. Butterfield and E. Wattie, *Public Health Rept.*, **61**, 157 (1946).
9. E. W. Moore, *Water Sewage Works*, **98**, 130 (1951).
10. *Standard Methods for the Examination of Water and Waste Water*, 12th ed., American Public Health Association, New York, 1965, pp. 90–112.
11. H. C. Marks and R. R. Joiner, *Anal. Chem.*, **20**, 1197 (1948).
12. H. C Marks, R. R. Joiner, and F. B. Strandskov, *Water Sewage Works*, **95**, 175 (1948).
13. H. M. Webber and E. A. Wheeler, *Analyst*, **90**, 372 (1965).
13a. *Standard Methods for the Examination of Water and Waste Water*, 12th ed., American Public Health Association, New York, 1965, p. 116–122.
14. J. W. Ellms and S. J. Hauser, *Ind. Eng. Chem.*, **5**, 915 (1913).
15. W. M. Clark, B. Cohen, and W. D. Gibbs, *Public Health Rept., Suppl. No. 54*, 1926.
16. D. Tarvin, H. R. Todd, and A. M. Buswell, *J. Am. Water Works Assoc.*, **26**, 1645 (1934).
17. A. T. Palin, *J. Inst. Water Engrs.*, **3**, 100 (1949).
18. N. S. Chamberlin and J. R. Glass, *J. Am. Water Works Assoc.*, **35**, 1065 (1943).
18a. *Carcinogenic Substances Regulations 1966*, H.M. Stationery Office, London, 1966.
19. N. S. Chamberlin and J. R. Glass, *J. Am. Water Works Assoc.*, **35**, 1205 (1943).
20. H. W. Adams and A. M. Buswell, *J. Am. Water Works Assoc.*, **25**, 1118 (1933).
21. H. F. Muer and F. E. Hale, *J. Am. Water Works Assoc.*, **13**, 50 (1925).
22. R. D. Scott, *Water Works Sewerage*, **82**, 399 (1935).
23. G. Dragt and M. G. Mellon, *Ind. Eng. Chem. Anal. Ed.*, **10**, 256 (1938).
24. R. W. Aitkin and D. Mercer, *J. Inst. Water Engrs.*, **5**, 321 (1951).
25. E. S. Hopkins, *Ind. Eng. Chem.*, **19**, 744 (1927).
26. R. D. Scott, *J. Am. Water Works Assoc.*, **26**, 1234 (1934).
27. R. Hulbert, *J. Am. Water Works Assoc.*, **26**, 1638 (1934).
28. R. D. Scott, *J. Am. Water Works Assoc.*, **26**, 634 (1934).

29. F. J. Hallinan, *J. Am. Water Works Assoc.*, **36**, 296 (1944).
30. F. W. Gilcreas and F. J. Hallinan, *J. Am. Water Works Assoc.*, **36**, 1343 (1944).
30a. *Standard Methods for the Examination of Water and Waste Water*, 12th ed., American Public Health Association, New York, 1965, p. 375–380.
31. C. Lea, *J. Soc. Chem. Ind.*, **52**, 245T (1933).
32. F. B. Strandskov, H. C. Marks, and D. H. Horchler, *Sewage Works J.*, **21**, 23 (1949).
33. H. Heukelekian and R. V. Day, *Sewage Ind. Wastes*, **23**, 155 (1951).
34. R. V. Day, D. H. Horchler, and H. C. Marks, *Ind. Eng. Chem.*, **45**, 1001 (1953).
35. W. Rudolfs and J. V. Ziemba, *J. Bacteriol.*, **27**, 419 (1934).
36. G. E. Symons and R. W. Simpson, *Sewage Works J.*, **13**, 1149 (1941).
37. C. H. Connell, *J. Am. Water Works Assoc.*, **39**, 209 (1947).
38. P. C. Laux, *J. Am. Water Works Assoc.*, **32**, 1027 (1940).
39. P. C. Laux and J. B. Nickel, *J. Am. Water Works Assoc.*, **34**, 1785 (1942).
40. D. B. Williams, *Water Sewage Works*, **98**, 429 (1951).
41. A. T. Palin, *Water Sewage Works*, **101**, 74 (1954).
42. R. N. Aston, *J. Am. Water Works Assoc.*, **42**, 151 (1950).
43. A. T. Palin, *J. Inst. Water Engrs.*, **2**, 61 (1948).
44. M. A. Post and W. A. Moore, *Anal. Chem.*, **31**, 1872 (1959).
45. C. Brandt and H. A. Reffes, unpublished work, 1962.
46. W. Olszewski and H. Radestock, *Pharm. Zentralhalle,* **68**, 733 (1927).
47. K. Alfthan, *J. Am. Water Works Assoc.*, **20**, 407 (1928).
48. R. C. Stratton, J. B. Ficklen, and W. A. Hough, *Ind. Eng. Chem. Anal. Ed.*, **4**, 2 (1932).
49. R. Belcher, A. J. Nutten, and W. J. Stephen, *Anal. Chem.*, **26**, 772 (1954).
50. G. Milazzo and L. Paolini, *Mikrochem. Mikrochim. Acta*, **36/37**, 255 (1951).
51. K. L. McEwen and G. A. Grant, *Can. J. Technol.*, **30**, 66 (1952).
52. L. M. Kul'berg and L. D. Borzova, *Zavodsk. Lab.*, **21**, 920 (1955).
53. N. J. Nicolson, *Analyst*, **90**, 187 (1965).
54. I. M. Kolthoff, *Chem. Weekblad (Neth.)*, **23**, 203 (1926).
55. D. H. Byers and M. G. Mellon, *Ind. Eng. Chem. Anal. Ed.*, **11**, 202 (1939).
56. A. T. Palin, *Analyst*, **70**, 203 (1945).
57. L. W. Haase and G. Gad, *Z. Anal. Chem.*, **107**, 1 (1936).
58. A. Moore, *J. Am. Water Works Assoc.*, **35**, 427 (1943).
59. A. Moore, S. Megregian, and C. C. Ruchhoft, *J. Am. Water Works Assoc.*, **35**, 1329 (1943).
60. A. T. Palin, *J. Am. Water Works Assoc.*, **49**, 873 (1957).
61. A. T. Palin, *J. Am. Water Works Assoc.*, **58**, 509 (1966).
62. J. G. Bjorklund and M. C. Rand, *J. Am. Water Works Assoc.*, **60**, 608 (1968).
63. A. T. Palin, *J. Am. Water Works Assoc.*, **60**, 847 (1968).
64. A. T. Palin, *Proc. Soc. Water Treat. Exam.*, **9**, (1), 81 (1960).
65. A. P. Black and G. P. Whittle, *J. Am. Water Works Assoc.*, **59**, 471 (1967).
66. A. P. Black and G. P. Whittle, *Water Sewage Works*, **114**, 437 (1967).
67. A. P. Black and G. P. Whittle, *J. Am. Water Works Assoc.*, **59**, 607 (1967).
68. L. W. Winkler, *Z. Angew. Chem.*, **28**, 22 (1915); *Z. Anal. Chem.*, **61**, 197 (1922).
69. K. Holwerda, *Mede. Dienst Volksgezondh. Ned.-Indië*, **17**, 251 (1927); **19**, 325 (1930).
70. Besemann, *Chem. Zeitung.* **52**, 826 (1928) (in German); through *Chem. Abstr.*, **23**, 792 (1929).
71. M. Taras, *Anal. Chem.*, **19**, 342 (1947).
72. F. W. Sollo, Jr., and T. E. Larson, *J. Am. Water Works Assoc.*, **57**, 1575 (1965).
73. G. Gad and E. Priegnitz, *Gesundh.-Ing.*, **68**, 174 (1947) in German; through *Chem. Abstr.*, **44**, 6991i (1949).

74. M. Taras, *J. Am. Water Works Assoc.*, **38**, 1146 (1946).

75. R. F. Milton, *Nature*, **164**, 448 (1949).

76. E. Asmus and H. Garshagen, *Z. Anal. Chem.*, **138**, 404 (1953).

77. H. A. L. Morris and P. K. Grant, *Analyst*, **76**, 492 (1951).

78. E. Asmus and H. Garshagen, *Z. Anal. Chem.*, **136**, 269 (1952).

79. I. Nusbaum and P. Skupeko, *Anal. Chem.*, **23**, 1881 (1951).

80. S. Katz and H. Heukelekian, *Sewage Ind. Wastes*, **31**, 1022 (1959).

81. L. M. Kul'berg, L. D. Borzova, I. S. Mustafin, A. I. Cherkesov, and V. F. Barkovskii, *Uch. Zap. Saratovsk. Gaz. Univ.*, **71**, 251 (1959) (in Russian); through *Chem. Abstr.*, **55**, 19590a (1961).

82. A. W. Hodgden and R. S. Ingols, *Anal. Chem.*, **26**, 1224 (1954).

83. W. Masschelein, *Anal. Chem.*, **38**, 1839 (1966).

84. F. W. Gilcreas and F. J. Hallinan, *J. Am. Water Works Assoc.*, **31**, 1723 (1939).

85. M. Hahn, F. Schutz, and S. Pavlides, *Z. Hyg.*, **108**, 439 (1928).

86. F. J. Hallinan, *Ind. Eng. Chem. Anal. Ed.*, **12**, 452 (1940).

87. A. M. Buswell and C. S. Boruff, *J. Am. Water Works Assoc.*, **14**, 384 (1925).

88. F. J. Hallinan and W. R. Thompson, *J. Am. Chem. Soc.*, **61**, 265 (1939).

89. W. V. D. Tiedeman, *J. Am. Water Works Assoc.*, **15**, 391 (1926).

90. D. Tarvin, *Sewage Ind. Wastes*, **24**, 1130 (1952).

91. M. L. Koshkin, *J. Am. Water Works Assoc.*, **29**, 1761 (1937).

92. I. Nusbaum and L. A. Meyerson, *Sewage Ind. Wastes*, **23**, 968 (1951).

93. E. R. Woodward, G. A. Petroe, and G. P. Vincent, *Trans. Am. Inst. Chem. Engrs.*, **40**, 271 (1944).

94. E. K. Rideal and U. R. Evans, *Analyst*, **38**, 353 (1913).

95. C. W. Foulk and A. T. Bawden, *J. Am. Chem. Soc.*, **48**, 2045 (1926).

96. I. M. Kolthoff and Y. D. Pan, *J. Am. Chem. Soc.*, **61**, 3402 (1939).

97. H. C. Marks and G. L. Bannister, *Anal. Chem.*, **19**, 200 (1947).

98. J. F. Haller and S. S. Listek, *Anal. Chem.*, **20**, 639 (1948).

99. W. A. Mahan, *Water Sewage Works*, **96**, 171 (1949).

100. J. Holluta and H. Meissner, *Z. Anal. Chem.*, **152**, 112 (1956).

101. J. J. Morrow, *J. Am. Water Works Assoc.*, **58**, 363 (1966).

102. H. C. Marks, D. B. Williams, and G. U. Glasgow, *J. Am. Water Works Assoc.*, **43**, 201 (1951).

103. C. K. Banks and J. A. Sultzaberger, *J. Am. Chem. Soc.*, **69**, 1 (1947).

103a. *Standard Methods for the Examination of Water and Waste Water*, 12th ed., American Public Health Association, New York, 1965, p. 440.

104. H. C. Marks and N. S. Chamberlin, *Anal. Chem.*, **24**, 1885 (1952).

105. H. Heukelekian, R. V. Day, and R. Manganelli, *Ind. Eng. Chem.*, **45**, 1004 (1953).

106. H. C. Marks, *J. New Engl. Water Works Assoc.*, **66**, 1 (1952).

107. J. V. Feuss, *J. Am. Water Works Assoc.*, **56**, 607 (1964).

108. J. R. Baylis, H. H. Gerstein, and K. E. Damann, *J. Am. Water Works Assoc.*, **38**, 1057 (1946).

109. G. J. Hazey, *J. Am. Water Works Assoc.*, **43**, 292 (1951).

110. E. L. Streatfield, *Chem. Ind. (London)*, **52**, 1208 (1951).

111. W. B. Murray, *J. Am. Water Works Assoc.*, **49**, 795 (1957).

112. K. Heller and E. N. Jenkins, *Nature*, **158**, 706 (1946).

113. F. C. Schmelkes, *J. Am. Water Works Assoc.*, **25**, 695 (1933).

114. S. L. Chang, *J. New Engl. Water Works Assoc.*, **59**, 79 (1945).

115. C. E. Keefer and J. Meisel, *Sewage Ind. Wastes*, **25**, 759 (1953).

116. D. Backmeyer and K. E. Drautz, *J. Water Pollution Control Federation*, **33**, 906 (1961).
117. V. H. Remington and H. M. Trimble, *J. Phys. Chem.*, **33**, 433 (1929).
118. G. A. Rohlich, *Sewage Works J.*, **20**, 650 (1948).
119. J. W. Hood, *Sewage Works J.*, **20**, 640 (1948).
120. R. A. Hoot, *Water Sewage Works*, **96**, 267 (1949).
121. R. Rosenthal, *Am. Soc. Testing Mater. Spec. Tech. Publ. No. 130*, Philadelphia, Pa. 1953, p. 12.

Chapter **23** Analysis for Pesticides and Herbicides in the Water Environment†

Samuel D. Faust

DEPARTMENT OF ENVIRONMENTAL SCIENCES
RUTGERS, THE STATE UNIVERSITY
NEW BRUNSWICK, NEW JERSEY

Irwin H. Suffet

ENVIRONMENTAL ENGINEERING AND SCIENCE PROGRAM
DREXEL UNIVERSITY
PHILADELPHIA, PENNSYLVANIA

I. Introduction

In this chapter, the major analytical problems are discussed that arise from the recovery, separation, and identification of organic pesticides from aquatic environments. First, however, the problem area is defined, the evidence is evaluated, and the significance is ascertained.

Since 1945, organic pesticides and herbicides have been reported in drinking, recreational, irrigational, and fish and shellfish water, and in sediments and bottom muds; the evidence has been obtained, in the main, from physiological responses of aquatic organisms. Since 1961, however, direct analyses by gas–liquid chromatography and other techniques indicate that most natural waters and their equilibrium solid phases contain trace amounts of these organic pollutants.

II. Occurrences in Natural Waters and Bottom Muds

Organic pesticides and herbicides may enter natural waters from (a) direct application for control of aquatic weeds, trash fish, and aquatic insects, (b) percolation and runoff from agricultural lands, (c) drift from aerial and land applications, (d) discharge of industrial waste waters, and (e) discharge of waste waters from cleanup of equipment used for pesticide formulations and application. Suspected instances of pollution from these sources have been reviewed(1–4) for the 1945–1964 period. These papers mentioned the specific organic pesticides that have been detected in water, sediments, and bottom muds from "500 compounds in more than 54,000 formulations"(5). By the same token, many of the so-called incidents of pollution may be labeled suspect because (a) direct evidence was not offered due to the unavailability of analytical techniques, (b) there was a lack of collateral analytical evidence, and (c) the analytical data were incorrectly interpreted.

Some of the more recent confirmed occurrences of organic pesticides and herbicides in aqueous and solid phases are presented in Table 1. These reports are judged to be reliable by the criteria of specific methodology accompanied by confirmatory techniques. For example, Nicholson (13) tentatively identified parathion as the etiological agent for a fish kill by electron capture and microcoulometric gas chromatography. Confirmation of the parathion molecules was obtained through infrared

TABLE 1

Some Confirmed Incidents of Organic Pesticide and Herbicide Pollution of Aquatic Environments of the United States

Pesticides	Concentration[a]	Method(s)[d]	Year	Reference
DDT	1-20 μg/liter	IR	1953, 1960	6,7
DDT	0.005–0.346 μg/liter	GLC, IR	1961	8
DDT	0.02–16.0 μg/liter	GLC, PC	1961, 1962	9
DDT	0.7–144 μg/kg[b]	GLC, PC	1961, 1962	9
DDT	110.0 ng/liter	GLC[e]	1967	10
Aldrin	0.04–1.6 μg/liter	GLC, PC	1961, 1962	9
Aldrin	1.2–2.3 μg/kg[b]	GLC, PC	1961, 1962	9
Aldrin	0.085 μg/liter	TLC, GLC	1964	11
Endrin	0.094 μg/liter	TLC, GLC	1964	11
Endrin	40.0 ng/liter	GLC[e]	1967	10
Endrin	0.013 ppm[b]	GLC[e]	1966	12
DDD	15.0 ng/liter	GLC[e]	1967	10
Dieldrin	0.118 μg/liter	TLC, GLC	1964	11
Dieldrin	0.005 ppm[b]	GLC[e]	1966	12
2,4-D esters	Trace–10.0 μg/liter	GLC[e]	1961, 1962	9
Parathion	2.0–51.0 μg/liter[c]	GLC, IR	1966	13

[a]Ground waters and surface waters unless noted otherwise.
[b]Bottom sediments or silts.
[c]Waste water.
[d]See Appendix A for abbreviations.
[e]Multicolumn technique.

spectroscopy. It is evident from Table 1 that most aquatic environments (aqueous and solid phases) contain trace concentrations of organic pesticides, especially the chlorinated hydrocarbons. The national surveys of Weaver et al.(11) and Brown and Nishioka(10) revealed this information.

III. Water Quality Criteria

Man's priority use of water for ingestion always raises the health issue. An advisory committee was established by the U.S. Public Health Service in 1962 to consider standards for insecticide levels in drinking water (*14*). Insufficient information, at that time, was cited as the reason for not establishing specific limits. Where information was available, however, the concentrations of organic pesticides were found to be "below levels which would constitute a known health hazard" (*13*). Nevertheless, the committee recommended continued surveillance of the problem.

A later report by the National Technical Advisory Committee's subcommittee on public water supplies recommended some criteria for permissible concentrations of organic pesticides in raw surface waters (*15*):

> These values were derived by an expert group of toxicologists as those levels which, if ingested over extensive periods, could not cause harmful or adverse physiological changes in man. In the case of aldrin, heptachlor, chlordane, and parathion, the Committee adopted even lower than physiologically safe levels; namely, amounts which, if present, can be detected by their taste and odor.

It was noted in the report "that limits for pesticides and herbicides have been set with relation only to human intake directly from a related domestic water supply." The recommended criteria for organic pesticides in water are listed in Table 2.

TABLE 2

Permissible Concentrations for Organic Pesticides in Public Water Supplies[a]

Pesticide	Permissible concentrations, μg/liter
Aldrin	17.0
Chlordane	3.0
DDT	42.0
Dieldrin	17.0
Endrin	1.0
Heptachlor	18.0
Heptachlor epoxide	18.0
Lindane	56.0
Methoxychlor	35.0
Organic phosphates and carbamates	100.0[b]
Toxaphene	5.0
2,4-D; 2,4,5-T; and 2,4,5-TP	100.0

[a]From Ref. *15*.
[b]As parathion equivalents in cholinesterase inhibition.

It would appear from a comparison of Tables 1 and 2 that the levels of organic pesticides in natural waters are lower than the permissible criteria. This observation is made, however, from the very few data cited in Table 1 and must be considered as a generalization. There may be isolated and localized occurrences where concentrations of organic pesticides exceed the recommended concentrations for public drinking water supplies.

IV. Chemical and Physical Properties of Organic Pesticides

A. Classes of Organic Pesticides

Organic pesticides are generally classified by use, i.e., insecticide, miticide, nematocide, rodenticide, fungicide, herbicide, etc. A more appropriate classification would be by chemical species. A list of some of the more important organic pesticides appears in Table 3. Generic names appear in Appendix B.

TABLE 3

Classes of Organic Pesticides[a]

A. Chlorinated hydrocarbons
   1. Class I: oxygenated compounds
     Dieldrin, methoxychlor, endrin
   2. Class II: benzenoid, nonoxygenated compounds
     BHC, DDD(TDE), DDT, perthane
3. Class III: nonoxygenated, nonbenzenoid compounds
     Aldrin, chlordan, heptachlor, strobane, toxaphene

B. Organophosphorus compounds
   1. Aliphatic derivatives
     Demeton, dimethoate, ethion, malathion, phosdrin, phorate
   2. Aromatic derivatives
     Trithion, diazinon, EPN, fenthion, parathion, ronnel

C. Herbicides, fungicides, nematocides, etc.
   1. Phenoxyalkyl acids: 2,4-D; 2,4,5-T; 2(2,4,5-TP)
   2. Substituted ureas: fenuron, monuron, diuron
   3. Substituted carbamates: IPC, CIPC, EPTC, sevin
   4. Symmetrical triazines; simazine, atrazine
   5. Substituted phenols: PCP, DNBP, DNC

---

[a]In part from Ref. 16.

B. Physical Constants

Selected physical constants of 32 organic pesticides are included in Appendix C for the convenience of the analyst. These constants were compiled from several sources(*16–21*). Gunther et al. have compiled the solubilities of organic pesticides in water(*22*).

C. Metabolites, Hydrolysis Products, and Oxidation Products

In the analysis of natural waters for organic pesticides, the analyst must recognize the possible presence of metabolites, hydrolysis products, and oxidation products. Many of these metabolites and products are compiled in an excellent treatise by Menzie(*23*). The importance of isolation and identification of these derivatives should not be overlooked, for they may (a) be more toxic than the parent molecule, (b) cause taste and odor problems, and (c) provide some historical data concerning the occurrence and/or persistence of an organic pesticide in water. A few examples are provided herein. For detailed information, the analyst is referred to Menzie(*23*).

1. CHLORINATED HYDROCARBONS

Perhaps the most extensively studied system is the metabolism of DDT in various insects and mammals. Some of the metabolites and metabolic pathways are given in Fig. 1. Apparently the DDT molecule remains stable with the exception of alterations of the ethane linkage between the two *p*-chlorophenyl groups.

The in vivo conversion of aldrin to dieldrin has been studied extensively also. It has been noted in cattle, pigs, sheep, rats, and poultry and has been observed in soils and on plants. In other studies, dieldrin was administered to rabbits, and six metabolites were isolated and purified via TLC (see Appendix A for abbreviations used in the text). Rosen et al. were able to isolate and identify photochemical decomposition products of dieldrin and endrin(*24*).

γ-BHC (lindane) also may undergo metabolism in the rat to pentachlorocyclohexene and then to several chlorothiophenols. These conversions are shown in Fig. 2. The end products of 2,3,5- and 2,4,5-trichlorophenol are especially pertinent in view of their taste and odor potential in drinking waters(*25*).

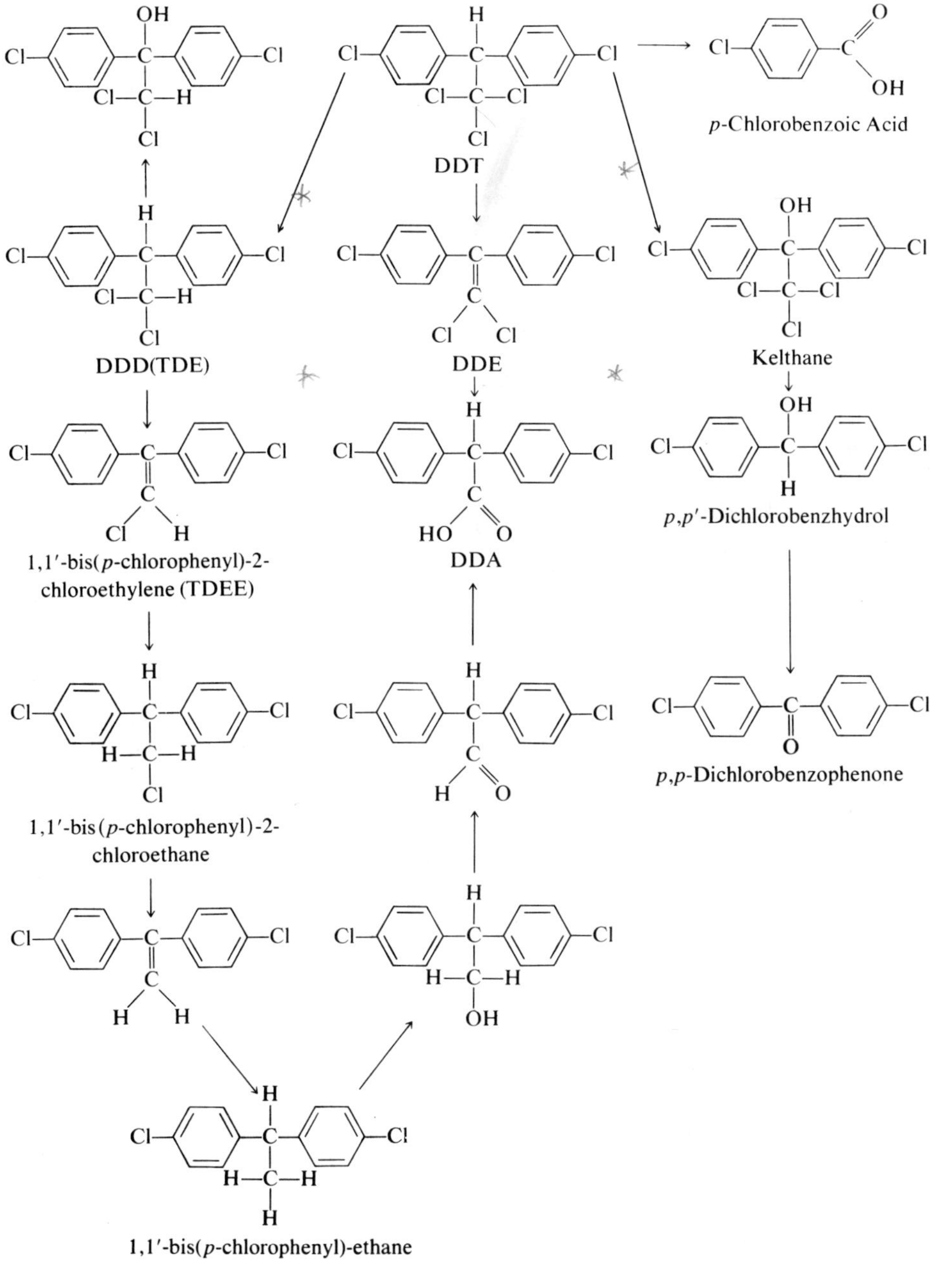

Fig. 1. Metabolites of DDT.

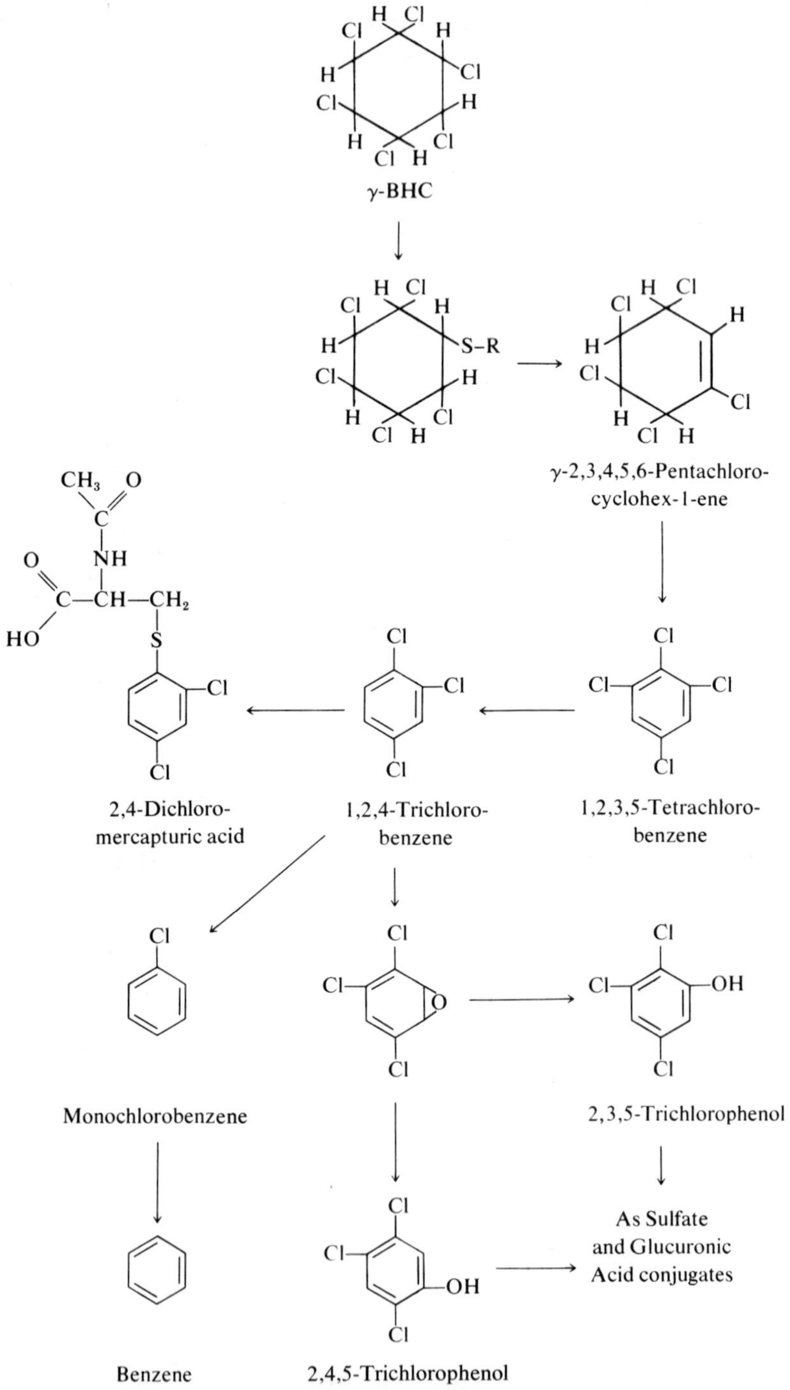

Fig. 2. Metabolites of γ-BHC.

2. ORGANOPHOSPHATES

Most organic phosphorus pesticides have the general structure:

$$\begin{array}{c} R{-}O \\ \diagdown \\ P{-}O{-}X \\ \diagup \\ R{-}O \end{array}$$

where R is an alkyl group and X is an organic radical. Four types of general structure are observed when either oxygen or sulfur is substituted in the indicated position.

A chemical characteristic of organophosphate pesticides is their ability to hydrolyze. This is expressed by the general equation:

$$\begin{array}{c} R{-}O \;\; O(S) \\ \diagdown \; \parallel \\ P{-}O{-}X \\ \diagup \;\; (S) \\ R{-}O \end{array} + {}^{-}OH = \left[\begin{array}{c} R{-}O \;\; O(S) \\ \diagdown \; \parallel \\ HO\cdots P{-}O{-}X \\ \diagup \;\; (S) \\ R{-}O \end{array}\right] = \begin{array}{c} R{-}O \;\; O(S) \\ \diagdown \; \parallel \\ HO{-}\!\!-\!\!P \\ \diagup \\ R{-}O \end{array} + \begin{array}{c} XOH \\ (S) \end{array} \quad (1)$$

The hydrolytic reagent approaches the reaction site and interacts with it. This step is considered to be rate controlling. The second step is very fast and may even be concurrent with the first(26).

Activation reactions, i.e., the conversion of a poorer to a stronger cholinesterase inhibitor, have been described for organophosphate pesticides, the major one being the $>P{=}S$ to $>P{=}O(26)$. This conversion may occur by chemical or enzymic mechanisms. An example of the latter has been observed in mice and insects where diazinon was converted to diazoxon(23).

In addition to oxons and hydrolysis products, organophosphates may yield other metabolites. For example, sulfoxides and sulfones of fenthion (Baytex) have appeared as metabolites in bean plants(23). Other metabolites of fenthion are shown in Fig. 3. Other schemes of metabolism are provided by Menzie(23) for parathion, malathion, demeton, and phorate. Special note should be made here of analytical schemes for fenthion and its metabolites provided by Bowman and Beroza(27) and by Suffet et al.(28).

3. PHENOXYALKYL ACIDS

Extensive studies have been conducted on the metabolic fate of 2,4-D and its organic esters in bean plants, cows, peas, tomato plants, soil and aquatic microorganisms, etc.(23). Several metabolites have been isolated

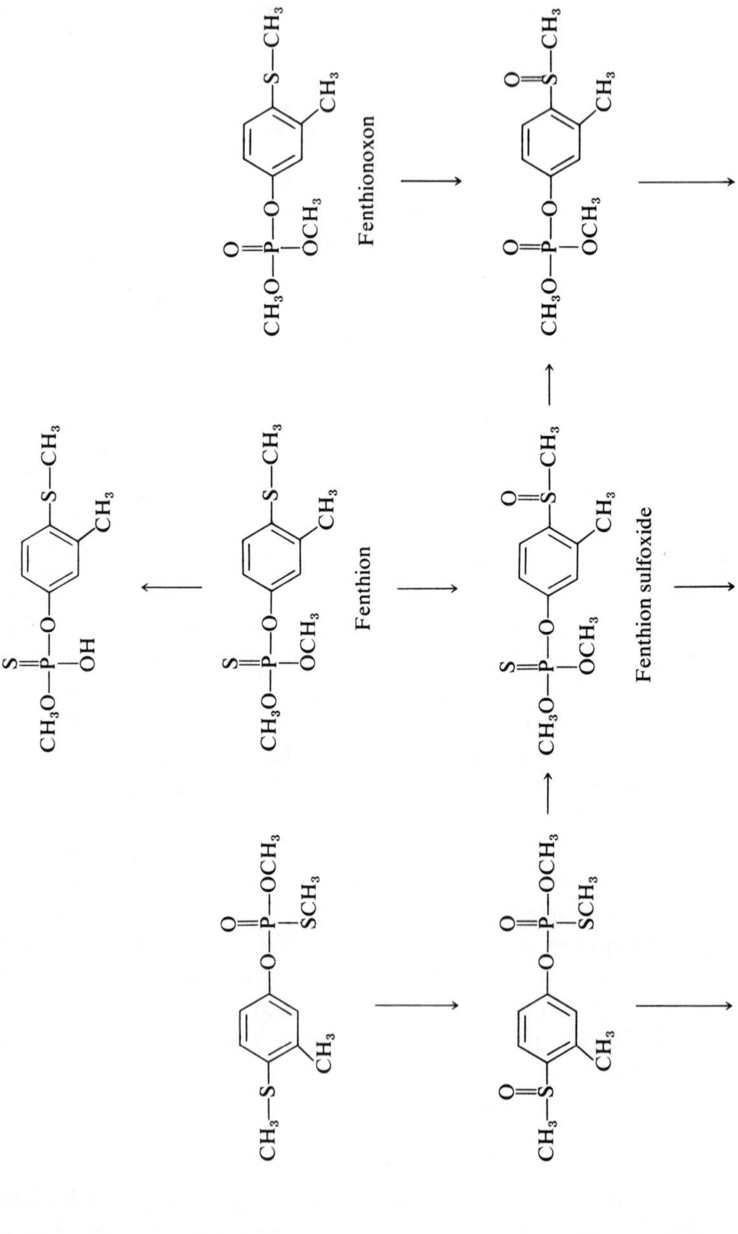
Fenthionoxon
Fenthion
Fenthion sulfoxide

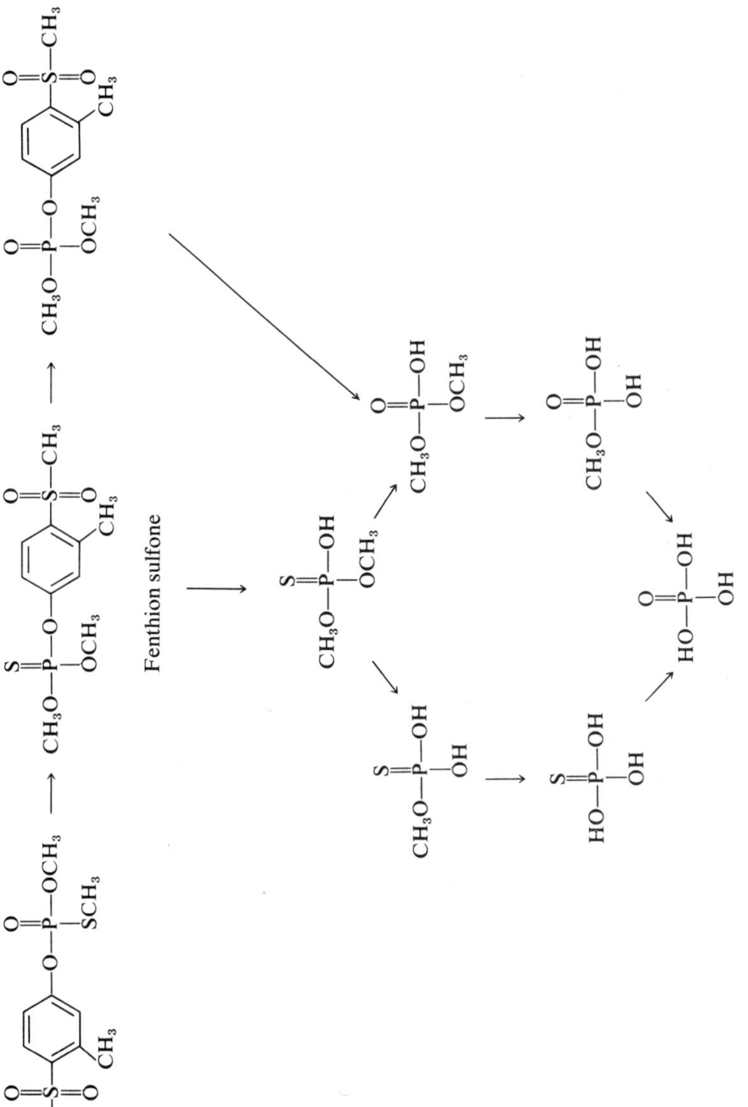

Fig. 3. Metabolites of fenthion.

and identified, as shown in Fig. 4. One of these metabolites, 2,4-dichloro-phenol, has special significance because of its taste and odor potential in drinking waters (25).

4. GENERAL COMMENTS

Natural waters may provide the enzymic opportunities for the conversion of organic pesticides to metabolites by aquatic insects, fish, and microorganisms. In turn, these metabolites may represent a potential water quality problem. Another consideration is the possible conversion of parent molecules into metabolites via chemical oxidation. For example, Gomaa (29) observed the conversion of parathion to paraoxon in dilute aqueous potassium permanganate solutions. Other conversion opportunities may be provided by dissolved oxygen and chlorine. The latter oxidative agent is used at conventional water treatment plants. Accordingly, the same justification must apply for the recovery and confirmation of metabolites, oxidative products, and hydrolysis products from natural and treated waters.

V. Significant Levels of Detection

The residue analyst should be concerned with a lower limit of detection, i.e., the true analytical sensitivity which is the reproducible, minimum detectable concentration for a given procedure. Frequently, confusion reigns from the interpretation of sensitivity in the detection of organic pesticides in aquatic environments.

Several factors determine the analytical sensitivity. First, the sample source is pertinent. Holden and Marsden (30) suggested that the minimum level for detection of dieldrin in "clean waters" was 0.5×10^{-6} mg/liter. Confirmation at this level was possible only by variants of their GLC technique. In domestic sewage effluents, they felt confident in reporting dieldrin levels in the $100-300 \times 10^{-6}$ mg/liter range. Epps et al. (31) stated that "problems of cleanup and confirmation in relation to number of samples necessitated that detection levels be established." These analysts did not attempt to realize the ultimate sensitivity of their analytical techniques. For water, 1 μg/liter was the objective for minimum detection, but in some cases, where interferences were great, concentrations less than 100 μg/liter were not reported with confidence.

Secondly, statistical variability in the analytical technique is important (32). Variance around a mean value, whether it be widespread or tight, influences the confidence with which minute levels of pesticides are

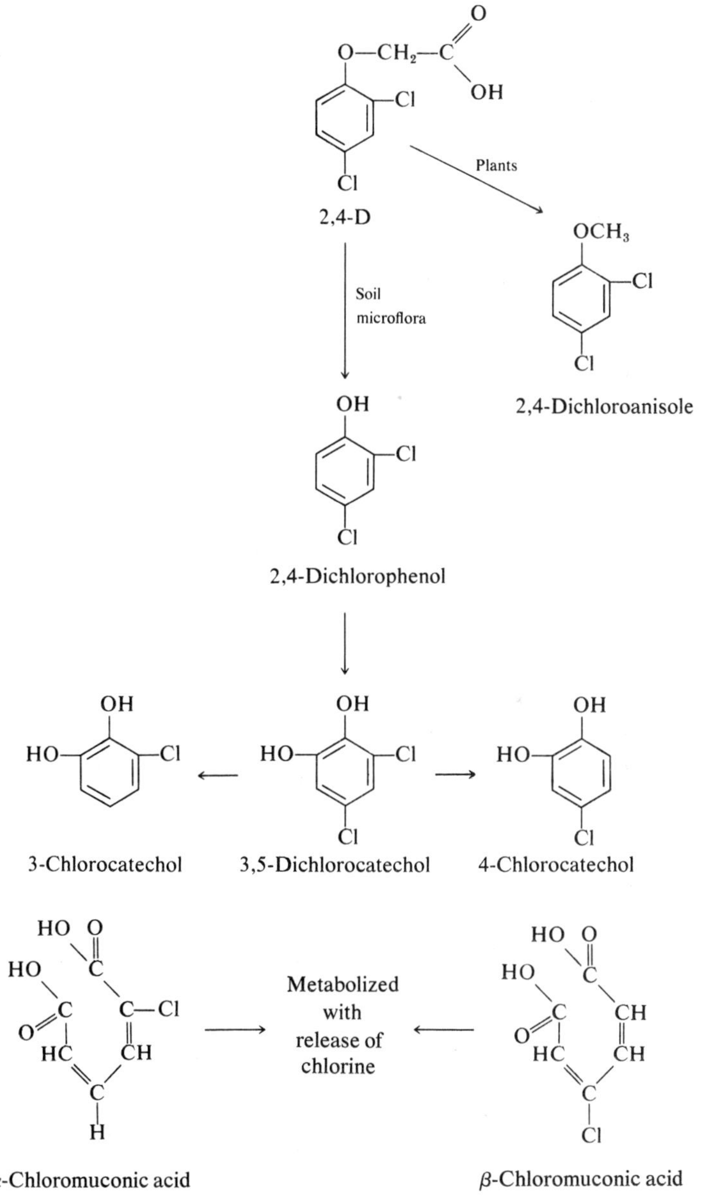

Fig. 4. Metabolites of 2,4-D.

reported. There is also the variance between operators and between laboratories.

Thirdly, a problem may arise from the required quantities for detection and for confirmation. GLC techniques, for example, are quite sensitive in the 10^{-9} to 10^{-12} g range, whereas such confirmatory techniques as infrared, ultraviolet, and mass spectroscopy require 10^{-3} to 10^{-6} g.

Nicholson(13) suggested minimum detectable limits for selected chlorinated hydrocarbon insecticides based on determination by gas chromatography with an electron capture detector and for a sample volume of 3.8 liters (1 gal). These limits are shown in Table 4 and are less that 10^{-6} g/liter. Brown and Nishioka(10) also suggest minimum detect-

TABLE 4
Suggested Minimum Detectable Limits for Selected Organic Pesticides in Aquatic Environments

Pesticide	Amount[a], μg/liter	Amount[b], μg/liter
BHC	0.005	—
Lindane	0.005	0.005
Heptachlor and/or heptachlor epoxide	0.010	0.005
Aldrin	0.010	0.005
Dieldrin	0.010	0.005
Endrin	0.020	0.005
Chlordane	0.040	—
Toxaphene	0.080	—
Strobane	0.080	—
DDT	0.020	0.010
DDD (TDE)	0.020	0.005
2,4-D	—	0.100
2,4,5-T	—	0.005
2 (2,4,5-TP)	—	0.005

[a]Reprinted from Ref. 13, p. 875, by courtesy of the American Association for the Advancement of Science, Washington, D.C. Based on determination by gas chromatography with an electron capture detector for a sample volume of 3.8 liters, under "average conditions," i.e., in a sewage effluent with associated suspended matter.

[b]Reprinted from Ref. 10 (for 1-liter samples).

able limits for several compounds, based upon their surveys. The reader should use the values in Table 4 as guidelines and not as absolute minimums.

A statistical approach to this question has been provided by Sutherland(32), who defines limit of detectability as "the concentration level of pesticide or drug above which a given sample of material can be said, with a high degree of assurance, to contain the chemical analyzed." This involves the correction of an apparent concentration of pesticide in the sample for interferences, via the equation:

$$C_{mc} = C_{ma} - \bar{y}c \tag{2}$$

where C_{mc} is the corrected limit of detectability, C_{ma} is the minimum apparent concentration, and $\bar{y}c$ is the average value of control samples. If a sufficient number of treated samples are run, then C_{ma} values can be calculated for various confidence limits (99%, for example). C_{mc} values can be generated if $\bar{y}c$ values are known or have been determined for aqueous samples.

VI. Collection of Samples

Several factors must be considered whenever a surface water or bottom mud is sampled for subsequent analysis for organic pesticides. The concern is to obtain and to preserve a sample that is representative of the environment from which it is taken. The U.S. Geological Survey and the Federal Water Pollution Control Administration have considered this problem at great length in their monitoring programs(33).

These two agencies recommend the following procedure for the collection of either a 1- or 4-liter grab sample:

All samples are to be collected in glass bottles. Prior to collection, scrupulous cleansing of sample containers is required. Chromic acid cleaning solution or other suitable cleansing agents are to be used, followed by several rinsings with organic-free distilled water. Containers are to be further treated as necessary to destroy remaining traces of organic matter; heat treating of containers at 300°C has been found satisfactory. Bottles are to be stoppered immediately to prevent air-borne contamination. The sample must have no contact with rubber, cork, and most plastics; Teflon, however, will not contaminate the sample. Rubber or cork stoppers may be used if wrapped carefully with a double layer of organic-free tinfoil or aluminum foil, taking care to avoid rupture of the foil cover when stoppering the bottle.

The sample is to be collected in the prepared glass bottle by lowering it in a weighted bottle holder in a vertical section of the stream which is representative of the stream cross section. The bottle is to be lowered as nearly as possible to the bottom of the stream and returned to the surface so that all points in the vertical section are represented in the sample.

It is also recommended that the samples be shipped in suitable packing cases for analysis within 2 days after collection. If storage is required beyond this 2-day period, then a cool, dark place is required.

Most of these recommendations were made with the more stable chlorinated hydrocarbon and phenoxyalkyl acids in mind. If, however, the presence of an organophosphate, carbamate, or urea is suspected, then the analytical work should follow the sample collection as quickly as possible. As these pesticides may hydrolyze in water, determination of a total concentration would not be possible if the sample were neglected or stored for an appreciable period of time.

Losses of chlorinated hydrocarbons, especially DDT, are quite high in the collection and transportation of samples. Bowman et al. (34) found that a high proportion of DDT was deposited on the inner walls of glass, aluminum, or paper containers. Another loss was found when more than 50% of a 0.01 ppm suspension of DDT was volatilized from the liquid phase within 24 hr.

Another consideration of importance is the loss of pesticide during sample preparation for the determinative step. Chiba and Morley (35) have provided an excellent study of this problem for γ-BHC, p,p'-DDT, and dieldrin. Although losses were small in the steps of filtration, partitioning, and washing, they were cumulative. Evaporation to dryness caused the most significant loss. For example, γ-BHC recovered to the extent of only 80.4% when a petroleum ether solution was evaporated almost to dryness by an air stream. If the evaporation was not included, then a 100% recovery of γ-BHC was effected.

VII. Efficiencies of Recovery

A 100% efficiency is a desirable goal for recovery of organic pesticides from aquatic environments. As cited above, some losses occur during each analytic step in a given procedure. It is general practice to report efficiencies of recovery by techniques of fortification, i.e., the addition of a known quantity of a pesticide to water in a laboratory test prior to processing the sample through each step of the analytic procedure. Fortification provides data only on the theoretical efficiency of recovery of the total analytical procedure.

The question remains: Do the percentages of recoveries from the fortification represent the actual efficiencies from field samples? The pitfalls of fortification have been discussed for plant and soil residues(36). Gunther and Blinn(37) concluded that fortification is "illusory except in a few instants." A completely homogeneous system as a completely dissolved solution, i.e., "pesticide residues in olive oil," may approach actual field recovery. Field samples are "weathered" (i.e., subject to physical, chemical, and metabolic transformations) and may be in aggregates or in molecular form in or on plant tissue or soil, etc. Therefore, the pesticidal residue in field samples may not be in the same form as the fortified sample. The actual recovery of the field sample can be determined by isotope tracers.

Recovery efficiencies from field samples will approach laboratory fortification if a pesticide is dissolved completely, not adsorbed on particulate matter, and not dissociated, and if the natural water characteristics of pH, temperature, and ionic strength can be duplicated. In another approach, natural water quality characteristics can be altered to those of laboratory fortified samples in order to compare efficiencies of recovery.

VIII. Extractive Techniques

Extraction, concentration, and cleanup techniques must be employed prior to quantification. Grossly polluted waters contain many organic compounds that would interfere with subsequent confirmation and quantification. On the other hand, relatively unpolluted waters may not require an extensive cleanup since organic interferences would be at a minimum. This section offers a critique of extraction and preliminary cleanup techniques that have been employed mainly for natural water systems.

A. Carbon Adsorption Method (CAM)

Carbon has been employed for extraction of organic pesticides from water in two different analytical techniques, as a continuous device in the field and as a batch-wise concentration and preliminary cleanup in the laboratory.

The continuous system was developed and introduced by the U.S. Public Health Service(38) as the carbon adsorption method (CAM) for isolation of synthetic organic contaminants affecting taste and odor qualities of surface waters. Middleton and co-workers(6,7) extended the CAM to organic pesticides because contemporary serial liquid–liquid extractive methods could not yield the necessary quantities for existing infrared identification techniques.

The CAM has many serious limitations. First, the adsorption and subsequent recovery by solvent extraction are not quantitative. Other errors in the CAM may arise from (a) the formation of a microbiological slime layer on the carbon surface, (b) removal of organics in the sand prefilter, (c) preferential adsorption for certain types of organics, as influenced by type of carbon, pH, temperature, and mineral constituents in the water, (d) change of selectivity during the course of a carbon column run (i.e., some organics with a high affinity for carbon will displace compounds with lower affinities), (e) improper pretreatment of the carbon, which leads to high blanks, and (f) decomposition of the parent compound during the adsorption and desorption steps(39).

In spite of the aforementioned and well-documented limitations, the CAM may have qualitative usefulness for continuous surveillance of groundwater and surface water for organic pesticides. It does carry two major advantages as an extraction technique: (a) It is a continuous process, and (b) large volumes of water are sampled. The CAM also might be utilized in the continuous monitoring of treated municipal waters, as noted by Nicholson et al.(40). Pesticide recoveries may be greater from treated waters since most interfering suspended and organic matter should have been removed.

B. Liquid–Liquid Extraction

In liquid–liquid extraction, a solute expresses its preference between two available solvents. The state of the art for prediction of the extraction capabilities of a binary solvent system has been expressed recently by Peppard(41): "The preference is determined by chemical and physical principles which are inviolable but in most instances are identified in a qualitative fashion only." Semiempirical knowledge about the polar–nonpolar character of the solute and each solvent, and solute–solvent interactions as dipole or hydrogen bond interactions, have been used to choose a solvent(42). Peppard concluded: "Consequently, the behavior of a given solute in a given extraction system can not be predicted in any but the roughest manner. However, this behavior once determined, can be relied upon in any future experiments"(41).

After selection of a solvent with the proper polarity, the analyst must decide between serial and continuous extraction. Then the analyst must consider such environmental factors as pH, ionic strength, and temperature. Often these decisions that were based upon percentage of recovery have been arbitrary and empirical. These recoveries, in turn, varied considerably. For example, Kawahara et al.(43) reported recoveries of 61.1 and 62.6% for dieldrin and endrin, respectively, extracted with hexane in a semiautomatic technique, whereas LaMar et al.(44) reported recoveries of 97 and 102%, respectively, when these two pesticides were added first to the aqueous phase as emulsifiable concentrates and then extracted serially with hexane. Cueto and Biros(45) averaged an 80.5% recovery of dieldrin via hexane extraction from fortified human urine.

1. SERIAL EXTRACTION

a. Chlorinated Hydrocarbons

The hydrocarbon solvents of hexane and petroleum ether have found the most utilization in recovery of chlorinated hydrocarbons from water via serial extraction. As seen in Table 5, extraction efficiencies are quite inconsistent. Also, the number of serial extractions varies from one to five, and the ratio of sample volume to solvent volume varies as well. It is, therefore, difficult to cite a universal procedure for the liquid–liquid extraction of chlorinated hydrocarbons from water.

b. Organophosphates

Many solvents appear to be suitable for recovery of the organophosphates from water, as suggested in Table 6. Which is the best solvent for a particular organophosphate: a group of organophosphates or an organophosphate and its degradation products? This is still not clear in most cases. Many analysts failed to report recoveries of water extracts and the water characteristics of the water extracted. An adjustment of pH is extremely important also since the organophosphate can hydrolyze under acid or alkaline conditions. The hydrolytic stability or "half-life" plays an important role in recovery efficiencies if this variable is not recognized. Some analysts prefer a neutral or slightly alkaline medium in the 7.0–8.0 pH range(53,54). On the other hand, some analysts utilize an acid medium in order to determine a hydrolysis product as indicative of the original organophosphate's concentration(55,56). Percent recoveries vary also, as suggested by column 6, Table 6.

c. Phenoxyalkyl Acids

The phenoxyalkyl acid herbicides have received little attention in the general concern over pesticidal pollution of natural waters. Consequently, little has been reported in the analytical literature. As seen in Table 7, chloroform and ether appear to be the preferred solvents for

TABLE 5

Conditions Used for Serial Liquid–Liquid Extraction of Chlorinated Hydrocarbons from Aqueous Samples

Pesticide(s)	Solvent(s)	Sample volume, ml	Solvent volume, ml	pH	Mean recovery, %	Reference
DDT	3:1 Ether:n-hexane	1500	250, 200, 150, 150	4.0–5.0	88[a], 61[b]	46
Dieldrin and metabolites	n-Hexane	100	4 × 50	—	90–95	47
Five CH	1:1 Ether:pet. ether, $CHCl_3$	1000	100, 4 × 50	—	88	48
Eight CH	n-Hexane	1000	2 × 25	—	80–115	44
Aldrin	1:1 Benzene:hexane	850	100, 4 × 50	—	43.4	43
Aldrin	1:1 Ether:hexane	850	100, 4 × 50	—	69.6	
Eleven CH	n-Hexane	12	3 × 2	7.0	2.3–106.0	45
Heptachlor	2:1 Pet. ether: isopropyl alcohol	1000	150, 3 × 75	—	80	48
Five CH	n-Hexane	1000	50	—	95	30
Nine CH	Benzene	250	25	—	84.6–101.8	49
Three CH	n-Hexane	1000	30	—	95	50
Four CH	1:1 n-Hexane:ether	100	3 × 100	—	97–100	51
Three CH	n-Hexane	1000	100	Acid	>90	52

[a]Interferences absent.
[b]Interferences present.

TABLE 6

Conditions Used for Serial Liquid–Liquid Extraction of Organophosphate Pesticides from Aqueous Samples

Pesticide(s)	Solvent(s)	Sample volume, ml	Solvent volume, ml	pH	Mean recovery, %	Reference
Parathion and diazinon	1:1 Ether:pet. ether or CHCl$_3$	1000	100, 4 × 50	—	90	57
Dipterex and DDVP	Ethyl Acetate	50	50, 25	8.0	94.5	53
Malathion	Dichloromethane	?	3 × 30	7.0–8.0	100	54
Diazinon	Benzene	100	100	—	?	43
Abate	Chloroform	1000	50, 25, 25	1.0	70	55
Parathion	Benzene	1000	500	Acid	99–100	56
Methyl parathion, diazinon, malathion, azinphos–methyl	Benzene	250	25	?	95.3–99.7	49
Parathion, methyl parathion, baytex	n-Hexane	1000	100	Acid	>90	52,57
Parathion, methyl parathion	1:1 Hexane:ether	100	3 × 100	—	98	51
Dursban	Dichloromethane	50	100, 50	—	92	58

TABLE 7

Conditions Used for Serial Liquid–Liquid Extraction of Phenoxyalkyl
Acids from Aqueous Samples

Pesticide(s)	Solvent(s)	Sample volume, ml	Solvent volume, ml	pH	Mean recovery, %	Reference
2,4-D; MCPA; 2,4,5-T	CHCl$_3$	25	3 × 10	2.0	93	59
2,4-D	1:3 Ether:CHCl$_3$	250	3 × 50	2.0	99.9–102	60
2,4-D; 2,4,5-T; MCPA; MCPB; 2,4-DB; dalapon	Ether	100	100	2.0	96	61
2,4-D; 2,4,5-T; 2,4,5-TP	Ether	800	150, 2 × 50	2.0	96	62

the phenoxyalkyl acids, with a pH adjustment to 2.0 necessary. Recoveries range from 93 to 102%.

2. CONTINUOUS EXTRACTION

Few continuous extractive techniques have been reported in the literature (Table 8). This is surprising since continuous extraction has an
obvious advantage over serial extraction, as larger sample volumes may
be extracted. Kahn and Wayman(63) described a continuous, multi-

TABLE 8

Conditions Used for Continuous Extraction of Organic Pesticides
from Aqueous Samples

Pesticide(s)	Solvent(s)	Sample volume	Mean recovery, %	Reference[a]
Six CH	Pet. ether	20 liters	97	63
2,4-D; dieldrin, perthane	Pet. ether	18 liters	93, 87, 93	64
Parathion	1:1 Benzene: n-hexane	200 liters	97	65
Dieldrin	n-Hexane	850 ml	61.1	43
Endrin	n-Hexane	850 ml	62.6	
Dieldrin	4% Benzene in n-hexane	850 ml	94.7	
Endrin	4% Benzene in n-hexane	850 ml	77.8	
Aldrin	1:1 Benzene: n-hexane	850 ml	63.6	

[a]No pH adjustment was mentioned in these references.

chamber liquid–liquid system with an internal solvent recycle, for extraction of several chlorinated hydrocarbons from natural waters. Three extraction chambers are connected in series and are charged with petroleum ether (38.7–57.7°C). Water is pumped into the chambers at rates of 0.5–1.0 liter/hr. The solvent is recycled at various rates depending upon Variac settings. Maximum recovery efficiencies were obtained with an average residence time of 45 min per chamber at a 1:1 solvent-to-aqueous phase ratio. Extraction efficiencies ranged from 96 to 100% for aldrin, isodrin, dieldrin, endrin, and two intermediates recovered from the waste water stream of a manufacturing facility. Detection and identification were accomplished by gas chromatographic and infrared techniques. Sanderson and Ceresia(64) modified the above continuous extraction system by replacing the peristaltic pump with gravity feed from an 18-liter carboy.

Sumiki and Matsuyama(65) continuously extracted parathion from 200 liters of water with a mean recovery of 97%. A 1:1 mixture of benzene and n-hexane was used.

C. Limitations of Liquid–Liquid Extractive Techniques

Most limitations of liquid–liquid extractive techniques are concerned with the practical aspects of extracting natural waters. Recoveries are less than 100%. No single solvent can extract all organic pesticides. The technique is not selective since organic interferences are coextracted from polluted waters. Therefore, an additional cleanup step usually must be included. The pH value of the water sample is an important variable if inclusive extraction is attempted. Volatilization losses may occur during the extraction and subsequent evaporation steps. In conclusion, liquid–liquid extraction techniques offer two major advantages: (a) Recoveries are greater than 85% in most systems, and (b) relatively little time is required to recover and to prepare the pesticide for the separation and quantification steps.

D. Quantitative Liquid–Liquid Extraction (LLE)

1. GENERAL CONSIDERATIONS

Several LLE techniques for recovery of organic pesticides from aqueous environments were reviewed critically above. These methods were concerned with one compound, a related group of compounds, or many groups of compounds. Each serial procedure utilized a particular solvent and specified the number of extractions and the solvent-to-water ratios. In some cases, per cent recovery as determined by a fortification technique was the experimental justification for selection of the solvent.

Many questions can be asked of these procedures: What is the advantage of the solvent used? What effects do such variables as pH, ionic strength, and temperature have on the extraction efficiency? What is the correct pH for sample storage between the time of sampling and extraction? What is the optimum pH for extraction? What is the most efficient solvent-to-water ratio? How many times must the sample be extracted for maximum recovery? For the most part, answers to these questions were not reported. It is entirely possible that the most efficient parameters were not used.

General criteria may be cited as an aid in the selection of a solvent for the extraction of organic pesticides from water. These are: (a) limited solubility of the solvent in water, (b) character of the solvent: polar or nonpolar, aromatic or aliphatic, (c) ability to be used with a gas-liquid chromatographic detection system, (d) volatility, (e) "pesticide quality" grade (i.e., purity), (f) ease of handling, (g) toxicity, and (h) flammability.

The final choice of solvent will depend on the desired analysis, for example: (a) the extraction of only one component for a mixture from water, (b) the extraction of many components from water, or (c) the use of different solvents, as necessitated by the determinative step.

2. THEORETICAL CONSIDERATIONS

It is possible to quantitate a LLE method based upon thermodynamic parameters (66). Gibbs' phase rule states:

$$F = C - P + 2$$

This equation shows the relation between the number of components, C, the number of phases, P, and the number of independent variables, F (degrees of freedom), necessary to characterize a system. Gibbs' phase rule helps to predict the behavior of a LLE multiphasic system and elucidate the thermodynamic state. LLE is a two-phase system of three components. Therefore, there are three degrees of freedom. If two degrees of freedom, temperature and pressure, are kept constant, then only concentration need be specified to define the system completely. The phase rule is specific for only a single molecular species.

The thermodynamic partition coefficient, D, is:

$$D = \frac{\gamma_a [A]_a}{\gamma_b [A]_b} = K \frac{\gamma_a}{\gamma_b} \tag{3}$$

and

$$K = \frac{[A]_a}{[A]_b} \tag{4}$$

where the γ's are the activity coefficients, $[A]_a$ is the concentration of the solute in the solvent phase, $[A]_b$ is the concentration of the solute in the water phase, and K is the extraction coefficient or distribution ratio. When γ_a/γ_b approaches 1.0, the analytical concentration of A approaches zero and K approaches D. This occurs in dilute solutions. The experimentally determined K value will remain constant and independent of the relative amounts of solutes if chemical equilibra, i.e., association or dissociation of solute species, are minimized, and if the activity coefficients in the two phases are equal regardless of concentration. From the above, it is possible to develop dependable quantitative LLE parameters based upon theory.

A general equation may be derived for the distribution coefficient, K, for binary solvent systems:

$$K = \frac{p}{\alpha(1-p)} \tag{5}$$

where p is the fractional amount of a solute partitioning into the nonpolar phase of an equal volume, two-phase solvent system (67,68). The α value is a volume correction factor (69):

$$\alpha = \frac{V_n}{V_p} \tag{6}$$

in Eq. (6), V_n and V_p are the equilibrated volumes of the nonpolar and polar phases, respectively, after extraction. When $\alpha = 1.0$, the distribution coefficient becomes:

$$K = \frac{p}{(1-p)} = \frac{p}{q} \tag{7}$$

where q is now the fractional amount of solute partitioning into the polar phase of an equal volume, two-phase solvent system. Beroza and Bowman have utilized p values rather than the distribution ratios for cleanup procedures by partition of organic pesticides in two-phase solvent systems (70–72). These investigators showed that p values can be used as an aid in the confirmation of pesticide identity at the nanogram quantity level.

3. CALCULATION OF THE p VALUE

The following passages describe a method for calculation of p values from a one-step LLE from water for unequilibrated solvents (73). In these calculations the aqueous phase is considered the polar phase and any other solvent is considered the nonpolar phase. The p value is calculated from the following equation (67,68):

$$p = \frac{E}{\alpha - E(\alpha - 1)} \tag{8}$$

where E is the fractional amount (of pesticide) extracted into the nonpolar phase of an unequal, unequilibrated, two-phase solvent system. For a one-step LLE, tedious calculations can be eliminated by plotting E versus p at several α values. This plot is shown in Fig. 5 for solvents insoluble in water at ratios of 1:1, 5:1, and 10:1 solvent:water. Alpha value corrections for solvents soluble in water would fall between 0.10 and 1.00 in Fig. 5.

4. DETERMINATION OF THE E VALUE

In order to solve Eq. (8), it is necessary to examine quantitatively either the nonpolar (i.e., solvent) phase or the polar phase for the organic pesticide after extraction. Often the solvent phase is examined because of analytical convenience. Two common techniques are gas–liquid chromatography and ultraviolet spectroscopy.

a. Gas–Liquid Chromatography

Here, the nonpolar solvent phase is analyzed. The E value is calculated from (68):

$$E = \frac{A_n}{A_s} \frac{V_n}{V_s} \tag{9}$$

where A_n is the amount (grams) in the nonpolar phase after extraction and equilibration, A_s is the amount (grams) in the water phase before extraction and equilibration or the maximum amount possibly extracted into the nonpolar phase, and V_n/V_s is a correction factor due to unequilibrated phase volumes and/or unequal original phase volumes. V_n is the volume of the nonpolar phase after extraction and equilibration. V_s is the original volume of the nonpolar phase before equilibration and extraction.

The number of micrograms that are actually extracted into the nonpolar phase and injected onto a GLC column (A_n) is the only experimental variable. A_n is determined from a standard gas–liquid chromatographic curve.

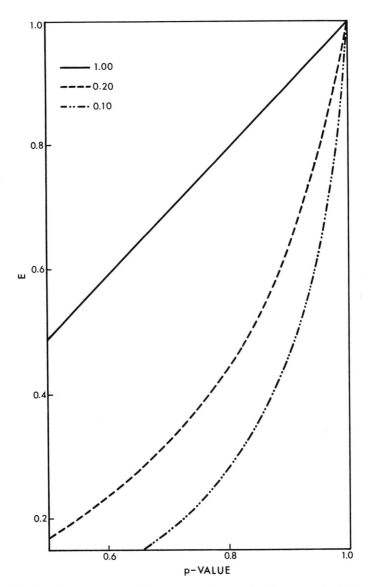

Fig. 5. Fractional amount extracted (E) versus the p value for unequilibrated water: solvent systems for α values of 0,10, 0.20, and 1.00.

b. Ultraviolet Spectroscopy

$$E = 1 - \frac{A_s - A_n}{A_s} \frac{V_p}{V_h} \tag{10}$$

where $A_s - A_n$ is the amount in the water phase after extraction and equilibration. V_h is the water volume phase to be extracted. The rest of the terms have been defined above. If the solvent absorbs at the wavelength maxima, it must be stripped from the water phase, whereupon

$$E = 1 - \frac{A_s - A_n}{A_s} \tag{11}$$

The number of micrograms left in the water phase $(A_s - A_n)$ is the only experimental variable. A_s is known. A_n is determined from a standard UV curve.

5. Determination of the Volume Correction Factors: V_n/V_s, V_p/V_h, V_n/V_p

The admixture of two unequilibrated solvents at constant temperature will show that each solvent's volume will change after equilibration. This is caused by mutual solubility (74).

Beroza and Bowman's p values were determined originally with equal volume solvent pairs that were equilibrated overnight at a constant temperature (71). This procedure eliminated volume changes that occurred during LLE and allowed direct comparisons of p value data. Later experiments showed that p values determined with unequal volumes of unequilibrated solvent phases were within experimental error of p values determined with equal volumes of equilibrated phases (68). The mutual solubility phenomenon requires that appropriate correction factors be determined and applied.

6. p Values of Some Organic Pesticides

Suffet (73) reported p values for the parent molecule, oxon, and hydrolysis products of diazinon, Baytex, parathion, and malathion. Some of these p values are given in Table 9.

7. Reproducibility of the p Value Determination

Bowman and Beroza have stated that for single extractions, "the p value deviated by 0.02 or less from those carefully determined in a six-tube countercurrent distribution" (72) and "we can reproduce p values consistently within 0.02" (71). In other words, the precision of the one-step p value determination is 1.00 ± 0.02.

TABLE 9
p Values of Some Organophosphates[a,b]

Compound	pH	Hexane	Benzene	Ethyl acetate	Ether
Diazinon	7.4	0.95	0.99	0.95	0.99+
Diazoxon	7.4	0.84	0.99	0.95	0.95
Baytex	3.4	0.93	0.97	0.93	0.95
Bayoxon	3.4	0.91	0.99	0.98	0.97
Parathion	3.1	0.89	0.88	0.84	0.93
Paraoxon	3.1	0.77	0.99	0.98	0.99
Malathion	6.0	0.98	0.99	0.99	0.99
Malaoxon	6.0	<0.20	0.99	0.99	0.97

[a]From Ref. 73.
[b]Extractive conditions were: $25 \pm 0.5°C$, $0.2\,M$ phosphate buffers to establish pH; p values are the average of three trials from unequilibrated solvents and with one-step extraction.

8. CRITERIA FOR QUANTITATIVE LIQUID–LIQUID EXTRACTION

The ultimate goal of the LLE step is 100% recovery of the solute from water. Some general criteria to achieve this goal are presented in Table 10(73). The selection of a solvent system which yields a high p value will enable the use of smaller sample volumes and/or smaller solvent volumes. The choices of the proper solvent-to-water ratio and the number of successive extractions for maximum recovery are considered below.

TABLE 10
General Criteria for the Quantitative Extraction of Organic Pesticides from Aqueous Systems

(1) The choice of a solvent with the highest p value.

(2) The setting of aqueous conditions to stabilize the solute, i.e., stop hydrolysis.

(3) The setting of aqueous conditions for best recovery, i.e., pH.

(4) The choice of the smallest aqueous volume which will give a sufficient amount of pesticide for quantitative analysis.

(5) The choice of a solvent-to-water ratio to give maximum recovery in minimum volume and/or minimum steps.

(6) The choice of the minimum number of times to reextract the sample and give the maximum recovery in the minimum volume and/or minimum steps.

a. Selection of the Solvent-to-Water Ratios

Beroza and Bowman(67) demonstrated that when single extractions of unequal volumes were exercised, the differences in E values became greater for solutes with low p values when the α values were increased. For solutes with high p values, the differences in E values became greater when α values were decreased. These relationships between p, E, and α values are shown in Fig. 5.

b. Selection of the Number of Extractions

Equation (8) may be rearranged to:

$$E = \frac{\alpha p}{\alpha p - p + 1} \tag{12}$$

This form can be used for a single extractive step in unequal, unequilibrated, two-phase solvent systems. If however, multiple extractions are desired, then:

$$F_n = E_1 + E_2 + E_3 + \cdots \tag{13}$$

or

$$F_n = \frac{\alpha_1 p}{\alpha_1 p - p + 1} + \frac{\alpha_2 p}{\alpha_2 p - p + 1}(1 - E_1) + \frac{\alpha_3 p}{\alpha_3 p - p + 1}(1 - (E_1 - E_2)) + \cdots \tag{14}$$

where F_n is the total fraction of solute extracted after n extractions. When the E or the p value is known and the α value remains constant, the number of extractions necessary to reach a desired F_n value can be determined from Fig. 6(73).

9. DISCUSSION

As noted above, the p value concept is an extremely useful tool in the liquid–liquid extraction of organic pesticides from water. Choice of solvent, solvent-to-water ratio, pH value of extraction, and number of extractions may be made from p value determinations. Unfortunately, few p values have been reported for water-solvent systems. Some have been determined for a few organophosphates and are listed in Table 9(73).

At a pH value of 7.4, benzene, ethyl acetate, and ether are excellent solvents for diazinon and diazoxon. Diazinon has been extracted from water by 1:1 ether:petroleum ether and chloroform(57), and by benzene (49). Per cent recoveries for the whole analytical procedures were 90 and 84%(57), and 98.4%(49), respectively. These recoveries by benzene are explained on the basis of p values reported for diazinon in Table 9.

At a pH value of 3.1, ether appears to be the best solvent for extrac-

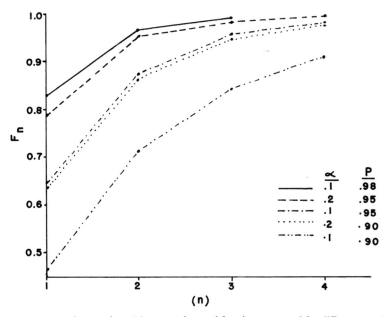

Fig. 6. Number of extractions (*n*) versus the total fraction extracted for different combinations of *p* and α values.

tion of parathion from water when based upon the *p* value concept. Other investigators have utilized 1 : 1 ether : petroleum ether and chloroform (57), benzene (49,56), hexane (52), and 1 : 1 ether : hexane (51), as noted in Table 6.

At a pH value of 3.4, benzene and ether are excellent solvents for the recovery of Baytex from water. Hexane and ethyl acetate yield slightly lower recoveries with a *p* value of 0.93, but may be used also. Baytex was extracted by hexane from acidified waters with a reported recovery greater than 90% (57).

At a pH value of 6.0, hexane, benzene, ethyl acetate, and ether are excellent solvents for the recovery of malathion from water in accord with the *p* values in Table 9. Pionke et al. (49) reported an extractive recovery of 99.7% for malathion with benzene from distilled water and seven natural waters.

The pH values cited in Table 9 have been determined as optimum for the extraction of diazinon, parathion, baytex, and malathion from water (73). As the organophosphate hydrolyzes in water, the hydronium ion concentration affects the kinetics of this process. Thus, the recovery of the unhydrolyzed parent molecule from water systems is affected by the pH value of the water either as sampled or as adjusted for solvent extraction. Diazinon is most stable in neutral pH ranges(75); thus, a pH of 7.4 was chosen as the extractive value. Parathion is most stable in acid solutions(76); thus, an extractive pH value of 3.1 was selected for parathion. Subsequently, the pH values of 3.4 and 6.0 were chosen for Baytex and malathion(73).

IX. Cleanup Techniques

A. General Requirements

Coextracted organic interferences may arise from domestic and industrial pollution, formulation impurities and solvents, and naturally occurring substances in surface waters and bottom muds. For example, the carbon chloroform extracts (CCE) from the CAM contain oxygenated organics that interfere with the identification of chlorinated hydrocarbons. o-Nitrochlorobenzene, phenyl ether, phenols, ketones, aldehydes, and numerous aromatic compounds in microgram per liter concentrations have been isolated from various rivers by the CAM(7). Also, a cleanup step is necessary when grossly polluted waters are extracted.

B. General Application

1. CHROMATOGRAPHIC TECHNIQUES

Partition chromatography, with columns of a silica-type support, appears to be the prevalent cleanup technique for natural water and bottom mud extracts. Table 11 summarizes these efforts. These supports have been investigated in order to standardize cleanup procedures(83,84). Very few evaluations have been reported for separation efficiencies. Furthermore, little effort was given to identification of interferences. The criterion for successful separation came from recovery studies on the entire analytical procedure. Apparently, the investigators were satisfied if the detection system exhibited minimum interference when compared with a reference standard. In some cases, heavy reliance was placed upon the identification step for additional separation, as exemplified by the procedures of paper, thin layer, and gas–liquid chromatography. In any event, it would seem prudent to include an efficient cleanup step prior to the steps of identification and quantification if gross pollution is encountered.

TABLE 11
Column Chromatography Cleanup Systems Applied to the Separation of Organic Pesticides from Interferences in Natural Waters and Bottom Muds

Pesticide(s)	Support	Elution solvent	Confirmatory technique	Reference
CH	Alumina	CHCl₃	IR	6
Toxaphene	MgO–celite	Pet. ether : ether	PC	77
Parathion	Alumina	CHCl₃	Color	78
2,4-D	Silicic acid	n-Hexane	Color and UV	60
OP	Florisil–water	—	PC	9
2,4-D esters	Aluminum silicate-muds	—	MC	9
Toxaphene, BHC	Florisil	Pet. ether : ether	MC	40
DDT	Florisil	Pet. ether : ether	MC	8
CH	Silica gel	Isooctane, benzene, and mixture of CHCl₃ and CH₃OH	IR	79
CH and triazines	Silica gel	Trimethyl pentane, benzene and mixture of CHCl₃ and CH₃OH	PC	80
CH	Silica gel	Hexane and 2, 40, 60, 70% benzene in hexane	GLC	81
Parathion, diazinon, malathion	Silica gel	Benzene and ethyl acetate	GLC	81
Dursban	Silicic acid	n-Hexane : methylene chloride	GLC	82

Some support for the above thesis is provided by Hindin et al. (9) and Epps et al. (31). The former group observed that "most water samples from surface and subsurface sources and stream sediment samples contained a considerable amount of organic matter other than pesticides." Hindin et al. were unable to cleanup water and mud samples sufficiently with Florisil, since streaking occurred on paper chromatograms of chlorinated hydrocarbons. No difficulties were encountered, however, with samples containing organophosphates. Epps' group encountered problems of cleanup and confirmation in a monitoring study of waters and bottom muds for chlorinated hydrocarbons. These investigators found it necessary to establish detection levels: "No attempt was made to realize the ultimate sensitivity of the analytical techniques. For water, the minimum detection objective was set at 0.001 ppm. For most other samples the target level was 0.01 ppm. In some cases interferences were so great that amounts less than 0.1 ppm could not be reported with confidence" (31).

2. Separation from Natural Color Compounds

Some mention should be made of the special problem of separating organic pesticides from naturally occurring color compounds in water. This is especially pertinent to lakes, ponds, and reservoirs. Black and Christman (85) have characterized natural color bodies in water as negatively charged colloids of vegetable origin. Chemically, they are fulvic acids or aromatic polyhydroxymethoxycarboxylic acids.

Aly and Faust (60) reported successful separation of 2,4-D, 2,4-dichlorophenol, and organic color (65 units) in a natural water by partition chromatography on a silicic acid column. n-Hexane served as the mobile solvent, whereas a 1:9 mixture of glacial acetic acid and methanol was employed as the immobile solvent. Recoveries of 100% were observed for 2,4-D when measured by visible and ultraviolet spectrophotometric techniques. Faust and Hunter (86) applied a rather unique technique to separate natural organic color from diquat and paraquat in water. These aquatic herbicides are highly polar, cationic, quaternary dipyridylium compounds that cannot be extracted from water by organic solvents. Complete removal of organic color was effected by passage through an anion exchange column which permitted a mean recovery of 100% of diquat and paraquat from several natural waters.

3. Suggested Techniques

Several chromatographic and extraction cleanup techniques from other residue systems may have some application to natural waters. For example, Crosby and Laws (87) used thin layer chromatography for separation of organic interferences in the CCE from aldrin, DDD, DDE, DDT, dieldrin, endrin, heptachlor, heptachlor epoxide, and lindane. This study was designed to clean up these chlorinated hydrocarbon pesticides prior to identification and quantification by gas-liquid chromatography. An average recovery of 91% (85.0–97.7% range) was claimed for six of the nine pesticides.

Storherr and Watts(88) describe a single-sweep codistillation method for cleanup of crude crop extracts. A sample of extract is injected into a heated glass tube filled with glass wool. A nitrogen flow carries the gaseous sample through the tube to a cooling bath and collection tube. The organic interferences remain on the glass wool and the pesticides are collected for analysis. The cleanup compared favorably with an adsorption column chromatographic method. The method "eliminates the need for specialized absorbents and equipment, large volumes of costly purified solvents and laborious cleanup methods."

Hamence et al. (89) describe a cleanup procedure by which interferences are eliminated with an alumina column subsequent to an acetone extraction of chlorinated hydrocarbons from animal viscera. Identification of the pesticide was by a derivative technique, with resolution by gas–liquid chromatography. Hartman(90) describes a gas–liquid chromatographic procedure for the cleanup of pesticide residues prior to their TLC determination. Recoveries ranged from 85 to 105% for 25 chlorinated hydrocarbons and organophosphates.

Beroza and Bowman(70) suggest a Craig countercurrent type of distribution system for extraction and cleanup of organic pesticides from foods prior to identification. Data are given on the extraction behavior of 25 insecticides in 19 binary solvent systems in terms of a p value.

C. Limitations

It is somewhat difficult to cite limitations of cleanup techniques that have been applied to natural water extracts. Very few data have been offered that indicate specific effectiveness of a given cleanup system. Efficiency has been measured solely by per cent of recovery of a particular pesticide. This may have resulted from a lack of information about the exact chemical species of the organic interferences, which made their separation from organic pesticides more difficult to measure directly. Also, some of the cleanup responsibility was shifted to the identification technique. This is not always acceptable, since some interferences may have the same R_f values or retention times as the pesticide. It is obvious that the current philosophy is to employ some sort of a silica column for gross cleanup of natural water extracts, with heavy reliance on techniques for identification and quantification to complete the separation.

X. Chromatographic Separation and Tentative Identification of Organic Pesticides and Their Metabolites

The chromatographic literature is filled with numerous techniques for the separation, tentative identification, and estimation of organic pesticides from soil, water, food, animal, and plant extracts. In essence, any methodology reduces to the technologies of column construction and detector specificity. Thin layer and gas–liquid chromatographic techniques offer the analyst several advantages: (a) cleanup prior to other methods of separation and confirmation, (b) separation from other pesticides, (c) separation of parent molecules from their metabolities, and (d) tentative identification of the pesticide and/or metabolite. The latter three advantages are emphasized below.

A. Thin Layer Chromatography

Although this technique has a high capacity to resolve pesticides from extraneous matter, it has rarely been applied to aquatic extracts. Abbott et al.(61) presented a TLC procedure for separation and identification of phenoxyalkyl acids from soil and water extracts. Identification was accomplished by comparison of the R_f value with a known standard. An average recovery of 92.5% from spiked soils and waters was claimed. Interferences in the aromatic fraction of the CCE from the CAM were resolved from five chlorinated hydrocarbons(90a). These spots were eluted with 1:1 ether:petroleum ether with subsequent identification by GLC. Table 12 summarizes pertinent conditions required for TLC separation of several classes of organic pesticides; some of these conditions have been applied to aquatic extracts.

B. Gas–Liquid Chromatography

Most applications of gas–liquid chromatography to extracts of aqueous samples have been concerned with separation and identification of parent compounds. Very little attention has been directed toward hydrolytic and metabolic products. Furthermore, most efforts have been applied to chlorinated hydrocarbons. The pertinent information gathered from aquatic situations is summarized below.

TABLE 12
General Conditions Required for Thin Layer Chromatographic Separation of Organic Pesticides

Pesticides	Absorbents	Solvents	Developers	Sensitivities	Reference
Nine CH	Silica gel G	CCl$_4$	AgNO$_3$–UV; Br$_2$–fluoroscein-AgNO$_3$	—	90a[a]
CH	Aluminum oxide G, Silica gel G	n-Heptane: acetone	AgNO$_3$-2-PE	0.01, 0.05, 0.1 µg	91
Five phenoxy acids	Silica gel G and Kieselguhr G	Liquid paraffin, benzene, gl. acetic acid	AgNO$_3$-2-PE	—	61
14 OTP	Aluminum oxide G	15 and 20% DMF/MCH	AgNO$_3$-citric acid-TBPP	0.05, 0.1 µg	92
14 OTP	Alumina-neut., silica gel G	Benzene: acetone	BPB	0.1–0.2 µg	93
Diazinon, parathion	Silica gel G	Acetone	0.7% EtOH CHCl$_3$	—	94
2 OTP; nine CH	Silica gel G	CCl$_4$	AgNO$_3$-UV	—	95
Eight triazines	Silica gel G	CHCl$_3$. Acetone	Br$_2$-Brilliant green	—	96[a]
10 OTP	Kieselguhr G	15% Acetone in hexane	Br$_2$-enzymes	5.0 ng	97
Seven OTP	Silica gel G	1,2-DCE: benzene	4-NP	2.0 µg	98
44 CH	Aluminum oxide G	n-Heptane, acetone	AgNO$_3$-2-PE	—	99

[a] Used on aqueous extracts.

1. CHLORINATED HYDROCARBONS

These pesticides and/or their metabolites are usually separated on columns of such relatively nonpolar liquid phases as DC–200, SE–30, and Dow–11 grease and on such polar phases as QF–1 and DEGS. [Relative polarity of liquid phases is classified by Brown(*100*).]

Table 13 summarizes the conditions used to separate chlorinated hydrocarbons extracted from water and bottom muds. These data were selected on the basis of prevalent use in monitoring studies and apparent success in resolution of a considerable number of parent molecules and metabolites on a given column. The appropriate sensitivities are also given. Appendix D lists the several chlorinated hydrocarbons (with retention times) that were resolved on the columns of LaMar et al.(*44*), Cueto and Biros(*45*), and Breidenbach et al.(*79*). These resolutions and retention times should serve as guidelines for the analyst.

2. ORGANOPHOSPHATES

There is little consistency in the use of either a nonpolar or a polar liquid phase for separation of the relatively polar organophosphates and their metabolites. Such liquid phases as DC–200, SE–30, Carbowax 20 M, and Reoplex-400 have been used with apparent success.

Operational conditions for the gas–liquid chromatographic separation of the organophosphates are cited in Table 14. These data were chosen on the basis of the criterion of apparent success in resolution of several parent molecules, metabolites, oxons, and hydrolysis products on a specific column. Appendix E lists the several compounds that were resolved on the columns of Suffet et al.(*102*), Ruzicka et al.(*103*), and Hariguchi et al.(*104*).

Special note should be made of a gas–liquid chromatographic method for determination of the hydrolytic metabolites of organophosphorus compounds in cow urine. St. John and Lisk(*105*) were able to separate the dimethyl or diethyl esters of phosphate, thiophosphate, or dithiophosphate on a 2% Ucon polar liquid phase at a column temperature of 105°C. The method includes an ether extraction of the acidified urine followed by methylation of the phosphoric acids. A thermionic detector was employed. These hydrolytic metabolites have not been reported under the conditions described in Table 14 and must be determined separately.

3. PHENOXYALKYL ACIDS AND ESTERS

Very few techniques have been conceived for gas–liquid chromatographic separation of the phenoxyalkyl acids and their esters in extracts of

TABLE 13

Summary of Conditions Used to Separate Chlorinated Hydrocarbons by Gas–Liquid Chromatography for Extracts of Water

Column construction	Column length, ft	Column temp, °C	Detector	Sensitivity claimed	Reference
1a. 5% Dow-11 on 60/80 Chromosorb W	5	185	EC	5.0 ng/liter	*10,44*
1b. 5% QF-1 on 60/80 Chromosorb W	5	185	EC	5.0 ng/liter	*10,44*
2a. 3% QF-1 on 70/80 Chromosorb W	6	180	EC	0.1 μg/liter	*45*
2b. 2% Dow-200 on silanized Chromosorb G	4	180	EC	0.1 μg/liter	*45*
2c. 3% DEGS on 60/80 Chromosorb G	6	180	EC	0.1 μg/liter	*45*
3a. 5% Dow-200 on 60/80 Chromosorb P	4	180	EC	0.001 μg/liter	*79*
3b. 5% Dow-200 on 60/80 Chromosorb P	4	190	MC	0.002 μg/liter	*79*
Other systems where retention times were not reported					
4. 3.8% SE-30 on 80/100 Diatoport S	3,4	190,200	EC	0.2 μg/liter BHC	*30*
5. 5.0% DC-200 on gas Chromosorb Q	6	210	EC	0.1 μg/liter DDT	*101*

TABLE 14

Summary of Conditions Used to Separate Organophosphates by Gas–Liquid Chromatography

Column construction	Column length	Column temp, °C	Detector	Sensitivity claimed	Reference
1. 15% Reoplex-400 on 80/100 Chromosorb W	3,4 ft	180	FID, EC	–	102
2. 10% SE-30, 1% Epikote on 80/100 Chromosorb W	150 cm	165	NaTh	0.5 ng	103
3a. 1% SE-30 on 80/100 Chromosorb W	2 m	125	EC	0.2 μg	104
3b. 1% QF-1 on 80/100 Chromosorb W	2 m	130	EC	–	104
3c. 1% DEGS on 80/100 Chromosorb W	2 m	145	EC	0.2 μg	104
Other systems					
4. 5% DC-200 on 60/80 Chromosorb P	6 ft	195	MC	50–100 μg/liter	57
5. 25% Carbowax 20m on 80/90 Anakrom ABS	5 ft	195	NaTh	0.01 μg	53
6. 5% Dow-11 on 60/80 Chromosorb W	6 ft	270	FID	0.05 ng/liter	55
7. 2% Ucon polar on 80/100 Gas Chromosorb Q	6 ft	105	NaTh	0.05 ng/liter	105
8. 5% SE-30 on 80/90 Anakrom ABS	4 ft	200	EC	0.1 μg/liter	82

natural waters and bottom muds. Gutenmann and Lisk(106) reported a column for the resolution of 2(2,4,5-TP) from interferences. The operational conditions were: (a) a column of 5% Dow silicone on 80/100-mesh Chromosorb W, (b) column length of 6 ft, (c) column temperature of 200°C, and (d) an electron capture detector. A sensitivity of 0.05 mg/liter was claimed.

Goerlitz and LaMar(62) provided some data on the application of microcoulometric detection of 2,4-D, 2,4,5-T, and 2(2,4,5-TP) in water extracts. These herbicides, free acid form, were extracted from acidified water with 1:1 (v/v) ether:benzene and were methylated before resolution under the following column conditions: (a) 10% QF-1 on 60/80-mesh, acid-washed Chromosorb W, (b) column length of 5 ft, and (c) column temperature of 165°C. A sensitivity of 10 ng/liter was claimed. Recoveries of these three herbicides from distilled water ranged from 73 to 95%.

C. Limitations of Chromatographic Techniques

Thin layer and gas–liquid chromatographic techniques are not without limitations and problems(107). Neither technique alone can suffice for the positive and absolute identification of parent compounds, metabolites, and hydrolysis products extracted from any medium. Each technique should be utilized in conjunction with some other form of analytical identification.

It may not be possible to develop a single, universally applicable system for isolation and separation of parent compounds, metabolites, and hydrolysis products because of widely different polarities of individual compounds. Therefore, the use of a single-solvent system for TLC or a single column for GLC for resolution is very unlikely. The residue analyst should run several trials before satisfactory conditions are established for his particular system. R_f values and relative retention times should be determined for each system under consideration. The analyst will discover that sample aliquots to give minimum detectable amounts of a particular pesticide will not always clean up to the point of no interference throughout the chromatogram. The collection of developed zones or gas–liquid chromatographic peaks is often arbitrary and is done with considerable error.

Thin layer chromatography has some virtues. Kovacs(91) claims that "thin layer chromatography is superior to paper chromatography because spots are more compact and resolution is much sharper." Smith and Eichelberger(90a) state: "The most important advantage of the method is that it provides corroborative evidence of the presence or absence of a particular pesticide. The first part of the evidence is that the pesticide occurs in a specific TLC-separated section. The second part is that the peaks of the gas chromatogram obtained from a particular section possess the required retention times of pesticides known to migrate to this section."

The application of gas–liquid chromatographic techniques for positive and absolute identification of unknown pesticides extracted from aqueous environments is a most serious fallacy. The authors are in firm agreement with Robinson and Richardson (108), who state:

> It is quite invalid to infer, on the basis of the gas chromatographic retention time determined with only one stationary phase, that a particular insecticide is present.

Benyon and Elgar (109) reemphasized this point:

> In the absence of a valid control sample, GLC analysis cannot provide positive identification of a particular pesticide when only one retention time value is obtained. Many naturally occurring products respond to GLC detectors and can be mistaken for pesticides, and often such natural products are not completely removed by the cleanup procedures that are used.

Confirmatory data must be obtained from such auxiliary procedures as thin layer and paper chromatography and various spectrophotometric and spectroscopic techniques.

In addition, there is every reason to suspect the occurrence of a metabolite or a hydrolytic product in the event of a pesticide water quality problem. There is no one column that will resolve pesticides and their degradation products, nor is there one detector sufficiently selective and specific. This suggests the multicolumn and detector approach for identification.

As an example of GLC operational difficulties, Hindin et al. (9) evaluated several columns, temperature conditions, and three detectors for separation of organic pesticides from one another and from interferences in lake sediments. Although extracts of these samples were passed through an activated aluminum silicate column, multiple leaks on the chromatograms prevented interpretation and detection of any pesticides. This was observed with a hydrogen flame detector using a 4-ft, 10% Dow-11 plus 2% Epon-1001 on 60/80-mesh Chromosorb W column programmed at 20°/min from 200 to 260°C. Multiple peaks were also observed with an electron capture detector using a 5-ft, SE-30 silicone oil on 60/80-mesh Chromosorb W column operated isothermally at 200°C. Subsequently, a microcoulometric detector was subjected to sediment extracts wherefrom three peaks were confirmed as the three isomers of DDT by relative retention times and by thin layer chromatography. Hindin and co-workers concluded that the microcoulometric detector would be more specific but less sensitive for field samples containing interferences than an electron capture detector. The latter could be used only in situations "for extremely clean extracts."

Holden and Mardsen(30) provide another excellent example of the many problems that may be encountered in the examination of surface waters and sewage effluents for γ-BHC, dieldrin, p,p'-DDE, p,p'-TDE, and p,p'-DDT. Column construction was the first major problem: A 3.8% SE-30 silicone oil on Diatoport S was unable to separate dieldrin from p,p'-DDE. A 3.0% apiezon L grease + 0.3% Epikote resin 1001 on Chromosorb W/HMDS was able to effect this separation, but some degradation of p,p'-DDT to p,p'-TDE was observed. Sewage effluent hexane extracts required an alumina column cleanup prior to GLC or TLC. Dieldrin, DDE, TDE, and DDT were found and confirmed in sewage extracts, but 10–12 other peaks frequently appeared on the chromatograms. These unknown peaks were suspected to be other chlorohydrocarbons, but confirmation was not obtained.

In summary, chromatography is a powerful analytical tool in which organic pesticides can be separated from one another and from gross quantities of interferences. Furthermore, these separations occur where quantities of the pesticides are usually less than 10^{-3} g. On the other hand, no one chromatogram can separate, nor can one detector detect all organic pesticides; i.e., there is no universal procedure. Confirmation of GLC peaks and TLC spots is another important consideration, especially when the original sample is obtained from an extremely polluted aqueous environment. Thus, the analyst must be prepared for a trial and error approach when attempting organic pesticide analysis by GLC and TLC.

D. Detectors for Gas–Liquid Chromatography

Many devices are available for the detection of organic pesticides following separation on gas–liquid chromatographic columns. The choice of detector will depend upon sensitivity and specificity. It is often assumed that the detector of choice can differentiate an organic pesticide molecule from an interfering molecule. Also, it is assumed that cleanup operations have separated the interferences completely from the organic pesticide molecule. This is a misconception, as cleanup operations are not perfect. These are some of the fallacies that surround many of the efforts to detect organic pesticide residues in water. Westlake and

Gunther(*110*) have compiled a detailed review of the chronological development of several gas chromatographic detectors. Excerpts of this paper follow as well as recent developments.

1. NONSPECIFIC DETECTORS

Harley et al.(*111*) introduced the flame ionization detector in 1958. It has found application in pesticide residue analysis where specificity is not required, i.e., extremely pure extracts, laboratory development of analytical techniques, etc. One major advantage is base line stability. Practical minimum detectability lies in the microgram area, although nanogram sensitivity has been reported.

An extensively used nonspecific detector is the electron capture detector of Lovelock and Lipsky(*112*). This has a very low ideal response in the picogram area, although a more practical minimum detectability would lie in the nanogram range. This device is often misused for specific detection of organic pesticide residues. Westlake and Gunther(*110*) emphatically state:

> For pesticide residue analysis, the electron–capture detector is the next-to-last choice, primarily because of its nonspecificity. Data obtained by use of this detector can be considered valid only if they are confirmed by at least one other method.

Some theoretical aspects and practical difficulties of electron capture detectors have been discussed by Gaston(*113*). For example, care must be exercised to avoid injection of extractives or bleeding of liquid phases that will destroy the detector's efficiency and necessitate cleaning the electron source.

2. SEMISPECIFIC DETECTORS

The detectors reviewed in this section have been labeled in the literature as specific for certain organic pesticides. This stems from the specificity with which the device detects an atom, a molecule, or a group peculiar to the pesticide. By the same token, there is the possibility that an interference would have the same responding group and the same elution time as a pesticide. Hence, these detectors are semispecific in a strict interpretation.

a. Coulometric Detectors

Coulson and associates introduced a halogen-detecting system to the pesticide residue analysis around 1960(*114*). Later, a sulfur cell(*115*) and a nitrogen cell(*116*) were developed. The principle of microcoulometric detectors lies in the oxidative combustion of effluent gases from the chromatographic column to produce a halide ion, sulfur dioxide, or ammonia which are titrated automatically. This system claims to be specific (certainly more so than the electron capture) but is, on occasion, less sensitive than the electron capture.

b. Thermionic Detectors

The flame ionization detector was modified by Karmen(117) and Karmen and Giuffrida(118) to yield an enhanced response to phosphorus and chlorine. The essentials of these modifications lie in the suspension of a circular electrode coated with sodium sulfate above the flame. This development signified the first time that gas chromatographic techniques could selectively measure organophosphorus pesticides that did not contain sulfur or halogens. Several modifications of the original thermionic detector have appeared, of which the most notable are the "stacked flame" of Abel et al.(119) and the cesium bromide cell of Hartmann(120). All of these modifications of the flame ionization detector claim increased stability, sensitivity, minimum detectability, and simplicity. Selectivity for organophosphorus compounds is based upon the hydrogen flow rate of 14–16 ml/min where organochlorine and normal hydrocarbons do not respond.

c. Electrolytic Conductivity Detector

Coulson(121) exploited electrolytic conductivity as a detection system for nitrogen-containing organic pesticides. This system involves the reductive combustion of gas chromatographic effluents whose ammonia product is dissolved in distilled water. Any change in electrical conductivity of this aqueous solution is recorded. The Coulson system was the first practical detector for nitrogen compounds. Nanogram sensitivity is claimed. Also, the detector is reported to be insensitive to nitrogen-free compounds, which suggests that minimal or no cleanup of extracts is required. Under oxidative combustion conditions, halogen- or sulfur-containing pesticides can also be determined with good selectivity and sensitivity.

d. Flame Emission Detector

Brody and Chaney(122) developed, independently from other researchers, a flame emission detector specific for phosphorus and sulfur. This can be considered a modification of the flame ionization system in which interference filters are used to isolate specific spectral emissions from the flame. These emissions are recorded through photomultiplier systems. This detector is especially selective and sensitive to organic pesticides containing phosphorus and sulfur atoms.

TABLE 15

Detectors Available with Gas–Liquid Chromatographic Systems

Detector	Major developers		
	Responder	Minimum detectability[a]	Reference
FID	Carbon	μg	*111*
EC	Electron transfer	ng	*112,113*
MC	Cl, S, N	ng, μg, μg	*114–116*
Thermionic	Cl, P	ng	*118, 118*
Electroconductivity	N	ng	*121*
Flame emission	S, P	ng	*122*

[a]In a practical sense — from extracts after some cleanup.

A summary of the detectors available for gas–liquid chromatographic systems appears in Table 15.

3. OTHER SYSTEMS

Several other detectors are in the development stage but should be mentioned herein because they may find application for the detection of pesticide residues in aqueous extracts. The microwave-excited spectral systems of McCormack et al. (*123*) and Bache and Lisk (*124*) are selective for Cl-, S-, and P-containing pesticides. This system has found application by Bellet et al. (*125*).

Even some dual detector systems have been developed. For example, Giuffrida et al. (*126*) examined a dual system of potassium thermionic (KTh) and electron capture detectors arranged in parallel. Wessel (*127*) extended this concept to arrangements whereby the detectors were in series.

XI. Confirmatory Techniques

A. General Considerations

An editorial appeared in the June, 1968, issue of the *Pesticides Monitoring Journal* (*128*) entitled, "The Need for Confirmation." The lead paragraph read:

With the increased amount of analytical work being done in the surveillance and monitoring of pesticides in our environment and on pesticide residues in general, it seems appropriate to reiterate the importance of confirming the identity of pesticides being reported. Confirmation is particularly important with samples from the environment when little or no information is available concerning the pesticides likely to be present.

Many techniques and innovations are available for confirmation of the parent molecule or any one of numerous metabolites, oxidative products, or hydrolysis products. The analyst may choose from various combinations of TLC, GLC, PC, IR, UV, NMR, MS, etc. Identification, therefore, becomes a function of the analyst's imagination and ingenuity in devising an appropriate scheme. As shown in Table 16, some attempts have been made to identify pesticide residues extracted from water and bottom muds. The majority of these have been directed at the chlorinated hydrocarbons.

An excellent effort was made by Smith and Eichelberger[90a] to identify chlorinated hydrocarbons by a GLC procedure *following* a separation step by TLC. Multiple GLC peaks were observed in unseparated aromatic fraction of a CCE. A TLC technique was able to resolve gross quantities of unknown interferences from p,p'-DDE, o,p'-DDT, o,o'-DDT, and p,p'-DDT. Subsequent solvent elution of the TLC spot led to identification by relative retention times on a Dow-11 silicone grease column operating at 175°C. A similar approach was utilized by Kawahara et al. [132] for 14 chlorinated hydrocarbons extracted from a waste water.

Teasley[130] was able to identify a cholinesterase inhibitor, S,S,S-

TABLE 16
Summary of Attempts to Confirm Pesticide Residues from Water and Bottom Mud Extracts[a]

Pesticide(s)	Extraction	Separation	Confirmation	Reference
DDT	CAM	GLC	TLC	9
CH	Hexane	GLC	Two columns-GLC	44
CH	CAM	TLC	GLC	90a
CH	Hexane	GLC	TLC	30
CH	Pentane: ether	TLC	GLC	129
CH	Hexane	GLC	Three columns-GLC	45
CH	—	GLC	TLC, IR	31
OTP	CHCl₃	CC, GLC	IR, NMR, MS	130
Parathion	—	GLC	IR	9
Heptachlor	Hexane-IPA	GLC	TLC	48
Parathion	Ether: pet. ether	GLC	TLC	131
CH	Hexane: benzene	TLC	GLC	132

[a]Reprinted from Ref. *133*, p. 39, by courtesy of the American Society for Testing and Materials, Philadelphia, Pa.

tributyl phosphorotrithioate, extracted from water and bottom muds through a combination of column chromatography, GLC, IR, NMR, and MS. The general scheme was $CHCl_3$ extraction, cleanup on an activated Florisil column, elution with 6% ether and 94% petroleum ether and with 15% ether and 85% petroleum ether, and chromatography of the CC eluants on a column of 5% DC-200 oil on 30/60-mesh Chromosorb P at 180°C. Aliquots of the column chromatographic eluants were prepared concurrently for the IR, NMR, and MS analyses.

Many options are open to the analyst for verification of pesticide

TABLE 17
Flow Diagram for Analytical Routes Leading to Positive Identification of Organic Pesticides[a]

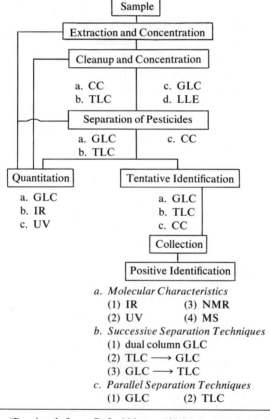

a. Molecular Characteristics
  (1) IR   (3) NMR
  (2) UV   (4) MS
b. Successive Separation Techniques
  (1) dual column GLC
  (2) TLC ⟶ GLC
  (3) GLC ⟶ TLC
c. Parallel Separation Techniques
  (1) GLC   (2) TLC

[a]Reprinted from Ref. *133*, p. 40, by courtesy of the American Society for Testing and Materials, Philadelphia, Pr.

identity. These are presented in Table 17. The usual approach has been to extract, clean up, and split samples to obtain confirmatory evidence from parallel procedures of GLC, TLC, IR, UV, etc. Another option would be to extract, clean up, separate by GLC, and collect the effluent peaks for identification. This option would offer an advantage wherein the fraction from the GLC is actually examined.

B. Confirmation from Molecular Characteristics

Here, the key for confirmation lies (a) in the rigorous cleanup of the aqueous extracts, (b) in obtaining sufficient quantities of the pesticide for examination by IR, UV, MS, and NMR spectrometry, and (c) in making sure that the extract is free from interferences prior to the confirmatory step. This is usually accomplished by a sequence of steps of extraction and chromatographic techniques of column, thin layer, or gas–liquid.

1. SPECTROPHOTOMETRIC TECHNIQUES

Spectrophotometric confirmation of an organic pesticide by ultraviolet and infrared techniques is possible because the radiation which is absorbed is specific and characteristic for a given organic molecule. Thus, a spectrum can be produced from infrared or ultraviolet absorption that can be matched with spectra of known compounds. Some compounds are extremely sensitive to infrared or ultraviolet energy absorption; hence they can be detected in very low quantities. Since the absorbing molecules are not destroyed, they may be retrieved and used for further examination by other confirmatory techniques (see Chaps. 20 and 30).

a. Infrared Spectra and Techniques

Numerous infrared spectra of chlorinated hydrocarbons have been published (6,134–136,138). A simplified chart of some of the more important infrared bands of several chlorinated hydrocarbons appears in Fig. 7 (6). Infrared spectra of organophosphate pesticides have also been published (87,133,139–141). Table 18 shows the characteristic absorption bands of organophosphate pesticides.

Blinn (137) describes several infrared techniques that are useful in measurement and identification of organic pesticides. One of the major problems is instrumental sensitivity. In general, only a few micrograms of the pesticide can be obtained in pure form by conventional techniques of separation. Therefore, microtechniques of infrared spectrophotometry must be employed, of which the two most common are: (a) collection and

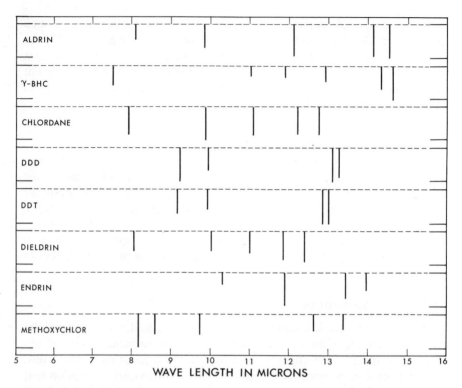

Fig. 7. Principal infrared bands of some chlorinated hydrocarbons. Reprinted from Ref. 6, p. 1730, by courtesy of the American Chemical Society.

concentration into such transparent solvents as carbon disulfide and carbon tetrachloride, and (b) collection by adsorption upon KBr pellets. Sensitivity of an infrared instrument may be increased by an increase in cell pathlength and ordinate scale expansion. Sensitivities of 5–10 μg (137) have been reported for the solution microtechnique and as low as 1 μg(138,142) for the KBr pellet microtechnique.

The organic pesticide molecule must be isolated in as pure form as possible. These isolations are affected usually by some chromatographic technique – column, thin layer, or gas–liquid. For example, Payne and Cox(138) used a combination of paper, column, and thin layer chromatography to isolate dieldrin and endrin for infrared examination in KBr pellets. Crosby and Laws(87) separated by gas–liquid chromatography several organophosphorus pesticides from which the effluent was collected in dichloromethane, evaporated to dryness, and dissolved in carbon disulfide for infrared examination.

An extremely useful system for the collection of GLC peaks was

TABLE 18

Characteristic Infrared Absorption Bands of
Organophosphate Pesticides

Structural group	Wavelength, $\mu^{a,d}$	Wavelength, $\mu^{b,d}$
P=O free[c]	7.9 m	8.0 s
P—O—C aromatic[c]	8.1 m	
P—O—C aromatic[c]	10.7 s	
P—O—C aliphatic[c]	9.7–9.8 s	9.6–9.9 s
P—O—ethyl	8.6 w	8.6 w, 10.2–10.5 s
P—O—methyl	8.4 w	8.4 w
P=S[c]	12–13 m	Near 12 wm
C—NO$_2$ aromatic[c]	7.4 s	7.5 s
C=O	5.7–5.9 s	5.8 s
P—O—C aromatic		
or		
P—O—C aliphatic		10–11[e] s

[a]From Ref. 87.
[b]From Ref. 141.
[c]Suitable for quantitative measurements.
[d]s, m, and w denote strong, medium, and weak intensities, respectively.
[e]Affected by type of group attached to C.

evolved from the KBr technique of Chen(143) and the dual column operation of Giuffrida(144). This system was employed by Faust and Suffet(133) to isolate diazinon, diazoxon, and 2-isopropyl-4-methyl-6-hydroxypyrimidine (IMHP) for infrared examination. This is illustrated in Fig. 8. This system was chosen for the following reasons:

(1) There is direct GLC to microspectrophotometric operation with minimum intermediate handling. This eliminates the contamination problem in handling microgram quantities.
(2) The collection of several peaks during one chromatographic run is possible.
(3) The ease of operation and repeatability.
(4) Collected fractions can be dissolved and rechromatographed.
(5) Recoveries of 50–75% have been obtained with this procedure.

b. Ultraviolet Spectra and Techniques

The characteristic ultraviolet spectra of organic pesticides may be used as supplemental information in confirmatory techniques. In other words, the match of an isolate's ultraviolet spectrum with a standard's spectrum would add another piece of evidence to such other confirmatory

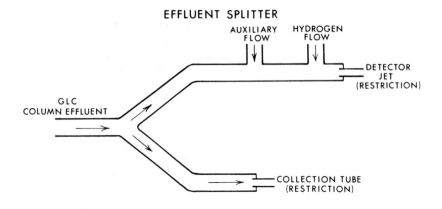

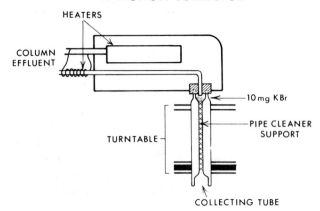

Fig. 8. Effluent splitter and fraction collector (KBr) for a gas–liquid chromatograph.

techniques as infrared spectroscopy, etc. Blinn and Gunther(*134*) cite a number of references for the ultraviolet spectral characteristics of several organic pesticides.

As in the case in infrared spectroscopy, microtechniques are often employed for the ultraviolet examination of organic pesticides. The separation and isolation techniques are usually column, thin layer, or gas–liquid chromatography. The pesticide is then dissolved in a suitable solvent for the UV region. Faust and Suffet(*133*) obtained the ultraviolet spectra of diazinon, diazoxon, and IMHP from GLC effluents that were described above. Suffet et al.(*28*) have reported the ultraviolet spectra of baytex and seven of its metabolites.

2. MASS SPECTROMETRY

Mass spectrometry is another tool for the confirmation of an organic pesticide. This instrumental technique involves the production of ions via bombardment of vaporized molecules with a beam of electrons. These ions are then separated according to their mass-to-charge ratio. A detection system measures and records the presence and intensity of the ionic fragments that correspond to a given mass-to-charge ratio.

This information leads to the molecular weight and structure of the original molecule. Once a mass spectrum is obtained for a molecule, it is matched with the spectrum of the standard or known compound. Quantities as little as 0.1 μg may be examined.

The mass spectra of several pure organic pesticides(145–149) have been reported. An example of the application of mass spectrometry is the one of Teasley(130) cited above. McFadden(150) describes the introduction of gas chromatographic samples into a mass spectrometer (see Chap. 29).

3. NUCLEAR MAGNETIC RESONANCE

A useful spectroscopic tool for the confirmation of organic pesticides is that of nuclear magnetic resonance. This technique determines the relative electronic environment of a proton or some other NMR-active element within the molecule, using radio frequency. An atom has NMR activity if the nucleus contains an odd number of protons and/or neutrons. A hydrogen atom attached to a carbon atom could have a different environment than one attached to a nitrogen atom. Even hydrogens attached to a carbon atom interact with each other, and a —CH$_3$ group has a different spectrum than a —CH$_2$— group. The area under an NMR signal is directly proportional to the number of nuclei giving the signal. Organic compounds have characteristic NMR spectra and those of several organophosphorus pesticides(151–153) have been reported. Teasley(130) used NMR to establish the identity of a chlorinesterase-inhibiting compound from an industrial effluent (see Chap. 20).

4. COMBINED TECHNIQUES

Several attempts have been made to combine directly the separatory and confirmatory steps. Several prototypes have been reported for GLC-IR(154), GLC-MS(155), and GLC-NMR(156). At present, it is expensive and often inconvenient to utilize these combinations on a routine basis. It is more convenient to collect the effluents from GLC, TLC, and CC for confirmatory analysis.

XII. General Discussion and Summary

In an unknown situation, the analyst faces a challenging problem in the recovery, separation, and identification of organic pesticides and their metabolites in aquatic environments. This problem is compounded by the several hundreds of species of pesticides and metabolites, all of which have different chemical and physical characteristics.

As the organic pesticide is potentially toxic when it is not used properly, the confirmation of its role as an etiological agent in fish kills, taste and odor problems, etc., becomes an extremely important consideration. The analyst must obtain a sufficient quantity of sample, which, in turn, should yield a sufficient quantity of the pesticide for quantitation and confirmation.

In aquatic situations, it is relatively easy to obtain a large enough sample for a complete residue analysis. Problems do arise, however, in the transportation and storage of the sample prior to analysis. In any event, it is recommended strongly that any sample of water or bottom mud should be analyzed as soon as possible after collection. This is especially true for the more unstable pesticides: organophosphates, carbamates, and ureas.

As direct analysis of the pesticide in the aqueous phase is virtually impossible, the analyst must employ some sort of an extractive system. In the course of extraction a concentration of the pesticide is usually effected. It is recommended that liquid–liquid extraction be the analyst's choice for recovery of the pesticide from the aqueous phase. The several advantages of this technique are cited above as well as guidelines for the proper selection of the solvent, the number of extractions, and the pH conditions of the water. Liquid–liquid extraction may be quantitative if the p value concept is utilized.

It may be stated that most aquatic environments contain analytic interferences. Consequently, a cleanup of the solvent extracts is recommended. This is usually accomplished by column chromatography, although it must be pointed out that TLC and GLC may be used primarily as cleanup procedures. Some cleanup is effected also in liquid–liquid extraction through selective solubility in various solvents. The inclusion of a cleanup step becomes imperative when confirmation of the pesticidal molecule is attempted later in the analytic scheme.

After rigorous cleanup, separation of organic pesticides from each other and from their metabolites is attempted through GLC and TLC. The former technique is used predominantly because of extremely low sensitivities. GLC and TLC provide tentative identification as well as separation of the molecules through retention times on a specific column or through R_f values. A word of caution must be mentioned here: GLC or TLC alone cannot suffice for the positive identification of an organic pesticidal molecule. An ancillary technique must be employed. Collective techniques are mentioned above for GLC peaks and for the elution of TLC spots.

Confirmatory techniques usually involve the separation and collection of a sufficient quantity of the pesticide in a purified form, whereupon some molecular characteristic is obtained. To date, the techniques of infrared and ultraviolet spectrophotometry have been utilized the most. However, mass spectroscopy and nuclear magnetic resonance techniques are being perfected and will be used more and more in future pesticidal analysis. Some attempts have been made to confirm pesticides through detector specificity, dual column GLC, TLC to GLC, GLC to TLC, and parallel use of GLC and TLC. It would be prudent to supplement these separation techniques with data from molecular characteristics.

Permissible concentrations have been suggested for organic pesticides in public water supplies (Table 2). These criteria are at the microgram per liter level. It appears that current analytic methodology is capable of detection and quantitative determination of microgram quantities of organic pesticides.

ACKNOWLEDGMENT

This publication was supported in part by Research Grant ES-00016, Office of Resource Development, Bureau of State Services, U.S. Public Health Service, Washington, D.C.

Appendix A. Abbreviations Used in This Chapter

BPB   Bromophenol blue
CAM   Carbon absorption method
CC   Column chromatography
CCE   Carbon chloroform extract
CH   Chlorinated hydrocarbons
DCE   1,2-Dichloroethane
DMF   Dimethylformamide
EC   Electron capture
FID   Flame ionization detector
GLC   Gas–liquid chromatography
IPA   Isopropyl alcohol
IR   Infrared
KTh   Potassium thermionic detector
LLE   Liquid–liquid extraction

MC   Microcoulometer
MCH   Methylcyclohexane
MS   Mass spectrometer
NaTh   Sodium thermionic detector
ng   Nanogram; 10^{-9} g
NMR   Nuclear magnetic resonance
4-NP   4-(p-Nitrobenzyl)-pyridine
OTP   Organothiophosphate
PC   Paper chromatography
PE   Phenoxyethanol
TBPP   Tetrabromophenolphthalein ethyl ester
TLC   Thin layer chromatography
UV   Ultraviolet

Appendix B. Chemical Names of Pesticides and Related Compounds Mentioned in Text

Common name	Chemical name
Abate	O,O,O',O'-Tetramethyl-O,O'-thiodi-p-phenylene phosphorothioate
Aldrin	1,2,3,4,10,10-Hexachloro-1,4,4a,5,8,8a-hexahydro-1,4-endo, exo-5,8-dimethanonaphthalene
Amizol	2-Amino-1,2,4-triazole
Atrazine	2-Chloro-4-ethylamino-6-isopropylamino-5-trazine
Azinphos-methyl	See guthion
Baytex (fenthion)	O,O-Dimethyl-O-[4-(methylthio)-m-tolyl]] phosphorothioate
Bayoxon	O,O-Dimethyl-O-[4-(methylthio)-m-tolyl)] phosphate
BHC	Mixture of stereoisomeric 1,2,3,4,5,6-hexachlorocyclohexanes
Chlordane	2,3,4,5,6,7,8,8-Octachloro-2,3,3a,4,7,7a-hexahydro-7-methanoindene
CIPC	Isopropyl N-(3-chlorophenyl) carbamate
Co-Ral	O,O-Diethyl O-3-chloro-4-methyl-2-oxo-2H-1-benzopyran-7-yl phosphorothioate
2,4-D	2,4-Dichlorophenoxy acetic acid
Dalapon	2,2-Dichloropropionic acid
2,4-DB	2,4-Dichlorophenoxy butyric acid
2,4-D esters	Butyl, butoxethanol, isooctyl, isopropyl, propylene-glycol-butyl ether
DBP	p,p'-Dichlorobenzophenone
DDD	1,1-Dichloro-2,2-bis(p-chlorophenyl) ethane
DDDE	1-Chloro-2,2-bis(p-chlorophenyl) ethylene
DDE	1,1-Dichloro-2,2-bis(p-chlorophenyl) ethylene
DDT	1,1,1-Trichloro-2,2-bis(p-chlorophenyl) ethane
DDVP	O,O-Dimethyl-2,2-dichlorovinyl phosphate
Delnav	2,3-p-Dioxanedithiol-bis(O,O-diethyl phosphorodithioate)
Demeton-O	O,O-Diethyl O-ethyl-2-thioethyl phosphorothioate
Demeton-S	O,O-Diethyl S-ethyl-2-thioethyl phosphorothiolate
Diazinon	O,O-Diethyl-O-(2-isopropyl-4-methyl pyrimidinyl) phosphorothioate
Diazoxon	O,O-Diethyl-O-(2-isopropyl-4-methyl-6-pyrimidyl) phosphate
Dibrom	1,2-Dibromo-2,2-dichloroethyl dimethylphosphate

Dieldrin	1,2,3,4,10,10-Hexachloro-exo-6,7-epoxy-1,4,4*a*,5,6,7,8,8*a*-octahydro-1,4-endo, exo 5,8-dimenthanonaphthalene
Dimethoate	O,O-Dimethyl S-(N-methylacetamide) phosphorodithioate
Dipterex	O,O-Dimethyl-trichloro-hydroxyethyl phosphonate
Diquat	1,1'-Ethylene-2,2'-bipyridylium dibromide
Di-syston (disulfoton)	O,O-Diethyl S-(2-ethylthioethyl) phosphorodithioate
Diuron	3-(3,4-Dichlorophenyl)-1,1-dimethylurea
DNBP	2-sec-Butyl-4,6-dinitrophenol
DNC	Dinitrocresol
Dursban	3,5,6-Trichloro-2-pyridylphosphorothioate
Endothal	2,3-Dicarboxylic acid, 7-oxobicyclo(2,2,1)heptane
Endrin	1,2,3,4,10,10-Hexachloro-6,7-epoxy 1,4,4*a*,5,6,7,8,8*a*-octa hydro-1,4,5,8-endo,endo-dimethanonaphthalene
EPN	O-Ethyl O-*p*-nitrophenyl phenyl phosphonothioate
Ethion	O,O,O',O'-Tetraethyl S,S'-methylene-bis-phosphonothioate
Fenac	2,3,6-Trichlorophenyl acetic acid
Fenchlorophos	See ronnel
Fenuron	3-Phenyl-1,1-dimethylurea
FW-152	4,4'-Dichloro-α-dichloromethylbenzhydrol
Guthion	S-(3,4-Dihydro-4-oxo-1,2,3-benzotriazin-3 ylmethyl)O,O-dimethyl phosphorodithioate
Heptachlor	1,4,5,6,7,8,8-Heptachloro-3*a*,4,5,5*a*-tetrahydro-4,7-endo-methanoindene
Heptachlor-epoxide	1,4,5,6,7,8,8-Heptachloro-2,3-epoxy-2,3,3*a*,4,7,7*a*-Hexahydro-4,7-methanoindene
IMHP	2-Isopropyl-4-methyl-6 hydroxypyrimidine
IPC	Isopropyl N-phenylcarbamate
Isodrin	1,2,3,4,10,10-Hexachloro-1,4,4*a*,5,8,8*a*-hexahydro-1,4,5,8-endo, endo-dimethanonaphthalene
Kelthane	4,4'-Dichloro-α-(trichloromethyl) benzhydrol
Lindane	γ-isomer of 1,2,3,4,5,6-hexachloro cyclohexane
Malaoxon	O,O-Dimethyl S-(1,2-dicarbethoxyethyl) phosphorothioate
Malathion	O,O-Dimethyl S-(1,2-dicarbethoxyethyl) phosphorodithioate
MCPA	4-Chloro-2-methyl phenoxyacetic acid
MCPB	α-(4-Chloro-2-methylphenoxy)butyric acid
Mecarbam	S-((Ethoxycarbonyl)methylcarbamoyl)methyl O,O-diethyl phosphoro-dithioate
Methoxychlor	1,1,1-Trichloro-2,2-bis(*p*-methoxyphenyl) ethane
Methyl parathion	O,O,-Dimethyl O-*p*-nitrophenyl phosphorothioate
Methyl systox	See demeton-O and -S, substitute methyl for ethyl
Mevinphos	See phosdrin
MMTP	3-Methyl-4-methylthiophenol
Morphothion	O,O-Dimethyl S-(morpholinocarbonylmethyl) phosphorothioate
Nankor	See ronnel
Nemacide	O-(2,4-Dichlorophenyl)O,O-diethyl phosphorothioate
OMPA	Octamethylpyrophosphoramide
Paraoxon	Diethyl *p*-nitrophenyl phosphate
Paraquat	1,1'-Dimethyl-4,4'-dipyridilium
Parathion	O,O-Dimethyl-O-*p*-nitrophenyl phosphorothioate
PCP	Pentachlorophenol

Perthane	2,2-Dichloro-1,1-bis(*p*-ethylphenyl) ethane
Phenkapton	O,O-Diethyl S-(2,5-dichlorophenylmercaptomethyl) dithiophosphate
Phorate	O,O-Diethyl S-ethylthiomethylphosphorodithioate
Phosdrin	Dimethyl 1-carbomethoxy-1-propen-2-yl phosphate
Phosphamidon	Dimethyl diethylamido-1-chlorocrotonyl-(2)-phosphate
Rhothane	See DDD
Rogor	See dimethoate
Ronnel	O,O-Dimethyl O-2,4,5-trichlorophenyl phosphorothioate
Rotenone	Cube and derris roots
Sevin	I-Napthyl-N-methylcarbamate
Simazine	2-Chloro-4,6 bis(ethylamino)-S-triazine
Strobane	Mixture of chlorinated terpenes
Sulfotep	Bis-O,O-diethylphosphorothionic anhydride
Sumithion	O,O-Dimethyl O-3-methyl-4-nitrophenyl phosphorothioate
Systox	See demeton
2,4,5-T	2,4,5-Trichlorophenoxy acetic acid
Telodrin	Octachlorohexahydromethanoisobenozofuran
TEPP	Bis-O,O-diethylphosphoric anhydride
2,4,5-T ester	Butoxyethanol
Thiodan	6,7,8,9,10,10-Hexachloro-1,5,5*a*,6,9,9*a*-hexahydro-8,9-methano-2,4,3-benzodioxathiepin-3-oxide
2,4,5-TP	2,4,5-Trichlorophenoxy propionic acid
Toxaphene	Essentially a mixture of isomers of octachlorocamphene
Trithion	8-(*p*-Chlorophenylthiomethyl) O,O-diethylphosphorodithioate
Zinophos	O,O-Diethyl O-(2-pyrazinyl) phosphorothioate

Appendix C. Physical Constants of Selected Organic Pesticides

Compound	Molecular weight	Physical state	Melting or boiling point, °C	Vapor pressure, mmHg (°C)	Solubility in water, ppm (°C)
Aldrin	364.93	Solid	104–104.5	6×10^{-6} (25)	0.20 (25)
Atrazine	215.7	Solid	173–175	—	70 (27)
γ-BHC	290.83	Solid	112.9	9.4×10^{-6} (20)	10 (20)
CIPC	213.7	Solid	36–40	—	108 (20)
cis-Chlordane	409.8	Liquid	175 at 2 mm	—	—
2,4-D	221	Solid	140.5	0.4 (160)	620 (25)
DDD (TDE)	320.05	Solid	109	—	—
DDT	354.49	Solid	108.5–109	1.5×10^{-7} mm (20)	0.0017 (25)
DDVP	221.0	Liquid	84 at 1 mm	0.032 mm (31)	10000 (24.5)
Demeton	258.0	Liquid	134 at 2 mm	10^{-3} mm (33)	100 (?)
Diazinon	204.3	Liquid	83–84 at 0.002 mm	1.4×10^{-4} (20)	40 (room)
Dieldrin	380.91	Solid	175–176	1.8×10^{-7} (25)	0.14 (?)
Dimethoate	229.3	Solid	51–52	0.025 (25)	7000 (?)
Diuron	233.1	Solid	150–155	3.1×10^{-6} (50)	42 (25)
Endrin	380.91	Solid	>200	2×10^{-7} (25)	0.19 (?)
Fenthion	278.3	Liquid	105 at 10^{-2} mm	—	5.3–6.0 (20)
Fenuron	164.2	Solid	127–129	1.6×10^{-4} (60)	2900 (24)
Heptachlor	373.34	Solid	95–96	3×10^{-4} (25)	—
IPC	179.2	Solid	66.5–68.5	—	20–25 (25)
Malathion	330.36	Liquid	156–157 at 0.7 mm	4.0×10^{-4} (30)	145 (25)
Methoxychlor	345.65	Solid	89	—	—
Parathion	291.26	Liquid	157–162 at 0.6 mm	0.03 (24)	24 (25)
Phorate	260.4	Liquid	118–120 at 0.8 mm	3.9×10^{-3} (36)	85 (25)
Phosdrin	224.0	Liquid	99–103.3 at 0.03 mm	2.9×10^{-3} (21.1)	—
Ronnel	321.4	Solid	35–37	8×10^{-3} (25)	800 (?)
Simazine	201.7	Solid	225–227	6.1×10^{-9} (20)	5.0 (20)
2,4,5-T	255.5	Solid	158	—	251 (25)
2(2,4,5-TP)	269.5	Solid	179–181	—	140 (25)
PCP	266.4	Solid	190–191	0.12 (100)	14 (20)
DNBP	240.2	Solid	39.4	—	1000 (25)
DNC	198.1	Solid	85.8	1.05×10^{-6} (25)	125 (?)
Sevin	201.2	Solid	145	—	<1000 (?)

Appendix D. Relative Retention Times of Several Chlorinated Hydrocarbons

Pesticide	LaMar et al. (44)		Cueto and Biros (45)			Breidenbach et al. (79)	
	5% Dow-11	5% QF-1	3% QF-1	2% Dow-200	3% DGS	5% Dow-200[a]	5% Dow-200[b]
α-BHC	—	—	0.29	0.18	0.40	—	—
γ-BHC	—	—	0.37	0.23	0.59	—	—
β-BHC	—	—	0.45	0.20	1.81	—	—
δ-BHC	—	—	0.50	0.23	1.44	—	—
Lindane	0.46	0.84	—	—	—	0.43	0.50
Heptachlor	0.79	0.89	—	—	—	0.77	0.81
Aldrin	1.00	1.00	—	—	—	1.00	1.00
Heptachlor epoxide	1.30	1.94	0.91	0.65	0.84	1.33	1.24
γ-Chlordane	—	—	—	—	—	1.52	1.42
o,p'-DDE	—	—	0.75	0.81	0.78	—	—
p,p'-DDE	—	—	1.00	1.00	1.00	2.10	1.79
Dieldrin	2.06	3.06	1.45	1.45	1.23	2.08	1.83
Endrin	2.28	3.53	—	—	—	3.06–4.64	2.53–3.59
o,p'-DDT	2.72	2.63	1.29	1.33	1.36	2.89	2.39
p,p'-DDT	3.48	3.89	1.95	1.88	3.08	3.78	2.99
p,p'-DDD	—	—	1.80	1.33	2.50	—	—
DDD	—	—	—	—	—	2.79	2.33

[a] 180°C.
[b] 190°C.

Appendix E. Relative Retention Times of Several Organophosphates

Pesticide	Suffet et al.(102) 15% Reoplex-400	Ruzicka et al.(103) 10% SE-30, 1% Epikote	Harigichi et al.(104) 1% SE-30
Mevinphos	—	0.14	—
Phorate	—	0.25	—
Dibrom	—	—	0.39
Demeton-S-methyl	—	0.33	—
Disulfoton	—	0.41	—
Diazinon	1.00	0.38	0.78
Dimethoate	—	0-78	0.58
Methyl parathion	—	—	1.00
Nankor	—	—	1.17
Sumithion	—	—	1.31
Diazoxon	1.45	—	—
IMHP	1.77	—	—
MMTP	2.67	—	—
Malathion	5.09	0.91	1.62
Baytex	5.98	—	—
Parathion	7.14	1.00	1.62
Malaoxon	—	1.02	—
Bayoxon	7.66	—	—
Paraoxon	8.25	1.20	—
p-Nitrophenol	11.70	—	—
Phosphamidon	—	1.21	—
Mecarbam	—	1.42	—
EPN	—	—	8.20

REFERENCES

1. H. P. Nicholson, J. Am. Water Works Assoc., 51, 981 (1959).
2. S. D. Faust and O. M. Aly, J. Am. Water Works Assoc., 56, 267 (1964).
3. S. D. Faust, Clin. Pharmacol. Therap., 5, 677 (1964).
4. I. West, Arch. Environ. Health, 9, 626 (1964).
5. Report on the Use of Pesticides, President's Science Advisory Committee, U.S. Government Printing Office, Washington, D.C., 1963.
6. A. A. Rosen and F. M. Middleton, Anal. Chem., 31, 1729 (1959).
7. F. M. Middleton and J. J. Lichtenberg, Ind. Eng. Chem., 52, 99A (1960).
8. A. R. Grzenda, H. P. Nicholson, J. I. Teasley, and J. H. Patrick, J. Econ. Entomol., 57, 615 (1964).
9. E. Hindin, D. S. May, and G. H. Dunstan, Residue Rev., 7, 130 (1964).
10. E. Brown and Y. A. Nishioka, Pesticides Monit. J., 1, 38 (1967).
11. L. Weaver, C. G. Gunnerson, A. W. Breidenbach, and J. J. Lichtenberg, Public Health Rept. (U.S.), 80, 481 (1965).
12. R. J. Moubry, J. M. Helm, and G. R. Myrdal, Pecticides Monit. J., 1(4), 27 (1968).
13. H. P. Nicholson, Science, 158, 871 (1967).

14. *Public Health Service Drinking Water Standards*, *Publ. No. 956*, U.S. Government Printing Office, Washington, D.C., 1962.
15. *Water Quality Criteria*, National Technical Advisory Committee Report to the U.S. Dept. Interior, Washington, D.C., April 1, 1968.
16. L. E. Mitchell, *Organic Pesticides in the Environment*, *Advances in Chemistry Series No. 60*, American Chemical Society, Washington, D.C., 1966, p. 1.
17. G. W. Baily and J. L. White, *Residue Rev.*, **10**, 97 (1965).
18. V. H. Freed, *Pesticides and Their Effects on Soils and Waters*, *Spec. Publ. No. 8*, Soil Science Society of America, 1966, p. 25.
19. D. E. H. Frear, *Chemistry of the Pesticides*, Van Nostrand, New York, 1955.
20. J. W. Biggar, G. R. Dutt, and R. L. Riggs, *Bull. Environ. Contamination Toxicol.*, **2**, 90 (1967).
21. *Pesticide Chemicals Official Compendium*, Association of American Pesticide Control Officials, College Park, Md., 1966.
22. F. A. Gunther, W. E. Westlake, and P. S. Jaglan, *Residue Rev.*, **20**, 1 (1968).
23. C. M. Menzie, *Metabolism of Pesticides*, *Spec. Sci. Rept.—Wildlife No. 127*, U.S. Dept. Interior, Washington, D.C., 1969.
24. J. Rosen, D. J. Sutherland, and G. R. Lipton, *Bull. Environ. Contamination Toxicol.*, **1**, 133 (1966).
25. R. A. Baker, *J. Am. Water Works Assoc.*, **55**, 913 (1963).
26. R. D. O'Brien, *Insecticides—Action and Metabolism*, Academic, New York, 1967, p. 104.
27. M. C. Bowman and M. Beroza, *J. Agr. Food Chem.*, **16**, 399 (1968).
28. I. H. Suffet, G. Dosza, and S. D. Faust, *Water Res.*, in press (1971).
29. H. A. Gomaa, private communication, 1969.
30. A. V. Holden and K. Mardsen, *Inst. Sewage Purif.*, *J. Proc.*, Part 3, 1966, p. 3.
31. E. A. Epps, F. L. Bonner, L. D. Newsom, R. Carlton, and R. O. Smitherman, *Bull. Environ. Contamination Toxicol.*, **2**, 333 (1967).
32. G. L. Sutherland, *Residue Rev.*, **10**, 85 (1965).
33. R. S. Green and S. K. Love, *Pesticides Monit. J.*, **1**, 13 (1967).
34. M. C. Bowman, F. Acree, Jr., C. H. Schmidt, and M. Beroza, *J. Econ. Entomol.*, **52**, 1038 (1959).
35. M. Chiba and H. V. Morley, *J. Assoc. Offic. Agr. Chem.*, **51**, 55 (1968).
36. W. B. Wheeler and D. E. H. Frear, *Residue Rev.*, **16**, 86 (1966).
37. F. A. Gunther and R. C. Blinn, *Analysis of Insecticides and Acaricides*, Vol. 6, Wiley (Interscience), New York, 1955, p. 476.
38. H. Braus, F. M. Middleton, and G. Walton, *Anal. Chem.*, **23**, 1160 (1951).
39. G. F. Lee, G. W. Kumke, and S. L. Becker, *Intern. J. Air Water Pollution*, **9**, 69 (1965).
40. H. P. Nicholson, *Limnol. Oceanog.*, **9**, 310 (1964).
41. D. F. Peppard, *Advan. Inorg. Chem. Radiochem.*, **9**, 1 (1966).
42. E. W. Berg, *Physical and Chemical Methods of Separation*, McGraw-Hill, New York, 1963, p. 50.
43. F. K. Kawahara, J. W. Eichelberger, B. H. Reid, and H. Stierly, *J. Water Pollution Control Federation*, **39**, 572 (1967).
44. W. L. LaMar, D. F. Goerlitz, and L. M. Law, *Geol. Surv. Water Supply Paper 1817-B*, U.S. Dept. Interior, Washington, D.C., 1965.
45. C. Cueto and F. J. Biros, *Toxicol. Pharmacol.*, **10**, 261 (1967).
46. B. Berck, *Anal. Chem.*, **25**, 1253 (1953).
47. C. Cueto and W. J. Hayes, Jr., *J. Agr. Food Chem.*, **10**, 366 (1962).
48. W. M. Weatherholtz, G. W. Cornwell, R. W. Young, and R. E. Webb, *J. Agr. Food Chem.*, **15**, 667 (1967).

49. H. B. Pionke, J. G. Konrad, G. Chesters, and D. E. Armstrong, *Analyst*, **93**, 363 (1968).
50. G. A. Wheatly and J. A. Hardman, *Nature*, **207**, 486 (1965).
51. M. Beroza and M. C. Bowman, *J. Assoc. Offic. Agr. Chem.*, **49**, 1007 (1966).
52. S. L. Warnick and A. R. Gaufin, *J. Am. Water Works Assoc.*, **57**, 1023 (1965).
53. H. R. El-Refai and L. Giuffrida, *J. Assoc. Offic. Agr. Chem.*, **48**, 374 (1965).
54. D. I. Mount and C. E. Stephan, *Trans. Am. Fisheries Soc.*, **96**, 21 (1967).
55. F. C. Wright, B. N. Gilbert, and J. C. Riner, *J. Agr. Food Chem.*, **15**, 1038 (1967).
56. M. S. Mulla, *J. Econ. Entomol.*, **59**, 1085 (1966).
57. J. I. Teasley and W. S. Cox, *J. Am. Water Works Assoc.*, **55**, 1093 (1963).
58. J. R. Rice and H. J. Dishburger, *J. Agr. Food Chem.*, **16**, 867 (1968).
59. K. Erne, *Acta Chem. Scand.*, **17**, 1663 (1963).
60. O. M. Aly and S. D. Faust, *J. Am. Water Works Assoc.*, **55**, 639 (1963).
61. D. C. Abbott, H. Egan, E. W. Hammond, and J. Thomson, *Analyst*, **89**, 480 (1964).
62. D. F. Goerlitz and W. L. LaMar, *Geol. Surv. Water Supply Paper 1817-C*, U.S. Dept. Interior, Washington, D.C., 1967.
63. L. Kahn and C. H. Wayman, *Anal. Chem.*, **36**, 1340 (1964).
64. W. W. Sanderson and G. B. Ceresia, *J. Water Pollution Control Federation*, **37**, 1176 (1965).
65. Y. Sumiki and A. Matsuyama, *Bull. Agr. Chem. Soc. Japan*, **21**, 329 (1957).
66. H. Irving and R. J. P. Williams, in *Treatise on Analytical Chemistry*, Part I, Vol. 3, Wiley (Interscience), New York, 1961, p. 1309.
67. M. Beroza and M. C. Bowman, *Anal. Chem.*, **38**, 837 (1966).
68. M. C. Bowman and M. Beroza, *Anal. Chem.*, **38**, 1427 (1966).
69. T. P. King and L. C. Craig, in *Methods of Biochemical Analysis* (D. Glick, ed.), Wiley, New York, 1962, p. 201.
70. M. Beroza and M. C. Bowman, *J. Assoc. Offic. Agr. Chem.*, **48**, 358 (1965).
71. M. C. Bowman and M. Beroza, *J. Assoc. Offic. Agr. Chem.*, **48**, 943 (1965).
72. M. Beroza and M. C. Bowman, *Anal. Chem.*, **37**, 291 (1965).
73. I. H. Suffet, Ph.D. Thesis, Rutgers University, January, 1969.
74. C. Marsden and S. Mann, *Solvent Guide*, 2nd ed., Cleaver-Hume, London, 1963.
75. H. Gomaa, I. H. Suffet, and S. D. Faust, *Residue Rev.*, **29**, 171 (1969).
76. R. Muhlmann and G. Schrader, *Z. Naturforsch.*, **12b**, 196 (1957).
77. B. J. Kallman, O. B. Cope, and R. J. Navarre, *Trans. Am. Fisheries Soc.*, **91**, 14 (1962).
78. H. P. Nicholson, H. T. Webb, G. T. Lauer, R. E. O'Brien, A. R. Grzenda, and D. W. Shanklin, *Trans. Am. Fish. Soc.*, **91**, 213 (1962).
79. A. W. Breidenbach, J. J. Lichtenberg, C. F. Henke, D. J. Smith, J. W. Eichelberger, and H. Stierli, *The Identification and Measurement of Chlorinated Hydrocarbon Pesticides in Surface Waters*, Public Health Serv. Publ. WP-22, U.S. Dept. Health, Education, and Welfare, Washington, D.C., 1966.
80. A. Goodenkauf and J. Erdei, *J. Amer. Water Works Assoc.*, **56**, 600 (1964).
81. A. M. Kadoum, *Bull. Environ. Contamination Toxicol.*, **2**, 264 (1967).
82. J. R. Rise and H. J. Dishberger, *J. Agr. Food Chem.*, **16**, 867 (1968).
83. J. A. Burke and B. Malone, *J. Assoc. Offiic. Agr. Chem.*, **49**, 1004 (1966).
84. P. A. Mills, *J. Assoc. Offic. Agr. Chem.*, **51**, 29 (1968).
85. A. B. Black and R. F. Christman, *J. Am. Water Works Assoc.*, **55**, 753 (1963).
86. S. D. Faust and N. E. Hunter, *J. Am. Water Works Assoc.*, **57**, 1028 (1965).
87. N. T. Crosby and E. Q. Laws, *Analyst*, **89**, 318 (1964).
88. R. W. Storherr and R. R. Watts, *J. Assoc. Offic. Agr. Chem.*, **48**, 1154 (1965).
89. J. H. Hamence, P. S. Hall, and J. T. Caverly, *Analyst*, **90**, 649 (1965).
90. K. T. Hartman, *J. Assoc. Offic. Agr. Chem.*, **50**, 615 (1967).

90a. D. Smith and J. Eichelberger, *J. Water Pollution Control Federation*, **37**, 77 (1965).

91. M. F. Kovacs, *J. Assoc. Offic. Agr. Chem.*, **46**, 884 (1964).

92. M. F. Kovacs, *J. Assoc. Offic. Agr. Chem.*, **47**, 1097 (1964).

93. P. J. Bunyan, *Analyst*, **89**, 615 (1964).

94. C. W. Miller, B. M. Zuckerman, and A. T. Charig, *Trans. Am. Fisheries Soc.*, **95**, 345 (1966).

95. F. K. Kawahara, J. J. Lichtenberg, and J. W. Eichelberger, *J. Water Pollution Control Federation*, **39**, 446 (1967).

96. D. C. Abbott, J. A. Bunting, and J. Thomson, *Analyst*, **90**, 365 (1965).

97. C. E. Mendoza, P. J. Wales, H. A. McLeod, and W. P. McKinley, *Analyst*, **93**, 34 (1968).

98. N. A. Smart and A. R. C. Hill, *J. Chromatog.*, **30**, 626 (1967).

99. E. J. Thomas, J. A. Burke, and J. H. Lawrence, *J. Chromatog.*, **35**, 119 (1963).

100. I. Brown, *J. Chromatog.*, **10**, 284 (1963).

101. H. Cole, D. Barry, and D. E. H. Frear, *Bull. Environ. Contamination Toxicol.*, **2**, 127 (1967).

102. I. H. Suffet, W. F. Carey, and S. D. Faust, *Environ. Sci. Tech.*, **1**, 639 (1967).

103. J. Ruzicka, J. H. Thomson, and B. B. Wheals, *J. Chromatog.*, **30**, 92 (1967).

104. M. Hariguchi, M. Ishida, and N. Higosaki, *Chem. Pharm. Bull.*, **12**, 1315 (1964).

105. L. E. St. John, Jr., and D. J. Lisk, *J. Agr. Food Chem.*, **16**, 48 (1968).

106. W. H. Gutenmann and D. J. Lisk, *J. Am. Water Works Assoc.*, **56**, 189 (1964).

107. R. A. Conkin, *Residue Rev.*, **6**, 136 (1964).

108. J. Robinson and A. Richardson, *Chem. Ind. (London)*, **11**, 1460 (1963).

109. K. I. Benyon and K. E. Elgar, *Analyst*, **91**, 143 (1966).

110. W. E. Westlake and F. A. Gunther, *Residue Rev.*, **18**, 175 (1967).

111. J. Harley, W. Nel, and V. Pretorius, *Nature*, **181**, 177 (1958).

112. J. E. Lovelock and S. R. Lipsky, *J. Am. Chem. Soc.*, **82**, 431 (1960).

113. L. K. Gaston, *Residue Rev.*, **5**, 21 (1964).

114. D. M. Coulson, L. A. Cavanagh, J. E. DeVries, and B. Walther, *J. Agr. Food Chem.*, **8**, 399 (1960).

115. D. M. Coulson, *Advan. Pesticide Control Res.*, **5**, 153 (1962).

116. R. L. Martin, *Anal. Chem.*, **38**, 1209 (1966).

117. A. Karmen, *Anal. Chem.*, **36**, 1416 (1964).

118. A. Karmen and L. Guiffrida, *Nature*, **201**, 1204 (1964).

119. K. Abel, K. Lanneau, and R. K. Stevens, *J. Assoc. Offic. Agr. Chem.*, **49**, 1022 (1966).

120. C. H. Hartmann, *Bull. Environ. Contamination Toxicol.*, **1**, 159, (1966).

121. D. M. Coulson, *J. Gas Chromatog.*, **4**, 285 (1966).

122. S. S. Brody and J. E. Chaney, *J. Gas Chromatog.*, **4**, 42 (1966).

123. A. J. McCormack, S. C. Tong, and W. D. Cooke, *Anal. Chem.*, **37**, 1470 (1965).

124. C. A. Bache and D. J. Lisk, *Anal. Chem.*, **37**, 1477 (1965).

125. E. M. Bellet, W. E. Westlake, and F. A. Gunther, *Bull. Environ. Contamination Toxicol.*, **2**, 255 (1967).

126. L. Giuffrida, N. F. Ives, and D. C. Bostwick, *J. Assoc. Offic. Agr. Chem.*, **49**, 8 (1966).

127. J. R. Wessel, *J. Assoc. Offic. Agr. Chem.*, **51**, 666 (1968).

128. M. S. Schechter, *Pesticides Monit. J.*, **2**, 1 (1968).

129. *Monitoring Agricultural Pesticide Residues, ARS 81-13*, U.S. Dept. Agriculture, Washington, D.C., December, 1966.

130. J. I. Teasley, *Environ. Sci. Tech.*, **1**, 411 (1967).

131. C. W. Miller, W. E. Tomlinson, and R. L. Norgren, *Pesticides Monit. J.*, **1**, 47 (1967).

132. F. K. Kawahara, R. L. Moore, and R. W. Gorman, *J. Gas Chromatog.*, **6**, 24 (1968).

133. S. D Faust and I. H. Suffet, in *ASTM STP 448*, American Society for Testing and Materials, Philadelphia, 1969, p. 24.

134. R. C. Blinn and F. A. Gunther, *Residue Rev.*, **2**, 99 (1963).

135. R. C. Blinn, *Residue Rev.*, **5**, 130 (1964).

136. W. W. Morris and E. O. Haenni, *J. Assoc. Offic. Agr. Chem.*, **46**, 964 (1963).

137. R. C. Blinn, *J. Assoc. Offic. Agr. Chem.* **48**, 1009 (1965).

138. W. R. Payne and W. S. Cox, *J. Assoc. Offic. Agr. Chem.*, **49**, 989 (1966).

139. D. F. McCaulley, *J. Assoc. Offic. Agr. Chem.*, **48**, 659 (1965).

140. J. A. A. Ketelaar and H. R. Gersmann, *Rec. Trav. Chim.*, **78**, 190 (1959).

141. D. F. McCaulley and J. W. Cook, *J. Assoc. Offic. Agr. Chem.*, **43**, 710 (1960).

142. A. K. Klein, E. P. Lang, P. R. Datta, J. O. Watts, and J. T. Chen, *J. Assoc. Offic. Agr. Chem.* **47**, 1129 (1964).

143. J. T. Chen, *J. Assoc. Offic. Agr. Chem.*, **48**, 380 (1965).

144. L. Giuffrida, *J. Assoc. Offic. Agr. Chem.*, **48**, 354 (1965).

145. T. R. Kantner and R. O. Mumma, *Residue Rev.*, **16**, 138 (1966).

146. L. A. Cavanagh, *SRI Pesticides Res. Bull.*, **3(i)**, 1 (1963).

147. J. N. Damico and W. R. Benso, *J. Assoc. Offic. Agr. Chem.*, **48**, 344 (1965).

148. J. N. Damico, *J. Assoc. Offic. Agr. Chem.*, **49**, 1027 (1966).

149. R. O. Mumma and T. R. Kantner, *J. Econ. Entomol.*, **59**, 491 (1966).

150. W. H. McFadden, *Separation Sci.*, **1**, 723 (1966).

151. L. W. Keith, A. W. Garrison, and A. L. Alford, *J. Assoc. Offic. Agr. Chem.*, **51**, 1063 (1968).

152. H. Babad, W. Herbert, and M. C. Goldberg, *Anal. Chim. Acta*, **41**, 259 (1968).

153. H. Babad, T. N. Taylor, and M. C. Goldberg, *Anal. Chim. Acta*, **41**, 387 (1968).

154. M. J. D. Low and S. K. Freeman, *Anal. Chem.*, **39**, 194 (1967).

155. "Unit Combines GLC and Mass Spectroscopy," *Chem. Eng. News*, June, 1966, p. 48.

156. E. G. Bramer, Jr., *Anal. Chem.*, **37**, 1183 (1965).